Two Manuals Combined into One Book

Marine Operations on Urbanized Terrain

And

Military Role in Civil Disturbance Operations

Department of Defense

Pentagon
Publishing

Military Operations on Urbanized Terrain

Table of Contents

Chapter 6. Noncombatant Considerations in Urban Operations

Chapter 7. The Urban Environment and Restrictions to Operations

Appendix A. Organization for Urban Combat and Fundamental Combat Skills

Chapter 1

Modern Urban Warfare

"... the likelihood is high that in the future, the National Command Authorities will again commit Marines to missions in urban areas."

A Concept for Future Military Operations on Urbanized Terrain

1001. Importance of Urban Areas. Throughout history, military planners have viewed cities as centers of gravity. As such, in war, cities are something to be either protected or taken away, depending upon one's perspective (MCDP 1, *Warfighting*). Cities house the population centers, transportation hubs, seats of government, sources of wealth, centers for industry, information networks, and key nodes of communication within a nation. Recent forecasts based on population statistics and the worldwide migration trend from agrarian to industrialized societies predict that 85 percent of the world's population will reside in urbanized areas by the year 2025. As the world trend toward urbanization increases, the military significance of cities is likely to increase proportionally.

Urbanized areas, themselves, may be significant sources of future conflict. Cities historically are where radical ideas ferment, dissenters find allies, mixtures of people cause ethnic friction, and discontented groups receive media attention. Adversaries may focus on the capture of radio and television stations in an attempt to influence public opinion and attain their political goals. Our political leaders may take advantage to neutralize or stabilize some extremely volatile political situations, or to provide assistance to allies in need of support, by deploying U.S. forces into urban environments,

1002. The Marine Corps Role in Urban Warfare. As the Nation's force in readiness, forward deployed with expeditionary forces, Marines must be prepared to fight on urbanized terrain. In the past two decades, MAGTFs ranging in size from MEFs (Saudi Arabia, Desert Shield/Desert Storm; Somalia, Restore Hope) through Marine expeditionary units (MEUs) (Beirut, Lebanon; Grenada, Urgent Fury; Somalia, Eastern Exit and Restore Hope) have participated in MOUT. The task-organization and combined-arms aspect of the MAGTF makes it well suited for combat on urbanized terrain.

The results of geographical studies show that 60 percent of politically significant urban areas outside allied or former Warsaw Pact territory are located along or within 25 miles of a coastline; 75 percent are within 150 miles; 87 percent are within 300 miles; 95 percent are within 600 miles; and all are within 800 miles. U.S. embassies and diplomatic facilities are primarily located in cities where the host country's political and economic leadership is concentrated. The Marine Corps will continue to play a prominent role in future evacuations of U.S. citizens, as well as the conduct of peace, counterinsurgency, and contingency operations centered on urbanized areas.

Today's Marine air-ground task forces (MAGTFs) are deployed as part of naval expeditionary forces (NEFs) that maintain a *global forward presence for rapid crisis response*. These integrated combined-arms forces are part of the Nation's proven contingency and naval power projection force. Therefore, Marines may find themselves rapidly deployed and employed in actions across the spectrum of military operations. Many of these trouble spots will likely be located in or around large urban centers.

In the years since World War II, the United States has employed military force more than 200 times. Of these, four out of five involved naval forces, and the majority of the naval efforts included Marines embarked in amphibious ships. The reasons are straightforward: availability and adaptability. Availability derives from the loiter time of forward deployed forces embarked on amphibious shipping. Adaptability comes from the Marine Corps' MAGTF organization, doctrine, training, and equipment, which prepare us for expeditionary missions from the sea in support of a variety of missions, including forcible entry. Enhancing our adaptability are the maritime prepositioning forces (MPFs). MPFs provide a rapid buildup of combat and logistics equipment that is joined with Marines on a distant shore, creating a substantial combat force. Despite our availability and adaptability, the prospect of urban warfare combined with an amphibious assault is a complex task which requires special preparation.

At the outset of a developing situation, forward-deployed expeditionary forces can move quickly within range of a crisis that threatens the political stability of a country. Urban intervention operations must often be planned and executed in a matter of hours or days (rather than weeks or months) to take advantage of the internal turmoil surrounding a developing crisis. Navy and Marine forces should anticipate deployment to urbanized areas on a day-to-day basis.

1003. Distinguishing Features of Urbanized Terrain. Urbanized terrain is a complex and challenging environment. It possesses all of the characteristics of the natural landscape, coupled with manmade construction, resulting in an incredibly complicated and fluid environment that influences the conduct of military operations in unique ways.

Military operations on urbanized terrain (MOUT) is defined as *all military actions planned and conducted on a topographical complex and its adjacent terrain where manmade construction is the dominant feature. It includes combat in cities, which is that portion of MOUT involving house-to-house and street-by-street fighting in towns and cities* (Marine Corps Reference Publication (MCRP) 5-12A, *Operational Terms and Graphics*). MOUT effects the tactical options available to a commander. A built-up area is *a concentration of structures, facilities, and populations, such as villages, cities, and towns,* that form the economic and cultural focus for the surrounding area. (MCRP 5-12A)

a. Cities. Cities are centers of finance, politics, transportation, communication, industry, and culture. They generally have large population concentrations ranging from tens of thousands to millions of people. Because of their psychological, political, or logistical value, control of cities have often been the scenes of pitched battles.

(1) Operations in built-up areas are normally conducted to capitalize on the operational or tactical significance of a particular city. In developing nations, control of only a few cities is often the key to the control of national resources. The side that controls a major city usually has a psychological advantage, which can be enough to significantly affect the outcome of a countrywide conflict.

(2) The abundance of guerrilla and terrorist operations conducted in built-up areas (e.g., Santo Domingo, Caracas, Belfast, Managua, and Beirut) demonstrates the importance many insurgent groups place on urban warfare.

(3) In the past 40 years, many cities have expanded dramatically, losing their well-defined boundaries as they extended into the countryside. New transportation systems (highways, canals, and railroads) have been built to connect population centers. Industries have grown along those connectors, creating "strip areas." Rural areas, although retaining much of their farmlike character, are connected to the towns by a network of secondary roads.

b. Multiple Avenues of Approach. Urbanized terrain is a unique battlespace that provides both attacker and defender with numerous and varied avenues of approach and fields of fire. The urban battlespace is divided into four basic levels: *building*, *street*, *subterranean,* and *air*. Operations can be conducted from above ground, on ground level, inside buildings, or below the ground. Most operations will include fighting on all levels simultaneously.

(1) Building Level. Buildings provide cover and concealment; limit or increase fields of observation and fire; and canalize, restrict, or block movement of forces, especially mechanized forces. They provide optimum perches for snipers and antiair weapons. Buildings also provide antitank weapons optimum positioning to allow engagement from above, exploiting an inherent weakness found in most armored vehicles.

(2) Street Level. While streets provide the means for rapid advance or withdawal, forces moving along streets are often canalized by buildings and have little space for off-road maneuver. Because they are more difficult to bypass, obstacles on streets in urbanized areas are usually more effective than those on roads in open terrain.

(3) Subterranean Level. Subterranean systems are easily overlooked but can be important to the outcome of operations. These areas may be substantial and include subways, sewers, cellars, and utility systems (Figure 1-1 on page 1-4). The city of Los Angeles alone has more than 200 miles of storm sewers located under the city streets. Both attacker and defender can use subterranean avenues to maneuver to the rear or the flanks of an enemy. These avenues also facilitate the conduct of ambushes, counterattacks, and infiltrations. (See Appendix E)

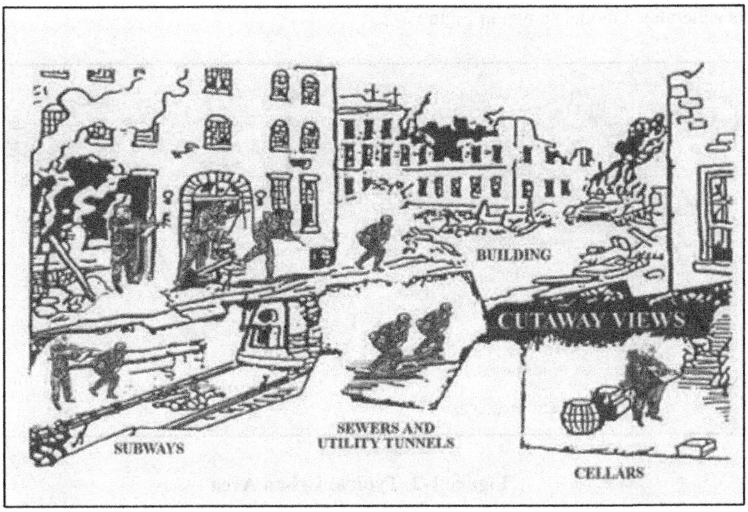

Figure 1-1. Subterranean Systems

(4) Air Level. The air provides another avenue of approach in urbanized areas. Aviation assets can be used for high speed insertion or extraction of troops, supplies, and equipment. While aviation assets are not affected by obstacles on the streets, they are affected by light towers, signs, power lines, and other aerial obstructions. They are also vulnerable to the man-portable surface-to-air missile threat, crew served weapons, and small arms fire.

c. Categories of Built-Up Areas. Built-up areas are generally classified as:

- Villages (populations of 3,000 or less)
- Strip areas (industrialized zones built along roads connecting towns or cities)
- Towns or small cities (populations of up to 100,000 and not part of a major urban complex)
- Large cities with associated urban sprawl (populations in the millions, covering hundreds of square kilometers).

d. Characteristics of Urbanized Areas. A typical urban area consists of combinations of the *city core, commercial ribbon, core periphery, residential sprawl, outlying industrial areas*, and *outlying high-rise areas*. Each of the urban area's regions has distinctive characteristics that may weigh heavily in planning for MOUT. Most urbanized areas resemble the generalized model shown in Figure 1-2.

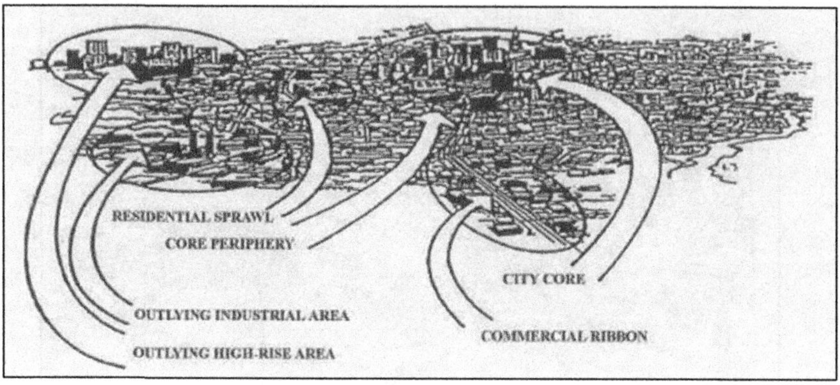

Figure 1-2. Typical Urban Area

(1) City Core. In most cities, the city core has undergone more recent development than the core periphery. As a result, the two regions are often quite different. Typical city cores are made up of high-rise buildings which vary in height. Modern urban planning for built-up areas allows for more open spaces between buildings than in old city cores or in core peripheries (Figure 1-3).

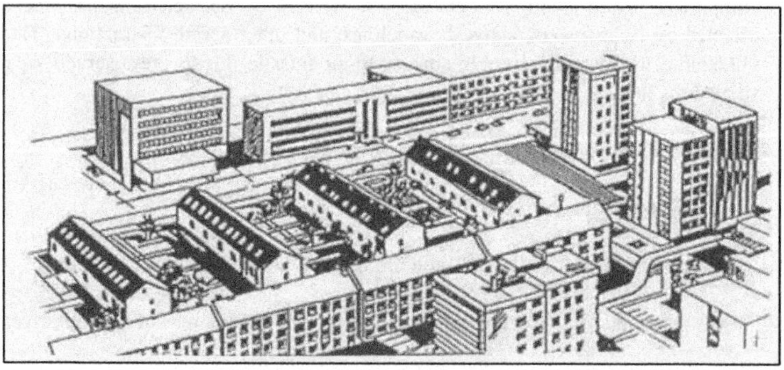

Figure 1-3. City Core

(2) Commercial Ribbon. Commercial ribbons are composed of rows of stores, shops, and restaurants that are built along both sides of major streets through built-up areas.

Typically, such streets are 25 meters wide or more. The buildings in the outer areas are uniformly two to three stories tall—about one story taller than the dwellings on the streets behind them (Figure 1-4).

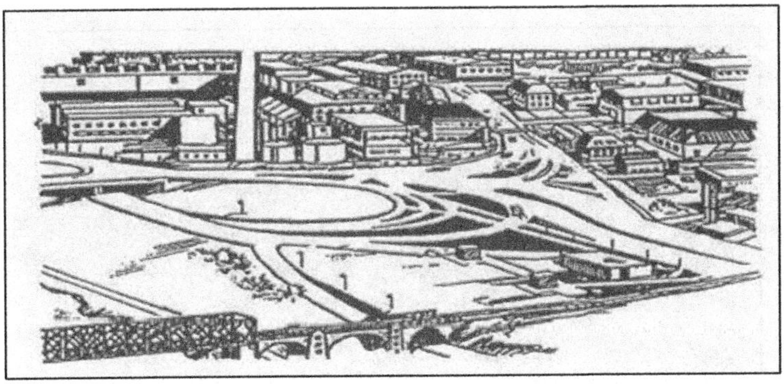

Figure 1-4. Commercial Ribbons

(3) Core Periphery. The core periphery generally consists of streets 12 - 20 meters wide with continuous fronts of brick or concrete buildings. The building heights are fairly uniform—2 or 3 stories in small towns, 5 to 10 or more stories in large cities (Figure 1-5).

Figure 1-5. Core Periphery

(4) Residential Sprawl. Residential sprawl areas consist mainly of low houses or apartments that are one to three stories tall. The area is primarily composed of detached dwellings that are usually arranged in irregular patterns along streets, with many smaller open areas between structures (Figure 1-6).

Figure 1-6. Residential Sprawl

(5) Outlying Industrial Areas. These areas generally consist of clusters of industrial buildings varying from one to five stories in height. Buildings generally vary dramatically in size and composition to match the needs of the particular businesses they house. Industrial parks are good examples of this category (Figure 1-7).

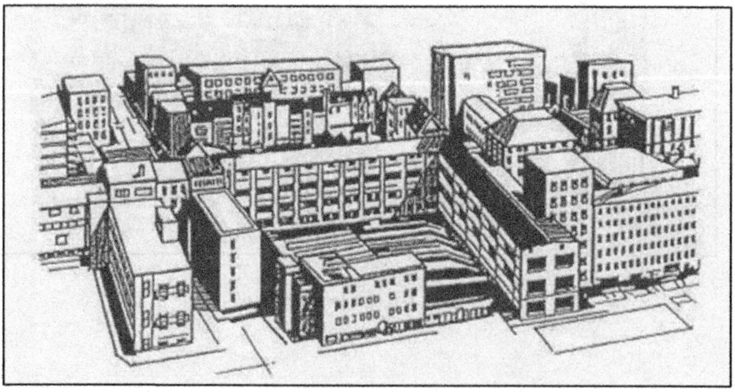

Figure 1-7. Outlying Industrial Areas

(6) Outlying High-Rise Areas. These areas are similar in composition to city core areas, but may be composed of clusters of more modern multistory high-rise buildings in outlying parts of the city. Building height and size may vary dramatically (Figure 1-8). Generally, there is more open space between buildings located in the outlying high-rise areas than is found within the city core area.

Figure 1-8. Outlying High-Rise Area

1004. Modern Battles for Urbanized Terrain. Urban warfare is as old as war itself. Since man began building villages, he has fought battles in and around them. Geography, politics, and economics dictate that cities will continue to be an objective of armies in warfare. From the armies that invaded and liberated Europe twice during the 20th century, to the forces that fought in Korea and Vietnam, to our most recent urban battles in the Middle East and Southwest Asia, *the basic principles of combat in built-up areas have essentially remained unchanged in this century*. While the principles remain the same, the introduction of helicopters, fixed-wing aircraft, armor, and precision-guided munitions (PGMs) has altered some of the techniques associated with urban combat.

Twenty two modern urban battles are discussed to illustrate the trends, dominant factors, and principles of combat in urbanized areas. (The majority of this material is extracted from the 1987 study, *Modern Experience in City Combat,* produced by Abbott Associates.)

• Aachen	• Manila
• Arnhem	• Ortona
• Ashrafiyeh	• Quang Tri City I
• Ban Me Thuot	• Quang Tri City II
• Beirut Port/Hotel (I)	• Seoul
• Beirut 1982 (II)	• Sidon
• Berlin	• Stalingrad
• Cherbourg	• Suez City
• Hue	• Tel Zaatar
• Jerusalem	• Tyre
• Khorramshahr	• Zahle

A brief description of each battle is provided to illustrate that battle's significance.

a. Stalingrad (1942 - 1943). The tenacious Soviet defense of Stalingrad cost the attacking Germans dearly in every way and set up conditions for a decisive counteroffensive. This classic urban battle involved large forces and resulted in innovative urban combat techniques and the creation of the highly successful storm groups (task-organized assault units). (Length of battle: greater than 30 days) (Casualties: 1,630,000+)

b. Ortona (1943). In this Italian town, determined resistance by a battalion of the elite German 3rd Parachute Regiment against Canadian Army attackers demonstrated the difficulty of overcoming a well-prepared defense. The Canadians were unfamiliar with urban combat and had to develop urban fighting techniques during the battle. After the town was largely destroyed and the defender had extracted a high cost in time and casualties from Canadian forces, the German parachute battalion withdrew. (Length of battle: 6 - 13 days) (Casualties: estimated in the hundreds)

c. Aachen (1944). The battle for Aachen, Germany, in the fall of 1944 developed during the U.S. First Army's offensive to breach the Westwall fortifications and the vaunted Siegfried Line. Aachen, the ancient capital of Charlemagne, had symbolic political and psychological significance to both the Germans and Americans. Furthermore, it was the first city on German soil to face an assault by the Allies. This first major battle on German soil foreshadowed bitter resistance against the American attackers in subsequent battles. The German defenders surrendered only after the city was destroyed. Although the U.S. Army had achieved a clear tactical victory, the German defense of Aachen cost the U.S. First Army valuable time and delayed the planned attack to the Rhine river. (Length of battle: 14 - 30 days) (Casualties: 8,000+)

d. Arnhem (1944). On September 17, 1944, Operation Market-Garden, the largest airborne operation in history, was launched in the Netherlands. The plan was to land three airborne divisions to seize key bridges along a 100-kilometer-long corridor through which allied mechanized forces would pass as the first step in the final offensive into Germany. The British 1st Airborne Division made a surprise landing near the Dutch city of Arnhem in order to seize a bridge over the Rhine river for advancing British forces. An unexpected German armor force counterattacked and eliminated all footholds, virtually destroying the British division before a linkup could occur. (Length of battle: 6 - 13 days) (Casualties: estimated in the thousands)

e. Cherbourg (1944). By June 17, 1944, U.S. forces advancing toward Cherbourg from the Normandy beachhead succeeded in cutting off defending German forces in the Cotentin Peninsula. Four German divisions withdrew to a perimeter surrounding Cherbourg. After much fighting, particularly in strongpoints outside the city, the German garrison surrendered to the Americans. Unfortunately for the Americans, the port facilities were destroyed which denied their early use by Allied forces. (Length of battle: 6 - 13 days) (Casualties: estimated in the thousands)

f. Berlin (1945). The long, bloody Soviet offensive to seize the German capital city effectively concluded the last battle of World War II in Europe. Bitter fighting occurred, but the defense was never well coordinated due in part to poor preparation by the Germans. (Length of battle: 14 - 30 days) (Casualties: estimated in the thousands)

g. Manila (1945). Japanese Army troops evacuated Manila under pressure from advancing American forces, but the local Japanese naval commander independently decided to hold the city at all costs. Despite defending Manila with poorly trained and equipped personnel, the determined resistance resulted in a high number of casualties to attacking U.S. forces as well as the destruction of the city and much of its population. (Length of battle: 14 - 30 days) (Casualties: 22,000+)

h. Seoul (1950). Following the Inchon landing, U.S. and Republic of Korea (ROK) forces recaptured the South Korean capital from the North Koreans. The fighting was unusual in that combat was largely centered on seizure of street barricades rather than buildings. (Length of battle: 6 - 13 days) (Casualties: Marines, 2,383; others, estimated in the thousands)

i. Jerusalem (1967). Israeli forces seized Jerusalem in a well prepared and well executed operation. Despite an uncoordinated Jordanian defense, Israeli casualties in this battle were the highest of those encountered during the Six Day War. Regular Jordanian forces withdrew during the latter stages of the battle, effectively ending organized resistance. (Length of battle: 48 hours to 5 days) (Casualties: Israeli forces, 400+; Jordanian forces, estimated in the hundreds)

j. Hue (1968). On January 31, 1968, the North Vietnamese Army (NVA) and Viet Cong (VC) forces launched their Tet Offensive at targets throughout South Vietnam. As part of this operation, two NVA/VC regiments and two sapper battalions conducted a surprise attack and seized part of the walled city (Citadel) of Hue. The NVA/VC held this part of the walled city for about three weeks against determined U.S. and South Vietnamese forces before finally succumbing. The battle for Hue is considered one of the most intense and savage battles of the Vietnam War. (Length of battle: 14 - 30 days) (Casualties: Marines, 433; others, 5000+)

k. Quang Tri City I and II (1972). An objective of the North Vietnamese 1972 winter-spring offensive was the capture of Quang Tri, the northernmost major city in South Vietnam. The NVA overwhelmed the Army, Republic of Vietnam (ARVN) defenders (I). Later, the city was recaptured (II) by a smaller ARVN force using extensive artillery and air support. The large conventional forces involved on both sides made Quang Tri I and II the major urban battles of the Vietnam War. (Length of battle: Quang Tri I, 6 - 13 days; Quang Tri II, 30 days or greater) (Casualties: battles combined, 30,000+)

l. Suez City (1973). Israeli Defense Forces (IDF) attempted to seize this Egyptian city before the anticipated United Nations (U.N.) cease-fire to end the Yom Kippur War. IDF armored shock tactics led to disaster against a well-prepared Egyptian defense. High casualties forced the IDF to withdraw. (Length of battle: less than 24 hours) (Casualties: Israeli forces, estimated 100 - 500; others, unknown)

m. Ban Me Thuot (1975). This South Vietnamese highlands town was the first strategic city to fall in the final North Vietnamese general offensive in 1975 South Vietnamese forces were surprised and overwhelmed. The fall of Ban Me Thuot resulted in a rout that the North Vietnamese exploited to achieve total victory in Vietnam. (Length of battle: 24 - 48 hours) (Casualties: estimated in the hundreds)

n. Beirut I (1976). When the Lebanese civil war broke out in the spring of 1975, combat in the capital city of Beirut assumed a central role. The battle for Beirut I was a series of small, local operations between largely irregular Christian and Muslim forces fighting over control of the hotel and port districts. Combat was not decisive, but led to changes in the boundary, called the "Green Line." This separated the antagonists and lead to the stagnation of the Lebanese conflict. (Length of battle: greater than 30 days) (Casualties: estimated in the hundreds)

o. Tel Zaatar (1976). Lebanese Christian attackers encircled and methodically besieged this Palestinian camp before overcoming its defenders with a final assault. (Length of battle: greater than 30 days) (Casualties: estimated in the hundreds)

p. Ashrafiyeh (1978). The Syrian forces occupying portions of Lebanon faced a complex political situation in which the power of the Christian militia was seen as a clear threat to stability. In an attempt to weaken the Christian militia by an attack on their center of power, the Syrians laid siege to the Christian militia stronghold of East Beirut (Ashrafiyeh). This urban battle was essentially an artillery bombardment without air attacks. Syria failed to break the will of the defenders and final positions remained unchanged. (Length of battle: greater than 30 days) (Casualties: estimated in the hundreds)

q. Khorramshahr (1980). Iranian regular forces initially evacuated this port city in the face of an Iraqi offensive. Irregular Iranian forces, however, continued to fight. They offered prolonged resistance and inflicted heavy casualties. Iraq eventually won this battle, but at a high cost in time and resources that ultimately served to halt the entire offensive against Iran. The intensity of fighting during the battle for Khorramshahr earned the city the nickname, "City of Blood." (Length of battle: 14 - 30 days) (Casualties: Iraqi, 3,000 - 9,000; Iranian, estimated in the thousands)

r. Zahle (1981). Syria laid siege to the Lebanese regular forces and militia in this key crossroads town. Fighting was inconclusive and ended in a negotiated settlement whereby the Lebanese defenders evacuated the town. (Length of battle: greater than 30 days) (Casualties: estimated in the hundreds)

s. Beirut II (1982). The siege of Beirut culminated the Israeli campaign to evict the Palestine Liberation Organization (PLO) from Lebanon. Fighting under domestic and world political pressures, the IDF besieged the PLO, selectively applying heavy ground and air firepower in conjunction with psychological warfare and limited-objective ground operations. The fighting resulted in a negotiated PLO evacuation from the city. (Length of battle: greater than 30 days) (Casualties: 2,300+)

t. Sidon (1982). Israeli forces easily seized this PLO southern headquarters during the invasion of Lebanon. The IDF was fully prepared for major urban combat using lessons learned from earlier battles, but resistance was unexpectedly light as PLO forces had largely withdrawn from the city. (Length of battle: 48 hours to 5 days) (Casualties: unknown)

u. Tyre (1982). The Israeli attack on the PLO in this Lebanese coastal city was well planned, with excellent intelligence on the target. All branches of the IDF participated in an operation that included naval fire support and amphibious landings. PLO resistance was uncoordinated and easily overcome. (Length of battle: 48 hours to 5 days) (Casualties: Israeli forces, 120+; others unknown)

1005. Modern Urban Battle Analysis and Observations. The 20th-century urban warfare experience has caused us to reevaluate old factors and consider new developments that affect the way we fight in this environment. The following discussion, based on the historical analysis of these select urban battles, illustrates the importance of maneuver warfare and combined arms philosophies in the urban environment. Factors that have had an impact on the manner in which urban warfare has been conducted are:

Intelligence is imperative to success in urban warfare. "Maneuver warfare requires a firm focus on the enemy" (MCDP 2, *Intelligence*). Few subsequent tactical changes can overcome the far-reaching impact of a major intelligence error.

a. Intelligence. The historical review of modern urban battles discloses that the attacker will usually win. Failures to win generally reflect classic military errors not characteristically unique to cities. *However, of the battles studied, battles lost were attributed to errors in* *initial intelligence.* The battles of **Arnhem** and **Suez City** probably would never have occurred had the attacker known the strength and locations of the defender's forces. At **Stalingrad**, the attacking Germans were aware of the defending forces facing them in the Sixth Army's zone, but the buildup of Soviet forces in other areas of the front was not anticipated and resulted in tactical surprise at those points, diluting the German offensive push to seize the city.

Surprise is a combat multiplier and can substantially reduce the cost of urban warfare.

b. Surprise. Surprise is a combat multiplier and can substantially reduce the cost of urban warfare. It can be achieved through deception, stealth, and ambiguity (MCDP 1-3, *Tactics*). Surprise was achieved by the attacker at **Aachen** and **Ban Me Thuot** and by the defender at **Suez City**. Surprise can be an important asset to increase leverage, but, as the failure at **Arnhem** shows, not necessarily a decisive one. When surprise is employed as a means to overcome other disadvantages, it is important to maintain accurate intelligence. In urban areas, tactical surprise by the attacker can be used to preempt effective defensive preparation of a city.

c. Combined Arms. The MAGTF must capitalize on one of the key means for gaining advantage in maneuver warfare — the use of combined arms (MCDP 1-3, *Tactics*). The use of combined arms places the enemy in a dilemma. Any action the enemy takes to avoid one combat arm makes him more vulnerable to another. An analysis of categories of weapons systems found in a MAGTF helps to illustrate it is a warfighting organization well suited for MOUT:

Combat in urban areas is primarily a small unit, infantry intensive operation

(1) Infantry. Combat in urban areas is primarily a small-unit, infantry intensive operation. Restrictions on maneuver, particularly for mechanized units, increases oportunities for infiltration. Urban combat requires small-unit leadership, initiative, and skill. Decentralized actions and difficulties in command , control, and communications are typically encountered. Built-up areas, like close terrain found in other operational environments, are generally considered to be most suited for operations conducted by infantry. Infantry units can be organized, trained and equipped to negotiate urbanized terrain that restricts observation, fields of fire, and mechanized movement.

(2) Armor. The role of armor in urban warfare can be significant. Of the 22 battles studied, armor participated in 21. In three-fourths of these battles, organic tank support was a central element when special assault teams were employed. *Overall, special assault units supported by tanks were more successful than any other task organization.*

The use of tanks to the attacker inside a city has been effective only when they were protected by infantry. Tanks in support of infantry act as an "assault gun" that delivers concentrated, sustained fires to reduce held strongpoints.

The U.S. experience in **Hue** demonstrated the key role that armor can play on a combined-arms team fighting inside the city. The Marines' most effective weapons during the battle were the M48A1 Patton tank and the M-50 Ontos. Both were protected by infantry. The M48A1, with its 90-mm main gun, was used extensively to reduce fortified positions. The Ontos, an armor-protected tracked vehicle mounting six 106-mm recoilless rifles, was highly effective against concrete and steel structures. The munitions of these armored systems provided breaches that the infantry could exploit.

Armor providing direct-fire support inside a built-up area requires protection by infantry.

In contrast, during the **Suez City** battle, Israeli armor forces attacked on "armor thrust avenues" into the city, outpacing their armored personnel carrier (APC) mounted paratroop/infantry support. The Egyptian defenders lacked organic artillery (except limited antiaircraft artillery (AAA) and mortars) and had no air support and virtually no armor support. The Egyptians prepared "kill zones" on the principal avenues down which the IDF armored forces attacked. As the lead IDF armor battalion entered the second of the three road intersection objectives, the Egyptians engaged with Sagger missiles, RPGs, ZSU-23 antiaircraft guns, antitank grenades thrown from balconies, and small arms. All of the tank commanders in the lead battalion were killed or wounded. Disabled vehicles blocked the road. Vehicles veering into the narrow side streets became trapped and were destroyed. *The lack of infantry to protect the armor proved disastrous to the attacking armor battalion.*

(3) Artillery. Artillery has played an important role in most major urban conflicts. At **Aachen,** U.S. forces combined infantry with antitank teams and artillery (in a direct-fire role) down to the squad and fire team level. Also, artillery firing shells with "delay" fuzes in an indirect-fire role were used to penetrate one or more floors before exploding, thus driving the enemy to the ground where infantry and armor could attack. Artillery was also positioned to fire *perpendicular* to the direction of movement of assaulting forces. Thus, fratricide from artillery range errors was alleviated.

Artillery has two distinct roles: outside the built-up area to isolate or prevent isolation with indirect-fire; and within the built-up area to provide direct-fire support.

Artillery proved most useful for interdicting enemy supplies, enemy evacuation, and the movement of reinforcements in the enemy's rear; for physically and psychologically harassing the enemy; and in direct-fire roles within a city.

Artillery employed in the indirect-fire role has been effective in disrupting defenders in half of the studied battles. However, some indirect-fire roles have proved more effective than others in urban combat. Artillery was most effective in the interdiction of supplies, enemy evacuation, movement of reinforcements in the enemy's rear (outside the city), and for indiscriminate physical and psychological harassment of the enemy. Artillery used as an indirect-fire siege weapon, as was done at **Ashrafiyeh** and **Zahle**, proved ineffective. Artillery can also cause problems for the attacker. The rubble resulting from indirect artillery fires can create considerable obstacles for the attacker while providing the defender with obstacles, materials, cover, and concealment.

Artillery employed in the direct-fire role proved useful in the reduction of strongpoints. Self-propelled artillery was used effectively inside **Aachen** and **Stalingrad** and more recently by the IDF at **Beirut II**.

(4) Mortars. The mortar is the most used indirect-fire weapon in urban combat. The mortar's high angle of fire allows the round to reach the street level accurately without being masked by surrounding structures. During the battle for **Hue** city, the most effective fire support provided to Marines was the indirect fire from 60-mm, 81-mm, and 4.2-inch mortars. Approximately 20,000 rounds of high explosive (HE) 60-mm and 81-mm mortar ammunition were expended during the battle. The 4.2-inch mortar was used primarily to deliver riot control agent munitions. It was discovered that 4.2-inch riot control agent munitions could be fired with great effectiveness through the tops of buildings to drive the enemy out. When the enemy emerged from their concealed positions, HE rounds were fired to complete the attack.

(5) Antiaircraft Artillery. Antiaircraft artillery (AAA) was extremely useful in a ground fire role in some urban battles. AAA was used only rarely in World War II, and then generally against assaulting personnel rather than against structures. AAA has been used more frequently in more recent events, but against buildings rather than people. *The high rates of fire of modern AAA make it excellent in terms of shock and destructive potential.* However, ammunition supply can be a problem because the volume of fire. During **Hue**, the Duster, an AAA vehicle mounting twin 40-mm guns, was provided to the Marines by the U.S. Army. These guns were extremely effective in the suppression of enemy positions due to the lethality of the 40-mm round and the quantities in which it could be delivered. During the **Suez City** battle, the Egyptians used the ZSU-23 antiaircraft gun in the direct-fire role against armored columns. This gun proved to be frightening and effective. Its high volume of explosive power created shock among IDF armor personnel akin to the shock that the IDF armor thrust tactics were designed to create. During **Sidon,** the IDF used 20-mm antiaircraft Vulcans in the direct-fire role against ground targets with great success.

(6) Aviation. Historically, aviation assets have played an important role in helping to isolate the objective and to interdict the flow of the defender's supplies and reinforcements. However, aviation has been relatively ineffective when not used in conjunction with ground forces. In past wars, bombing operations have been used in attempts to reduce the defender's will to resist and destroy their physical capabilities. In the majority of urban battles, aerial bombing, by itself, did not erode the defender's will to resist, nor did it significantly degrade the defender's military capabilities.

Aviation plays an important role in interdicting movement of the defender's supplies and reinforcement into the built-up area.

The Marine air-ground task force is employed as a combined arms team. As such, the aviation combat element will not usually be called upon to conduct independent operations in MOUT. It will function as part of a MAGTF and, regardless of operating on or over urbanized terrain, will still execute the six functions of Marine aviation (i.e., offensive air support, assault support, antiair warfare, electronic warfare, air reconnaissance, and control of aircraft and missiles) in support of the MAGTF.

In future urban warfare, aviation will be even more effective due to advances in fixed- and rotary-wing aircraft, unmanned aerial vehicles, precision guided munitions, improved munitions, communications, sensors, and targeting systems. Our battle study already indicates a trend toward more extensive aviation participation in MOUT. As an example, the IDF in **Beirut II** employed bombing by fixed-wing aircraft using cluster bomb units, "smart" bombs, phosphorous, and other munitions. Attack helicopters operated on the outskirts of the built up area with impunity, and medical evacuation (MEDEVAC) proceeded swiftly and efficiently using helicopter support.

d. Combat Forces. Whether attacking or defending, the size of the force relative to the enemy can be critical to success. When provided with adequate forces, the attacker can isolate and encircle the defender and prevent a breakout or linkup. The defender can use them to create a mobile defense or to create strong reserves for counterattacks.

Other factors that impact on the size of forces required are the degree of surprise achieved and the firepower utilized (aviation, armor, artillery, mortars, etc.). They should be weighed against the sophistication of the prepared defense. Ample consideration should be given to the local population, degree of external support, and utilization of existing services (communications, water, etc.).

Regardless of the size or quality of defensive forces, the defender usually extracts large costs from the attacker in time, resources, and casualties.

In the historical examples reviewed, the defender was usually outnumbered by the attacker, the quality of the defender's available forces was inferior, and defeat of the forces defending the city was usually certain. However, regardless of the size or quality of defensive forces, the defender can extract enormous costs to the attacker in time, resources, and casualties. As was seen at **Khorramshahr**, the Iranian defenders, outnumbered 4 to 1, still held the city for 26 days. Although the Iranian defenders eventually lost the city, its defense allowed the remaining Iranian forces time to organize and redeploy. Furthermore, the winter rains that followed the battle turned much of the region into a sea of mud and largely halted further Iraqi efforts. The Iraqi army's offensive thrusts into Iran lost momentum as a result of the defense of Khorramshahr.

e. Special Assault Teams. In these battle studies, "shock units" or "special assault teams" have been used by attackers (and often by defenders) with great success. These organizations were characterized by the integration of combined arms at the battalion level and below. Control was decentralized to lower echelons (down to the squad level in some cases). Assault teams typically contained infantry with various combinations of armor, artillery, and engineers. Hence, all combat arms should plan, train, and develop common tactics, techniques, and procedures for use on urbanized terrain. Before the battle of **Aachen**, U.S. ground forces conducted intensive urban training with antitank teams and artillery pieces down to the squad and fire team levels. In contrast, the defender, in most cases having to fight with whatever forces were locally available, had no opportunity to conduct combined-arms training.

The use of combined arms warfare on urbanized terrain is imperative.

f. Time. In most cases, successful conclusion of an urban battle took two to three times longer than the initial estimates. Consequently, the additional tome resulted in the expenditure of more logistics and the loss of more personnel then initially anticipated. This often had adverse affects on the overall campaign. Well-planned urban defense, even if the defender is isolated or lacking in aviation, armor, or artillery weapons, can be time consuming to the attacker. Time can allow the defender to reorganize, redeploy, or marshal resources in other areas.

Generally, urban warfare is time consuming.

Three battles where time played an unanticipated critical role in the attacker's strategic timetable were **Aachen**, **Khorramshahr**, and **Stalingrad**. In these battles, the defenders delayed the attacker longer than was estimated, resulting in the modification of operational or strategic plans.

g. Isolation. *The attacker won all urban battles where the defender was totally isolated.* Even the partial isolation of the defenders resulted in attackers enjoying a success rate of 80 percent. Conversely, attackers won only 50 percent of the battles in which defenders were not significantly isolated, and those victories came at great cost.

No single factor is more important to the attacker's success than isolation of the urban area.

In most urban battles, some form of isolation occurred as a result of the attacker's actions. While it is unrealistic to envision complete isolation of a city until enemy forces to the rear of the city are pushed far beyond its outermost boundaries, *total isolation does not appear necessary.* The key to the attacker's success is in stemming the *unimpeded* flow of manpower, supplies, and weapons to replace the defender's losses.

h. Cost. The cost of conducting urban warfare is relative to the percentage of total expended resources, the time elapsed, and the results achieved. The cost to the attacker was considered high in the majority of urban battles. A high cost does not necessarily imply that the results were not worth the price. *The attacker and defender must thoroughly evaluate the overall cost prior to committing to an urban battle.*

From the offensive point of view, several factors can be associated with cost. First, and most importantly, isolating the urban area is critical to the attacker's success. Second, overwhelming superiority is needed if all costs are to be minimized. Third, the operation should be carefully planned. Fourth, intelligence is invaluable. Knowing where and how the city has been prepared for defense is also important. Fifth, attacking forces should understand the unique nature of urban combat. Clear tactics, techniques, and procedures for urban combat is required. Every aspect, from taking a building to using destructive technology and coordinating combined arms, must be thoroughly understood. Careful consideration of these five factors can minimize the cost of urban warfare to the attacker.

Attacker cost was generally high in casualties, time, and resources in the majority of urban battles studied.

From the standpoint of the defender, the critical variable is the defensive preparation of the city. Defensive preparations should include measures to prevent isolation of the city by the attackers. The capture of a prepared city can be made to be extremely costly. Preparations can include creating kill zones, clearing fields of fire, constructing canalizing obstacles, establishing reinforcing and fall-back positions, decentralizing command and control, and organizing multiple movement routes above ground, between rooftops, and below ground in subways and sewers. Artillery and aviation support can also be valuable force multipliers in defense.

i. Rules of Engagement (ROE). The nature of the military operation may restrict our use of weapons. The majority of urban battles since 1967 (such as **Beirut II**, **Hue**, **Jerusalem**) have had one or more of the following restrictions imposed on the attacking force:

(1) Minimizing civilian casualties and/or collateral destruction in order to:

- Avoiding alienation of the local population
- Reducing the risk of adverse world or domestic opinion
- Preserving facilities for future use
- Preserving cultural facilities and grounds.

(2) Limiting the use of specific ground or air weapons.

j. Logistics. Timely combat service support, particularly in the areas of ammunition resupply and casualty treatment and evacuation, is a critical element in MOUT.

(1) Logistics Support. Historically, combat in urban environments has seen a dramatic increase in the amounts of Class IV (shoring, sandbags, concertina wire), Class V (ammunition), and Class VIII (medical material). Unique items, such as rope, grappling hooks, and ladders are required for operations on urbanized terrain. Intense close-quarter combat requires a continuous flow of ammunition, particularly small-arms, tank, antitank, mortar, and artillery ammunition, as well as mines, grenades, and demolition explosives. Medical supplies must be readily available to treat the anticipated increase in casualties. Once battle the has been initiated, combat forces will require continuous supply.

Urban warfare requires a flexible, balanced logistics system capable of sustaining the close-quarter fighting evolutions within the built-up area.

One method of providing continuous logistics support is to establish a "push system." Essentially, this system pushes supplies to fighting units without their having to request them. A "push system" should negate the inherent delays of a "pull system," which requires units to request supplies then await their arrival. Optimally however, a balanced approach should be taken to sustain engaged forces (MCDP 4.). Combat service support, in terms of its timeliness and anticipation of the true needs of the combat units, should sustain the tempo of operations. It should avoid the delivery of unwanted or unnecessary supplies. To achieve this balanced approach, logisticians must develop flexible support plans that readily adapt to the ebb and flow of urban combat.

Mobile aid stations and surgical teams should be positioned as far forward as the tactical situation allows and be provided with dedicated evacuation vehicles.

(2) Health Service Support (HSS). Responsive treatment and evacuation plans should be established to handle the expected increase in casualties. Both immediate treatment and evacuation are critical in maintaining the morale and confidence of forces engaged in urban warfare. To meet casualty treatment and evacuation needs, mobile aid stations with surgical teams should be placed as far forward as the tactical situation allows. Dedicated vehicles and aircraft for the evacuation of casualties should be provided.

In addition to casualties resulting from physical injury on the battlefield, there are often considerable psychological casualties in modern battles, particularly urban battles. This can be attributed to the intense, sustained, and close-quarter combat associated with urban warfare. In many battles, attacking troops were not properly trained for urban warfare and felt inadequate in their ability to fight on equal terms with the defender. Many felt moral or ethical dilemmas associated with collateral damage and civilian casualties. Finally, the constant threat of being killed from any quarter coupled with the extreme fatigue of fighting in an urban environment can create psychological casualties. Leaders at all levels should be attuned to the symptoms associated with psychological casualties in order to get effected individuals prompt treatment so that they can be returned to their units.

1006. Implications of Urban Warfare. The commander charged with making decisions needs to understand the operational and strategic implications of a tactical struggle in an urban area (MCDP 1-1, *Strategy*, and MCDP 1-2, *Campaigning*.). Three urban battles (**Stalingrad**, **Hue**, and **Beirut II**) illustrate the importance of seeing beyond the tactical nature of the battle.

Commanders must see beyond the immediate tactical implications of fighting an urban battle.

The battle for **Stalingrad** had major operational and strategic implications. The entanglement of German forces at Stalingrad bought time for the Soviets to mobilize and prevent a major combined effort of two German Army Groups in the south. While Stalingrad resulted operationally in the destruction of the German Sixth Army, the strategic results were even greater. Stalingrad caused a complete change of German strategy in the east. Hitler made major changes in his General Staff, and from this point on in the war, he was a man estranged from his military leadership. The loss of enough men and equipment to field one-fourth of the German Army shook the foundations of the Third Reich.

The Battle for **Hue**, although only one of over one hundred different attacks of the Tet Offensive of 1968, had a negative impact on the will of both the American people and their political leadership. Hue marked a revolution in the coverage of war by modern mass media. It was the first time Americans could and watch an ongoing battle from their living room on the evening news. Hue was a television bonanza for almost a month. When North Vietnamese leadership directed that Hue be held for at least seven days, it was clearly *not* their intent to win a tactical battle, but to strike at the *strategic center of gravity*—in this case, the will of the American people. Although the battle for Hue was a tactical victory for the U.S., the North Vietnamese had achieved their strategic goal of making the American public question the costs associated with the war.

During the battle of **Beirut II**, the IDF's objective was to drive the PLO from Lebanon. The ability of the PLO to leverage the media to gain an advantage was one of the most significant weapons in their arsenal. Despite an Israeli tactical victory, the costs in image, prestige, allies, and, most importantly, its own national will were enormous. The results of this battle eventually caused a change of political leadership at the highest levels of government.

1007. Key Insights. Analysis of modern urban battles provides insight into the major factors affecting the course and outcome of combat. Consideration of these key insights are required in order to effectively and efficiently plan and execute MOUT:

a. MOUT is infantry intensive.

b. A tactical battle may have far-reaching implications. No longer are the strategic, operational, and tactical levels overlap, creating the situation where tactical actions can have operational and possibly strategic repercussions (MCDP 1-1, *Strategy*, and MCDP 1-2, *Campaigning*).

c. Commanders at all levels must understand the impact that media representation will have on the accomplishment of operational and strategic objectives.

d. Maneuver warfare doctrine must be applied to the environment.

e. Intelligence is imperative to success in urban warfare.

f. Surprise is a combat multiplier.

g. Armor, artillery, and aviation are effective at the outer perimeter of built-up areas for causing isolation or preventing reinforcement.

h. Armor operating inside a built-up area must be protected by infantry.

i. Artillery providing direct fire inside a built-up area can be effective in the reduction of strongpoints.

j. As force ratio increases in favor of the attacker, combat duration decreases.

k. Urban warfare is time consuming.

l. Isolation of an urban defender ultimately ensures his defeat.

m. Attack of an urban area is costly to the attacker in terms of resources and casualties.

1008. Necessity for Preparation. Since MAGTFs are usually forward-deployed forces, it is inevitable that MAGTFs will be tasked to deploy to urban areas and conduct military operations. Fighting in this environment will be violent, close, and personal. Fighting in villages, towns, and cities are likely to become more frequent for military forces as the United States responds to global crises. Forces will be expected to exercise adaptability in responding to a wide variety of missions in the world's built-up areas. Marines may soon find themselves operating in the conceptual "three block battlefield" within a city grid: feeding refugees in one block, separating combatants in the adjoining block, and returning hostile fire in a third block.

MOUT is infantry intensive. The remaining chapters and appendices of this publication focus on how the GCE fights and integrates supporting elements in an urban environment.

Chapter 2

Offensive Operations

Urban combat isolates and separates units. Operations are reduced to a series of small-unit actions, placing a premium on small-unit leadership, initiative, and skill.

2001. Introduction. The Marine Corps' manuever warfare philosophy guides the conduct of offensive operations in MOUT. This warfighting philosophy serves to guide commanders through the development, planning, and execution of missions. A detailed discussion of maneuver warfare can be found in MCDP 1, *Warfighting*, MCDP 1-3, *Tactics*, MCWP 3-1, *Ground Combat Operations* (under development), and other Marine Corps doctrinal publications.

This chapter focuses only on offensive operations in an urban environment where collateral damage is of minimal concern. It describes tactics, techniques and procedures to be employed for seizing and clearing buildings and other urbanized areas. Chapter 6 addresses some considerations associated with fighting under constraints and restraints on urbanized terrain.

Section I

Planning

2101. Considerations. Identification of the adversary's centers of gravity and critical vulnerabilities will allow the commander to focus his efforts on those portions of the built-up area essential for mission accomplishment.

a. Reasons for Attacking a Built-Up Area. A commander considers the following before deciding to attack a built-up area:

(1) Tactical Advantage. Cities control key routes of commerce and provide a tactical advantage to the commander who controls them. Control of features such as bridges, railways, and road networks can have a significant impact on future operations. Urbanized areas may be used by the enemy as a base of operations from which they launch their own offensive operations. It may be advantageous to attack those bases and separate the enemy from their support infrastructure.

(2) Political Advantage. The political importance of a built-up area may justify the use of time and resources to liberate it. Capturing a city could destroy the seat of local and national government. At the very least, it could deal the enemy a decisive psychological blow.

(3) Economical Advantage. The destruction or capture of key industrial and commercial cities with the resulting denial of production and distribution of equipment and supplies strikes at the enemy's future ability to wage war. The requirement for a logistics base, especially a port or airfield, may play a pivotal role in the enemy's ability to continue the conflict. Capture of such cities may prove extremely beneficial to the attackers, who can use these resources to their advantage.

(4) Potential Threats to Operations

(a) Enemy Threat Too Great To Bypass. Though the terrain around a built-up area may facilitate its bypass, the enemy within that urbanized area may remain a threat capable of interdicting lines of communications. This may require the enemy force to be contained or destroyed.

(b) Terrain Does Not Allow Bypass. The urbanized area may sit between two natural slopes on the avenue of approach and thus require capture in order to secure the main supply route. Additionally, the urbanized area, itself, may sit on dominating terrain that threatens combat support and CSS elements.

b. Reasons for Not Attacking a Built-Up Area. The commander considers the following reasons for not attacking a built-up area:

(1) Built-Up Area Not Required To Support Future Operations. The attacker may have adequate supply support and resources established at another site from which combat forces can be supported. Therefore, during the estimate process, commanders may assess that the urbanized area is not necessary to support future operations.

(2) Enemy Not a Threat. The commander may decide to bypass upon determination that no substantial threat exists in the built-up area that could affect the unit's ability to accomplish its mission.

(3) Time or Risk Unacceptable. The commander's intent may dictate that speed of movement is essential to the mission. Because MOUT can be time-consuming, the commander may choose to bypass the urbanized area to maintain tempo. Furthermore, the potential for numerous casualties, the expenditure of critical resources, or the restrictions placed on attacking forces may result in unacceptable risks to the commander's primary mission.

(4) Declared an Open City. The area may have been declared an "open city" because it is undefended or of religious or historical significance. By international agreements, open cities are demilitarized and must be neither defended nor attacked. (See HR, 25, Annex to Hague Convention No. IV, *Embodying Regulations Respecting the Laws and Customs of War on Land*, 18 October 1907.) The attacking force must assume civil administrative control and treat the civilians as noncombatants in an occupied country. The defender must immediately evacuate and cannot arm the civilian population. A city can be declared open only before it is attacked. Other reasons for not defending could be the presence of large numbers of noncombatants, hospitals, or wounded personnel or the city's cultural, religious, or historical significance.

2102. Commander's Estimate. Once assigned the mission to attack a built-up area, a thorough mission analysis is conducted. Marines follow the same planning process in MOUT as for other operational environments by evaluating the mission, enemy, terrain and weather, troops and support available, and time available (METT-T). Based on METT-T and application of maneuver warfare concepts, the commander can decide on a plan that defeats the enemy by exploiting identified critical vulnerabilities. The commander may not have sufficient forces to fight everywhere and so must focus on the enemy and concentrate resources to decisive ends. This may involve making conscious decisions to accept risk at some place or time and to seize opportunities at others. The challenges of an urbanized environment dictate an even greater need to fully understand maneuver warfare.

This section addresses METT-T considerations as they pertain to the offense in urban warfare. METT-T is covered in detail in MCRP 3-11.5A, *Marine Troop Leader's Guide.*

a. Mission. Mission analysis is the first step of the estimate process. One of the important mission analysis considerations in urban conflict is the scope of clearance required to accomplish the mission. Commanders and planners should ask themselves the following questions:

- Do I need to clear every building?
- Should I clear only certain blocks?
- Should I only control certain areas?
- What level of protection is required for my lines of communication?

b. Enemy. Maneuver warfare focuses on the enemy. The intelligence preparation of the battlefield (IPB) process should be used to analyze the enemy and terrain in detail. (See Field Manual (FM) 34-130, *Intelligence Preparation of the Battlefield.*) The commander decides whether threat forces are conventional or unconventional.

(1) Conventional Forces. Most potential adversaries have adopted techniques of urban combat from either the United States or the former Soviet Union. Therefore, potential adversaries will build their urban defense to counter the attack of a combined arms force. Potential enemy forces will normally organize their defense in depth. Prepared strong points will usually form the perimeter of a larger defense while reserves locate in a separate position within the perimeter. You can expect ambushes to fill gaps in the perimeter while dummy strongpoints are established to deceive the attacker. Positions for securing the entrances to and exits from underground structures and routes are normally established. Security positions will normally be located forward of first-echelon defensive positions.

(2) Unconventional Forces. Urban areas have become a haven for unconventional forces. The large noncombatant population provides cover and concealment for unconventional force operations. Conventional forces operating in MOUT will normally be placed under restrictive ROE to minimize collateral damage. Unconventional forces will often use our restrictive ROE and the noncombatant population to their advantage when devising an urban defense. (See FM 34-130, FM 7-98, *Operations in a Low-Intensity Conflict*, and Chapter 6 of this manual.)

c. Terrain and Weather

(1) Terrain. Offensive operations should be tailored to the urban environment based on a detailed analysis of urbanized terrain (see Appendix I). Commanders and subordinate leaders use **KOCOA** (**k**ey terrain, **o**bservation and fields of fire, **c**over and concealment, **o**bstacles, and **a**venues of approach) to identify important terrain factors:

(a) K: Key Terrain. Key terrain surrounding an urban area can facilitate entry or deny escape. Within the city, airports or airfields, stadiums, parks, sports fields, school playgrounds, public buildings, road junctions, bridges, or industrial facilities may be key terrain.

Critical public buildings are identified during the terrain-analysis phase of an IPB. Hospitals, clinics, and surgical facilities are important because the laws of war prohibit their attack when not being used for military purposes other than medical support. The locations of civil defense, air raid shelters, and food supplies are critical in dealing with civilian affairs. Additionally, population size, locations, and density; density of the built-up area; firefighting capabilities; the location of hazardous materials; police and security capabilities; civil evacuation plans; and key public buildings should be identified.

(b) O: Observation and Fields of Fire. Urbanized terrain is characterized by restrictive observation and fields of fire. Weapon ranges can be greatly reduced because of buildings and other manmade structures. On the other hand, high ground or tall buildings can provide perches which enhance line of sight (LOS) for observation and communications as well as for individual and crew-served weapons. This includes man-portable surface-to-air missiles.

(c) C: Cover and Concealment. Buildings, sewers, and subways can provide excellent cover and concealment for enemy and friendly forces. They also provide covered and/or concealed maneuver routes within the built-up area. The civilian population can also offer cover and concealment to enemy forces.

(d) O: Obstacles. Natural or manmade obstacles restrict or deny maneuver within the urban area. Bridges, walls/fences, canals, streams, rivers, as well as rubble created by the effects of weapons should be thoroughly analyzed. Construction sites and commercial operations such as lumberyards, brickyards, steelyards, and railroad maintenance yards are primary sources of obstacle and barrier construction materials.

These sites can also supply engineers with materials to strengthen existing obstacles or to set up antitank hedgehogs or crib-type roadblocks.

(e) A: Avenues of Approach. Avenues to the city should support maneuver and be concealed either by terrain, darkness, smoke, or a combination of the three. Avenues that canalize or choke maneuver, due to the density of built-up area or natural terrain, should be identified and avoided. Roads, rivers, streams, and bridges provide high-speed avenues for movement. **Generally, military maps do not provide enough detail for urbanized terrain analysis.** They usually do not show the underground sewer system, subways, underground water system, mass transit routes, and power plants. Local maps of intracity road networks and subway systems and city hall/department of public works blueprints of all city buildings (if available), coupled with aerial photos, should be used.

(2) Weather. As in any military operation, weather affects equipment, terrain, and visibility, but its greatest impact is on the individual Marine. Snow, ice, dust, wind, rain, humidity, and temperature extremes reduce human efficiency. Weather extremes coupled with stress and the physical strain of urban combat can be minimized with effective small-unit leadership. Weather factors include:

(a) Precipitation. Rain or melting snow may flood basements and subway systems. This is especially true when automatic pumping facilities that normally handle rising water are deprived of power. Flooding makes storm and other sewer systems hazardous or impassable. In an NBC environment, chemical agents can be washed into underground systems by precipitation. As a result, these systems may contain chemical agent concentrations that are much higher than surface areas and thus become contaminated "hot spots." Hot spot effects become more pronounced as agents are absorbed by brick or unsealed concrete sewer walls.

(b) Fog. Many major cities experience fog, especially those located in low-lying areas and along canals or rivers. Industrial and transportation areas are the most likely to be affected by fog because of their proximity to waterways. Fog adversely affects vision and optical aids. It may also be used to help conceal friendly movement.

(c) Inversion Layers. Air inversion layers are common over cities, especially cities located in low-lying "bowls" or in river valleys. Inversion layers trap dust, smoke, chemical agents, and other pollutants, reducing visibility and often creating a greenhouse effect, which raises ground and air temperatures.

(d) Temperature. Built-up areas are often warmer than surrounding open areas during both summer and winter. This difference can be as great as 10 to 20 degrees hotter throughout the year.

(e) Wind Effects. Generally, wind chill is not as pronounced in built-up areas. However, the configuration of streets, especially in the city core and outlying high-rise areas, can cause wind tunneling. This increases the effects of the wind along streets that parallel the wind direction, while cross-streets remain relatively well protected. Wind tunneling can have a negative impact on aviation support within the urbanized area by creating dangerous wind effects down streets and between buildings.

(f) Day/Night Differences. Night and periods of reduced visibility favor surprise, which in turn, may facilitate infiltration, detailed reconnaissance, attacks across open areas, seizure of defended strongpoints, and reduction of defended obstacles. Because of the difficulties of night navigation in restrictive terrain (usually without reference points and in close proximity to the enemy) forces may consider using simple maneuver plans with easily recognizable objectives.

Rotary-wing aircraft are best utilized at night when operating on urbanized terrain. The most effective method of protecting helicopters from a successful engagement by hostile ground forces is to remain unseen. During night flights, the use of night vision goggles (NVGs) should be weighed carefully. Artificial lights can render NVGs partially ineffective and can increase the possibility of mid-air collisions by degrading the goggles to the point where nearby aircraft and aerial obstructions are not seen.

(g) Aviation Weather Considerations. Weather conditions can be a critical factor in determining the amount of aviation support the aviation combat element can provide in an urbanized area. Aviation forces should consider the following:

- Presence or absence of fog, industrial haze, low clouds, heavy rain, and other factors that limit visibility for pilots
- Illumination and moon phase/angle during night vision goggle operations
- Ice, sleet, and freezing rain that degrade aerodynamic efficiency
- Updrafts and downdrafts in the urban canyon that may present an unpredictable hazard
- High temperatures and/or high-density altitudes that degrade aircraft engine performance and lift capability
- High winds (large gust spreads) and crosswinds, to include the tunneling effect, that may create localized and unpredictable hazards to aviation
- Weather conditions that create hazards on pick-up zones and LZs, such as blowing dust, sand, or snow.

d. Troops and Support Available. The commander must be aware of all of his combat power (e.g., types of weapons systems, their numbers, and their capabilities) when performing analysis for operations in a built-up area. In the initial planning phase, force size must be evaluated in relation to urban size, enemy forces, and mission assignment. In the attack of a built-up area (population 100,000+), the GCE of a MEF would be a Marine division. However, in the initial stages of an amphibious or MPF operation, a reinforced infantry battalion or regiment may provide the initial forces. No matter what the size GCE, operations will be conducted by task-organized, reinforced battalions and their companies/platoons. Therefore, by using the infantry battalion as the basis for projecting the size of an overall force, the number of regiments or divisions required to secure a built-up area can be determined. Whatever size MAGTF is required for operations in an urbanized area, much of the fighting may be generally conducted by small teams of Marines. Consequently, our success in urban fighting may largely depend upon small unit and individual tactical skills. Specifically, leaders at all levels must analyze the following factors:

- Number and type of available units
- Task organization for urban combat (See Appendix A.)
- Availability of critical weapons systems
- State of training and discipline (training for urban warfare is imperative)
- Strength in terms of men and materiel in relation to enemy and size of built-up area
- Aviation support available
- State of maintenance and supplies
- Available combat support (based on availability and mission requirements)
- Available CSS (critical logistical and maintenance items)
- Host nation support available. (If the city's occupants are friendly to the attacker, then support in intelligence, deception, and diversion may be possible.)

e. Time Available. The following issues should be considered when analyzing the time available for an attack on urbanized terrain:

- A significant amount of time is required for clearing buildings, blocks, or axes of advance.
- Marines tire more quickly when clearing buildings because of stress and additional physical exertion.

- Adequate time should be allocated for thorough reconnaissance, planning, and rehearsals. Ultimately, this can result in higher tempo in the execution of the plan.

Once an estimate of the situation (using METT-T) has been conducted, planning for offensive operations can proceed. It is important to remember that the initial estimate of the situation must be continually updated throughout the planning process and the conduct of operations to account for changes in METT-T. (See MCWP 5-1 *Marine Corps Planning* [under development])

2103. Phases of the Attack. Attacks are categorized as either hasty or deliberate. Both hasty and deliberate attacks should take advantage of as much planning, reconnaissance, and coordination as time and the situation permit. Regardless of the size of the attacking force or of the objective to be secured, the phases of the attack (i.e. reconnoiter, isolate, secure foothold, and control the objective) remain constant.

a. Deliberate Attack. A deliberate attack *is a type of offensive action characterized by preplanned coordinated employment of firepower and maneuver to close with and destroy or capture the enemy* (Jt Pub 1-02). It is a fully coordinated operation that employs the MAGTF's combined arms team against the enemy's defense. It is used when enemy positions are well prepared, when the built-up area is large or severely congested, or when the element of surprise is lost. Given the nature of urbanized terrain, the deliberate attack of a built-up area is similar to the assault of a strongpoint (see MCWP 3-1, *Ground Combat Operations* [under development]). The deliberate attack of a built-up area is conducted in the following phases:

(a) Phase I: Reconnoiter the Objective. Intelligence gathering and reconnaissance/ surveillance are critical to the planning process and success of the operation. All available intelligence resources should be used to gain vital information on the enemy. Whenever possible, a leader makes a personal reconnaissance of the objective area to collect first-hand information regarding the area to be attacked. A thorough reconnaissance yields a wealth of information to use in the continued development of the plan. Avenues of approach, observation posts (OPs), supply routes, and the emplacement positions of direct- and indirect-fire weapons systems are all examples of information that may be amassed during the reconnaissance of the objective area. Composition and structure of buildings and roadbeds, cover and concealment opportunities, and other information not apparent in a map study may have a significant impact on the plan.

(b) Phase II: Isolate the Objective. The objective can be isolated by seizing natural and man-made features that dominate the area. Isolation may also be accomplished by coordinated use of supporting arms to seal off enemy lines of communication. This phase may be conducted simultaneously with Phase III (securing a foothold). Figure 2-1 shows a task-organized, reinforced infantry battalion isolating the objective.

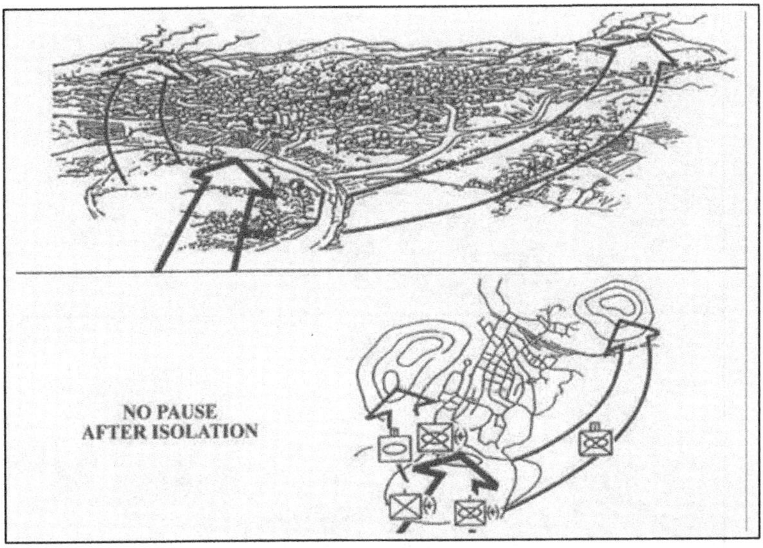

(c) Phase III: Secure a Foothold. Once the objective is isolated, a foothold should be secured as soon as possible in order to maintain tempo. The seizure of a foothold provides the attacking force with a position from which they can continue the assault through the objective area. The assault is supported by direct and indirect suppressive fires and smoke. Figure 2-2 (pg. 2-10) shows the battalion objective being isolated. One company is providing suppressive fires while another company seizes an initial foothold. Fire support assets are providing suppressive fires on the objective.

Figure 2-1. Isolation of the Objective Area

(d) Phase IV: Seizing the Objective. Once a foothold is seized and consolidated, supporting forces move to the built-up area to support the seizing of the objective area. To maintain tempo, the transition between the phases should be seamless. Once the foothold has been established, forward units continue the attack through the objective area. Supporting units assist as required. The momentum of the assault is continued until the objective area is cleared or controlled.

The assault force should establish limited objectives to ensure that the attacking forces do not get strung out along the axis of advance. Gaps may give the enemy the opportunity to infiltrate along the line of advance or make isolated friendly forces vulnerable to attack.

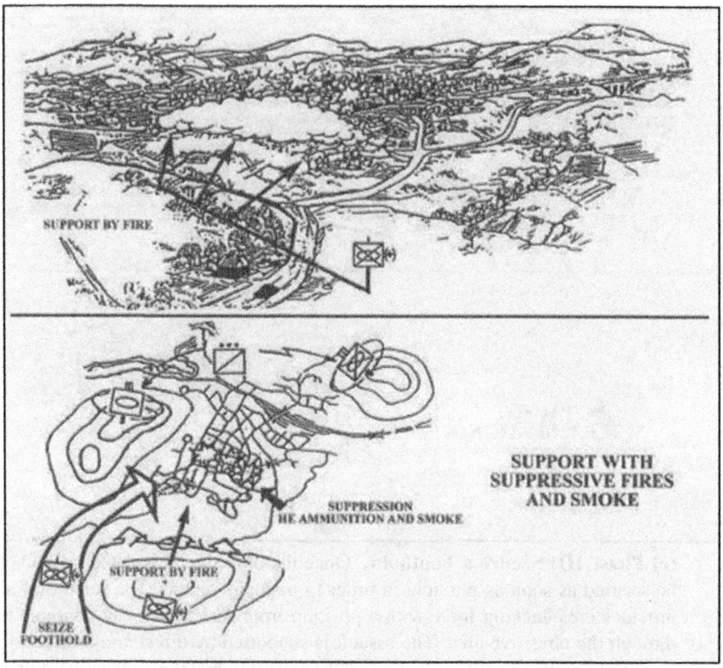

This phase also includes consolidation of the objective area and reorganization for future taskings. By setting limited objectives, the attacking forces have an opportunity to reorganize and defend against counterattacks while maintaining momentum. Once the limited objectives have been secured, the urbanized area may be sectored for detailed clearing. Clearing by sectors may help in the allocation of forces and in the prevention of fratricide. In addition to defensive preparation against counterattack, preparation should be made for follow-on missions (e.g., restoration of civilian facilities, marking and clearing minefields, etc.).

Figure 2-2. Seizing a Foothold

Seizing may consist of a systematic house-by-house, block-by-block advance through the entire zone of action, or it may be a rapid advance through a lightly defended area to seize a key objective. The difference between the two techniques is made by the commander on the basis of mission requirements and METT-T analysis. Seizing an objective in urbanized terrain requires detailed planning, coordination, decentralized command and contol, and small-unit execution. It is a continuation of the assault until all mission requirements are met.

(1) Rapid Advance. Rapid advance (Figure 2-3) may be used to maintain

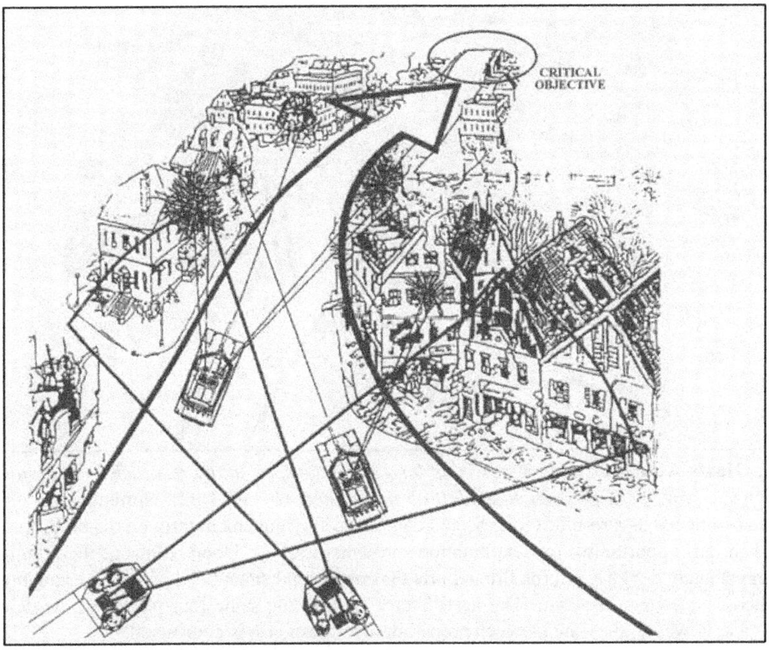

momentum through the zone of action and to seize key objectives. Some enemy positions may be isolated or bypassed in order to maintain tempo. This procedure may be used when:

- A key objective is identified (an enemy critical vulnerability or an objective that supports the friendly scheme of maneuver)
- Time is critical in reaching the objective
- Only selected buildings on the route of advance require clearing.

Figure 2-3. Rapid Advance En Route to a Key Objective

(2) Systematic Clearance. Systematic clearance (figure 2-4 on pg. 2-12) involves the deliberate reduction of all significant enemy positions throughout the objective area. It is usually used when time is not a critical factor.

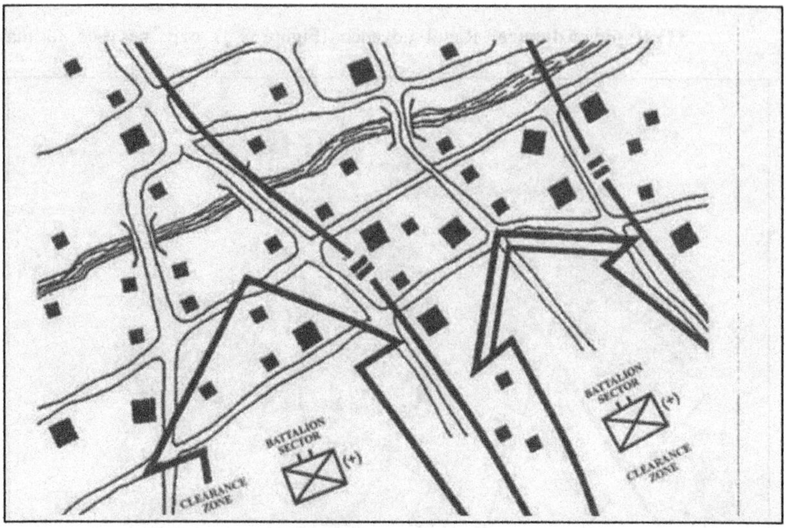

b. Hasty Attack. A hasty attack is, *in land operations, an attack in which preparation time is traded for speed in order to exploit an opportunity* (Jt Pub 1-02). Normally, hasty attacks are conducted as a result of unexpected contact during movement through the built-up area or when an opportunity for exploitation presents itself. Upon contact, the commander immediately deploys, returns fire, reports the contact and situation, develops the situation, and chooses a course of action. The hasty attack follows the same four phases as the deliberate attack; however, the time between preparation and response is compacted.

2104. Warfighting Functions. The warfighting functions (WFs) encompass all activities in the battlespace. Planners consider and integrate the warfighting functions when analyzing how best

Figure 2-4. Systematic Clearance Within Assigned Sectors

to accomplish the mission. The WFs include maneuver, intelligence, fires, logistics, command and control, and force protection.

a. Command and Control. The commander positions himself where he can best observe and influence the battle. Positioning is relative to the level of command. For example, a squad leader locates where he can observe and direct his fire teams, while the battalion commander locates where he can observe and direct his companies.

Close-quarter fighting in and around buildings makes command and control exceedingly difficult. Commanders and small-unit leaders rely on guidance and intent provided by higher

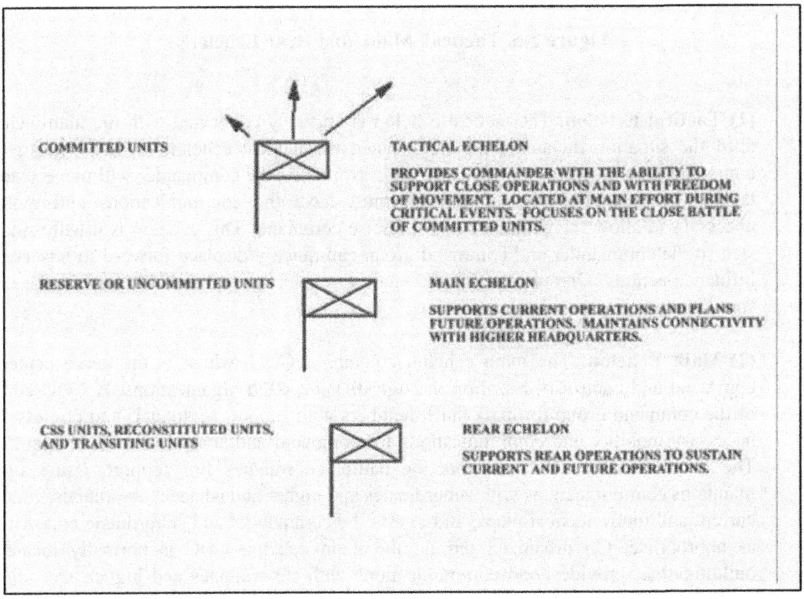

headquarters to facilitate decentralized execution while maintaining control and direction. In close-quarter fighting, commanders may have to rely upon messengers to communicate with their units and higher headquarters.

At battalion and higher levels, command and control is accomplished through three command echelons, which can be collocated or echeloned across the battlefield. The three echelons are tactical, main, and rear (see fig. 2-5). Each echelon may have a combat operations center (COC) capability. The COCs are the facilities (buildings, vehicles, or tents) used by the commander or staff at each echelon to plan, direct, control, and coordinate operations.

Figure 2-5. Tactical, Main, and Rear Echelons

(1) Tactical Echelon. The tactical echelon is normally collocated with the main echelon until the situation dictates separation. When the tactical echelon separates, designated command group members displace forward. Normally, the commander will move with the tactical echelon. The tactical echelon must have the communications and mobility necessary to allow the commander to exercise command. This echelon is usually small in size so the commander and command group can quickly displace forward to observe and influence actions. Organization of the tactical echelon should be addressed in the unit's standing operating procedures (SOPs).

(2) Main Echelon. The main echelon contains a COC, which is the nerve center for command and control of battalion through division-sized organizations. A COC consists of the command group (primary staff members with support personnel) and possesses the necessary mobility and communications for command and control of tactical operations. The main echelon COC monitors the battle, coordinates fire support, issues orders, maintains communications with subordinates and higher and adjacent commands, conducts current and future tactical plans, and assists the commander and subordinate commanders as appropriate. On urbanized terrain, the main echelon COC is normally located in buildings that provide good communications with subordinates and higher and adjacent commands. The command post (CP) is where the commander is physically located. Therefore, the CP can be located at any of the three echelons based on the commander's presence.

(3) Rear Echelon. The rear echelon's COC consists of the logistical and administrative personnel required to plan, coordinate, and execute logistics support. During urban warfare, the rear echelon COC is usually closer to combat forces than normal as it uses the city's captured buildings and facilities to provide cover and concealment for stocking supplies, conducting maintenance repair, and treating the injured.

Attack of a built-up area requires detailed planning due to restrictive terrain, close proximity to opposing forces, fire coordination requirements, and reduced communications capability. The GCE will frame a centralized plan of attack, fire support, and logistics support to conduct urban operations. However, the actual attack of a built-up area is a series of isolated, close-quarter battles carried out by small units. The execution of these attacks should be decentralized.

Communications planning procedures, as outlined in MCWP 3-40.3, *Communications and Information Systems* (under development), are applicable to the urban battle. Planners should be meticulous in their scrutiny of the environment and the effect it will have on communications. Manmade structures can create problems for single-channel radios. These structures inhibit LOS radio communications by absorbing or reflecting transmitted signals. However, the urban environment may have exploitable advantages such as the availability of electrical power, commercial telecommunications networks, and environmental control systems. Electrical power generation stations and other emergency power systems are normally found in protected structures and are probably usable. Due to their value to the enemy, these areas are more likely to be heavily defended. Enclosed areas offer excellent concealment and protection of communications and other command and control support equipment. Extensive commercial communications networks composed of miles of underground protected cable connecting central telephone exchanges are likely to be available, as well as a multitude of public service radio nets (police, fire, civil defense, taxi, etc.) complete with existing antenna systems and retransmission stations. To communicate effectively and continuously, commanders must minimize limitations imposed by the urban environment and maximize the advantage of existing civil communications.

b. Intelligence. Collection of intelligence information on urbanized terrain is difficult. Urbanized terrain provides overhead cover and concealment from observation by aircraft, reconnaissance personnel, and satellites.

(1) Information Requirements. In addition to information about the location of enemy units, nature of the terrain, trafficability of roads, etc., that all operations require, urban warfare requires detailed information on the following:

- Population density, characteristics, and culture
- Location and quality of water supply
- Width and construction of streets
- Layout of road networks
- Location, width, and load capacity of bridges
- Layout of underground networks such as subways, sewers, and utility ducts
- Types and construction of buildings
- Location of key communications and transportation facilities
- Location of canals and waterways.

Information on port facilities should include details on harbor sites, pier networks, unloading capabilities, warehouse/dockside facilities, harbor currents, and sandbars or other midstream obstructions.

(2) Collection. Accurate, detailed, and timely information is vital to the success of any urban operation. Collection operations conducted by national intelligence organizations may provide valuable information on the layout of the city, recent changes to transportation routes, political affiliations of various groups, and locations of military units. This information needs to be current and must be verified and updated as attacking forces close on the city. Reconnaissance activities such as the study of large-scale maps, aerial photographs, and background intelligence reports keep the commander's information current. The latest information on the enemy and the terrain should be continuously fed to the intelligence section using long-range reconnaissance patrols, human intelligence (HUMINT) reports, unmanned aerial vehicle (UAV) missions, and electronic intelligence. Use of reconnaissance assets should be tempered with a risk-versus-gain judgment. Reconnaissance assets should not be risked indiscriminately if the information can be collected by other means.

(3) Maps. Tactical maps should be supplemented by city street maps that contain street names and information on the location of important buildings and transportation terminals. These city street maps should be distributed to all units operating within the city. Although tactical maps are key in controlling fire support missions, simple street maps assist ground troops in maintaining their orientation within the city and in tracking buildings and areas that have been cleared.

c. Maneuver. Maneuver is the employment of forces on the urban battlespace through movement in combination with fire or fire potential to achieve a position of advantage (see MCWP 3-1). Maneuver is an essential element of combat power. Maneuver in itself cannot produce decisive results. Combined with mass, offensive action, economy of force, and surprise, maneuver provides favorable conditions for closing with the enemy. Maneuver contributes significantly to sustaining the initiative, exploiting success, preserving freedom of action, and reducing vulnerability. It is through maneuver that an inferior force can achieve decisive superiority at the necessary time and place. In many cases, maneuver is made possible only through the control of tempo and effective employment of firepower. The commander integrates supporting fires with the scheme of maneuver to create a dilemma for the enemy. Likewise, movement without fires exposes the force to effective enemy counteraction and risks losing the initiative and momentum. Maneuver that does not include violent action against the enemy will not be decisive. At all levels, successful application of this principle in the urban environment requires flexible and innovative plans and actions.

(1) Infantry/Armor Movement. Armored vehicles are vulnerable in built-up areas where streets and alleys provide ready-made fire lanes for defenders. Motorized traffic is greatly restricted, canalized, and vulnerable to ambush and close-range fire. Tanks are at a further disadvantage because their main guns cannot be depressed sufficiently to fire into basements or be elevated to fire into upper floors of buildings at close range (Figure 2-11). Whenever tanks are used in the direct-fire role, infantry must provide security against enemy ground attack. Movement of armor down narrow streets or down narrow paths through debris requires close coordination with infantry. Infantry and armor may use bounding overwatch techniques during movement (See MCWP 3-12, *Marine Tank Employment* [under development]). With this technique, infantry moves ahead of armored vehicles to clear the buildings on each side. Lead infantry units should not get so far ahead of armor that mutual support is lost. Conversely, if armor gets too far ahead of infantry, it may be quickly damaged or destroyed. Coordinated movement also contains a rear security element to protect armor from behind.

During movement down a street tanks should deploy with one tank on each side of the street focusing on the street and lower levels of buildings. *In MOUT, tanks are best deployed in pairs.* Other tanks of the tank platoon or section should move behind the infantry and fire at targets in the upper stories of buildings. When space is inadequate to deploy abreast, tanks may deploy in single file. Movement across open areas should be well coordinated to ensure mutual support. Suppression fire and smoke are used to cover force movements.

Armor can also be used in the reduction of obstacles. For example, grappling hooks and ropes attached to a tank can be used to pull down wire obstacles.

(2) Cover and Concealment. Walls and buildings can provide cover, concealment, and protection from enemy ATGM or heavy machine gun fire. Before armor enters a building, infantry should clear the building and check the ground floors to ensure that there is no basement into which an armored vehicle could fall and become trapped.

Tank and AAV generated smoke may be useful in concealing the location and movement of assaulting forces. Additionally, the thermal sights of the tank are very effective as they can use heat differentials to detect the movement or presence of concealed enemy soldiers, weapons, and vehicles.

(3) Tactical Considerations for Helicopterborne Movement. Helicopterborne operations are planned and conducted using the same combat fundamentals that apply to ground operations. However, there are considerations that apply only to helicopterborne operations. See MCWP 3-24, *Assault Support* (under development) for a detailed discussion on helicopterborne operations. These considerations include:

(a) Helicopterborne forces, once they have landed, lack vehicular mobility and heavy weapons. For this reason, it is important that the forces land on or near the objective. Such a force may be isolated for a time pending linkup with ground forces.

(b) Helicopters may be used for tactical deception. Helicopters may be used to make demonstration landings in several different zones to deceive the enemy as to the true objective of an operation.

(c) Helicopterborne assaults are normally conducted against undefended or lightly defended objectives. When attacking a well-defended objective, an LZ should be selected which provides cover and concealment for the assault force and is close to the objective.

(d) Helicopterborne forces are vulnerable to attack helicopters, fixed-wing aircraft, and surface-to-air missiles. To counter this vulnerability, friendly attack helicopters

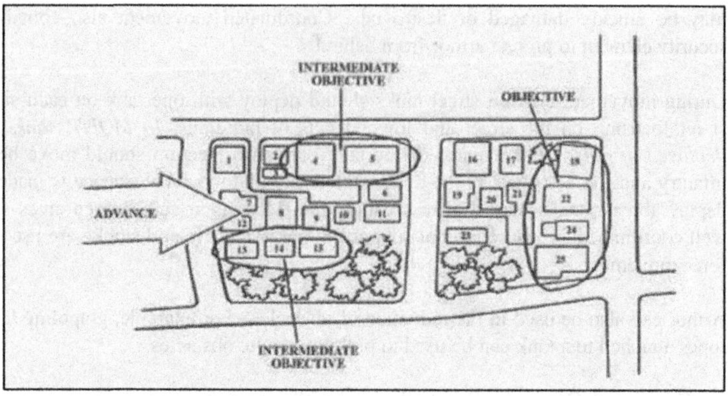

escort the helicopterborne forces during air movement, and indirect fires are used to suppress or neutralize enemy air defenses.

(4) Control Measures. The following control measures assist in the manuever of forces in MOUT:

(a) Objectives. When attacking to seize a foothold, the infantry unit may assign subordinate units the first block of buildings as their first objective. When an objective extends to a street, only the near side of the street is included. The final objective may be buildings or key terrain at the far edge of the built-up area. Key buildings, or groups

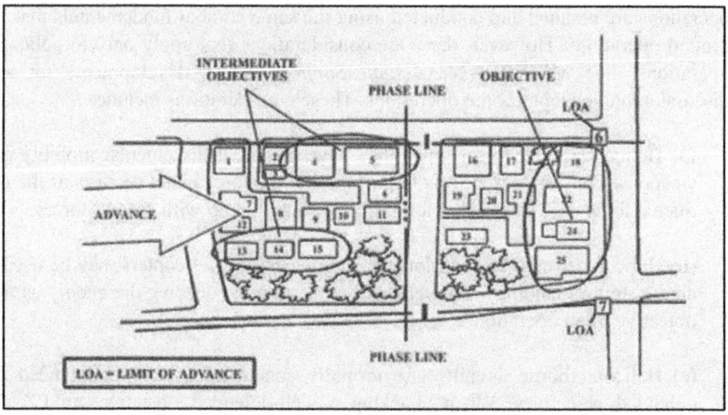

of buildings, may also be assigned as intermediate objectives. Buildings along the route of attack should be identified by numbers for clarification (Figure 2-6 on page 2-18).

During the seizing phase, bypassing buildings may increase the risk of attack from the rear or flank. It may be necessary for the unit to enter, search, and clear each building in its zone of action. A single building may be an objective for a rifle squad or, if the building is large, for a rifle platoon or company. When the commander's concept of operations is based on speed (rapid advance method) or when the force is conducting a hasty attack, a unit may be directed to bypass certain positions within its zone.

Figure 2-6. Objectives and Numbering System

(b) Phase Lines. Phase lines are control measures used to report progress or to control the advance of attacking units. Phase lines should be oriented on readily identifiable terrain features such as principal streets, rivers, and railroad lines. They should be placed on the near side of a street or open area. In an attack, a unit may have the mission to seize positions in its zone of action up to a particular phase line. (Figure 2-7).

Figure 2-7. Phase Lines

(c) Boundaries. Unit boundaries are used to define zones of action and are usually set within blocks so that a street is included in the zone. Both sides of a street should be included within the same unit's boundries.

(d) Checkpoints and Contact Points. Checkpoints aid in reporting locations and controlling movement. Contact points are used to designate specific points where units make physical contact. Checkpoints and contact points are designated by each unit as appropriate for command and control at street corners, buildings, railway crossings, bridges, or any other easily identifiable feature.

(e) Attack Position and Line of Departure (LD). A designated attack position may be occupied by forward units for last-minute preparation and coordination. The attack position is often behind or inside the last large building before crossing the LD. The LD should be located on the near side of an open area running perpendicular to the direction of attack, such as a street or rail line.

(5) Time of Attack. Ideally, the time of attack should be conducted during hours of darkness or limited visibility. Marines can exploit the poor visibility to cross open areas, to gain access to rooftops, to infiltrate enemy areas, and to gain a foothold. When attacking in unrestricted visibility, units should use smoke to conceal movement.

(6) Reserve. Reserves should be mobile and usually follow in trace of forward units. Battalion reserves normally follow one to two blocks to the rear of the lead company. If a company reserve is available, it follows within the same block so that it can immediately influence the attack. A unit with a reserve mission can be called upon to:

- Exploit an enemy weakness or friendly success
- Attack from another direction
- Attrit and/or clear bypassed enemy positions (normally requires designation of a new reserve)
- Secure the rear or a flank
- Maintain contact with adjacent units
- Counterattack

(7) Security. Each unit should ensure security of its flanks and rear (Figure 2-8 on page 2-20). Fighting in a built-up area is unique in that attacks can come from all three dimensions. Units assigned a mission to isolate can also provide security. Small units conduct security patrols and man OPs to supplement higher level reconnaissance and security operations.

Figure 2-8. Enemy Firing From Flank

(8) Engineers. Leading units should have engineers attached for immediate support. Tasks commonly assigned to engineers include:

- Preparing and using explosives to breach walls and obstacles
- Finding and exploding mines in place or helping to remove them

- Cratering roads and performing other countermobility measures
- Clearing rubble and obstacles

Requirements for engineer support in MOUT normally exceed capabilities. To compensate, all Marines should receive basic demolition training prior to conducting urban operations.

d. Fires. The attack of a built-up area may require extensive air and artillery preparation prior to the ground attack. Supporting fires suppress the defender's fire, restricts his movement, and possibly destroys his position. Consideration should be given to the rubbling effect produced by aerial and artillery bombardment. The assault should closely follow air and artillery fire to exploit its immediate effect on the defender. Maneuver units move near the final coordination line while the enemy is engaged by supporting fires. As the attacking force assaults, supporting fires lift and/or shift to block enemy withdrawal or to prevent the enemy from reinforcing their position. (See Chapter 4 for detailed discussion on air and artillery support during urban warfare.) Fire can be categorized into indirect and direct fires.

(1) Indirect Fire

(a) Artillery. Indirect artillery fire is planned to isolate objectives, to prevent reinforcement and resupply, to neutralize known and suspected CPs and OPs, and to suppress enemy defenders. Urbanized terrain requires that most indirect artillery fire use a high-angle trajectory.

(b) Mortars. Mortars are the most responsive indirect-fire weapon for the urban environment. They can be employed to hit targets of opportunity at the close ranges that are typical of combat in built-up areas.

(c) Other. Urban conflict is conducive to innovation. Many direct fire weapons can be employed in the indirect fire mode. Indirect fire weapons may also be effective at clearing the tops of buildings using variable-timed fuses.

(2) Direct Fire.

(a) Armor. Tanks are normally task organized with mechanized units for protection and security operations. The commander can employ tanks to take advantage of their long-range lethality, high-speed mobility, and survivability. They can be used outside the built-up area to cover high-speed armor avenues of approach. Tanks are normally employed in this role during the isolation phase. Tanks also support by fire the assault to seize a foothold and may support attacks to seize objectives within the built-up area. In MOUT, tanks can also be organized into special assault teams. The tank cannon's direct fire is very effective against structures. (Figure 2-9 on page 2-22 and Appendix B)

(b) Howitzers. Artillery rounds delivered by direct fire can be very effective for destroying targets in buildings. When used in this manner, the mobility, flexibility, responsiveness, and survivability of gun crews may be reduced. Artillery used in the direct fire role normally use HE shells.

(c) Infantry. Direct-fire systems, mainly machine guns, antitank guided missiles (ATGMs), shoulder- launched multipurpose assault weapons (SMAWs), and AT4s—are initially employed to support the seizure of a foothold. The positioning of antitank weapons in buildings must allow enough space for the effects of backblasts. Antitank weapons have limited as effectiveness when neutralizing targets behind walls. Designed primarily as armor penetrators, they neutralize a target only if that

Figure 2-9. Tank Direct-Fire Support Within the Built-Up Area

target is located directly behind the point of impact. Consideration must be given to weapons with minimum arming distances. Many of these factors can combine to limit the effects of weapons in the urban environment.

(d) Aviation. The vertical nature of large urban areas may block the pilot's line of sight with the target unless careful consideration is given to inbound headings and altitudes. Plan attack headings to take advantage of large open areas and street axis. Friendly ground forces must be clearly distinguishable from targets. The use of smoke, visible or infrared strobes, or other visual devices must be effectively employed by ground forces and clearly understood by friendly aircraft.

e. Logistics. Logistics support in urban operations should be timely, flexible, and responsive. Logistics planners, both in the supported units and the CSSE, should be fully integrated in the planning process. The final logistics plan must be thoroughly coordinated with the scheme of maneuver. Units may need special equipment such as ropes, grappling hooks, ladders, sandbags, and wire; medical supplies must be readied in anticipation to support increased casualties; and additional supplies must be prestocked. (See Chapter 5 for a detailed discussion on logistics and urban warfare, and MCWP 4-1, *Tactical Logistics*, for additional planning considerations.)

f. Force Protection. Force protection encompasses all those measures that preserve the fighting potential of the command. It is integrated throughout all the warfighting functions. It does not entail security operations alone. It involves the continuous implementation of active and passive measures by all units during all phases of an operation.

Some particular active and passive measures in MOUT include:

-Manmade structures in the urban environment provide cover and concealment for both the attacker and defender.

-Lines of communication can be more easily interrupted due to multiple avenues of approach.

-All units must be prepared to implement force protection in a 3-dimensional environment.

-Proximity of civilian population makes identification of non-combatants more difficult.

-ROE may be more restrictive

-Snipers and man-portable shoulder lauched air defense missiles are usually more difficult to detect.

Section II

Infantry Battalion Operations

2201. Overview. This section addresses some of the operations that a task-organized and/or reinforced infantry battalion could conduct in MOUT. These operations are seldom conducted independently and are usually part of a larger joint operation. In the example given below, the battalion has seized a foothold and is operating in its sector of a town of 100,000 people. Concurrent operations are on-going through other parts of the town. The threat in the battalion sector is a company (-) with a capability to reinforce.

2202. Seizure of a Key Objective. The tactical situation may require the immediate seizure of a building, industrial complex, or transportation site that is deemed decisive terrain. Examples of such objectives may include key bridges, intersections, railheads, or facilities that constitue the center of gravity for the enemy. As an example, a battalion may be assigned the mission of seizing the only bridge that is capable of supporting armored vehicles and resupply by heavy transports along a river bisecting a town. The intent is to isolate enemy forces and secure the bridge for follow-on forces. Intelligence indicates that the bridge is lightly defended at this time. In this case, speed of attack and surprise is critical to capture the objective. Some key intelligence information is required prior to the attack. This would include the number of enemy defending the bridge, their weapons, disposition, their ability to reinforce their defense, location of LZs near the bridge, whether the bridge has been set for demolition, and the location of enemy indirect fire assets that can range the objective.

The commander decides to conduct a penetration using a combination of helicopterborne attack to rapidly seize the bridge followed by a mechanized ground attack to link-up with the helicopterborne forces (Figure 2-10). The helicopterborne force can achieve surprise and speed of movement required while the mechanized force can thrust quickly along high-speed avenues to minimize the time the helicopterborne force must hold the objective against enemy counterattack.

For the ground attack, the battalion commander makes a determination that speed of movement outweighs the risks of rapidly extending his lines of communication. He directs, on the basis of his knowledge of the enemy situation and his mission, that his maneuver forces avoid decisive engagement with the enemy along the route to minimize the time it takes to link up with the helicopterborne force. The commander provides guidance on those enemy forces that should be bypassed and those that should be isolated and reduced.

The commander task organizes his battalion for rapid movement on two axes to allow flexibility in reacting to unexpected enemy contact. Lead and rear security is conducted throughout the operation. As lead units locate enemy positions, they fix them by fire, and quickly reduce or bypass them. They also report the contact so that follow-on forces can then deal with the threat.

As the ground attack force reaches the edge of the built-up area, Marines remain mounted so as not to slow the advance. All vehicle machine guns fire to flanks and into doors, windows, and

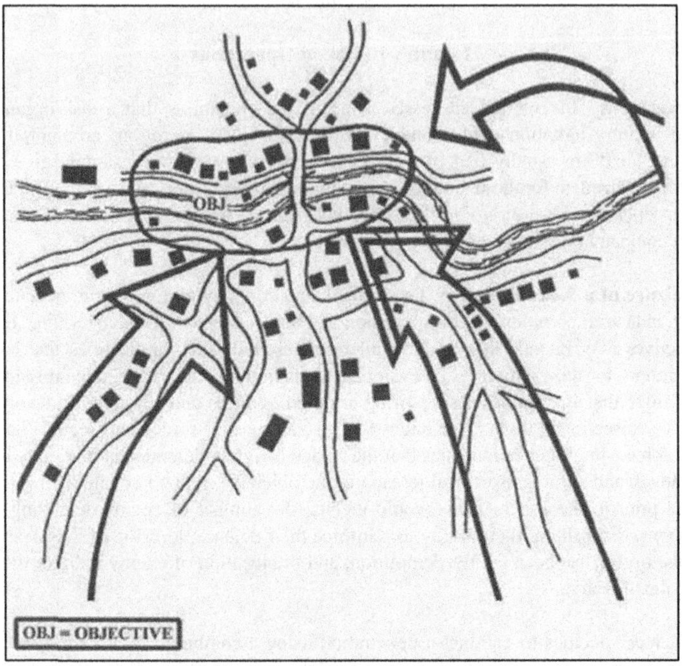

Figure 2-10. Seizure of a Key Objective

alleys to provide suppressing fires (Figure 2-11 on page 2-26). If heavy resistance is met and the column is halted, Marines dismount and the enemy and/or obstacles are cleared as quickly as possible. Once resistance and/or obstacles are cleared, the attack is immediately resumed. If necessary, units are dropped off to assume blocking positions and await follow-on clearing forces.

Once the objective is seized or linkup is accomplished, the battalion establishes a perimeter defense. The companies seize buildings around the objective and expand the size of the perimeter until it is large enough to secure the bridge. Attached engineers can then examine the bridge and clear it of any explosives.

2203. Infiltration. Infiltration is *"the movement through or into an area or territory occupied by either friendly or enemy troops or organizations. The movement is made, either by small groups or by individuals, at extended or irregular intervals. When the term is used in connection with the enemy, it infers that contact is avoided"* (Joint Pub 1-02).

Figure 2-11. Rapid Advance Movement

.3

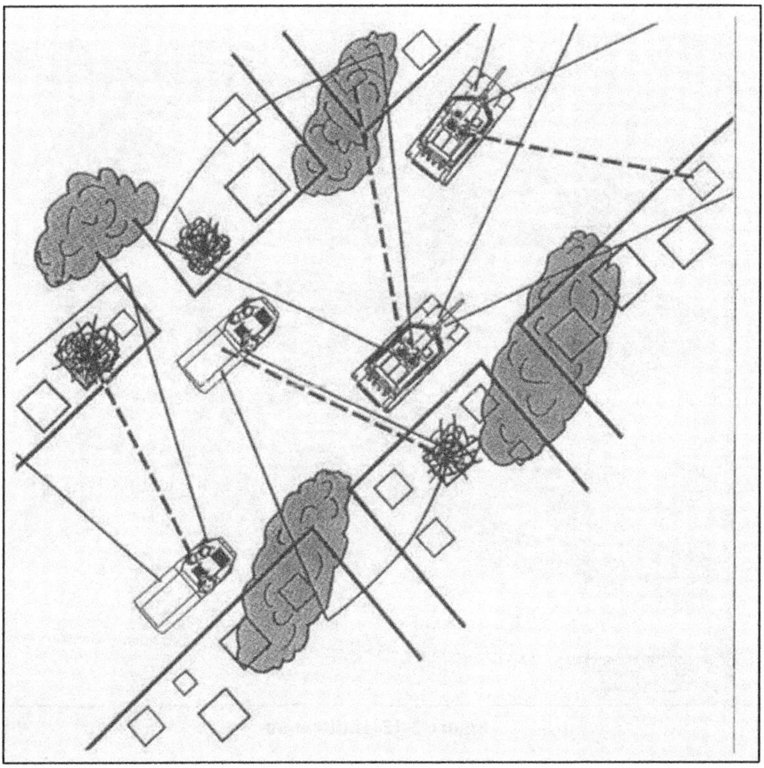

The outskirts of a town may not be strongly defended. The town's defenders may have only a series of antiarmor positions, security elements on the principal approach, or positions blocking the approaches to key features in the town. The enemy strongpoints and reserves may be positioned deeper in the city. In such a situation, an infantry battalion may be able to seize a part of the town by infiltrating companies and platoons between the enemy positions on the outskirts (Figure 2-12).

Moving by stealth on secondary streets, using the cover and concealment of alleys and buildings, the battalion may be able to seize key street junctions or terrain features, isolate enemy positions, and help following units move into the built-up area. Such an infiltration should be performed when visibility is poor.

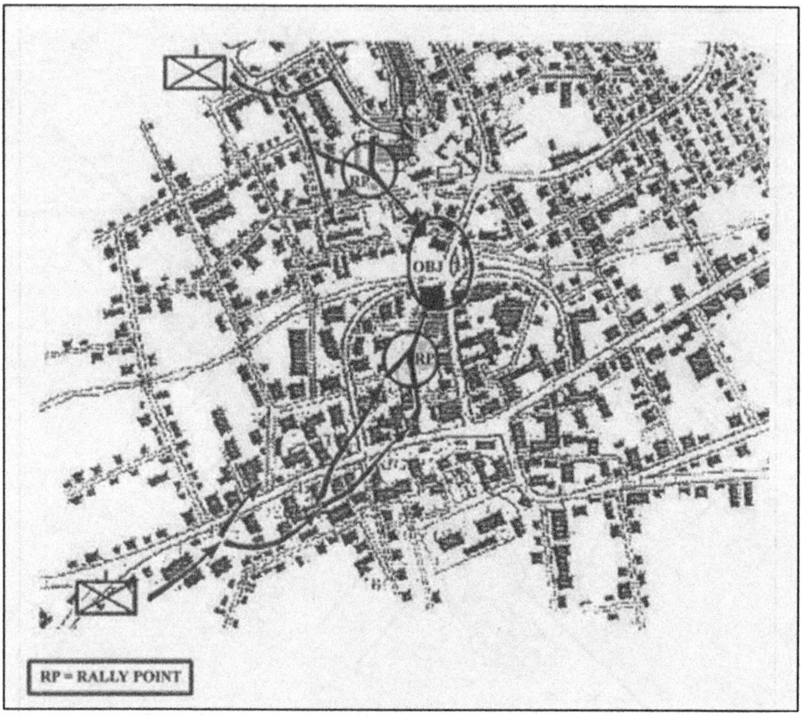

Figure 2-12. Infiltration

The infantry battalion is organized into infiltration companies with engineers attached and a designated reserve. Each company is assigned an infiltration lane that allows for infiltration by company-sized or smaller units. Depending on the construction of the built-up areas and streets, the infiltration lane may be 500 to 1,500 meters wide.

Mortar and artillery fire can be used to divert the enemy's attention and cover the sound of infiltrating forces. Armored vehicles and antitank weapons should be positioned to cover likely avenues of approach for enemy armored vehicles. Designated unit(s) provide isolation and security for infiltrating rifle companies. As the companies move into the built-up area, they secure their own flanks. Security elements may be dropped off along the route to warn of a flank attack. Engineers assist in breaching or bypassing minefields or other obstacles. Enemy positions are avoided but reported. The infiltrating companies proceed until they reach their rally points. At that time, they reorganize and make final preparations for attack of the objective.

2204. Route Security. In MOUT, a battalion may be tasked to provide route security. The process of securing the route may consist of a series of small-unit actions to clear buildings along the route. Fires should be planned to isolate the route as well. It may also require the reduction of obstacles and the emplacement of flank security along the entire route. In an outlying area, the forward units may proceed by bounds from road junction to road junction. Other units could augment flank security by moving down parallel streets and by probing the flanks.

The speed of movement required and the enemy situation determine whether the unit should be mounted or dismounted. For mechanized forces, narrow streets should be avoided if possible to avoid being trapped. Designated armored vehicles overwatch the squad to its front, keeping watch on the opposite side of the street. While armored vehicle movements and positioning provide mutual support, armored vehicles should be protected by dismounted infantry.

When contact is made with the enemy, armored vehicles seek covered positions and provide direct-fire support. Supporting fires fix and isolate enemy positions. Infantry dismount then maneuver to attack. Phase lines are used to coordinate movement along the route. For example, at each phase line, the forward companies might reestablish contact, reorganize, and continue clearing (Figure 2-13).

2205. Battalion Considerations. Some of the concerns for the commander to consider include:

a. Command and Control. CP location can be critical due to line of sight restrictions in the urban environment. Consideration should be given to maintaining a highly mobile CP with links to airborne retrans stations.

b. Intelligence. Organic collection capability resides primarily from units in contact with the enemy. The Scout Sniper Platoon also provides information but they may have obstructed visibility in the urban environment. Commanders may chose to employ them more often in a sniper role. Links to higher command intelligence sources may be key to maintaining current information on the enemy. The same line of sight restrictions encountered with command and contol will also restrict intelligence dissemination. Commanders may have to rely on messengers.

c. Manuever. Maneuver of mechanized forces in an urban environment may be restricted to the size, width, and load capabilities of the streets. Consideration must be given to plan for alternate routes for any attack. Significant manmade obstacles can be encountered throughout the battlespace. Additionally, the reserve is normally kept closer to the assault force. Movement of forces by aviation is less restrictive, but may also be vulnerable to surface to air missiles and small arms threat. Availability of LZs may also be a limiting factor.

d. Fires. A key consideration on the use of fires will be the ROE and the collateral damage that can result by their use. Rubbling of buildings can complicate an already restricted maneuver capability. Weapons effects can be more pronounced and can create a hazard for friendly forces. IPB determines potential firing positions for indirect fire assets that accounts

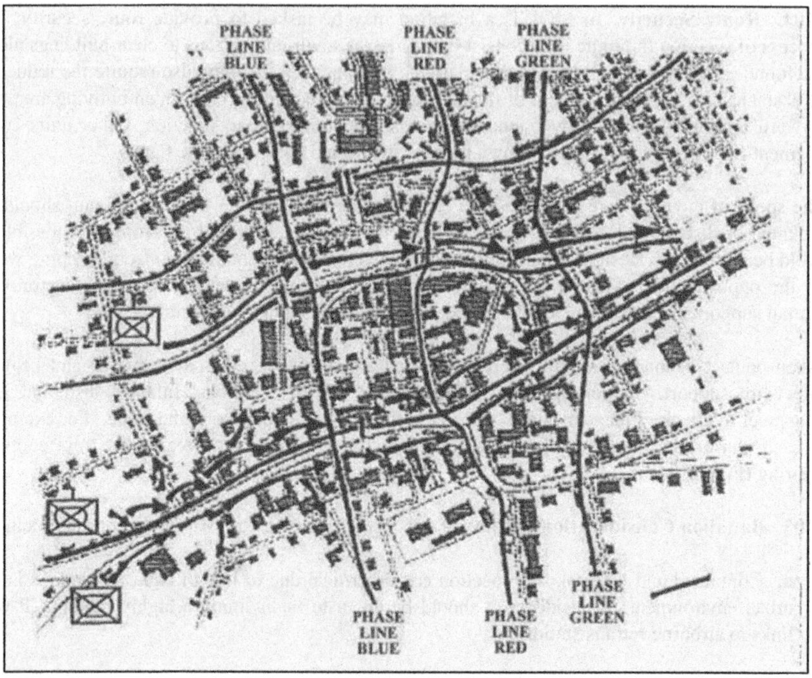

Figure 2-13. Clearing Along a Route

for the overhead masking by buildings and the ability to seat baseplates. Traditional indirect fire assets, such as artillery, may be used in the direct fire role.

e. Logistics. Historically in MOUT, the utilization of class I-V and VIII supplies is much greater. More vehicular assets should be allocated to the small unit level to transport ammunition and support the increased casualty evacuation requirements.

f. Force Protection. Lines of communication require extensive flank security in the urban environment. The commander must determine where his critical security requirements are and concentrate his force protection efforts in those areas while economizing in other areas. Manpower may further be depleted by security requirements for refugees and EPWs.

.3

Section III

Rifle Company Operations

2301. Overview. This section addresses some of the operations the infantry battalion could assign to the rifle company in MOUT. For our focus, the rifle company is a task-organized and/or reinforced unit. Company operations will normally be conducted as part of a larger battalion operation.

2302. Attacking within a Built-up Area. Attacking within a built-up area is a task the rifle company may be assigned (Figure 2-14). This task involves the isolation, attack, and clearing of a specified block of structures. It is usually conducted in conjunction with adjacent friendly attacks or security operations. (See Appendix A.)

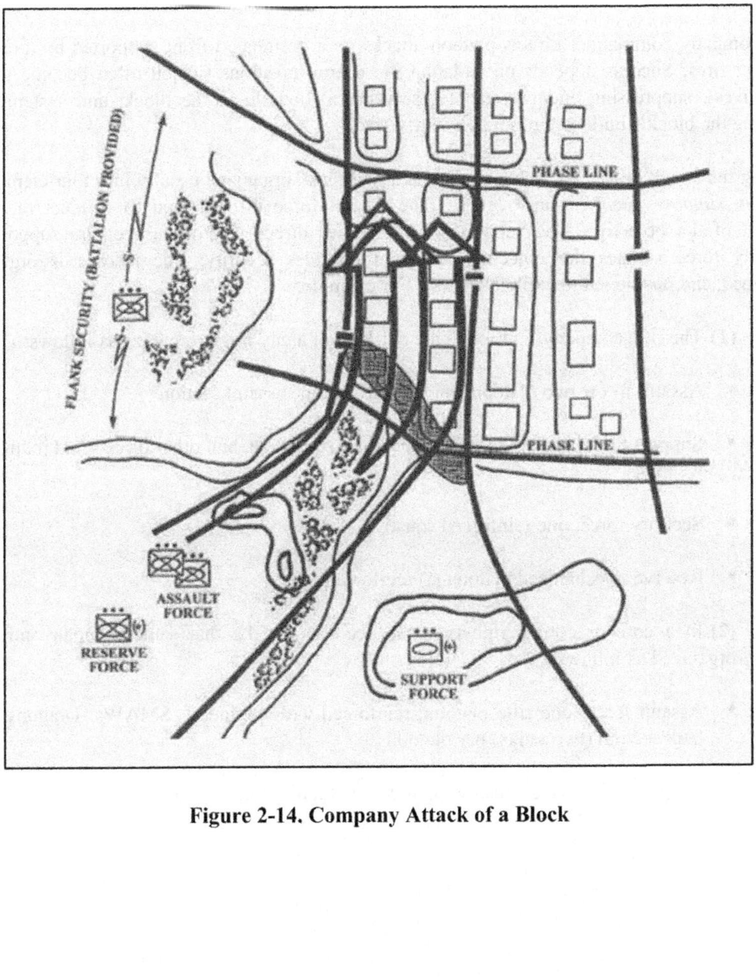

Figure 2-14. Company Attack of a Block

Before an assault, the company commander employs smoke to conceal the movements of assaulting platoons. He provides suppression fires to their flanks with direct-fire weapons. The battalion may provide additional security forces for flank protection.

(1) Concealed by smoke and supported by direct-fire weapons, an assaulting platoon attacks the first isolated building. The platoon must close on the building quickly while the enemy is still stunned by supporting fire. The company commander must closely coordinate the assault with supporting fire so that the fire is shifted at the last possible moment.

(2) The platoons continue to seize each building until the block is cleared. (See Appendix C for a detailed discussion on clearing of buildings.) After seizing the block, the company consolidates and reorganizes to repel a counterattack or to continue the attack.

The company commander directs platoon attacks on a designated front supported by direct and indirect fires. Success depends on isolating the enemy positions (which often become platoon objectives), suppressing enemy weapons, securing a foothold in the block, and systematically clearing the block's buildings room by room.

During the attack of a block, the company is usually task organized usually into four elements: *assault*, *support*, *security*, and *reserve*. The assault force is organized to conduct the actual assault of the objective. The support force provides direct- and/or indirect-fire support. The security force isolates the objective area and provides security. The reserve is organized, equipped, and positioned to exploit success. For example:

(1) The rifle company fighting in the outskirts of a city might organize as follows:

- Assault force: two platoons reinforced with the assault section
- Support force: armored vehicles, mechanized squad, and other direct- and indirect-fire weapons
- Security force: one reinforced squad.
- Reserve: mechanized platoon (-) reinforced.

(2) In a core or core periphery area (See Chapter 1), that same company might be organized as follows:

- Assault force: one rifle platoon, reinforced with engineers, SMAWs, Dragons, and a tank section (two tanks) per platoon
- Support force: One squad with direct- and indirect-fire weapons

.3

- Security force: one rifle platoon (-)

- Reserve: one rifle platoon, mechanized.

c. Armored vehicles, machine guns, and other direct-fire weapons under the company commander's control are assigned to isolate the objective and provide suppressive fires from covered positions. These weapons should not be fired for prolonged periods from one position. The gunners should use a series of positions and displace from one to another to gain better fields of fire and to avoid being targeted by the enemy. Direct-fire support tasks are assigned as follows:

- Machine guns fire along streets and into windows, doors, alleys, etc.; however, innovative use of direct fire weapons can be key in MOUT

- Tanks, ATGMs, and AT4s fire at enemy tanks and other armored vehicles.

- SMAWs can be used to make entrances in buildings and destroy identified bunker/fighting positions.

- Tanks can fire at targets protected by walls, make entrances in buildings, and engage enemy armored vehicles with their main gun and other targets with their machine guns.

- Riflemen engage targets of opportunity.

2303. Attack of an Enemy Outpost. A motorized or mechanized force may encounter enemy outposts, checkpoints, or strongpoints. In a hasty attack on an enemy outpost, the company commander makes a quick assessment based on the mission and situation, and then reacts in accordance with the commander's intent. If not able to bypass an enemy outpost, he may have to attack and reduce it.

The company occupies positions of advantage. The tanks or other armored vehicles support the dismounted infantry in the attack by assuming support-by-fire positions from which they can fire on the outpost, prevent the enemy from escaping, and destroy any reinforcements.

The infantry platoons dismount and move to engage the enemy (Figure 2-15). They should attack from a covered route to hit the outpost at a vulnerable point. As the platoons approach the outpost, smoke may be employed to screen their movement, and supporting fires are shifted. After platoons close on the outpost, they clear the buildings quickly and consolidate. The company is then ready to continue operations.

2304. Seizure of a Traffic Circle or Major Intersection. A company may have to seize a traffic circle or major road or traffic intersection (Figure 2-16 on page 2-34). This operation consists of seizing the buildings adjacent to the traffic circle or intersection while bringing direct-fire weapons into position to fire on likely avenues of approach.

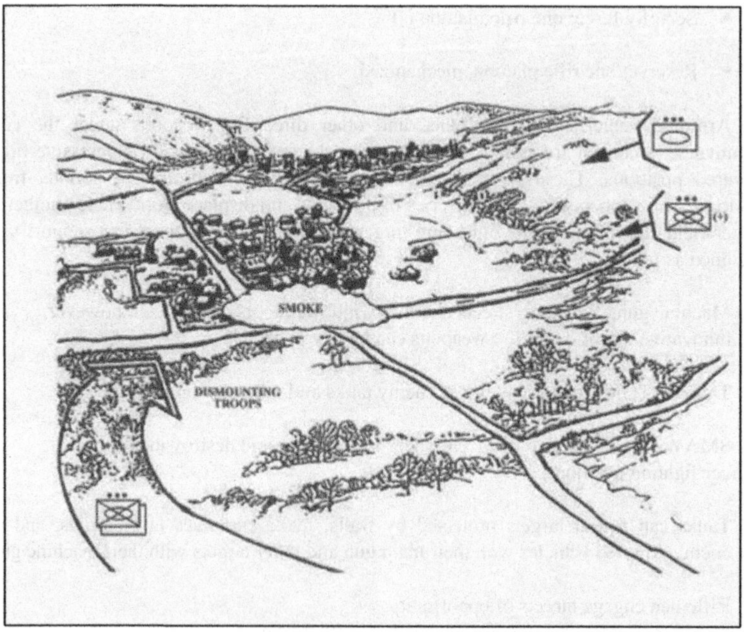

Figure 2-15. Hasty Attack of an Enemy Outpost

The company commander takes the following actions to seize a traffic circle or major intersection:

- Isolate the objective area.
- Seize and clear the buildings overlooking the traffic circle or major intersection under cover of direct-fire weapons.
- Secure the flanks against enemy attack.
- Consolidate and prepare for counterattack.

2305. Seizure of Key Terrain. The seizure of certain key terrain such as a bridge, intersection, facility, or structure may be critical to the success of an operation. Seizing key terrain for friendly use is a possible mission for a rifle company.

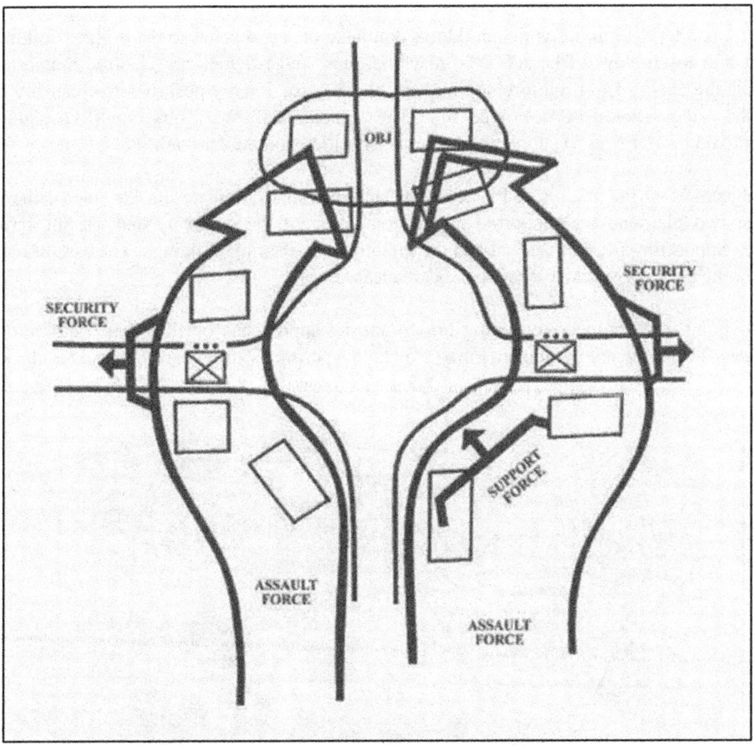

Figure 2-16. Seizure of a Traffic Circle or Major Intersection

To seize a bridge, a company should:

- Clear the buildings on the *near bank* that provide clear fields of fire for supporting weapons
- Suppress enemy weapons on the *far bank*
- Use screening smoke to obscure enemy observation
- Clear the buildings on the *far bank* that dominate the bridge
- Secure a perimeter around the bridge so that the engineers can clear obstacles or remove demolitions from the bridge.

The commander determines which buildings dominate the approaches to the bridge. Buildings that permit him to employ AT4s, ATGMs, machine guns, and riflemen are cleared. Supporting fire prevents the enemy from reinforcing units on the far bank. Enemy positions that can fire directly onto the bridge should be destroyed first. The company's forward observer (FO) should plan artillery and mortar fires to suppress the enemy's positions on the far bank.

The objectives of the assaulting platoons are buildings that dominate the far side of the bridge. One or two platoons are supported by the remainder of the company and assault across the bridge. Supporting fires are shifted and the infantry start clearing buildings. The assault continues until all the buildings in the objective area are cleared.

At this point, the engineers clear the bridge and its approaches of all mines, demolitions, and obstacles. The company commander may expand his perimeter to prepare for counterattack. Once the bridge is cleared, the armored vehicles are brought across to the far bank (Figure 2-17).

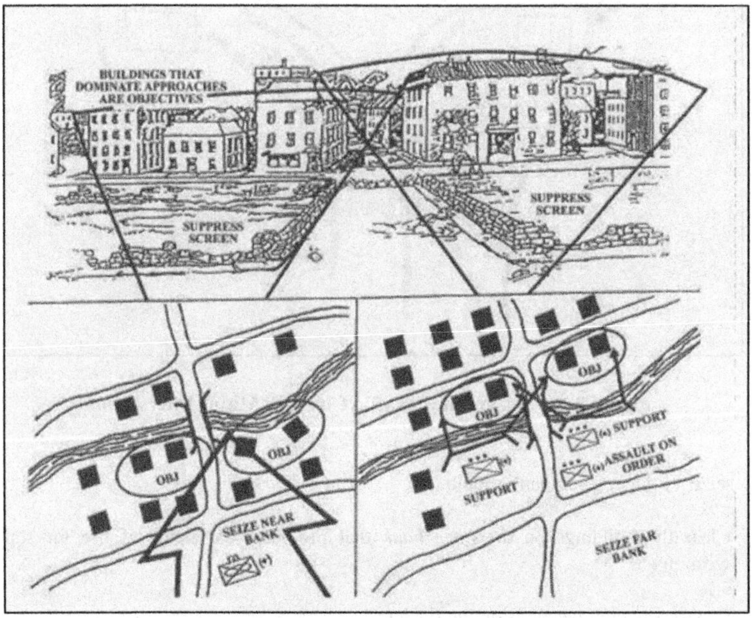

Figure 2-17. Seizure of a Bridge

.3

2306. Movement to Contact. A rifle company may conduct a movement to contact along a road. Similarly, the company may have to reconnoiter such a route to prepare for a battalion attack. This type of mission can best be accomplished by a mechanized company with appropriate attachments. (Figure 2-18 on page 2-36).

Figure 2-18. Movement to Contact Along a Highway (Commercial Ribbon)

The speed of movement depends on the mission, terrain, and enemy situation. In open areas where rapid movement is possible, tanks should lead. In close terrain, the infantry may lead while being overwatched by the tanks. Artillery fire should be planned along the route. Engineers may accompany the lead platoon on the main route to help clear obstacles and mines as they may be encountered. Between key terrain, when contact is not likely, the company remains on their vehicles. At key terrain or when enemy contact is likely, the company moves on foot to clear enemy positions or to secure the key terrain. Armored vehicles support the infantry.

Section IV

Rifle Platoon Operations

2401. Overview. This section discusses some of the various tactical missions that might be assigned to a rifle platoon.

2402. Attack of a Building. One offensive mission assigned to a rifle platoon in MOUT is the attack of a building. The attack requires isolating the building and entering the building at the least-defended point or through a breached hole (Figure 2-19 on page 2-38). When possible, clear the building from top to bottom. Close coordination between the assault and support forces is required.

When attacking a building, a platoon should be organized into an assault force, a support force, and a security force. The assault force provides the forces that actually conduct the assault. A support force provides supporting fires and efforts that assist the assault force. A security force covers the platoon's flanks and rear. (See Appendix A.) The platoon may be augmented by armored vehicles and from the company's weapons platoon or the battalions weapons company. If a reserve is required it should be provided by the company.

The assault consists of three steps:

(1) Isolate the building and deny the enemy the ability to maneuver or reinforce.

(2) Enter the building to secure a foothold for further action.

(3) Clear the building systematically.

Clearing is performed by individual squads and teams, which leapfrog as rooms and floors are secured. Platoons tasked to clear buildings may be reinforced with combat engineers.

2403. Movement Down a Street. When moving in urban terrain, a rifle platoon normally follows the same principles of movement as in other environments. However, some movement techniques must be modified to adjust to the unique dangers encountered in an urban environment. (See Appendix A, Section VII.) This section describes the movement down the street of the lead platoon of a company.

The speed of movement depends on the enemy situation. In outlying or lightly defended areas, a mechanized platoon may proceed along the street mounted but send dismounted infantry forward to reconnoiter danger areas. In the center of a built-up area the platoon dismounts to move on foot with two squads leading—one on each side of the road, using all available cover. They move through the buildings, if feasible, to avoid exposure on the streets. The squads coordinate to provide each other with all-around observation and mutual support.

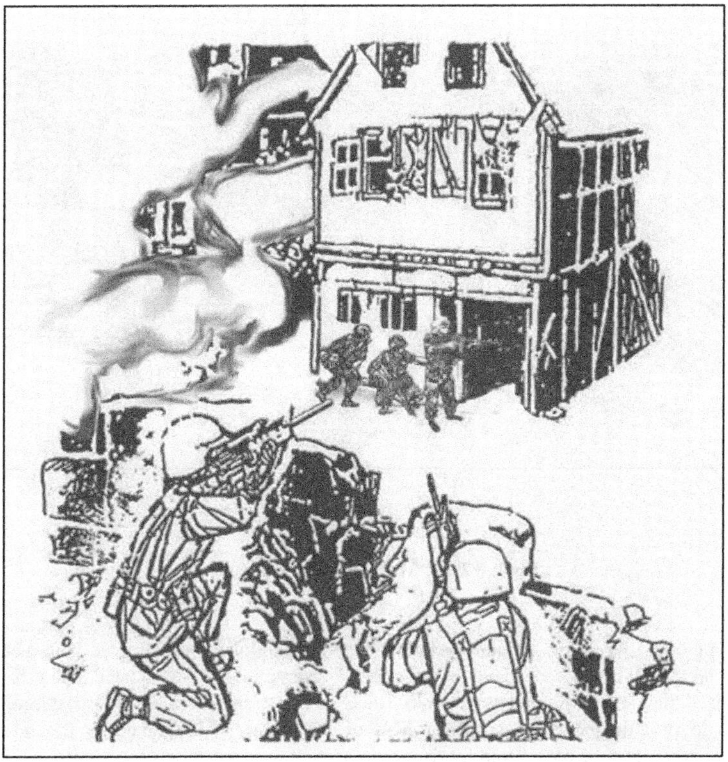

Figure 2-19. Attack of a Building

The platoon should organize and practice to move as two forces: a *maneuver* force and an overwatch force. The maneuver force (normally of one squad on narrow streets or two squads on wide streets) moves forward, scouts danger areas, and closes with the enemy. An *overwatch* force (the remainder of the platoon and its attached or supporting crew-served weapons), moves behind the maneuver force, secures the platoon's flanks and rear, and provides direct-fire support when needed. These two forces, can exchange roles from time to time to avoid overfatigue of the maneuver force (Figure 2-20).

2404. Utilization of the Reserve. In the offense, a rifle platoon may be assigned as a reserve. The reserve can be used to reinforce and attack or exploit success. Reserves are normally held at the company level.

Figure 2-20. Movement Down a Street

In outlying areas, where the terrain is relatively open, a tank platoon reinforced with mechanized assets (tank-heavy task force) may be tasked as the reserve. However, if inside the urban area, a mechanized platoon reinforced with tanks (mech-heavy task force) would most probably be directed to provide the reserve. Regardless of what unit is designated as the reserve, the following items must be considered for their employment in an urban environment:

- The reserve should plan to use a different axis of advance
- Route selection for the reserve may be more constructive since it may have to use building, side streets, sewers, etc.
- Reserves are kept closer to assault forces in MOUT than on open terrain
- Engineers should normally be attached
- Reserves should maintain a high degree of mobility

Chapter 3

Defensive Operations

The essence of defensive tactics is to place the enemy into a position that permits his destruction through the intelligent use of terrain and firepower.

3001. Introduction. In urban combat, the defender possesses key advantages over the attacker. The defender can shape the battlespace to his advantage by maximizing the natural restrictions and obstacles found in the urban environment. ROE for the attacker can add to the defender's advantage by placing restrictions on the application of force by the attacker. Knowledge of the terrain and time available for preparing defensive positions are advantages which may enable the defender to successfully resist a numerically superior force. A spirited and stubborn defense may persuade an attacker to abandon the attack. In some battles, urban defenders can be credited with repelling or decisively delaying an attacker which ultimately influenced the course of the war in their favor.

Marines may be called upon to defend a port city, an embassy located on or near a coastline, a transportation link, or the capital of a foreign government. They may also be tasked to defend from an urban area in an economy-of-force role in order to support operations elsewhere. In combat, offensive and defensive operations are inseparable. Marines need to be capable of conducting both offensive and defensive operations in an urban environment.

The fundamentals of defense do not change in an urban environment. The defenders of a city, however, usually have detailed knowledge of the terrain. This will allow them to establish an extensive defensive network that is designed to force an attacker to expend exorbitant amounts of time, supplies, equipment, and manpower. Commanders need to recognize both the advantages and disadvantages of defensive operations in an urban environment.

3002. Decision to Defend

a. Reasons for Defending Built-Up Areas. Historically, the following are some of the reasons for defending built-up areas:

(1) Denial of Important Strategy/Political Objectives. Capitals and cultural centers can be defended strictly for psychological or national morale purposes even if they do not offer a tactical advantage to the defender. Defending a city can cause an attacker to commit a significant amount of his forces which reduces his capability to attack elsewhere. The defense of a city can delay the overall offensive capability of the attacker.

(2) Retention of Key Economic Centers. In many countries, the entire nation's economic well-being may be tied to a few key cities. These key cities usually contain the country's primary industrial, transportation, and communications base. The capture of these key centers could result in the overthrow of the current government, or deny that government the ability to adequately support combat operations against enemy regular or insurgent forces.

(3) Control of Avenues of Approach. Most avenues of approach to large cities are straddled by small towns every few kilometers which must be controlled by defending forces. These areas can be used as battle positions or strongpoints to control the avenues of approach. For the attacker to utilize these avenues of approach he must sacrifice speed and expend resources. The defense of these cities or towns may constitute a large part of the overall defense of a city.

(4) Economy of Force. The tactical advantages provided by urbanized terrain allows the defender to engage a numerically superior force. The defender can conserve the bulk of his combat power so that it is available for use in other operations.

(5) Concealment of Forces. Reconnaissance and combat identification is more difficult in an urban environment. CPs, reserves, CSS complexes, and combat forces emplaced in built-up areas are much harder to detect.

3003. Reasons for Not Defending Built-Up Areas. The commander considers the following reasons for deciding not to defend built-up areas.

a. Unnecessary to the Defensive or Offensive Plan. If the built-up area is too far forward or back in a unit's defensive sector, is isolated, or is not astride an enemy's expected avenue of approach, the commander may choose not to defend it.

b. Bypassable. If the nature of nearby terrain allows the enemy to bypass the city, then it will not make a good blocking position. Some built-up areas, mainly smaller ones, are easily bypassed by existing main road and highway systems. A built-up area that can be easily bypassed normally will be, thereby effectively isolating that area's defenders from the remainder of their forces.

c. Inadequate Structures for Defense. Extensive areas of lightly built or flammable structures offer little protection to the defender. Built-up areas near flammable or hazardous industrial areas, such as refineries or chemical plants, also may not lend themselves to successful defense.

d. Adjacent Dominating Terrain. If the built-up area is small and dominated by close, prominent terrain, the commander may choose to defend on that terrain rather than in the built-up area.

e. Better Fields of Fire Elsewhere. The commander may choose to base all or part of his defense on the better fields of fire that exist outside built-up areas. An example would be an armor-heavy force defending in sectors from multiple, small, built-up areas surrounded by open or farm-type areas.

f. Open City. See chapter 2.

Section I

Defensive Planning

3101. Commander's Estimate. As in any other defensive operation, planning begins with the commander's estimate of the situation. The complxities involved with the defense of a built-up area require detailed and centralized planning. This section details the general procedures and principles for planning and organizing the defense of an urban area.

3102. METT-T

a. Mission. Mission analysis is the first step in conducting the estimate of the situation. When conducting mission analysis commanders and their staffs should analyze their assigned tasks and commander's intent to determine the scope of the defense. For example:

- Do I need to control every building or can I control the frontage from key buildings?

- Are certain parts of the urban area more significant? (i.e. main transportation centers, communication nodes government buildings)

b. Enemy. The commander must focus on the enemy and build his defense to concentrate his strength against enemy weakness. The IPB process is used to analyze the terrain and enemy capabilities. (See FM 34-130.) For example:

- If the attacker relies heavily on aviation for mobility, the defender may defend potential LZs and emplace manpad teams to counter the air threat

- If the enemy relies heavily on mechanized movement, the defender may employ additional countermobility assets.

c. Terrain and Weather

(1) Terrain. Analysis of all manmade and natural terrain features is critical for planning a defense in built-up terrain. A defending commander has the advantage of occupying the terrain on which the fight will occur, and therefore usually knows it better than the attacker.

The defender must make best use of the battlespace that he controls: ground level (streets and parks), above ground (buildings), and below ground (subways and sewers). Defenders must be aware that built-up areas are normally interdependent. A reservoir supplying water to a city may be many miles away, electricity can come from remote power stations, and the control points for these services and others, such as communications lines, may be vulnerable.

.3

(a) K: Key Terrain. Primary examples of key terrain in the urban environment are ports, airfields, power grids, communication nodes, bridges, government building complexes, or parks. We identify key terrain so that we can better select our defensive positions and also better determine the enemy's objectives.

1 **Villages.** Villages often dominate high-speed avenues of approach. If the buildings are constructed of stone, brick, or concrete, providing good protection against both direct and indirect fires, a formidable defense can be constructed

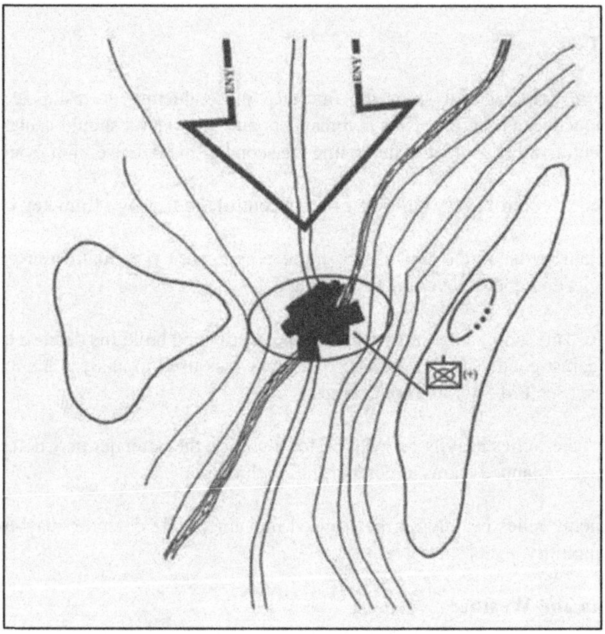

(Figure 3-1). Company-sized battle positions can be established in these small villages to block approaches into the larger urban area.

Figure 3-1. Village Defense (Chokepoint)

2 **Strip Areas.** Strip areas are built along roads or down valleys between towns and villages. They afford the defender the same advantages as villages. If effective fields of fire are available, a unit can occupy a few key positions within the strip to create an effective defense. (Figures 3-2A and 3-2B).

3 **Towns and Cities.** When a small town or city is positioned on a key enemy avenue of approach, a force can defend from the urban area while other forces

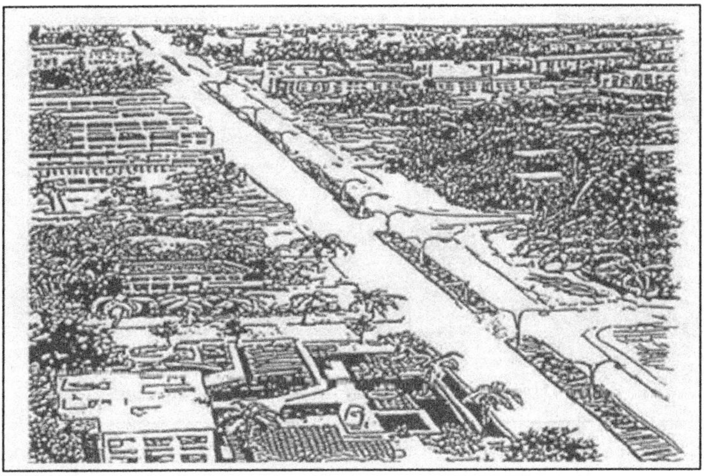

deny the enemy the ability to bypass. The town or city can be made into a strongpoint defense. (Figures 3-3A and 3-3B on page 3-6).

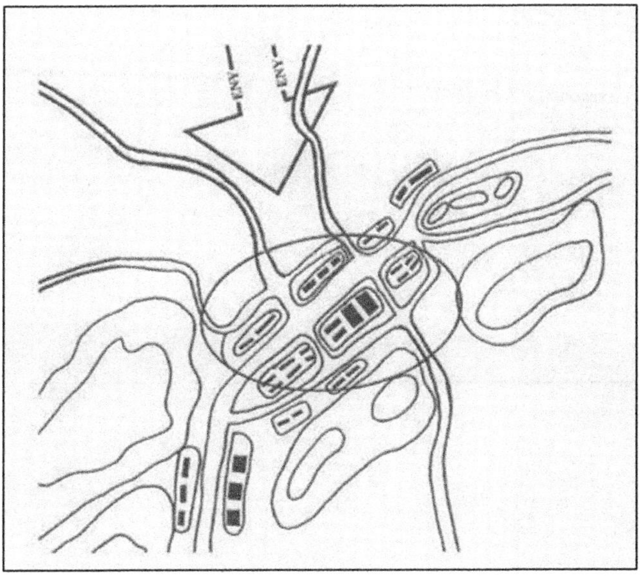

Figure 3-2A. Strip Area

Figure 3-2B. Strip Area Defense

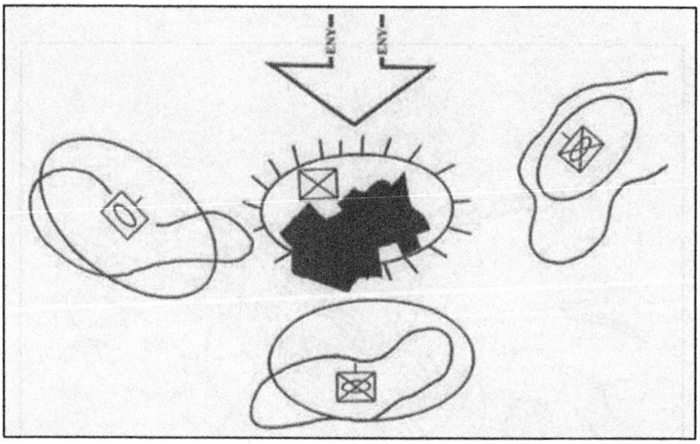

Figure 3-3A. Towns and Cities

Figure 3-3B. Defense of a Town or City

4 **Large Built-Up Areas.** Large buildings that are close together may require more forces and smaller defensive sectors than natural open terrain. The density of buildings, rubble, and street patterns will dictate the frontage of the unit.

(b) **O: Observation and Fields of Fire.** Attackers must generally advance by crossing streets and open areas between buildings where they are exposed to fires from concealed positions.

1 **Weapons and Range.** The defender must position weapons to obtain maximum effect and mutual support. FACs and FOs should be placed well above street level for increased observation. Fires and final protective fires should be preregistered on the most likely approaches.

2 **Limited Visibility.** The commander can expect the attacker to use periods of limited visibility to sustain or gain momentum. (See Appendix H for more detailed discussion.) The commander considers the following during periods of limited visibility:

- Unoccupied areas that can be observed and covered by fire during daylight may have to be occupied or patrolled at night.

- Remote sensors should be emplaced in dead space.

- Mines, noise-making devices, tanglefoot, and OPs should be positioned on avenues of approach to provide early warning.

- Artificial illumination to include street lamps, stadium lights, and flares should be integrated into the overall defense.

- When defenses are probed, indirect-fire weapons, grenade launchers, and hand grenades can be used to avoid disclosure of defensive position locations.

(c) C: Cover and Concealment. Although walls, floors, and ceilings can be used as protective cover, Marines should always improve these positions by using sandbags, rubble, etc. A defender can reduce his exposure by establishing routes between positions using:

- prepared breaches through buildings

- reconnoitered and marked underground systems
- trenches and sewage systems
- concealment offered by smoke and darkness to cross open areas.

(d) O: Obstacles. A city is an obstacle because it canalizes and impedes an attack. Avenues of approach should be blocked by obstacles and covered by fire. (See Paragraphs 3404.d., 3404.f., 3405.e., and Appendix F.)

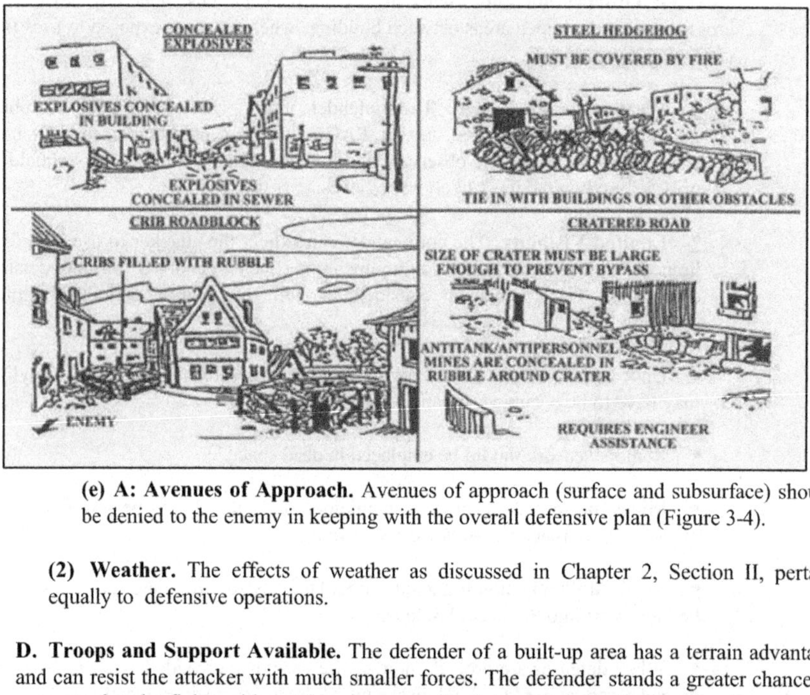

(e) A: Avenues of Approach. Avenues of approach (surface and subsurface) should be denied to the enemy in keeping with the overall defensive plan (Figure 3-4).

(2) Weather. The effects of weather as discussed in Chapter 2, Section II, pertain equally to defensive operations.

D. Troops and Support Available. The defender of a built-up area has a terrain advantage and can resist the attacker with much smaller forces. The defender stands a greater chance of success when he fights with combined arms and task organizes his forces. The discussion of troops and support available in Chapter 2, Section II, also pertains to defensive operations.

Figure 3-4. Denial of Avenues of Approach

e. Time Available. The commander establishes priorities of work, which depend on the time available. Many tasks can be accomplished at the same time, but priorities for work are accomplished according to the commander's order. An example priority-of-work sequence follows:

- Establish security and communications
- Assign sectors of responsibility, final protective fires
- Clear fields of fire
- Select and prepare initial fighting positions
- Establish and mark routes between positions (including alternate and supplementary routes)
- Emplace obstacles and mines
- Improve fighting positions.

3103. Organizing for a Defense

a. Types of Defense. The defense of a built-up area should be organized around key terrain features that preserve the integrity of the defense and provide the defender ease of movement. There are two types of defense: position and mobile. (See MCWP 3-1, *Ground Combat Operations* [under development] for details on defensive operations.) Most defenses will include a combination of position and mobile defense. The type of defense chosen is predicated on commander's intent and METT-T.

(1) Position Defense. The type of defense in which the bulk of the defending force is disposed in selected tactical localities where the decisive battle is to be fought. Principal reliance is placed on the ability of the forces in the defended localities to maintain their positions and to control the terrain between them. The reserve is used to add depth, to block, or restore the battle position by counterattack. (Joint Pub 1-02.) The position defense is usually characterized by defending key terrain. In the urban environment, this equates to the physical occupation of key public buildings, avenues of approach, transportation centers, industrial parks, etc. This defense focuses combat power to repel the attacker's advance and/or penetration while retaining the terrain. This type of defense is common for battalion-sized units and below.

The commander conducting a position defense organizes forces utilizes the same fundamentals that apply to any defense. Forces can be assigned sectors to defend and/or battle positions or strongpoints based on the commander's analysis of METT-T. Reserves are identified and located to respond quickly to exploit success.

(2) Mobile Defense. Defense of an area or position in which maneuver is used with organization of fire and utilization of terrain to seize the initiative from the enemy. (Joint Pub 1-02.) The mobile defense is characterized by its focus on the enemy, not on terrain. Terrain is still important. However, this defense is organized with a mobility capability to exploit enemy weakness. In order to conduct the mobile defense, the defender must have enough forces to defend the built-up area, and to rapidly conduct an enveloping or flanking attack.

As in the position defense, the sector, battle position, strongpoint, or any combinations thereof are also used in shaping the battlespace by canalizing the enemy into a designated location where he can be destroyed.

b. Defensive Options

(1) Defense Outside the City. The commander may defend from outside the city. This option is often selected when the terrain surrounding an urban area offers an advantage to a defender and when his defending force is of sufficient strength (See Cherbourg paragraph in Chapter 1). The presence of a large, friendly, civilian population, or buildings of historic significance or specific cultural value, would also contribute to deciding on an outside defense. A situation in which the built-up area has a preponderance of building structures that offer little protection to defenders or terrain that favors the attacker.

A commander organizes his defenses around strongpoints selected on the outer edges of the urban area when these may be suitable structures for siting weapons to gain maximum firepower range. The organization of an outer-city defense is also heavily influenced by the number and type of weapons with which the hostile force is armed. A defense against a predominantly infantry threat would differ from that planned to defeat an armored threat. The decision to defend or concentrate assets solely on the outer edge of a city are weighed against the enemy's ability to mass fires, infiltrate, or bypass strongpoints.

(2) Defense Inside the City. The commander may decide that his best defense is to defend within the city. This option is usually chosen when hostile forces possess strong, accurate, long-range fires and the defender wishes to minimize their effectiveness (See Quang Tri City example in Chapter 1). The strength of the defending force relative to the attacker may also lead to this option as an economy-of-force measure.

Structures on the outer edge of an urban area that obstruct fields of fire or that are likely to be used by an attacker should be prepared for demolition or demolished ahead of time. All defensive positions should be supplemented or covered by direct-fire weapons. Indirect fires should be planned for most likely avenues of approach. Additionally, close air support should be planned and integrated with the defense.

Security forces are placed outside the defended area to gain early enemy contact, inflict maximum casualties, and attempt to deceive the enemy as to the location of the friendly positions.

(3) Defense of a Key Sector. This option is usually employed when only a portion of an urban area be held or when strength is insufficient to defend an entire area (See the Japanese defense of Manila in Chapter 1). This option is often employed in situations where the mission requires clearing and holding a key sector, or sectors, in order to facilitate other operations.

(4) Entrapment and Ambush. This option is preferred when the defender lacks requisite strength to man organized, in-depth, defensive positions (See the highly successful Egyptian defense of Suez City example in Chapter 1). It requires the preparation of a series of blocking and/or ambush positions along major avenues of approach through an urban area. The defending force uses mines and obstacles to block alternate or possible detour routes in order to canalize the attack force into prepared ambush sites. The defender employs security forces forward of the main battle area and establishes inner or outer defensive strongpoints with extensive obstacles designed to assist in canalize the enemy into kill zones.

The entrapment and ambush option requires detailed planning and rehearsal by all echelons. Alternate plans should be prepared in the event the ambush is only partially successful or if the enemy fails to enter prepared ambush routes. Alternate ambush positions or counterattacks are planned.

Ambush planning in an urban environment does not differ significantly from planning in other environments. However, the urban environment offers an abundance of kill zones that afford the enemy little or no opportunity to escape. Ambush positions are established in selected buildings in the kill zone which offer optimum fields of fire for all weapons. Indirect fires and close air support should be planned to obtain maximum effect in the kill zone. The kill zone should also be covered with targets for indirect fire and CAS. Entrances into buildings not used as firing points should be blocked to entrap hostile forces in the kill zones.

(5) Defense in Depth. Defense in depth should be incorporated into any defensive plan. Units should be given enough terrain to allow for primary, supplementary, and alternate fighting positions as well as successive battle positions or strongpoints. This allows enough room for multiple ambush/entrapment kill zones and the positioning of a reserve.

c. Defensive Organization. The defensive battlespace is divided into three areas: the *security area*, *main battle area*, and *rear area*. Depending on the mission of the unit, forces defending a built-up area may have missions in any one or more of these areas. (Figure 3-5 on page 3-12).

.3

(1) Security Area. The defensive battle begins with a combined-arms force conducting security operations well forward of the main battle area. Security area missions can include screen , guard, and cover. Based on their assigned mission, security forces use all

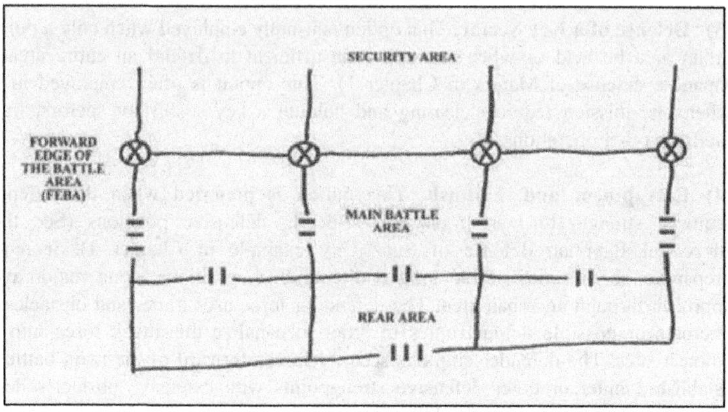

available combat power to gain time, destroy the enemy, or slow his momentum. The initial battle is fought using aviation assets and indirect fire weapons.

(a) Value of Security Area Operations. Security area operations complement to the defense by:

- Alerting the defense to the strength, location, and general direction of the enemy's main and supporting attacks
- Delaying enemy first-echelon detachments
- Initiating early engagement of enemy forces
- Deceiving the enemy as to the true location of the defender's main forces and main effort.

Figure 3-5. Organization of the Battlefield

(b) Withdrawal of Security Area Forces. The urban environment may complicate battle handover from the security force to the main battle area force. However, this transition must be accomplished smoothly to prevent the enemy from gaining momentum. The withdrawal of the security area forces must not result in an easing of pressure on the enemy.

(2) Main Battle Area. The decisive battle is fought in the main battle area. However, the

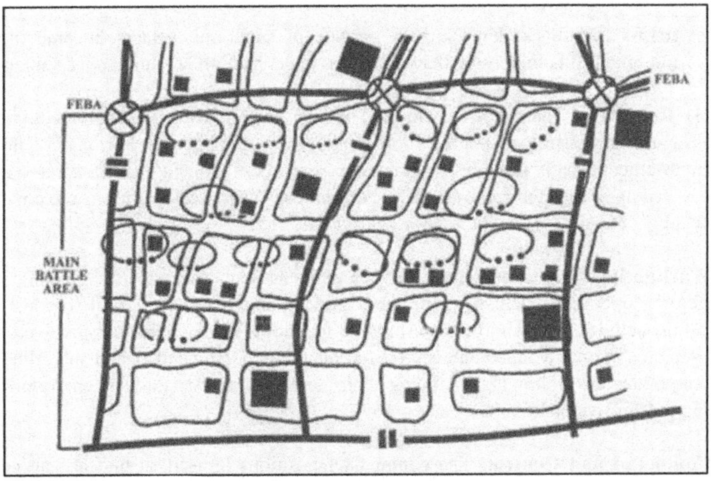

commander may deploy units on the forward edges of the city or in battle positions in depth. The defensive scheme should include forces to defend along the flanks. The commander normally employs security forces to the front to provide early warning and to deny the enemy intelligence on the unit's defensive dispositions (counterreconnaissance) (Figure 3-6).

(a) The size and location of battle positions depend on METT-T and the type of defense prepared. A sophisticated defense will often include a combination of both position and mobile defense but will always be planned in depth.

(b) A unit assigned battle positions on the forward edge of a city or town should:

- Provide early warning of the enemy's advance
- Engage the enemy at long range
- Deceive the enemy as to the true location of the defense.

Figure 3-6. Main Battle Area

(c) The defender employs all available fires to destroy and suppress the direct-fire weapons that support the ground attack. Depending on the concept of operations,

.3

weapons may engage targets at their maximum effective range or engage simultaneously for an ambush.

(d) As the attack develops, the actions of small-unit leaders become increasingly important. It is imperitive that all leaders understand their commander's intent.

(3) Rear Area. The rear area is located behind the main battle area and it contains the CSS and administrative support. Protection of these units is vital. Fortunately, the urban environment usually provides sufficient cover and concealment for CSS assets as well as covered and protected routes to the forward areas. Rear area operations are covered in detail in MCWP 3-41.1, *Rear Area Operations*.

3104. Warfighting Functions. The defensive planning sequence remains the same as in any other defensive operation. Defense planning considerations, based on METT-T, must take into account all the activities in each of the warfighting functions. The complexities involved in defense of a built-up area require detailed and centralized planning. Commanders and subordinate leaders must incorporate the following planning considerations for an urban environment when conducting a defensive operation.

a. Command and Control. The commander positions himself so that he can control the action. In an urban environment, this can be more difficult because of the close nature of fighting, prolific obstacles, poor visibility, and difficulty in communication.

Graphic control measures oriented on prominent terrain features are also used in planning and conducting combat in built-up areas. Prominent streets are ideal for use as phase lines and boundries. These and other control measures assist the commander in controlling manuever and fires throughout the battlespace.

COCs should be located in secure locations. (These can be inside or underground when possible.) (See Chapter 2, Section IV, Paragraph 2402.)

Commanders consider the effects of built-up areas on communications when they allocate time to establish communications. LOS limitations affect both visual and radio communications. Wire laid at street level is easily damaged by rubble and vehicle traffic. Also, the noise of urban combat is much louder than in other areas, making sound signals difficult to hear. Therefore, the time needed to establish an effective communications system may be greater than in more conventional terrain.

Wire is the primary means of communication for controlling the defense of a city and for enforcing security. However, wire can be compromised if interdicted by the enemy. Radio communication in built-up areas is normally degraded by structures and a high concentration of electrical power lines. Nonetheless, radio is an alternate means of communication. Messengers can be used as another means. Visual signals may also be used but are often not effective because of the screening effects of buildings, walls, and so forth. Signals must be planned, widely disseminated, and understood by all assigned and attached units. Commanders consider the following techniques when planning for communications:

(1) If possible, lay wire through buildings for maximum protection.

(2) Use existing telephone systems. (Remember, telephones are not secure.)

(3) Emplace radios and retransmission sites in buildings on top floors or atop buildings that are higher than surrounding structures. Ensure that antennas are placed on the side of buildings where the radio signal can be directed at friendly units. This will enhance LOS radio communications.

(4) Use messengers at all levels because they are the most reliable and secure means of communications.

b. Intelligence. The discussion in Chapter 2 pertains equally to a defensive operation. Intelligence gathering for defensive operations is not limited to studying only the enemy. Commanders must emphasize obtaining and using all intelligence. Commanders must emphasize the utility of the urban infrastructure to enhance his intelligence gathering and dissemination effort. The defender usually has the advantage of being familiar with the intricacies of the urban terrain that he is defending. For example, the:

- The sanitation or public works department will have knowledge of the subterranean sewer lines
- Civil engineers will have extensive knowledge of water and power distribution nodes
- Law enforcement representatives may have access to lists of people sympathetic to the attacker
- Local journalists may agree to use local sources to assist in intelligence gathering effort.

c. Manuever

(1) Positioning of Forces. Finding fighting positions in urban areas that provide both good fields of fire and cover is often difficult. The forward edges of a town usually offer the best fields of fire but can be easily targeted by enemy observation and supporting fire. These areas often contain residential buildings constructed of light material. Factories, civic buildings, and other heavy structures that provide adequate cover and are more suitable for a defense are more likely to be found deeper within the town, but they have limited fields of fire.

Because the forward edge of a town is the obvious position for the defender, it should generally be avoided. However, the defender may choose to use the edge of the town if the terrain limits the enemy's ability for engagement or if strongly constructed buildings that give defenders protection are available. The mission of such a force is to provide early warning of the enemy's advance, to engage the enemy at long range, and to deceive the enemy as to the true location of the defense. This force should withdraw in time to avoid decisive engagement. If there is limited observation from the forward edge, a force should be positioned on more favorable terrain to gain better observation and to engage the enemy at long range.

Reconnaissance by all defending elements should help select routes for use by defenders moving between positions. Movement is crucial in fighting in built-up areas. Early selection and marking of routes adds to the defender's advantages.

The commander covers probable landing and drop zones such as parks, stadiums, or large rooftops and heliports with obstacles or fire to prevent air assault. (See Appendix F.) In a large built-up area, a unit is given a sector to defend and normally establishes a series of defensive positions. Although mutual support between positions should be maintained, built-up terrain often allows the enemy to infiltrate between positions. Therefore, the defender must identify the following:

- Positions that allow suppressive fires on infiltrating routes
- Covered and concealed routes, such as subways and sewers, for friendly forces to use to move between positions
- Structures that dominate large areas
- Areas such as parks, boulevards, rivers, highways, and railroads where antiarmor weapons have longer fields of fire
- Firing positions for mortars
- Command locations that offer cover, concealment, and ease of command and control
- Protected storage areas for supplies.

Buildings that enhance the general plan of defense are selected. Mutual support between these positions is vital to prevent the attacker from maneuvering and outflanking the defensive position. Buildings chosen for occupation as defensive positions should:

- Offer good protection
- Have strong floors to keep the structure from collapsing under the weight of debris

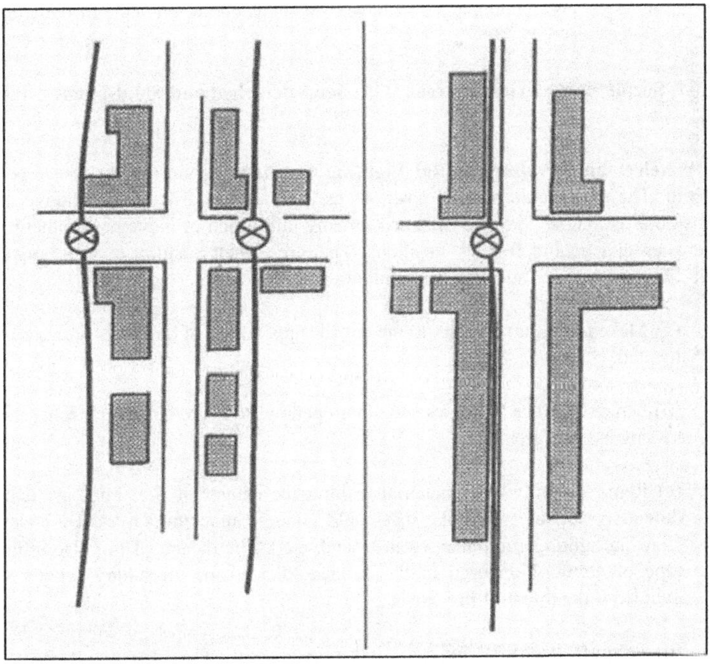

- Have thick walls
- Be constructed of nonflammable materials
- Be strategically located (corner buildings and prominent structures)
- Be adjacent to streets, alleys, vacant lots, and park sites (these buildings usually provide better fields of fire and are more easily tied in with other buildings)
- Provide for 360-degree protection and fields of fire
- Be stocked with adequate supplies as positions may become isolated and cut off for extended periods.

(2) Assign Sectors of Responsibility. Boundaries define sectors of responsibility. Sectors include areas where units may fire and maneuver without interference or coordination with other units. Responsibility for primary avenues of approach should never be split among units. In areas of semi-detached construction, where observation and movement are less restricted, boundaries should be established parallel to alleys or streets so that both sides of a street will be in a single sector. Where buildings present a solid

.3

front along streets, a boundary may have to be extended to one side of the street (Figure 3-7).

Figure 3-7. Sector Boundaries in Areas With Semi-detached and Solid-Front Construction

(3) Select and Prepare Initial Fighting Positions. Commanders select positions in depth. The unit should prepare positions as soon as they arrive and continue improving positions as long as they are occupied. Enemy infiltration or movement sometimes occurs between and behind friendly positions. Therefore, each position must be organized for all-around defense. The defender should also:

(a) Make minimum changes to the outside appearance of buildings where positions are located.

(b) Screen or block windows and other openings to keep the enemy from seeing in and tossing in hand grenades.

(c) Remove combustible material to limit the danger of fire. Fires are dangerous to defenders and create smoke that could conceal attacking forces. Defenders should store firefighting equipment (water, sand, etc.). The danger of fire also influences the type of ammunition used in the defense. Tracers or incendiary rounds should be avoided if the threat of fire exists.

(d) Turn off electricity and gas. Both propane and natural gas are explosive. Natural gas is also poisonous and is not filtered out by a protective mask. Propane gas, although not poisonous, is heavier than air. If it leaks into an enclosed area, it displaces the oxygen and can cause suffocation. Gas mains and electricity should be shut off at the main distribution nodes.

(e) Battle positions should be located so as not to establish a pattern. The unit should avoid obvious firing locations like church steeples.

(f) The unit should rehearse movements between positions.

(g) Camouflage positions.

(h) Reinforce positions with available materials, such as beds, furniture, etc.

(i) Block stairwells and doors with wire or other material to prevent or delay enemy movement. Create holes between floors and rooms to allow for covered movement within a building.

(j) Prepare range cards, fire plans, and sector sketches.

(k) Emplace machine guns in basements windows, where the guns can provide grazing fires. When basements are not used, access to them should be sealed to prevent enemy entry.

(l) Stockpile V (ammunition) and class VIII (medical supplies) items.

(4) Clear Fields of Fire. Commanders may need to rubble certain buildings and structures to provide greater protection and fields of fire. If the ceiling of a lower story room can support the weight of the rubble, collapsing the top floor of a building before the battle starts may afford better protection from indirect fires. Rubbling an entire building can increase the fields of fire and create an obstacle to enemy movement. Rubbling buildings too soon (or rubbling too many) may give away defensive locations and may destroy cover from direct fire. Rubbled buildings may also interfere with planned routes of withdrawal or counterattack.

(5) Improve Fighting Positions. When time permits, all positions, including supplementary and alternate positions, should be reinforced with sandbags and provided

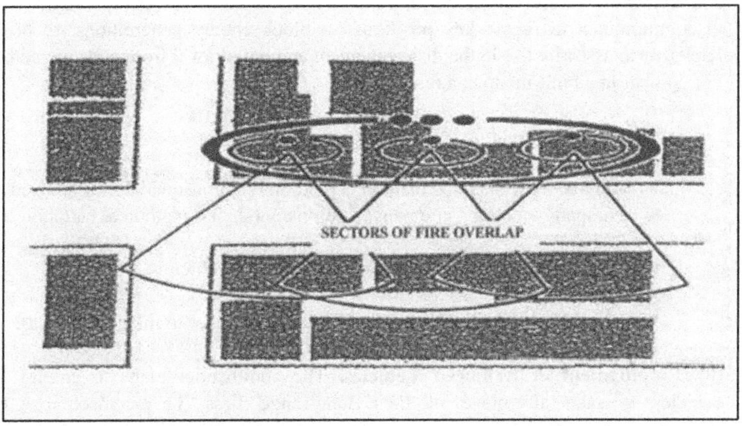

with overhead cover. Obtain support from attached engineers. (See Appendix D).

(6) Employment of Infantry

(a) The infantry battalion is normally assigned a sector to defend. However, depending on METT-T, it may be assigned to defend from a battle position or strongpoint. The battalion will usually assign its companies to either a battle position, a strongpoint, or a sector. These battle positions, strongpoints, or sectors are placed along avenues of approach to block or restrict the enemy's movement.

Depending on the type of built-up area, a company may be employed on the forward edge of the flanks of the area. This forces the enemy to deploy early

.3

without decisive engagement because it deceives the enemy as to the true location of the main defense. Other companies may then be employed in a series of battle positions and/or strongpoints in the center of the city or town. In all cases, mutual support between positions is vital. Companies and platoons should have designated alternate and supplementary positions.

(b) Once the rifle platoon commander has received his battle position or sector, he then selects the positions for his squads and crew-served weapons. Squad positions within the built-up area may be separated by rooms within buildings or be deployed in different buildings. Squad positions must be mutually supporting and allow for overlapping sectors of fire (Figure 3-8).

Figure 3-8. Sectors of Fire

(7) Employment of the Reserve. The commander's defensive plan always considers the employment of a reserve. The reserve force is organized and prepared to exploit success, to counterattack to regain key positions, to block enemy penetrations, to protect the flanks, or to assist by fire in the disengagement and withdrawal from endangered positions. For combat in a built-up area, a reserve force:

- Must be as mobile as possible
- May be a mechanized platoon, a platoon (-), or a squad at mechanized company level or a company, a company (-), or a platoon at battalion level
- May be supported by tanks or other armored vehicles
- Should be centrally located to facilitate manuever to threatened areas.

(8) Employment of Armored Vehicles. The commander plans to employ armored vehicles to take advantage of their long-range fires, heavy machine gun fires, self-generating smoke, and mobility. However, built-up areas restrict the mobility of armored vehicles and make them vulnerable to the antiarmor weapons of the enemy. Both the LAV and AAV are lightly armored and can be penetrated by heavy machine gun and antiarmor fires. Armored vehicles provide the commander with a mobile force to respond quickly to enemy threats on different avenues of approach.

(a) When armored vehicles are employed in the defense of a city, infantry should be positioned to provide them with security against close antitank fires and to detect targets for the armored vehicles. Armored vehicles should be assigned *primary*, *alternate*, and *supplementary* positions, as well as *primary* and *alternate* sectors.

(b) Armored vehicles should be located on likely avenues of approach to take advantage of their long-range fires. They may be:

- Positioned on the edge of the city in mutually supporting positions
- Positioned on key terrain on the flanks of towns and villages
- Used to cover barricades and obstacles by fire
- Placed with the reserve.

(c) Armored vehicles are normally employed as a platoon. However, sections of armored vehicles may be employed with rifle platoons or squads. This provides the armored vehicles with the close security of the infantry.

(9) Employment of Helicopterborne Forces

(a) Helicopterborne Task Force in the Defense. A helicopterborne force can defend against an infantry-heavy threat by utilizing its mobility to achieve a maneuver advantage. This force can be key in the execution of a mobile defense in both the security area and main battle area.

(b) Security Area. The helicopterborne force may be able to conduct security force operations for a larger force. Helicopters can position combat power and combat support quickly during rapidly changing situations.

(c) Main Battle Area. The mobility advantage that the helicopterborne force has over enemy infantry-heavy units may allow it to defend in greater depth in a large city. The helicopterborne force focuses on the destruction of advancing enemy forces and fights a series of battles in depth, it can attack the enemy from the front, flanks, and rear. Battle positions are selected and prepared throughout the main battle area along likely avenues of approach. Primary and alternate LZs and pick-up zones should be selected.

d. Fires. Fire support in the defense is as important as it is in the offense and has many of the same characteristics. The proximity of buildings to targets, minimum range restrictions, and repositioning requirements are all factors in the utilization of fire support in the defense. Additionally indirect fires are planned on top of and immediately around defensive positions for close support. Some of the indirect fire support considerations for the defense include:

(1) Historically, artillery fire has been used in some unique and innovative ways in an urban environment. Artillery fire support may be used in the direct- or indirect-fire mode. Artillery fire should be used to:

- Suppress and blind enemy overwatch elements

.3

- Disrupt or destroy an assault
- Provide counterfire
- Support counterattacks
- Provide direct fire when necessary
- Provide illumination in hours of darkness.

(2) Mortars at the infantry battalion and rifle company level are employed to maximize the effect of their high-angle fires. Mortars are one of the most effective indirect fire weapons in the urban environment. They should be used to engage:

- Enemy overwatch positions
- Enemy infantry before they seize a foothold
- Targets on rooftops
- Enemy reinforcements within range.

(3) Final protective fires are planned to stop dismounted assaults in front of the defensive positions. Fires within the city are planned along likely routes of advance to delay, disrupt, and destroy the enemy as he attempts to deepen a penetration.

(4) Commanders establish priorities of fire on the basis of enemy avenues of approach and threat systems that present the greatest danger to the defense. For example, during the attacker's initial advance, tanks, IFVs, and overwatching elements are the greatest threat to the defense. ATGMs should concentrate on destroying tanks first, then IFVs. Indirect fires should suppress and destroy enemy ATGMs and overwatch positions and/or elements. If enemy formations secure a foothold, priority is shifted to the destruction of enemy forces within the penetration.

(5) As the enemy attack progresses into the city, fires are increased to separate enemy infantry from its supporting armored vehicles. Friendly artillery concentrates on attacking infantry, conducting counterfire missions, and destroying reinforcements.

(6) Ample consideration must also be given to the rubbling effects of supporting arms. These effects often create massive rubbling of structures and buildings, making obstacles to both friendly and enemy movement.

(7) Mortars may be initially positioned forward in support of security forces. After their withdrawal, mortars are repositioned where they can support the in the main battle area.

(8) The commander assigns final protective fires and machine gun final protective lines (FPLs). Machine gun positions provide interlocking grazing fire and mutual support between adjacent units. FPLs are fired on planned signals. Proposed FPLs must be "walked out" to determine the extent of grazing fire available and to locate dead space. Dead space can be covered by:

- Rifle and light machine gun fire
- Grenade launchers
- Mines and boobytraps
- Indirect fires.

(9) Long-range antiarmor weapons are employed in the forward areas of the main battle area and the security area. Antiarmor weapons must be carefully positioned to prevent terrain and obstacles from reducing their range. Dismounted and manpacked antiarmor weapons should be positioned in buildings and along routes where engagement distances are reduced but where surprise and ambush are ideal.

e. Logistics. Just as in the offense, the defense utilizes a greater amount of class V and class VIII supplies in the urban environment. The defender needs to plan for multiple routes and means of supply. These can include:

- Using the city's existing transportation infrastructure to move supplies
- Prepositioning supplies in the defense in depth
- Using helicopter assets for rapid resupply
- Establishing and protecting supply routes
- Plan for utilization of existing repair and maintenance facilities
- Identify potable water sources

f. Force Protection

(1) Employment of Engineers. Engineers are employed at the battalion level or attached to companies and integrated down to squad level. Commanders must consider and assign the priority of work for engineer tasks that enhance survivability, mobility, and countermobility. Tasks that engineers can accomplish in the defense of a built-up area include:

- Constructing obstacles and rubbling for countermobility

- Clearing fields of fire
- Laying mines
- Preparing routes to the rear
- Preparing fighting positions.

(a) Employment of Obstacles. Obstacles are used to shape the battlespace. (See FM 5-102, *Countermobility*, and FM 20-32, *Mine/Countermine Operations*, for details.) The four types of obstacles are:

- **Disrupting.** Disrupting obstacles are used to disrupt the enemy's timetable and force him into conducting a time-costly bypass or breach, or cause his march columns to become staggered or "telescoped."

- **Turning.** Turning obstacles are used to canalize the enemy into engagement areas or kill zones where combined-arms fires can be concentrated and flanking shots are increased.

- **Fixing.** Fixing obstacles increase the time the enemy can be fired on in an engagement or target area and to cause increased confusion.

- **Blocking.** Blocking obstacles prevent or delay the enemy from moving past a certain point.

(b) Obstacle Belts. In defensive obstacle planning, obstacles may be emplaced in a series of belts. This enhances the overall effectiveness of the obstacle plan with successive belts designed for specific purposes. The following example lays out a three-belt obstacle plan.

1 First Obstacle Belt. The first obstacle belt is enplaced at the nearest buildings across from and parallel to the main defensive position. This belt consists of wire and improvised barriers (including inside buildings, in subterranean avenues of approach, and outside in open areas), danger areas, and dead space. These barriers and obstacles should be heavily boobytrapped and covered by fires. This belt is designed to impede enemy movement, breaks up and disorganize attack formations, and inflicts casualties.

2 Second Obstacle Belt. The second obstacle belt is placed between the first belt and the main defensive position buildings, but out of hand grenade range. It is designed to impede movement, canalize the enemy into the best fields of fire, break up attack formations, and inflict casualties. This belt is not meant to stop the enemy permanently. It consists mainly of wire obstacles, improvised barriers, road craters, and minefields. It should be heavily boobytrapped. Triple-strand concertina wire is placed along the machine gun FPL to delay the enemy in the kill zone.

3 **Third Obstacle Belt.** The third obstacle belt is the defensive position's denial belt. It consists of wire obstacles placed around and through the defensive buildings, close-in minefields, and subterranean accesses. It is designed to impede the enemy's ability to gain a foothold in the defensive area. It should be heavily boobytrapped, and integrated with mines.

(c) Emplace Obstacles and Mines. While principles for employing mines and obstacles do not change in the defense of a urban area, techniques for their employment change. For example, burying and concealing mines in streets is usually not done. To save time and resources in preparing the defense, commanders must emphasize using all available materials (automobiles, railcars, rubble) to create obstacles. Civilian construction equipment and materials should be utilized. Obstacles must be integrated into buildings and rubble areas to increase their effectiveness and to canalize the enemy. The family of scatterable mines (FASCAM) may be effective on the outskirts of a city or in parks and open areas. (See Appendix F for further discussion on obstacles, mines, and demolitions.)

(2) Air Defense. Air defense (Stinger) weapons are positioned on rooftops and parking garages in order to obtain LOS coverage for the units they are protecting. Stinger teams can be assigned the mission of protecting specific positions or placed in general support (GS) of the battalion.

(3) Fire Hazards. All cities are vulnerable to destruction by fire, especially those with many wooden structures. The defender's detailed knowledge of the terrain permits him to avoid areas that are likely to be fire hazards. The defender can deliberately set fires:

- To disrupt and disorganize the attackers
- To canalize the attackers into more favorable engagement areas
- To obscure the attackers' observation.

.3

Section II

Infantry Battalion Defense of a Built-Up Area

This section addresses some of the missions a task-organized and/or reinforced infantry battalion could be assigned. These missions are normally conducted as part of a larger joint operation.

3201. Defense in Sector. An infantry battalion is normally given the mission of defending a sector in a city (Figure 3-9). The interior buildings in the battalion sector form battle positions for the battalion's fixed defense. The size of a sector varies, but may cover the outlying structures and can be used to provide early warning and to delay the enemy.

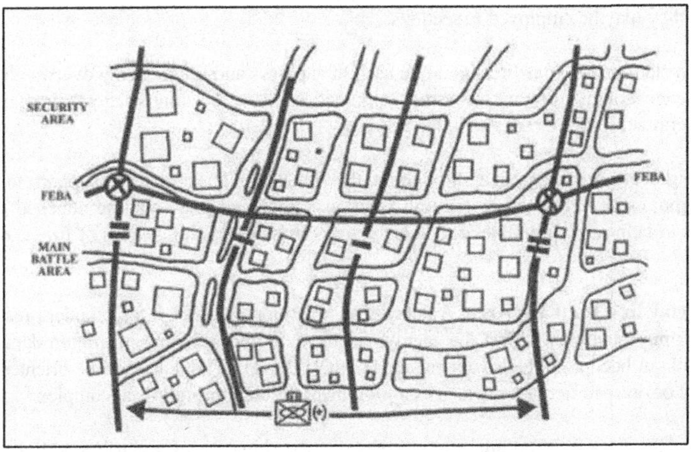

Figure 3-9. Defense of a Built-Up Sector

a. Defensive operations begin with the reconnoitering of the assigned sector to learn the terrain and to orient toward the enemy's expected advance. The main defensive positions should be formed on key terrain features, such as prominent buildings, road junctions, and positions which offer good fields of fire. Long-range antiarmor weapon systems should be deployed to engage the enemy as early as possible.

b. Depending on the situation, a reinforced rifle company may be used as a security force. The security force would operate forward of the FEBA and, depending on the mission, cause the enemy to deploy without without becoming decisively engaged. This can be done through maximum use of ambushes and obstacles while using covered and concealed routes. These routes are used for disengagement. Covered and concealed routes can be artificially made by making holes through buildings. The security force also attempts to detect the location of the enemy's main attack. Once the security force withdraws to the main battle area, it can be used as a reserve to reinforce or counterattack.

c. Based on METT-T considerations, the defense in sector (behind the FEBA in the main battle area) may consist of either company sectors, strongpoints, or battle positions. Strongpoints located on or covering decisive terrain are extremely effective in the defense.

d. Antiarmor vehicles and weapons should be used to engage enemy armored vehicles and/or cover obstacles with fire on likely avenues of approach.

e. Tanks are best used to engage enemy tanks, cover obstacles by fire, and engage in counterattacks. They should be employed in platoons where possible. However, in congested areas they may be employed in sections.

f. Artillery and mortar fire should be used to suppress and blind enemy overwatch elements, to engage enemy infantry on street approaches, provide counterbattery fire, and support counterattacks with direct and indirect fire.

g. Engineers are integrated throughout the defense. They provide support to forces in strongpoints to help prepare fighting positions. Engineers may also be attached to security forces to help in laying mines and constructing obstacles, clearing fields of fire, and preparing routes.

3202. Delay in a Built-Up Area. A delay is an operation in which a force under pressure trades space for time by slowing down the enemy's momentum and inflicting maximum damage on the enemy without becoming decisively engaged (MCWP 3-1). The delay can be oriented either on the enemy or on specified terrain such as a prominent building or industrial complex.

a. A delay in a built-up area may consist of a combination of ambushes and alternating or successive battle positions (Figures 3-10A and 3-10B on page 3-28).

(1) Ambushes are normally planned on observed obstacles. They require decentralized control and execution. Ambushes can be combined with limited-objective attacks on the enemy's flanks. They are most effective at the edge of open spaces, parks, wide streets, and so on. These attacks may be executed by dismounted infantry supported by tanks, LAVs/AAVs, or HMMWV-mounted machine guns and antitank vehicles.

(2) Battle positions should be placed where heavy weapons, antiarmor weapons, and machine guns will have the best fields of fire. Such locations are normally found at major street intersections, parks, and the edge of open residential areas. Battle positions should be carefully and deliberately prepared, reinforced by obstacles, and supported by indirect fire weapons. Battle positions are designed to inflict maximum casualties on the enemy and cause him to deploy for a deliberate attack.

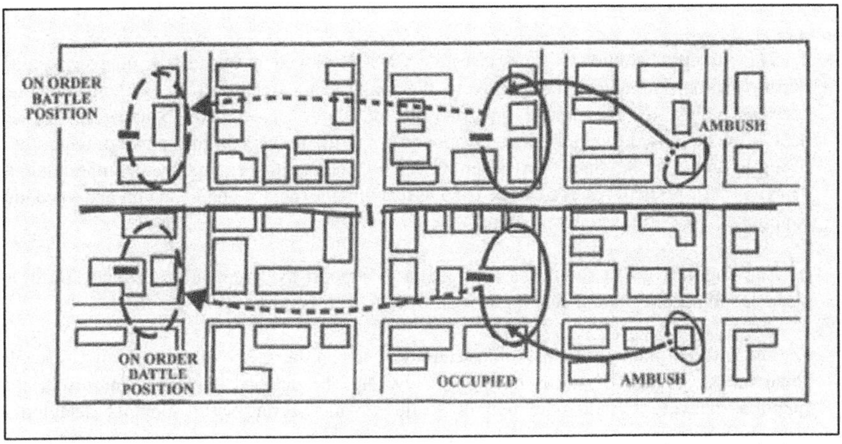

b. Tanks, LAVs/AAVs, and antiarmor weapons should prepare primary and alternate positions to improve their surivability and flexibility.

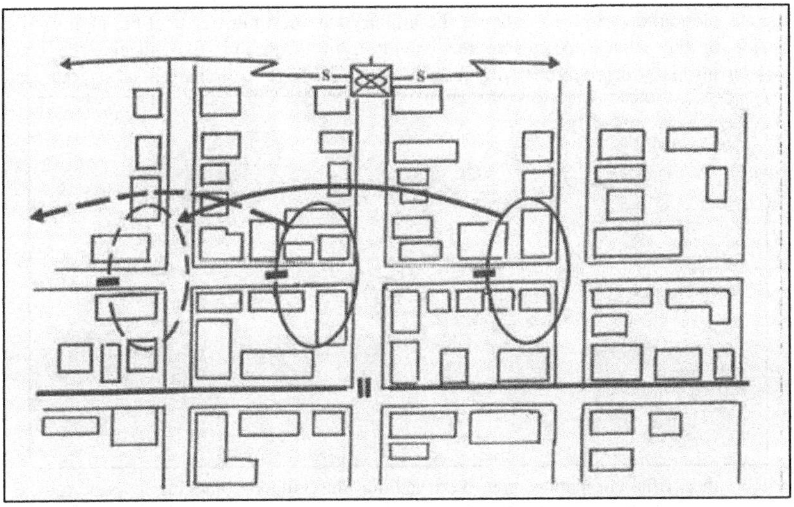

Figure 3-10A. Delay in a Built-Up Area (Successive Bounds)

Figure 3-10B. Delay in a Built-Up Area (Alternate Bounds)

c. The infantry battalion is most effective when deployed in successive delaying actions. Forces alternately engage the enemy to the maximum extent possible without becoming decisively engaged, then move on order in successive or alternate bounds. Smoke and demolitions are used to aid disengagement and cover the movement of the withdrawing force. Security units on the flank can be employed to prevent the enemy from outflanking the delaying unit. A small reserve can be used to react to unexpected enemy action and to conduct continued attacks on the enemy's flank.

d. The engineer effort should be centralized to support the preparation of battle positions, obstacle emplacement, and clearing routes.

e. The width of the infantry battalion sector depends on the nature of the buildings, obstacles along the street, and the time that the enemy must be delayed. Normally, narrow sectors supports delaying by alternate bounds while a wide sector better supports delaying by successive bounds.

3203. Defense of a Village. An infantry battalion assigned a defensive sector that includes a village may decide to incorporate the village as a strongpoint. This use is most common where the village stands astride a high-speed avenue of approach or lies between two difficult obstacles. To incorporate such an area into its defense, the infantry battalion must control the high ground on either side of the village to prevent the enemy from firing into the village. The battalion commander normally deploys one rifle company in the village to prepare a strongpoint defense

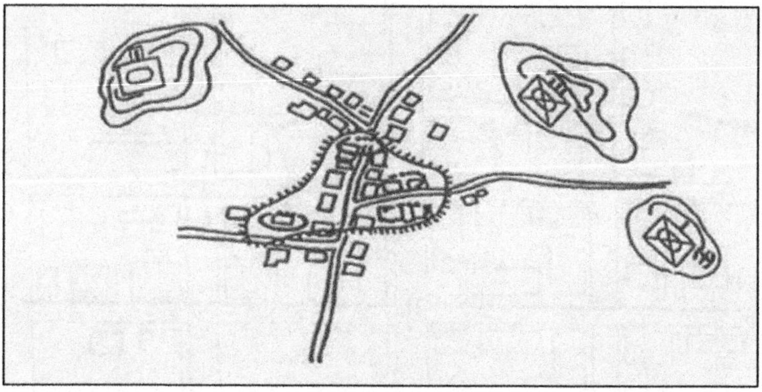

while the other rifle companies are overwatching the village, blocking bypass routes, and protecting the strongpoint's flanks. The strongpoint defense should provide a location where the enemy can be stopped and counterattacks can be launched (Figure 3-11).

Figure 3-11. Defense of a Village

a. Armored Vehicles and antiarmor weapons should be employed where the maneuver room is the greatest. This is usually on key terrain to the flanks of the village. As the security force withdraws and rifle companies become engaged, armored vehicles and weapons support by fire.

b. Although the infantry battalion's disposition should prevent large enemy forces from threatening the rear and flanks of the village, the danger of small-unit enemy infiltration requires the village be prepared for all-around defense.

c. Engineers should be assigned to the rifle company in the village strongpoint to provide continuous engineer support in preparation for the enemy's attack. The priority of engineer support should go to the strongpoint.

Section III

Rifle Company Operations

This section addresses some of the missions the infantry battalion could assign to the rifle company. For the purpose of this publication, the rifle company is a task-organized and/or reinforced unit. Assigned missions are normally part of a larger battalion operation.

3301. Defense of a Village. Once the commander has completed his reconnaissance of the village and the surrounding terrain, he develops his plan for the defense (Figure 3-12). The commander must decide whether to defend on the leading edge of the village or within the village.

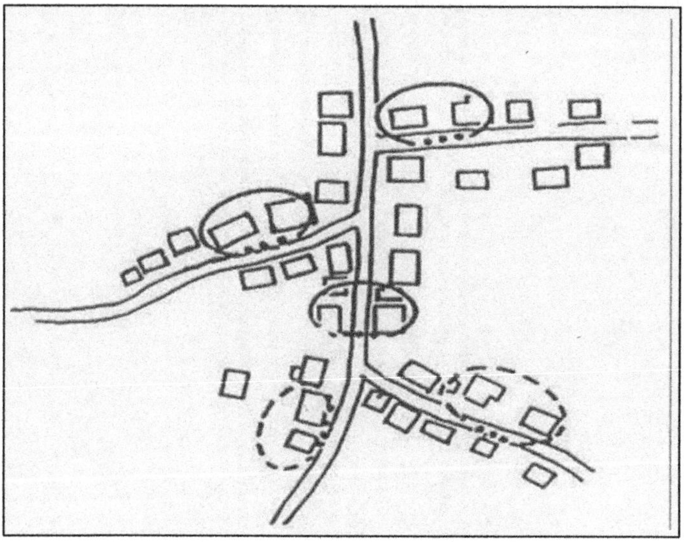

Figure 3-12. Defense of a Village

a. Several factors influence his decision. First, he must know the type of enemy that he will defend against. If the threat is mainly dismounted infantry, the greatest danger is in allowing them to gain a foothold in the town. If the threat is armor or motorized infantry, the greatest danger is that the company's defensive positions can be destroyed by direct fire. The commander must also consider the terrain forward and to the flanks of the village from which the enemy can direct fires against his positions.

b. The commander places platoons in a small group of buildings where each platoon leader establishes mutually supporting squad-sized positions. This positioning increases the area that the rifle platoon can control and hampers the enemy's ability to isolate or bypass a platoon. A rifle platoon can be responsible for the road through a village. The rest of the rifle company is positioned to provide all-around security and defense in depth.

c. Mortar positions must be protected from direct fire and allow for overhead clearance.

d. Combat vehicles should be assigned primary, alternate, and supplementary positions as well as primary and secondary sectors of fire. They should be positioned in defilade behind rubble and walls or inside buildings for cover, concealment, and movement into and out of the area.

e. If the company is reinforced with antiarmor weapons, the reinforcements should be positioned on high ground in and around the town to attain good fields of fire.

f. If the company is reinforced with tanks, the commander could place the tanks along the leading edge complement the other antiarmor weapons. The tank platoon leader selects exact firing positions and assign sectors of fire. If attacked by enemy infantry, the tanks should move to alternate positions protected by infantry. The alternate positions should allow the tanks to engage to the front and flanks with as little movement as possible. Tanks can be used as a mobile reserve after withdrawal from the leading edge of the town.

g. Obstacles are easily constructed in a built-up area. Commanders must stop enemy vehicles without interfering with his own movement in the village. Command detonating cratering charges at key street locations may be used to hamper enemy mobility. Mines should be laid on the outskirts of the town and along routes that the company will not use.

h. Engineers assist rifle platoons to prepare defensive positions. Engineers use explosives to make firing ports, mouseholes, and demolition obstacles. They also assist in the emplacement and recording of the minefields.

i. Ammunition expenditure is increased when fighting in an urban environment. Ammunition should be stockpiled at each squad position.

Other factors that the company commander may consider are:

- Resupply
- MEDEVAC
- Communications
- Firefighting

.3

- Sleep and alert plans
- Security
- Limited visibility
- Control of noncombatants.

j. The company should install a wire net and develop a plan for pyrotechnic signals to ensure adequate communications. Backup wire should be laid in case primary lines are cut by vehicles, fires, or the enemy. The commander also plans for the use of messengers .

3302. Delay in a Built-Up Area. A rifle company may be assigned the mission to delay in a built-up area.(Figure 3-13). To perform this mission, a rifle company can destroy enemy reconnaissance elements forward of the outskirts of the town, prevent their penetration into the built-up areas, and gain and maintain contact with the enemy to determine the strength and

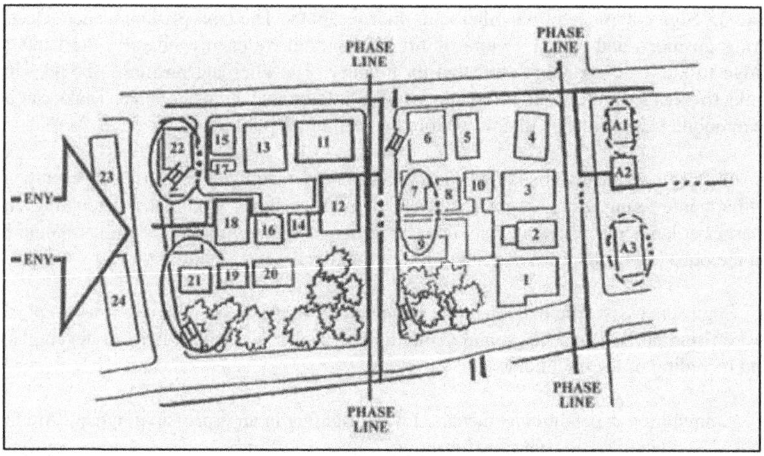

location of the main attack.

Figure 3-13. Delay in a Built-Up Area

a. Platoons delay by using patrols, OPs, and ambushes. Following each action, a platoon will disengage and withdraw.

b. The company's sector should be prepared with obstacles to enhance the delay. Engineers should prepare obstacles on anticipated enemy main routes of advance.

c. Antiarmor weapons may be positioned on the outskirts of the town to destroy the enemy at maximum range. They should be located in defilade positions or in prepared shelters. They fire at visible targets and then fall back or proceed to alternate positions.

d. Tanks in support may be used to engage enemy tanks, provide reinforcing fires, aid the disengagement of platoons, and cover obstacles by fire.

3303. Defense of a City Block. The rifle company may be assigned to defend a city block. The company actions should be coordinated with those of the security forces charged with delaying to the front of the company's position. The defense should take advantage of the protection of buildings that dominate the roads.

a. A well-organized company defense:

- Stops the attack of the enemy on the roads by using obstacles and enfilade fire
- Destroys the enemy by ambush and direct fire from prepared positions
- Ejects the enemy from footholds or fixes them for the battalion's counterattack.

b. The defensive forces should prepare ambushes on the avenues of approach, cover the obstacles by fire, and build a defense inside the buildings (Appendix D). The reserve can be tasked:

- To reinforce the fires of the defense
- To react to a danger on the flank
- To counterattack to throw the enemy from a foothold.

c. Engineers should be employed in GS to construct obstacles, prepare access routes, and assist in preparing defensive positions.

d. Tanks attached to the rifle company may be used to direct-fire support, engage enemy tanks, and support a counterattack.

3304. Defense of a Traffic Circle or Key Intersection. A rifle company may be assigned the mission to defend a key traffic circle or road junction (Figure 3-14).

a. The commander analyzes enemy avenues of approach and buildings that dominate those avenues. He plans to focus all possible firepower on the traffic circle and its approaches. He also plans for the all-around defense of the buildings that dominate the traffic circle to prevent encirclement. The commander prepares as many covered and concealed routes between these

.3

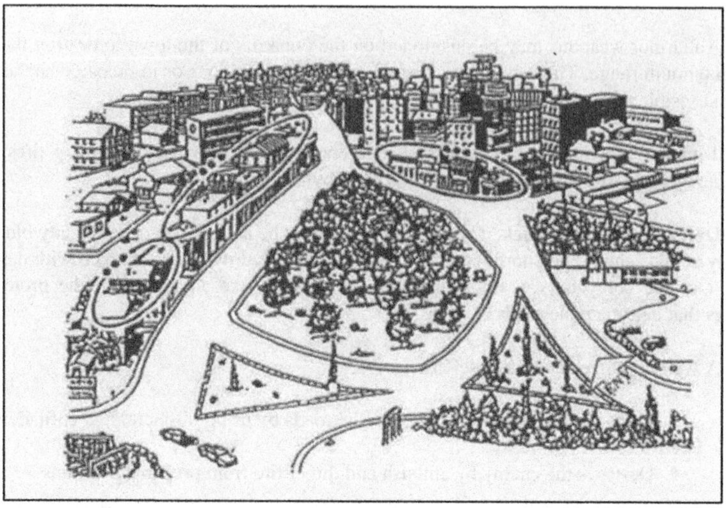

buildings as possible. These routes make it easier to mass or shift fires and to execute counterattacks.

Figure 3-14. Defense of a Traffic Circle

b. Obstacles can be used to deny the enemy the use of the traffic circle. In this case, obstacle planning should consider whether friendly forces require access to the traffic circle. Antiarmor weapons can be employed across the traffic circle if designated fields of fire exist. Tanks can be used to engage enemy tanks and provide direct-fire for counterattacks.

Section IV

Rifle Platoon Operations

This section discusses some missions the rifle company might assign to a rifle platoon.

3401. Defense of a Strongpoint. A rifle platoon may be given the mission to provide or conduct the strongpoint defense of a building, part of a building, or a group of small buildings (Figure 3-15). The rifle platoon's defense is normally integrated into the rifle company's defense. The rifle platoon must keep the enemy from gaining a foothold in its assigned buildings. It makes the best use of its weapons and supporting fires, organizes all-around defense, and counterattacks or calls for a rifle company counterattack to eject an enemy that has a foothold. The platoon leader analyzes his defensive sector to recommend to the company commander the best use of obstacles and supporting fires. (See Appendices B and D.)

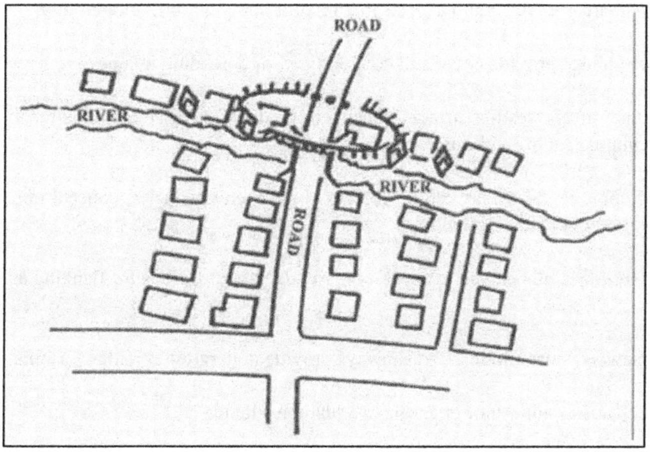

Figure 3-15. Defense of a Strongpoint

a. The rifle platoon should be organized into a series of firing positions located to cover avenues of approach, cover obstacles, and provide mutual support. Snipers may be located on the upper floors of the buildings. Unengaged units should be ready to counterattack, fight fires, or reinforce other elements of the rifle platoon.

b. Depending on the length of the mission, the rifle platoon should have available:

- Pioneer equipment (axes, shovels, hammers, picket pounders)

- Barrier material (barbed wire, sandbags)
- Munitions (especially grenades)
- Food and water
- Medical supplies
- Firefighting equipment.

3402. Defense Against Armor. Urbanized terrain is well suited to an infantry's defense against mechanized infantry and armored forces. Armored forces will usually avoid built-up areas but may be forced to pass through them.

a. Built-up areas have certain aspects that favor infantry antiarmor operations.

(1) Buildings provide cover and concealment for defending infantry.

(2) The streets restrict armor maneuver, fields of fire, and communications, thereby reducing the enemy's ability to reinforce.

(3) Rubble in the streets can be used to block enemy vehicles, conceal mines, and cover and conceal defending infantry.

(4) Rooftops, alleys, and upper floors provide good positions for flanking and overhead fire.

(5) Sewers, storm drains, and subways provide underground routes for infantry forces.

b. To prepare for antiarmor operations, a platoon should:

(1) Choose a Good Engagement Area. The best way for infantrymen to engage tanks is one at a time so that they can destroy one tank without being vulnerable to the fires of another. Good engagement areas include narrow streets, turns in the road, "T" intersections, bridges, tunnels, split-level roads, and rubbled areas. Less obvious locations can also be created by using demolitions or mines to create obstacles.

(2) Select Good Weapons Positions. The best weapons positions are places where the tank is weakest (tank deadspace) and the infantry is most protected. Forcing tanks to button up greatly reduces their ability to see and fire on close-in street and overhead targets. Figure 3-16 (on page 3-38) shows close-in tank weapon vision and engagement limitations (tank buttoned up) at street and overhead levels.

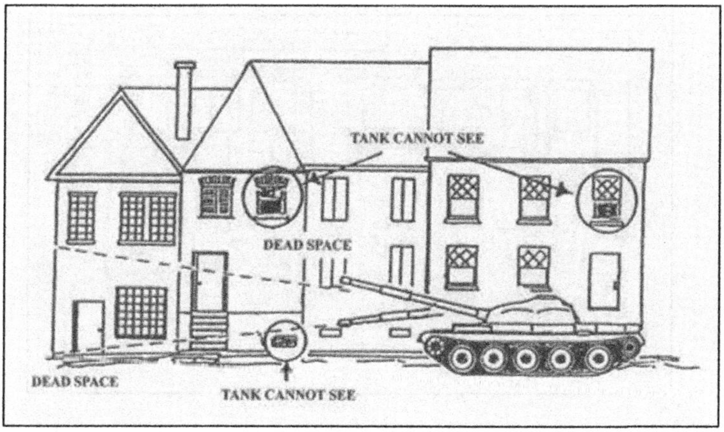

Figure 3-16. Close-in Tank Limitations at Street and Overhead Levels

(3) Assign Target Reference Points and Select Method of Engagement. After selecting the weapons positions, target reference points should be assigned to ensure adequate coverage and control fires. The target reference points should be clearly visible through the gunner's sights and should be relatively resistant to battle damage (for example, large buildings or bridge abutments, but not trees or cars). The three types of antiarmor engagement techniques are frontal, crossfire, and depth. Frontal fire is the least preferred because it exposes the gunner and it attacks the tank where the armor is the thickest. (For more information on target engagement techniques, see MCWP 3-15.5 *Antiarmor Operations*.)

Infantry forces should fire at tanks on the flanks and rear from ground level or at the top of tanks from an elevated position. A suitable antiarmor defense might be set up as shown in Figure 3-17. For a safe engagement from an elevated position, Marines should allow the tank to approach to a range three times the elevation of the weapons. However, engagement at longer ranges risks counterfire because the attacking weapon's position will not be in the tank's overhead dead space.

(4) Coordinate Target Engagement. Tanks are more vulnerable to infantry operations when they are buttoned up. The first task is to compel the tanks to button up by using direct and indirect fire. The next task is to coordinate the fires of the antitank weapons so that all targets in the engagement area are attacked simultaneously (Figure 3-18).

Figure 3-17. A Rifle Platoon's Antiarmor Defense

Figure 3-18. Coordinated Antiarmor Ambush

c. Infantry often accompany armored vehicles in built-up areas to provide them with close-in protection. Antiarmor weapons should be supported by an effective antipersonnel defense.

d. At a planned signal (for example, the detonation of a mine), all targets in the engagement area are attcked at the same time. If that is not possible, then the most dangerous target should be engaged first. Although tanks present the greatest threat, BMPs are also dangerous because their infantry can dismount and destroy friendly antiarmor positions. Rubble and mines should be used to reduce target mobility so that a target's time in the engagement area is increased.

3403. Conduct of an Armored Ambush. A rifle company can use an attached tank platoon to conduct an armored ambush (Figure 3-19). The tank platoon should be reinforced with infantry.

Figure 3-19. Armored Ambush

a. An armored ambush requires maneuver on obstacle-free roads. Obstacles outside the ambush area can be used to canalize and delay the enemy.

b. Ambushing tanks should hide approximately 1,000 meters from the expected enemy avenue of approach. A security post, located at a chokepoint, observes and reports the approach, speed, security posture, and activity of the enemy. When the enemy is reported at a trigger point or target reference point, the tank platoon leader orders the ambush.

c. The tanks move quickly from their hiding positions to firing positions, taking advantage of all available concealment. They attempt to fire flank shots on the approaching enemy at an average range of 300 to 400 meters. Firing at these ranges does not expose tanks to enemy infantry. After the engagement, the tank breaks contact and moves to a rally point with close security provided by infantry. They may then move to a new ambush site.

Chapter 4

Combat Support

Combat support units provide critical combat functions in conjunction with combat arms units to secure victory (MCRP 5-2A).

4001. Introduction. Success against a determined enemy is achieved in part by generating decisive combat power through the application of *combined arms*. This is especially true in MOUT. The MAGTF commander must effectively integrate all the capabilities of the command element (CE), GCE, ACE, and CSSE to win the battle in the shortest amount of time with the fewest casualties. The MOUT environment requires special considerations pertaining to combat support forces and capabilities.

Section I

Fire Support

4101. Indirect and Direct Fire Support

a. General. MOUT operations present unique fire support considerations. On urbanized terrain, buildings provide excellent cover and concealment to the enemy while limiting friendly observation and targeting efforts. Targets are generally exposed for brief periods of time and are often in close proximity to friendly forces. Observers will experience difficulty in finding OPs with adequate fields of observation. Terrain masking by tall buildings may restrict the delivery of indirect fires. Collateral damage and rubble effects must be considered during ammunition selection. The importance of effective communications, ROE, control measures, and procedures to prevent fratricide is magnified.

b. Offensive Support. In the offense, fire support plans should include fires to isolate the objective area, support the assault, and support the clearing action. Fires are delivered to isolate and fix the enemy and deny him the use of avenues of approach into and out of the built-up area. Most fires are normally planned and executed at the GCE or higher echelons. Fires are employed to rupture the enemy's established defenses and screen friendly maneuver, in order to maintain the momentum of the attack. Fire support is also allocated to units involved in clearing operations. Fire support plans should incorporate the employment of aerial observers and UAVs to compensate for restrictions to observation and to assist in the delivery of deep fires. Procedures for designating the forward line of own troops (FLOT), marking targets, shifting fires, and communicating in the urban environment should also be considered.

c. Defensive Support. In the defense, fire support plans address fires to disrupt and slow the enemy attack. Fires are delivered at maximum ranges along avenues of approach to separate

.3

armor and infantry forces, to canalize the enemy into killing zones, and to deny the enemy use of key facilities in the city, such as communications and transportation facilities. Most fires are planned and executed at the GCE or higher echelon. Defensive fires planned at lower echelons support fighting within the built-up area. Defensive fire support plans identify fire support coordination procedures necessary to execute the delivery of fire support.

4102. Artillery

a. Employment of Artillery. The mission of artillery does not change in MOUT. However, the method of employment is tailored to the unique requirements posed by the urban battlespace. MCWP 3-16.3, *Marine Artillery Support,* provides a detailed discussion on the employment of artillery. Considerations for MOUT include:

(1) Organization for Combat. Standard command relationships for tactical missions are used to organize for combat. During the initial phases of offensive MOUT, particularly isolation of the objective area and the initial assault to secure a foothold, control of artillery is more centralized than in other types of offensive operations. Centralized control provides better responsiveness in shifting and massing fires. As the attack progresses into the clearance phase, control of artillery becomes more decentralized with direct support (DS) artillery providing direct and indirect fires. In the defense, control of artillery should be sufficiently centralized to enable the rapid massing of fires.

(2) Movement and Positioning. Key considerations for the positioning of artillery in MOUT include:

- Mobility restrictions
- Limited availability of firing positions
- Masking of fires
- Security
- Enemy counterbattery fires

(a) Offense. Artillery is initially positioned outside or in the outskirts of the city. Artillery may displace to suitable firing positions within the city as the attack progresses. Positions should be selected that minimize masking, provide several routes for escape, and afford as much cover and concealment as possible. Potential firing positions can be found in parks, racetracks, sports fields, school yards, stadiums, parking lots, and similar areas. Special techniques for the emplacement of howitzers may be required if the ground is not suitable for normal emplacement. Some artillery positions may be maintained outside the city (see appendix B). Artillery generally makes few displacements in MOUT because it requires extensive route reconnaissance and security. If and when artillery is moved, it should be given priority of movement during displacements. Movement is conducted during night or periods of reduced visibility.

(b) Defense. Artillery is positioned on the friendly outskirts of the built-up area. Artillery displaces along predetermined routes to avoid traffic congestion.

(3) Delivery of Fires. The commander, in coordination with the fire support coordinator (FSC), considers the following when planning fire support in a built-up area:

(a) Target acquisition may be more difficult because of the increased cover and concealment afforded by the terrain. Ground observation is limited; therefore, FOs should be placed near tops of buildings.

(b) For targets in defilade, fire may be adjusted laterally so that the rounds impact on the street perpendicular to the observer's position. Airburst rounds are more suitable for this adjustment. Adjustments may have to be made by sound. When rounds impact on a perpendicular street, they are adjusted for range. When the range is correct, a lateral shift is made onto the target and the observer fires for effect.

(c) Special consideration must be given to shell and fuze combinations when effects of munitions are influenced by buildings. These include:

- Careful use of variable time is required to avoid premature arming.
- HE ammunition with impact or delay fuzes may create unwanted rubble.
- The close proximity of enemy and friendly forces requires careful coordination.
- White phosphorus (WP) may create unwanted fires and smoke.
- Fuze delay should be used to penetrate fortifications.
- Illumination rounds can be effective. However, friendly positions should remain in shadows and enemy positions should be highlighted. Tall buildings may mask the effects of illumination rounds.
- HE/variable time, HE/time fuze, and improved conventional munitions are effective for clearing enemy positions, observers, and antennas from rooftops.
- Swirling winds may degrade smoke operations.
- FASCAM may be used to impede enemy movements. FASCAM effectiveness is reduced when delivered on a hard surface.

.3

(4) Targeting is difficult on urbanized terrain because the enemy has many covered and concealed positions and movement lanes. The enemy may be on rooftops and in buildings and may use sewer and subway systems. Targets should be planned on rooftops to clear enemy FOs, communications, and radar equipment. Targets should also be planned on major roads, at road intersections, and on known or likely enemy fortifications. Employing artillery in the direct-fire mode to destroy fortifications should be considered. Also, restrictive fire support coordination measures (such as a restrictive fire area or no fire area) may be imposed to protect civilians and critical installations.

(5) The 155-mm howitzer can be effective in neutralizing concrete targets by direct fire. A 155-mm round can penetrate 36 inches of concrete at ranges up to 2,200 meters. Restrictions may be placed on the type of ammunition used in order to reduce rubbling on friendly avenues of approach.

(6) FOs must be able to determine the location and size of dead space. Dead space is the area in which indirect fires cannot reach the street level because of buildings. This area is a safe haven for the enemy because he is protected from indirect fires. For low-angle artillery, the dead space is about five times the height of the building. For mortars and high-angle artillery, the dead space is about one-half the height of the building.

(7) Aerial observers and UAVs are effective in providing observation behind buildings immediately to the front of friendly forces. They are extremely helpful in adjusting fire. Aerial observers can also relay calls for fire when communications are degraded because of power lines or building masking.

(8) Weapons locating radars are effective in locating enemy artillery and mortar targets in an urban environment because of their high percentage of high-angle fires. However, LOS must be maintained in order to exploit this advantage. Radars should not be placed in the midst of a city due to masking. They are best located on the outskirts of the city.

(9) The use of airburst fires is an effective means of clearing snipers and manpads from rooftops. HE shells with delay fuzes may be effective against enemy troops in the upper floors of buildings, however, because of the overhead cover provided by the building, such shells have little effect on the enemy in the lower floors. (The planning and employment of artillery in offensive and defensive operations is also addressed in Chapters 2 and 3.)

b. Local Security. Artillery crews and howitzers require close security when used in the direct-fire mode. Artillery is particularly vulnerable when units are employed in isolated firing positions.

c. Communications and Information Systems (CIS). CIS must be able to satisfy the unique C2 requirements for artillery in MOUT. CIS should provide the commander with the ability to collect, process, analyze, and exchange information. Requirements such as massing fires and call for fire must be carefully planned and adapted to meet the requirements of

MOUT. For example, small units conducting independent operations must have the communications capabilities and maps to allow them to call for fire.

4103. Mortars

a. Employment of Mortars. Mortars are often the primary indirect-fire weapon for forward units in the assault or defense of a built-up area. Their high rate of fire, short minimum range, and high trajectory give mortar units the ability to fire in the tight confines of the city. Mortars can provide obscuration, neutralization, suppression, or illumination fires.

(1) Organization for Combat. Normally, 81-mm mortars are employed in GS of the battalion. The 60-mm mortar is normally used at the discretion of the rifle company commander.

(2) Movement and Positioning. 81-mm mortars employed in GS normally occupy a position within or near the infantry battalion reserve. Key considerations for the selection of positions include:

- The minimum range of the weapon
- Mask and overhead clearance
- Terrain suitable for setting base plates
- Dispersion, and accessibility.

Special techniques can be used to position and lay the 81-mm mortar. These techniques include placing buffers under base plates, using curbs and sandbags to anchor or brace mortars, and filling cans with dirt or using sandbags to emplace aiming posts. Fixed objects may be used as distant aiming points. When the depth of the defensive position is shallow, or when suitable firing positions are not available, mortars may have to be positioned behind the reserve. Positioning of mortars behind buildings greatly enhances survivability. Displacement is often executed by section. Mortars are usually not placed on top of buildings because the location lacks cover and concealment and recoil/concussion could collapse roofs.

(3) Delivery of Fires. Mortar fires are effective in the offense as well as in the defense. The indirect fires are extremely responsive. Mortars are well suited for combat in built-up areas because of their high rate of fire, steep angle of fall, and short minimum range. Mortar fires can be used to inhibit enemy fires and movement, allowing friendly forces to maneuver to a position of advantage. Effectively integrating mortar fires with dismounted maneuver is key to successful combat in a built-up area.

.3

(a) HE Ammunition. HE is used more than any other type of ammunition. Although mortar fires are often targeted against roads and other open areas, the natural dispersion of indirect fires will result in many hits on buildings. Leaders must use care when planning mortar fires to minimize collateral damage. Considerations when using HE include:

- HE ammunition is particularly effective when used on lightly built structures within cities. However, it does not perform well against the reinforced concrete found in larger urban areas.

- Point-detonating fuzes should be used. The use of proximity fuzes should be avoided because the nature of built-up areas causes proximity fuzes to function prematurely. However, proximity fuzes are useful in attacking targets such as OPs on tops of buildings.

- During World War II, the Middle East conflicts, and most recently in Bosnia, mortar HE fires have been used extensively to deny the use of streets, parks, and plazas to enemy and civilian personnel.

(b) Incendiary Fires. Employment of these munitions is planned in detail to prevent the burning of buildings and surrounding terrain that may create obstacles to friendly forces.

(c) Illumination. In the offense, illumination rounds should burst above the objective to put enemy troops in the light. Buildings reduce the effectiveness of the illumination by creating shadows. If the illumination is behind the objective, the enemy troops would be in the shadows rather than in the light. In the defense, illumination is planned to burst behind friendly forces to put them in the shadows and place the enemy forces in the light. Continuous illumination requires close coordination between the FO and the fire direction center (FDC) to produce the proper effect.

(d) Other Considerations. When planning the use of mortars, commanders must consider the following:

- FOs should be positioned on tops of buildings so target acquisition and adjustments in fire can best be accomplished.

- Commanders must understand ammunition effects to correctly estimate the number of volleys needed for the specific target coverage. Also, the effects of using white or red phosphorus may create unwanted smoke screens or limit visibility, which could interfere with the tactical plan.

- Mortar sections should plan to provide their own security.

- Commanders must give consideration to where, when, and how mortars are to displace in order to maintain immediate indirect fire support. Combat in built-up areas may adversely affects the ability of mortars to displace because of rubbling.

b. Communications. An increased use of wire, messenger, and visual signals will be required. However, wire should be the primary means of communication between the FOs and the FDC because elements are close to each other. Also, frequency-modulated (FM) radio transmissions in built-up areas are likely to be erratic. Remote siting of antennas to upper floors or roofs may improve communications and enhance operator survivability. The use of radio retransmissions can also be effective. Both in the offense and defense, commanders consider using existing civilian systems to supplement the unit's capability.

c. Magnetic Interference. Anything that uses or generates electricity generates its own magnetic field. In an urban environment, all magnetic instruments are affected by surrounding structural steel, electrical cables, and automobiles. Minimum distance restrictions for the use of the M-2 aiming circle will be difficult to overcome. However, an azimuth may be obtained from a distant aiming point.

4104. Naval Surface Fire Support (NSFS) and Naval Gunfire (NGF)

a. General. When NSFS ships are available, naval surface fires and NGF can provide some fire support.

b. Employment of Naval Surface Fire Support. The urban environment severely limits NSFS, and NGF in particular, due to its flat trajectory. NSFS can be used to supplement existing fire support but should not be planned as the primary fire support asset for MOUT. NSFS is most effective when firing on targets to isolate the battlefield and engage targets located on the seaward side of a city. Only when the ROE allows extensive collateral damage can the full effect of NSFS be brought to bear on units occupying a city.

(1) Organization for Combat. NSFS is best employed under centralized control because of the limited use of NSFS within the confines of the city.

(2) Positioning. Many of the fires delivered by NSFS ships require a careful orientation of the gun target line and may limit the size of a fire support area.

(3) Delivery of Fires. NGF can deliver a variety of ammunition in MOUT. NGF can deliver HE, incendiary, and illuminating projectiles. The NGF high-capacity projectile is very effective in producing fragmenting and penetration effects. Offensive and defensive fire support plans can incorporate NSFS fires on bridges, roads, reinforcements, artillery positions, and command and control centers. NSFS can be effectively delivered on targets along the outskirts of the city to rupture established defenses. These fires are particularly effective on buildings and on targets with a linear axis such as streets and airfields. Because of its flat trajectory, NGF employment is limited on targets within the city. Firing reduced charges at high angles does enable NGF to hit more targets but will reduce its range and accuracy.

4105. Aviation

a. Employment of Aviation. Both fixed- and rotary-wing aircraft can provide fire support in MOUT. The aviation combat element (ACE) provides the MAGTF commander with a variety of options in ordnance delivery. Aerial delivered ordnance is particularly effective against hard targets. Cluster bomb units, rockets, cannons, laser-guided bombs, and electro-optically guided missiles can be used against a variety of targets. However, the presence of civilians, the requirement to preserve key facilities, and the residual effects caused by rubble and collateral damage must be taken into consideration when planning for and using aviation in an urban environment. (See MCWP 3-24 *Assault Support Helicopter Tactical Manual* (under development), MCWP 3-23.1 *Close Air Support* (under development),and NWP 55-3-AH1, *AH1 Tactical Manual,* for greater detail on MOUT considerations for aviation.)

(1) Fixed-Wing Aircraft

(a) Offensive/Defensive Operations. During offensive operations, fixed-wing aircraft can isolate, interdict, and reduce enemy strongpoints throughout the battlespace. During defensive operations, they attrite enemy forces as they approach, attack forces attempting to bypass, or canalize the enemy to take avenues of approach that are advantageous to the defender.

(b) Other Considerations

1 Fixed-Wing Attack of a Built-Up Area. Air strikes can reduce or destroy vital supplies and facilities supporting the enemy. The resulting shock, concussion, and psychological effects of such attacks can reduce the efficiency and fighting spirit of enemy forces.

2 Rubble and Debris. The rubble and debris resulting from air attacks may increase the defender's cover and create obstacles to the movement of friendly forces.

3 Proximity of Friendly Forces. The proximity of opposing forces to friendly forces increases the potential for fratricide. This may require the use of precision guided munitions (PGMs) or the temporary disengagement of friendly forces in contact.

4 Locations of Indigenous Civilians or Key Facilities. The use of aerial delivered ordnance may be restricted by the ROE because of the presence of civilians or the requirement to preserve key facilities within a city.

5 **Limited Ground Observation.** Limited ground observation may require the use of forward air controllers (airborne) (FAC(A)).

(2) Rotary-Wing Aircraft. Attack helicopters provide a distinct advantage to the commander in MOUT. They can rapidly maneuver within the built-up area to attack a variety of targets. Urban structures offer ideal cover for attack helicopters. The ability to deliver PGMs, rockets, and 20 mm cannon fire makes attack helicopters valuable assets in reducing strongpoints, breaking up enemy concentrations, and providing suppressive fires for attacking forces. Attack helicopters also provide a highly manueverable antiarmor capability.

(a) Attack Helicopter Fire Support During Offensive/Defensive Operations

1 Provide close air support in support of the commander's scheme of maneuver

2 Provide suppression and/or destruction of enemy forces within buildings and strongpoints

3 Conduct counterattacks

4 Isolate the built-up area or objectives within the built-up area

5 Conduct armed reconnaissance and patrols.

6 Conduct rapid concentration of fires.

4106. Armored Vehicles

a. Employment of Armored Vehicles. Tanks, LAVs, and AAVs are primarily mobility assets that function best in an unrestricted, mobile environment. However, these vehicles can play an important role inside the city by using their firepower in support of offensive and defensive operations. LAVs and AAVs highly are susceptible to armor-piercing rounds and manpacked antiarmor weapons. All vehicles are susceptible to top-attack munitions. Some of the fire support roles armored vehicles may provide include:

- Suppression and/or destruction of enemy forces within buildings and strongpoints (tanks/LAVs/AAVs)
- Provide machinegun suppression fires
- Reserve or counterattack suppression fires (tanks and/or AAVs)
- Creating entry/exit points in buildings
- Isolation of the built-up area or objectives within the built-up area (tanks/LAVs/AAVs)

- Artillery-towing vehicle (LAVs/AAVs)

- Antiarmor fires

- Breaching obstacles in a direct fire mode.

b. Special Consideration. The existence of waterways, canals, and rivers in and around urban areas can provide avenues of ingress into designated objective areas. If used in conjunction with supporting fires, smoke, and/or deception operations, AAVs can capitalize on their amphibious capability and make use of waterways previously viewed as obstacles.

4107. Antiarmor Weapons. Antiarmor weapons have more utility in the defense than in the offense. Their primary task is defeating enemy armor. Secondary tasks may involve blasting holes in structures and fortified positions. As with other weapons systems, antiarmor weapons are normally employed in a decentralized fashion at the small unit level. Enemy armor should be engaged in predetermined locations that afford the most advantage to the defender. Weapon positions should be selected that provide maximum range fires. TOWs and Dragons may be fired from an enclosed area if it is a sturdy ventilated structure of appropriate size. For a more detailed discussion of antiarmor weapons, see MCWP 3-15.5, *Antiarmor Operations*.

4108. Employment of Snipers

a. General. Snipers are highly trained in field skills and marksmanship and can deliver long-range precision fire at selected targets from concealed positions. The method by which snipers are employed in urban warfare will be governed by the nature of the terrain/weather, number of snipers available, and whether the enemy employs snipers. A sniper can play an important role in MOUT. Employment of snipers influences the action by:

- Creating adverse psychological affects on the enemy by negating feelings of security

- Inflicting casualties on enemy leadership

- Enabling the infantry to seem to cover a large area, regardless of whether the terrain is physically occupied or not

- Detailing enemy positions and activity through the use of advanced optical equipment and observation techniques

- Controlling fire support

- Conducting reconnaissance in conjunction with their sniper role

- Disrupting enemy movement, observation, and infiltration, and negating the possibility of surprise.

b. Offensive Employment. Snipers should be used to gain and maintain contact with the enemy. This enhances security and prevents surprise by keeping constant and unrelenting pressure on the enemy. Some of the considerations for sniper employment include:

(1) As the "eyes" of a commander, a sniper increases the commander's flexibility by gathering and transmitting information on fire support targets.

(2) Snipers attack those targets that affect the enemy's ability to wage battle and those that will cause the maximum amount of confusion on the battlefield. Potential targets for a sniper are:

- Officers and noncommissioned officers
- Enemy snipers and marksmen
- Engineer personnel
- Personnel manning crew-served weapons
- Communications personnel and equipment
- Fire support observers and/or equipment
- Commanders of armored vehicles.

c. Combat on Urbanized Terrain

(1) Snipers can be employed to operate for extended periods of time from hidden positions to dominate and establish a "no-man's land," screen flanks, protect the rear, and deny the enemy access to certain areas or avenues of approach. Snipers can operate with the covering element to deliver accurate fires in support of search elements. Optical devices enhances their ability to detect movement and engage it instantly.

(2) Snipers can also support (by fire) infantry movement across streets. They can provide immediate precision fire on enemy machine gun nests and enemy snipers. The best countersniper weapon is the sniper. Snipers are assigned the following supporting missions:

- Delaying and inflicting casualties on the advancing enemy
- Observing and reporting potential targets
- Covering (by fire) avenues of approach and obstacles

- Harassing the enemy and causing him to deploy prematurely
- Directing supporting arms fire.

The urban environment provides an abundance of secure firing positions that are not highly accessible to countersniping.

d. Defensive Employment. In defensive operations, the sniper is best employed forward of the FEBA to provide early warning of enemy approach, disrupt it, and if possible, cause the enemy to deploy prematurely. Snipers are employed to:

- Prevent enemy infiltration
- Detect and destroy infiltrators
- Protect unit at the FEBA from surprise attacks
- Protect patrols from ambush
- Screen the flanks and rear of the defensive positions.

More detailed information regarding the employment of snipers can be found in Marine Corps Warfighting Publication (MCWP) 3-15.3, *Scout Sniping* (under development).

Section II

Other Combat Support

4201. Engineers

a. General. The nature of the urban battlespace requires extensive use of engineers. In both the offense and defense, the proper employment of engineers can be decisive. Engineers must also be prepared to repair and maintain the internal support facilities of the city in order to support further operations. MAGTF commanders plan to use all available engineer assets, including engineers in the GCE, ACE, and CSSE, as well as any naval construction forces (NCFs).

b. Engineer Organization. Engineers are almost always task organized. The compartmentalized, closed-in nature of the city, which limits observation, mobility, and communications, will dictate the best organization for the mission. Teams should be formed to perform specific tasks for small units and must be positioned for quick employment. In MOUT, engineer support is key to maintaining the mometum of the attack. Engineer teams will usually move with the units they are supporting. Two-man demolition teams may be assigned to support a rifle platoon or squad to breach obstacles or to neutralize booby traps and improvised explosive devices (IEDs).

Commanders must carefully analyze mission requirements, the alignment of enemy forces, the condition of the battlespace, and available assets, to ensure an organization that will provide optimum engineer support. Most of the actual engineer work is done at the small unit level.

c. Engineer Tasks

(1) Offense. Urban combat is by nature a very slow, deliberate affair. Even if the enemy has not prepared the area for a defense, it will take time to secure the city. A well-constructed urban defense will slow or stop any attack unless impediments to movement can be quickly removed or bypassed. Properly organized and positioned engineers can reduce delays by performing a variety of tasks which include:

(a) Engineer reconnaissance to determine the level of effort required to penetrate outer and inner defenses, clear obstacles, and identify bypass routes.

(b) Bridge repair and construction on critical avenues of approach. Bridge construction may also open additional avenues of approach that the enemy is not prepared to defend.

(c) Breaching obstacles both outside and inside the city. This includes breaching minefields and neutralizing booby traps and IEDs.

(d) Assault breaching into buildings.

(e) Reducing enemy strongpoints using demolitions and heavy equipment.

(f) Preparing helicopter landing zones (HLZs).

(g) Assist in the preparation of defensive positions in the event of a counterattack.

(2) Defense. An urban area can be made nearly impregnable with proper organization of the battlespace and skillful preparation by engineers. Obstacles in outlying areas can delay/disrupt the enemy and canalize him into engagement areas. The streets and alleys inside a city can also be use to canalize enemy forces into engagement areas and kill zones. In preparing the defense of an urban area, engineers perform several critical tasks to include:

(a) Engineer preparation of the urban battlespace. This should begin well outside the city with complex obstacles in support of security forces. Selective demolition of bridges and roads and other countermobility measures can force the enemy into taking avenues of approach that favor the defense.

(b) Constructing obstacles forward of and between defensive positions to block, disrupt, fix, and turn enemy attacks. A majority of the obstacles will be directed at interfering with the movement of personnel and vehicles, but some effort must be directed at preventing vertical and subterranean movement. Material designed especially for obstacles and barricades, such as barbed wire and tetrahedron(s), is necessary, but improvised barriers can be just as effective and are readily available. Vehicles and building rubble are the most common improvised barriers.

(c) Enhancing mobility by opening and/or marking preplanned routes into attack positions for counterattacks or alternate positions to protect a flank.

(d) Providing technical advice on which buildings are more suitable for use as strongpoints or which rooftops are strong enough for helicopter landings can be vital. Any reinforcement of building roofs should be supervised by the engineers.

(e) Preparing tank positions that offer them cover and concealment.

(f) Assisting infantry units in clearing clear fields of fire, especially for longer range weapons like the TOW.

(g) Playing a potential role in the maintenance of critical urban services such as power, water, and communications. This function may also be performed by naval construction units.

(h) Assisting in the preparation and maintenance of railroads, airfields, and ports. This function may also be performed by naval construction units.

4202. Assault Support Helicopters

a. Combat Assault Transportation. The mobility provided by helicopters can be a significant advantage in the urban environment. In many modern cities, office buildings may have helipads on their roofs. Other buildings, such as parking garages, are usually strong enough to support the weight of a helicopter. The delivery of Marines onto a building can also be accomplished by rappelling, fastroping, or jumping out of the helicopter as it hovers just above the roof. The rooftops should be inspected to ensure that no obstacles exist, such as electrical wires, telephone poles, antennas, or mines and wire. The use of helicopters in this role is usually predicated on the antiair threat. Helicopterborne forces are highly vulnerable to small arms fire, AAA and manpads.

b. Small-Scale Helicopterborne Assaults. Small units may be helo-lifted onto the rooftop of a key building. Depending on the construction of the roof, rappelling or fastroping Marines from the helicopter may be more advantageous than landing them on the rooftop. Fastroping provides a rapid and safer means of insertion than jumping from a low hover. With practice, Marines can accomplish a fastrope insertion with minimum exposure.

c. Large-Scale Helicopterborne Assaults. For large-scale assaults, rooftop landings are not practical; however, open spaces (parks, parking lots, sports arenas) within the built-up area may be used. Depending on METT-T, helicopterborne forces may assault into an LZ in one wave or scheduled waves.

d. Mobility of Forces and Supplies. In the urban environment, movement of forces by helicopters may become a major requirement. Units engaged in house-to-house fighting normally suffer more casualties than units fighting in open terrain. These casualties must be evacuated and replaced quickly with new forces. At the same time, roads are likely to be crowded with resupply and evacuation vehicles and may also be blocked with craters or rubble. Helicopters provide a means of overcoming these obstacles.

e. Helicopterborne Assault Techniques. Helicopter assaults into enemy-held territory are extremely difficult (Figure 4-1 on page 4-16). One technique is to fly nap-of-the-earth down a broad street or commercial ribbon while attack helicopters and assault helicopter door gunners suppress buildings on either side of the street. Artillery preparations can be incorporated into the helicopter assault plan to suppress rooftop and building antiair threat. Feints and demonstrations in the form of false insertions can deceive the enemy as to the intended objective. Helicopters often deliver forces to the last covered position, short of the fighting, then return without exposure to enemy direct fire. Similar flight techniques can be used for transportation of supplies and casualties.

Figure 4-1. Helicopterborne Assault of a Built-Up Area

f. Combat Roles. Below are some of the roles that assault support helicopters may play in urban warfare:

- Conduct insertion and extraction of forces (See MCRP 3-11.7A, *Helicopter insertions and Extractions* [under development])
- Provide command/control and aerial retransmission
- Support CSS operations
- Provide air reconnaissance.

4203. Employment of Reconnaissance Forces. Reconnaissance units traditionally make maximum use of rural camouflage and terrain. Area, zone, and route reconnaissance are conducted from well-concealed vantage points through clandestine insertion by foot, vehicle, rubber boat, or helicopter. These methods of insertion may still apply in village and strip areas for collection of information, sensor employment, conducting OP radio relays, or a combination of missions. In major urban areas, undetected insertions are difficult to accomplish because of the high density civilian population.

The use of reconnaissance assets must be used wisely. Reconnaissance assets must not be risked indiscriminately if the information can be collected by other means. Consideration should be given to flank and rearward employment of ground reconnaissance units, in addition to close (0 to about 2,000 meters), distant (approximately 2,000 - 20,000 meters), and deep (from about 20,000 meters and beyond, plus preassault) missions. Reconnaissance of bypassed pockets of resistance, avenues of approach, canals, bridges, subways, and sewer systems (possibly in conjunction with combat engineers) is feasible. Observation, sniper, and listening posts manned by reconnaissance personnel are also missions that reconnaisance personnel perform. The limited combatant diving capability of division reconnaissance may be employed to advantage in many coastal cities.

4204. Military Police (MP). MPs can play a significant role in the urban environment. They may conduct battlefield circulation control, area security, enemy prisoner of war (EPW) and law and order operations. These operations require continuous coordination with MAGTF units as well as with host nation military and civilian police.

a. MP units support area damage control operations. With the increased possibility of rubbling, MP units report on and block off affected areas and reroute movement to alternate road networks.

b. When conducting EPW operations, MPs collect prisoners as far forward as possible and transport them to the rear. They operate temporary collecting points and holding areas to briefly retain EPWs and civilian internees. EPW operations are important in built-up areas because the rate of capture can be higher than normal.

c. Commanders must realize that MP support may not always be available and that infantry units may have to assume certain MP functions. These functions include:

(1) Providing convoy escort and security for lines of communications.

(2) Controling roads, waterways, and railroad terminals that are critical chokepoints in the main supply routes.

(3) Providing security of critical sites and facilities, including communications centers, government buildings, water and electrical supply facilities, command and control nodes, airfields, storage facilities, and other essential areas.

(4) Provide refugee control in close cooperation with host nation civil authorities.

(5) Collect and escort EPWs.

4205. Communications

a. General. Manmade structures create many challenges for CIS in an urban environment. Other CIS challenges include extensive commercial and communications networks and public service radio nets. Commercial communication networks are commposed of miles of underground protected cable which connect and control telephones and computer networks. Public service radio nets, (including police, fire, civil defense, and taxi) compete with existing communication systems. To communicate on an effective, continuous basis in the urban environment, commanders must minimize communication limitations imposed by this environment and maximize the advantage of existing civil communications capabilities.

b. Radio. Manmade structures will decrease the effectiveness of VHF radio communications. The physical locations of radios are critical for effective communications. Some specific locations to avoid include powerlines, underground areas, and steel bridges.

Very high frequency (VHF) radios provide the primary communication capability at the battalion level. It is imperative that these radios be employed as effectively as possible. Communicators can use the following techniques to enhance VHF communications equipment:

- Relocation of VHF radios to overcome phase cancellation due to numerous dead spots
- Use of the higher end (50 - 75 MHz) of the available frequency spectrum
- Use of retransmission facilities to improve the reliability of communications
- Use of radio remotes.
- Use of field expedient antennas.

In setting up retransmission stations, antennas should be camouflaged to blend in with urban surroundings and to prevent them from being targeted by enemy fires. Antennas can be concealed by such objects as water towers and steeples.

In larger cities with many tall buildings, it may be necessary to construct a terrain profile of the area of operations. A communications plateau can be deduced by calculating the average height of the structures in a zone. Operation of antennas above the median line of this plateau will result in better wave propagation and aid in retransmission.

c. Wire Communications. Wire may be used extensively in urban environments. It offers two primary advantages over radio communications:

- It is not susceptible to interference and absorption suffered by radio systems.
- Security of the overall system is greater because wire systems are essentially point-to-point.

Maximum use should be made of existing facilities for concealment. Sewers, subway tunnels, cable stairwells, elevator shafts, intact buildings, and pole lines offer ideal concealment for military wire systems. Precautions must be taken to lessen the possibility of vehicles tearing lines and disrupting communications. Existing telephone poles and tunnels can be used to keep wire above or below the street level.

Maximum use of existing commercial facilities should be made to supplement or replace military CIS. Telephone exchanges are linked together by underground cable systems and constitute a key resource in the urban communications system. These facilities are normally windowless buildings constructed of steel-reinforced concrete and provide ideal CPs and communications centers. The physical location of these exchanges, along with circuit diagrams and telephone directories, should be determined as soon as possible. Action should be initiated to secure these facilities early. Other facilities that contain extensive internal/external telephone systems may also be used as command centers. These include most municipal buildings, department stores, local police stations, and large business offices.

d. Messengers. Messengers provide a secure and flexible means of communication, particularly in an electronic warfare (EW) environment. However, careful selection of messenger routes is necessary to avoid pockets of enemy resistance. Disorientation in an urban environment occurs more often and more easily than in a rural environment. Marking facilities, transportation networks, and city streets on overlays often facilitates messenger service. Selected primary and alternate routes can often interface with designated logistics/resupply routes. Early restoration of street signs and recognizable landmarks are is essential.

e. Telephone Networks. Captured civil or military communications facilities can be utilized to augment organic communications capabilities. A civilian phone system can provide a reliable means of communication. If a telephone encryption device is not available, codes and authentication tables should be used.

Telephone technology differs from place to place, but there are universal concepts and practices in their operation. The most important concept to the military planner is network hierarchy.

Network hierarchy is a concept of organization by which a telephone company can rapidly and expeditiously allocate calls through a particular group of offices. As an ancillary function, it represents a control feature by which traffic can be rerouted to avoid congested areas or damaged facilities. This network hierarchy could play an important role in the planning and execution of the scheme of maneuver and fire support.

4206. Employment of Sensors

a. Sensor systems (seismic, infrared, acoustic, or magnetic) are normally employed in a rural context but also have traditional applications in village and strip area terrain. Use of sensors in urban areas is largely undocumented. In major urban areas, the capability for emplacement by air over select areas is reduced. Emplacement in, under, or around urban structures may also reduce microphone efficiency. Tactical employment will be degraded by such obstacles as concrete and steel structures, high-density population, and traffic.

b. Training in camouflage techniques for microphones and sensor terminals in an urban environment should be conducted. Operators should receive sufficient training to distinguish background-generated clutter from enemy foot, vehicular, and mechanized traffic. Use of and reliance on sensors may be seriously reduced during offensive operations in major urban areas. However, greater use of sensors will occur in subterranean defensive operations. Sensors should continue to be employed as far forward as possible and be centrally controlled.

c. Sensor employment should be coordinated with counterintelligence (CI) efforts for placement, readouts, and recovery. Sensors can also play a role in rear areas to augment rear security operations. Sensors, in themselves, are only component elements of an integrated intelligence system. Their employment is limited only by the imagination.

4207. Employment of Air Defense. Basic air defense doctrine does not change when units operate on urbanized terrain. The employment principles of mix, mass, mobility, and integration all apply to the employment of air defense assets. The commander must consider the following when developing his air defense plan:

a. Enemy air targets such as principal lines of communication and road and rail networks and bridges are often found in and around built-up areas.

b. Good firing positions may be difficult to find and occupy for long-range air defense missile systems in the built-up areas.

c. Movement between positions is normally restricted in built-up areas.

d. Long-range systems can provide air defense cover from positions on or outside of the edge of the city.

e. Radar masking and degraded communications reduce air defense warning time for all units. Air defense control measures must be adjusted to permit responsive air defense within this reduced warning environment.

f. Positioning of air defense weapons in built-up areas may be limited to more open areas without masking, such as parks, fields, and rail yards.

g. Stinger teams provide protection for infantry battalions the same as in any operation. When employed within the built-up area, rooftops normally offer the best firing positions.

h. Heavy machine guns emplaced on rooftops can also provide additional air defense.

Chapter 5

Combat Service Support and Legal Aspects of Combat

During combat in built-up areas, the terrain and the nature of operations create unique demands on the CSS system. Meeting these demands will require innovative techniques and in-depth planning.

Section I

Combat Service Support

5101. Introduction. CSS is "the essential capabilities, functions, activities, and tasks necessary to sustain all elements of operating forces in theater at all levels of war" (Joint Pub 1-02). The CSS organization throughout the MAGTF, down to battalion level, must provide a responsive CSS system that can perform all functions and tasks associated with meeting identified CSS requirements. This responsibility does not change during urban operations. Forward support for combat forces continues to be the basic concept governing CSS operations. No significant changes in doctrine or organization are required. However, the characteristics and nature of urban combat do affect how CSS is provided.

5102. Combat Service Support Resupply, Maintenance/Repair, and Replacement

a. Resupply of Ammunition. Combat in built-up areas is characterized by extremely high ammunition expenditure rates. Not only do individual Marines expend more ammunition, but they also use greater quantities of munitions such as smoke, concussion, and fragmentation grenades; AT4s; claymore mines; demolitions; and so on. The ammunition consumption rate for the first day of combat in a built-up area can be up to four times the normal rate. Even though this rate decreases during succeeding days, consumption remains high. Leaders should plan to meet these high consumption rates. The plan must include how ammunition and demolitions are to be moved forward to the companies. AAVs should be designated for the movement of ammunition if rubble or glass prevents wheeled-vehicle traffic. Marines may need to organize into carrying elements if streets are blocked by rubble.

b. Resupply of Fuel. The amount of bulk fuel needed by forces during combat in built-up areas is greatly reduced. Combat vehicles normally use less fuel in built-up areas because they travel shorter distances and perform less cross-country traveling. Engineer equipment and power generation equipment may use more fuel, but requirements are small. A unit may not use much fuel daily, but when it does need fuel, a problem exists in delivering bulk fuel to the vehicle. In open terrain, a vehicle that has run out of fuel can be recovered later. In built-up areas, the same vehicle is probably going to be lost quickly. Leaders should plan for and provide the means of moving limited amounts of bulk fuel forward to combat units.

c. Maintenance and Repair. Maintenance contact teams must operate well forward to support units fighting in built-up areas. Although some maintenance operations may be consolidated in civilian facilities, many vehicles will have to be repaired near fighting positions. Battle damage assessment (BDA) and repair procedures allow mechanics to analyze, repair, and return damaged vehicles to a serviceable condition. Other considerations:

(1) Combat in built-up areas generates a high demand for tires.

(2) The dust and rough handling that are characteristic of combat in built-up areas also place great strains on communications and night vision devices.

(3) The unit armorers and their small-arms repair kits provide only limited maintenance. S-4s should plan for increased weapons maintenance demands.

d. Replacements. Units conducting combat in built-up areas must expect high casualty rates. Casualty reports must be prepared scrupulously and forwarded via the S-1 to the battalion personnel officer located at the consolidated administration center (CONAD). Other considerations:

(1) The S-4s must plan to expedite the evacuation of wounded from the built-up area. Location of battalion aid stations and evacuation routes must be planned and disseminated to the lowest level. Higher casualty rates should be expected and may require the stockpiling of medical supplies and augmentation of additional HSS personnel.

(2) The personnel officer quickly processes replacements and coordinates their movement forward via the S-1 officer.

(3) Proper accountability of personnel at all levels is required. Timely and accurate personnel accountability and strength reporting support decisionmaking and initiate the replacement cycle. Leaders maintain accountability through the use of battle rosters and by establishing procedures for periodic reporting of numbers to higher commands. During combat, strength reports are provided on request or as significant changes occur.

5103. Critical Classes of Supply. Requirements for supply and resupply will vary from those encountered during operations in other tactical environments. The critical classes of supply listed below will assume greater importance during combat inside a built-up area.

a. Class I (Rations). The process of ordering and moving rations to forward units is complicated by the dispersed nature of combat in built-up areas and its increased caloric demands on Marines. Hot meals should be provided when practical.

(1) Combat in built-up areas not only causes great stress on Marines, but also requires great physical exertion. This combination of stress and exertion quickly causes dehydration. Unless potable water is continuously provided, Marines will seek local sources, which are usually contaminated by petroleum, oils, and lubricants (POL) runoff, sewage, bacteria, or unburied corpses. Marines who are not provided sufficient quantities of potable water become casualties as a result of drinking from contaminated sources or from dehydration. Waterborne contaminants can quickly render entire units combat ineffective.

(2) Water and other liquid supplements such as coffee, tea, or soup that must be forwarded to exposed positions may need to be backpacked at night.

b. Class II (General Supplies). Combat in built-up areas increases wear and tear on combat uniforms and footgear. Supply officers should increase on-hand stocks of uniforms, boots, and individual combat equipment such as protective masks and armored vests. Nuclear, biological, and chemical (NBC) protective suits become damaged quickly when worn in the urban environment. Extra stocks of these and protective mask filters should be kept on hand. (For further information on NBC considerations, see Appendix G.) Limited amounts of other Class II and IV (barrier materials) items may be available locally. These should be gathered and used if authorized and practical. Local shops may provide such items as hand tools, nails, bolts, chains, and light construction equipment, which are useful in preparing a defense or reducing enemy-held positions. The unit's organic wire communications net may be augmented with locally obtained telephone wire and electrical wire.

c. Class III (POL). Bulk fuel may have to be brought forward from fuel tankers by using 5-gallon cans. One man can carry a fuel can long distances, even over rubble, if it is lashed to a pack frame. Supplies of bulk Class III items and some prepackaged POL may be available at local gas stations and garages. These may be contaminated or of poor quality. The S-4s should coordinate to have fuel tests performed.

d. Class IV (Barrier Materials). If a unit is defending a built-up area, the requirements for Class IV materials are less than in other areas. This class of supply is probably the most available locally. After coordinating the effort with higher headquarters, S-4s and supporting engineer officers gather materials for use in strengthening the defense. Cargo trucks, wreckers, or recovery vehicles from maintenance platoons or engineer units can be used to load and move barrier materials. The defense of a built-up area may require concertina wire and/or barbed wire to restrict the enemy infantry's movements. Barriers can be built of abandoned cars and buses, which are dragged into position, turned on their sides, and chained together through the axles.

e. Class V (Munitions). Combat in built-up areas causes ammunition to be expended at extremely high rates. Commanders should plan for early resupply of explosives, grenades, and ammunition for small arms, direct fire, and indirect fire.

(1) In the defense, the S-4 should prestage as much ammunition as practical in dispersed storage areas. These storage areas should be protected, and they should be easily accessible to the forward defensive positions. In the offense, attacking forces should not be overburdened with excessive ammunition. Mobile distribution points may be set up as low as at company level.

(2) Leaders should plan to continuously deliver ammunition to the leading units as they advance. Ammunition may be carried by armored vehicles close behind the advancing forces or by designated carrying elements. Modern ammunition, particularly missiles, is characterized by extensive amounts of packing material. S-4s must remove the ammunition depot overpack before the ammunition is transported forward. Resupply by helicopter (prepackaged slingloads) may be feasible.

(3) Removing the overpack from large amounts of ammunition can be a time-consuming process. It may require being augmented by available Marines. If carrying elements are used to move ammunition forward, a Marine can carry about 75 to 90 pounds by using a pack frame. Bulky and heavier loads can be carried by lashing them to litters and using teams of two to four men. Loads of up to 400 pounds can be carried moderate distances using four-man teams.

Note: DO NOT use medical corpsmen to carry ammunition forward as described above—it is a violation of the Geneva Accords.

f. Class VIII (Medical Supplies). Because of the decentralized nature of combat in built-up areas, medical supplies should be dispersed throughout the unit, not just consolidated with the aid station and the individual corpsman. Marines should carry additional bandages, and units should have additional splints and stretchers.

5104. Health Service Support. The regimental and battalion surgeons are responsible for planning and executing HSS within their respective units. The most critical functions during combat in built-up areas include preventive medicine, trauma treatment, and evacuation. In addition, there should be a plan for the treatment and evacuation of NBC-related casualties that could occur in combat in built-up areas.

a. Combat in built-up areas exposes Marines not only to combat wounds, but also to the diseases endemic to the AO. Commanders must enforce preventive measures against the spread of infectious diseases. The unit surgeon advises the commander on how best to implement the use of prophylactics.

b. Corpsmen should be placed at strongpoints, at battle positions, and in units likely to be fighting in somewhat isolated positions.

c. Battalion aid stations must be placed farther forward than in operations in more open terrain. Protection offered by urban structures will permit this forward location, but the limited range of observation will require that personnel involved in evacuation be aware of their surroundings and the threat.

d. Corpsmen attached to rifle platoons are trained in the treatment of traumatic injuries; however, they can quickly become overwhelmed by the number of casualties needing care. If time and training are available before the conduct of urban operations, the commander should increase first aid training. Immediate first aid support by a buddy to an injured Marine may be the difference between life and death and will greatly assist the corpsmen. The aid station should plan to care for the increased casualties that are inherent to combat in built-up areas, as the incidence of crushing injuries, eye injuries, burns, shrapnel wounds, and fractures increases.

e. The difficulties encountered when evacuating casualties from urbanized terrain are many and require innovative techniques and procedures. The planning for evacuation on urbanized terrain must include special equipment requirements, use of litter teams, use of air MEDEVAC, use of an ambulance shuttle system, communications requirements, and techniques for locating casualties.

(1) Special equipment requirements include ropes, pulleys, skid litters, axes, crowbars, and other tools used to break through barriers.

(2) Although litter teams are labor intensive, they are required for evacuation from buildings, where casualties can occur on any level. Also, rubble in the streets, barricades, and demolition of roads will require a heavy reliance on litter teams.

(3) Communications presents one of the biggest obstacles to casualty evacuation. Urbanized terrain renders LOS radios ineffective. Also, individual Marines normally do not have access to radios. Therefore, a wounded Marine within a building may be difficult to find and evacuate. The unit SOP should contain alternate forms of communication such as colored panels or other forms of markers that can be displayed to hasten rescue when the battle is over. Also, a systematic search of the area after the battle may be required to recover casualties.

f. The use of local medical facilities, hospitals, professional medical help, and medical supplies may be available during combat in large, built-up areas. The commander must adhere to the guidelines established within the theater as to when and how these facilities can be used. If civilians are wounded in the unit's area, the commander is responsible for providing them with aid and protection without disrupting military operations. A commander cannot confiscate civilian medical supplies unless he makes provisions to provide adequate replacements if civilians are wounded.

g. The commander is responsible for the evacuation of deceased personnel to the nearest mortuary affairs collection point, whether they are U.S., allied, or enemy personnel, or civilians. (See Joint Pub 4-06, *JTTP for Mortuary Affairs in Joint Operations.*) Some general considerations for the handling of deceased personnel include:

(1) The theater commander is the approval authority for hasty burial.

(2) The deceased person's personal effects must remain with the body to assist in the identification of the body and to facilitate shipment of personal effects to the next of kin.

Retention of personal items is considered looting and is, therefore, punishable by the Uniform Code of Military Justice (UCMJ).

(3) When operating under NBC conditions, the bodies of deceased personnel should be decontaminated before removal from contaminated areas to prevent further contamination and casualties.

(4) Care must be exercised when handling deceased personnel. Improper handling of deceased personnel can result in a significant decrease in unit and civilian morale.

5105. Personnel Services. Timely and accurate personnel services are just as important during combat in built-up areas as in any other operation.

a. The S-1 plans for all personnel services that support and sustain the morale and fighting spirit of the battalion. Among the most important of these services are:

- Replacement operations
- Strength accounting
- Casualty reporting.

These functions allow the commander to maintain accountability of his forces and to redistribute personnel or change taskings for subordinate units to maintain combat power and tempo.

Chapter 6

Noncombatant Considerations in Urban Operations

Noncombatants can have a significant impact on the conduct of military operations.

Section I

6101. Introduction. Commanders must be well educated in the legal aspects of military operations involving noncombatants. The urban environment contains numerous legal pitfalls in dealing with the civilian population. Adherence to the ROE and the standards of the Geneva Accords will keep the commander on firm legal ground.

6102. Noncombatant Impact. Large concentrations of civilians can greatly impede tactical operations. Noncombatants may have the following effects on military operations:

a. Mobility. Noncombatants civilians, attempting to escape the battlespace, can block military movement. Commanders plan routes to be used by civilians and seek the assistance of the civil police in refugee control.

b. Firepower. The presence of noncombatants can restrict the use of firepower. Areas may be designated no-fire areas to prevent noncombatant casualties. Other areas may be limited to small-arms fire and grenades. The control of fire missions may be complicated by the requirement for positive target identification. Detailed guidance on the use of fires in the presence of noncombatants is promulgated by the MAGTF commander. In the absence of guidance, the general rules of the law of land warfare apply.

c. Security. The presence of noncombatants increases security requirements in an urban environment to preclude:

- Noncombatants entering defensive areas
- Pilferage of equipment
- Sabotage
- Terrorism

d. Obstacle Employment. The presence and movement of Noncombatants will influence the MAGTFs commander's obstacle plan. Minefields may not be allowed on designated refugee routes or, if allowed, must be guarded until the passage of refugees is completed. The use of boobytraps may be curtailed until noncombatants have been evacuated. Commanders need to review and understand current national and international treaties concerning the employment of mines and boobytraps.

6103. Command Authority. The limits of command authority over government officials and noncombatants must be established. The host government's responsibility for its populace and territory can affect the commander's authority in civil-military matters. In less secure areas, where the host government may be only partly effective, the commander may have to assume greater responsibility for noncombatants.

6104. Source Utilization. Operations in highly populated areas may require the diversion of men, time, equipment, and supplies to accomplish humanitarian tasks. If host government agencies collapse, the impact on military resources could be substantial.

6105. Health and Welfare. The disruption of noncombatant health and sanitary services may sharply increases the risk of disease among both noncombatant and military personnel.

6106. Law and Order. Marines may have to augment civilian forces to protect life and property and to restore order. Marines may also have to secure vital government facilities for the host nation. (For more information on how to control civilians violating civil law, see FM 9-15, *Civil Disturbances*.)

6107. Public Affairs Officer and Media Relations. Free media access to units in the field is desirable; however, operational security, existing guidelines, and ROE considerations have priority. All members of the media should have an escort officer. Escort officers may be detailed from line units when public affairs personnel are unavailable. Enforcing established guidelines helps prevent negative publicity which could jeopardize the operation or national and strategic objectives.

6108. Civil Affairs Activities and Psychological Operations (PSYOP).

a. General. CA and PSYOP are mutually supporting and may have essential roles in MOUT. They can be critical force multipliers. CA and PSYOP may offer the possibility of victory without the destruction, suffering, and horror of protracted urban conflict.

(1) CA activities are activities performed by commanders, staffs, department of defense elements and units, and foreign military forces that (1) embrace the relationship between military forces and civil authorities and populations in areas where military forces are present; and (2) involve application of civil affairs functional specialty skills, in areas normally the responsibility of civilian government, which enhance conduct of civil-military operations (Joint Pub 1-02). CA activities are inherently civil-military in nature. They assist in coordinating military efforts to strengthen host government legitimacy, and help prevent or reduce violence by bridging critical gaps between civil and military sectors. MAGTF responsibilities for CA activities are determined during planning. U.S. Army CA units can be requested to support the MAGTF. The Marine Corps Civil Affairs Group (CAG), located in the Marine Corps Forces Reserve, may be activated and assigned to the MAGTF.

(a) CA activities are characterized by applications of functional specialties in areas normally the responsibility of indigenous government or civil authority. CA activities may extend to assumption of government functions required in an occupied territory during or immediately subsequent to hostilities. Refer to Joint Pub 3-57, *Doctrine for Joint Civil Affairs*.

(b) CA support to a MAGTF would normally include procuring local goods and services for military use, reducing civil interference in military operations, and assisting a MAGTF commander in fulfilling moral and legal obligations to the local government and noncombatant populace. See MCWP 3-33.1, *MAGTF Civil Affairs*, for more detailed information.

(2) Psychological operations are planned operations to convey selected information and indicators to foreign audiences to influence their emotions, motives, objective reasoning, and ultimately, their behavior. The purpose of psychological operations is to induce or reinforce foreign attitudes and behavior favorable to the originator's objectives (Joint Pub 1-02).

PSYOPs are planned and conducted by the MAGTF. PSYOPs are integrated and coordinated down to the lowest tactical level. PSYOPs can weaken enemy resistance and reduce its combat effectiveness. PSYOPs can also help prevent civilian interference with military operations.

PSYOPs may include the use of television, radio, posters, leaflets, and loud speakers, to disseminate propaganda and information. Television, including video tapes, is one of the most effective mediums for persuasion. In areas where television is not common, televisions may be distributed to public facilities and selected individuals. See MCWP 3-36.2/FM 33-1, *Psychological Operations*, for more detail.

6109. Refugee Control. The control of refugee movements in combat operations is primarily a host nation function.

The most important principle in refugee control is reliance on host government forces (military/civilian) whenever possible. U.S. forces will have limited refugee control capabilities. If a host government is incapable of performing the function, MAGTFs and MPs may be required to conduct any or all of the following refugee control measures:

- Enforce standfast orders or prohibit other unauthorized movements
- Screen and take custody of enemy agents or other hostile personnel discovered among the refugees
- Screen refugees for suspicious activities (including detection of infiltration by guerrillas, enemy agents, and escaping members of hostile forces)

.3

- Perform traffic control on designated refugee routes
- Escort refugee columns
- Direct or divert refugee columns to secondary roads and through areas not used by military elements
- Perform police duties such as traffic control or security at refugee assembly areas, collecting points, and centers or camps
- Enforce prescribed restrictive measures such as curfews, identification and pass systems, restricted area restraining lines, no-passage lines, checkpoints and roadblocks, travel permits or manifests, prohibition on the use of vehicles, priorities of movement, and limits on personal belongings.

Section II

6201. Commander's Legal Authority and Responsibilities

a. General. Leaders at all levels of command are responsible for protecting noncombatants and minimizing collateral damage to their property. Looting, vandalism, and brutal treatment of civilians are strictly prohibited by the Geneva Accords. Individuals who commit such acts may be severely punished. Civilians, and their religions and customs, must be treated with respect. Women are especially susceptible to many forms of abuse, and special care must be taken to protect them. Urban warfare offers many opportunities for abuses of the civilian population. This section focuses on those civilian-military confrontations that are most common in the urban environment.

b. Control Measures. Commanders may enforce control measures to conduct operations, maintain security, or ensure the safety and well-being of the civilians.

(1) Curfew. A commander with the mission of defending a town could establish a curfew to maintain security or to aid in control of military traffic. However, a curfew would not be legal if imposed strictly as punishment.

(2) Evacuation. A commander can require civilians to evacuate towns or buildings if the purpose of the evacuation is to use the town or building for essential military purposes, to enhance security, or to safeguard those civilians. The commander must specify and provide security for those being evacuated. Food, clothing, and sanitary facilities should be provided at the destination.

(3) Forced Labor. The Geneva Accords prohibit the use of civilians in combat operations. However, they may be required to perform some forms of forced labor prior to and after the conduct of combat operations. Guidelines for use of civilian labor should be published. The commander may force civilians over 18 years of age to work if the work does not oblige them to take part in military operations. An example of forced labor that is permitted would be maintenance of public utilities as long as those utilities are not used in the general conduct of the war. Jobs can also include services to the local population such as care of the wounded and burial. Civilians can also be forced to help evacuate and care for military wounded, as long as doing so does not involve any physical danger. Prohibited forced labor include digging entrenchments, constructing fortifications, transporting supplies or ammunition, or acting as guards. Only civilians who are volunteers can be employed in such work.

c. Civilian Resistance Groups. Another situation that commanders might encounter is combat with a civilian resistance group. The combatant status of these groups varies with their organization, methods, and adherance to the Geneva Accords. For example:

(1) Civilians accompanying their armed forces (with an identity card authorizing them to do so) are treated as EPWs when captured. Examples would include civilian members of military aircraft crews, war correspondents, supply contractors, and members of labor units or of service organizations responsible for the welfare of the armed forces.

(2) Captured civilians of a nonoccupying territory who have taken up arms against an invading enemy without time to form regular armed forces are considered EPWs if they wear a distinctive insignia that can be seen at a distance, carry their weapons openly, and operate according to the rules and customs of warfare. Other civilians who provide assistance to such groups may not be entitled to status as combatants, depending on whether they are actually members of the resistance group. They are normally best treated as combatants until a higher authority determines their status.

(3) Armed civilian groups that do not meet the criteria of a legal resistance group or individuals caught in the act of sabotage, terrorism, or espionage are not considered legal combatants. If captured, they will be considered criminals under the provisions of the law of land warfare. They should be detained in a facility separate from EPWs and should be quickly transferred to the MPs. Reprisals, mass punishments, hostage taking, corporal punishment, pillaging, or indiscriminate destruction of property are prohibited.

(4) The law of land warfare lets the commander control the civil population under the conditions already described using his own organic resources. However, language and cultural differences between U.S. and foreign personnel make it good practice to use local authorities such as the police for such purposes. Use of the police does not relieve a commander of his responsibility to safeguard civilians in his area.

d. Protection of Property. Like civilian personnel, civilian buildings and towns normally have a protected status. As such, they are not legal targets. They lose their protected status if authorities determine that the enemy is using them for military purposes. All doubt as to whether a town or building is a military target must be resolved before it can be targeted. For example:

(1) If the enemy is using a building or a portion of the town for military purposes—such as a supply point or a strongpoint—that building or that portion of the town is a legal target. However, engagement of the target should be limited to the buildings or portion of the town being occupied by the enemy.

(2) Normally, religious, historical, and cultural objects and buildings are not legal targets. They are sometimes marked with symbols to signify their status. Medical facilities are identified under the internationally recognized Red Cross, Red Crescent, Red Lion, or Red Star of David symbols. The fact that such symbols are absent does not relieve a commander of his responsibility to protect objects he recognizes as having religious, cultural, medical, or historical value.

(3) The misuse of such objects by the enemy is grounds to disregard their protected status. A demand should be made for the enemy to stop his misuse of the protected object before any action is taken. However, if an enemy FO or sniper uses a church for an OP, a commander would be justified in destroying it immediately because of the immediate danger posed to his troops. Another example is a religious shrine that is used as a telephone switchboard; in this case a warning would be appropriate because it would take some time to dismantle the system. Attacks on these types of targets should be limited to the least amount necessary to neutralize the enemy installations.

(4) The destruction, demolition, or military use of other buildings is permitted under the law of land warfare if required by clear military necessity. Destroying a house to obtain a better field of fire would be a legal act. But destroying it as a reprisal would not. However, firing on any houses that are occupied or defended by an enemy force is usually legal.

Chapter 7

The Urban Environment and Restrictions to Operations

"The responsibility of officers . . . and the training necessary are of a very different order from their responsibilities and training in ordinary military duties. Instead of striving to generate the maximum power with the forces available, the goal is to gain decisive results with the least application of force and the consequent minimum loss of life."

USMC, 1940 (FMFRP 12-15, *Small Wars Manual*)

Section I

Military Operations Other Than War

7101. Introduction. One of the most likely missions that U.S. Marines will undertake abroad will be military operations other than war (MOOTW). These missions typically will take place in the Third World and are normally associated with urban areas. MOOTW focuses on deterring war, resolving conflict, promoting peace, and supporting civil authorities in response to domestic crises. It is important to remember that political considerations permeate at all levels. Although some types of MOOTW, such as peace enforcement, may have the same characteristics of war, they normally have more restrictive rules of engagement (ROE).

Urban environments are inherently complex, and somtimes unpredictable. Local governments may have limited functions. Services may be overburdened, leading to allocation on the basis of personal connections and corruption. Services may be provided by a mixture of formal, high-technology and informal, labor-intensive systems. Large portions of the population may be poor, young, and underemployed. Nevertheless, there may be a high degree of social organization and interaction. Word of mouth and informal organizations may be as important as mass media and formal organizations to the communication structure of the area. Indigenous populations may have experienced centuries of foreign interventions and become quite adept at innovative ways of communicating. These factors and others often combine to make MOOTW a challenging and frustrating endeavor.

7102. Types of MOOTW. MOOTW encompasses a wide variety of military operations, that may take place in an urban environment. Forward deployed MAGTFs are ideal forces to respond to contingency operations involving MOOTW. These operations include:

- Combatting Terrorism
- DOD support to counterdrug operations
- Enforcement of Sanctions/MIO
- Enforcing Exclusion Zones

- Humanitarian Assistance
- Military Support to Civil Authorities
- Nation Assistance/Support to Counterinsurgency
- NEO
- Peace Operations
- Recovery Operations
- Show of Force Operations
- Strikes and Raids.

7103. Recent Experience and Lessons Learned. Recent Marine Corps urban experiences— Lebanon (1982 - 1984); Grenada (1983); Panama (1989); and Somalia (1992 - 1994)— are examples of MOOTW. Although each operation was unique, the lessons learned are similar. For detailed information, commanders may obtain after action reports through the Marine Corps Lessons Learned System (MCLLS). Some of the lessons learned include:

(1) Demonstrate a powerful military presence. Where and when possible, use tanks, armored vehicles, heavy automatic weapons, fixed-wing aircraft, and attack helicopters.

(2) PSYOP, CI, HUMINT teams, and linguists, are essential to communications and intelligence gathering.

(3) Topographic detachment is useful to provide necessary sketches and maps.

(4) UAVs can be used for security, reconnaissance, and surveillance.

(5) Whenever entering an AO, isolate the area, patrol, and use intelligence to evaluate specific buildings or groups of buildings for search-and-clear operations. PSYOP, CI, HUMINT teams, and linguists can be used, as appropriate, to support patrols.

(6) Take advantage of the night by using night vision devices.

(7) The most effective means of dealing with snipers is with counter snipering operations.

(8) Snipers are also a tremendous asset to overwatch patrol movements, provide intelligence, and direct fire support. (See MCWP 3-15.3.)

(9) Random patrolling, traffic control points, random checkpoints/roadblocks, and ambushes disrupt enemy movements, reduce looting, seize curfew violators, and convince the population that order has been restored.

(10) Use armored vehicles to move quickly to new checkpoints and set up roadblocks or ambushes. They also increase protection for infantry. Their night vision sights also provide increased security and recognition during periods of limited visibility.

(11) Extensive patrolling provides current combat information used to maintain situational awareness. It helps to familiarize units with the AO and presents a strong military presence. (See Appendix A, Section VII)

(12) Tanks, armored vehicles, or attack helicopters increase military presence and can provide immediate, accurate gunfire.

(13) Before search and clear of a building, use loudspeakers to explain your intentions to noncombatants and enlist their cooperation.

(14) Use a minimum amount of force when dealing with noncombatants and their property. This will assist you in obtaining them to cooperate and may prevent turning them into active combatants. Assign engineers to search-and-clear forces, as necessary.

(15) MPs can instruct small unit leaders on the proper method of conducting vehicle, individual, and room search techniques. In some cases, it may be appropriate to attach MPs to search-and-clear forces. (See MCWP 3-34.1, *Military Police in Combat.*)

(16) Maintain a strong, mobile reserve.

(17) Avoid becoming predictable.

7104. Urban Terrorism. Terrorism is "the calculated use of violence or threat of violence to inculcate fear; intended to coerce or to intimidate governments or societies in the pursuit of goals that are generally political, religious, or ideological" (Joint Pub 1-02). Terrorism is a criminal act that is symbolic in nature. Its intent is to influence an audience other than the victim.

The likelihood of Marine involvement with terrorism increases in the urban environment. Terrorism in an urban environment can occur across the spectrum of conflict. Terrorists are an integral element in an insurgency and also play a major role in unconventional warfare. Terrorists can disrupt economic functions, expose a government's incompetence, eliminate opposition leaders, and elevate social anxiety. (See MCRP 3-02D, *Combating Terrorism*; and MCRP 3-02E, *The Individual's Guide for Understanding and Surviving Terrorism.*)

In countering this threat, Marines should determine whether it is internally or externally directed terrorism. Terrorism rooted externally must be severed from its roots. Against internal terrorism,

.3

Marines should attempt to penetrate the infrastructure and destroy the leadership of the terrorist groups.

The keys to defeating terrorists are awareness, education, and intelligence in order to deny, deter, delay, and detect terrorist acts. Rapid coordination among military units, military services, local police, and other government agencies is essential in denying the terrorists targets and refuge.

a. Nature of Urban Terrorism. Terrorism is generally an urban phenomenon. Incidents of terrorism are increasing in the international community, and in the United States, and pose a significant threat for U.S. military forces. Military personnel in noncombatant roles are considered terroist targets. For example, the attack on the Marine compound in Lebanon can be considered a terrorist attack instead of a military action because the Marines were officially in a noncombat activity (a multinational peacekeeping activity).

b. Terrorists in the Urban Area. Cities are target rich environments for terrorists. They offer ample cover and concealment. They contain large amounts of resources, transportation, communications, and potential recruits or sympathizers to their cause. Cities also provide the media attention essential for terrorists to accomplish their mission.

c. Combating Terrorism

(1) Counterterrorism. Counterterrorism includes the full range of offensive measures to prevent, deter, and respond to terrorism (Joint Pub 1-02). Population protection and control to combat the terrorist threat may be a major effort within the urban environment. The ability to protect the general population from attack and intimidation is essential.

(2) Antiterrorism. Antiterrorism includes defensive measures used to reduce the vulnerability of individuals and property to terrorist acts, to include limited response and containment by local military forces (Joint Pub 1-02). The extent of these measures varies based on assessments of the local threat. These include personnel awareness and knowledge of personal protection techniques. They also include crime prevention and physical security programs to "harden" the target, making installations and personnel less appealing as terrorist targets.

Section II

Restrictive Conditions

7201. Introduction. Commanders can expect to encounter restrictions on the use of force in MOOTW, especially in the urban environment. While we tend to focus on a traditional image of urban combat, MOUT extends beyond the house-to-house, close-combat effort to seize a city. Recent history provides many examples of the need for precise application of force to minimize collateral damage. MOOTW generally requires the application of force in varying degrees. Future missions in urban areas will likely follow this same pattern. While Marine forces must be prepared for more intensive urban combat, they must also train to apply only the level of force necessary to accomplish the mission. *Our tactics may have to rely more on innovation than firepower.*

Basic tenets of manuever warfare remain the same in MOOTW, but the tactics, techniques, and procedures may have to be modified to stay within established ROE and to avoid unnecessary collateral damage.

7202. Rules of Engagement. ROE are "directives issued by competent military authority which delineate the circumstances and limitations under which United States forces will initiate and/or continue combat engagement with other forces encountered" (Joint Pub 1-02). ROE for Marine forces are usually set by the JTF commander. They are based on National Command Authorities (NCA) guidance, the mission, the threat, the laws of war, and host nation constraints on force deployment.

MOOTW focuses on deterring war, resolving conflict, promoting peace, and supporting civil authorities in response to domestic crises. It is important to remember that political considerations permeate at all levels. Although some types of MOOTW, such as peace enforcement, may have the same characteristics of war, they normally have more restrictive rules of engagement (ROE). Restraint requires the careful balancing of the need for security, the conduct of operations, and the political objective. Excessive force antagonizes those parties involved, thereby damaging the legitimacy of the organization that uses it while possibly enhancing the legitimacy of the opposing party. Ensure all personnel understand the ROE and incorporate the practice into training exercises at all levels.

ROE must be fully integrated in the MAGTFs plans. As an example, table 7-1, presents the ROE used in Operation United Shield:

.3

ROE Used for Operation United Shield
Nothing in these Rules of Engagement limits your right to take appropriate action to defend yourself and your unit.
a. You have the right to use deadly force in response to a hostile act or when there is a clear indication of hostile intent.
b. Hostile fire may be returned effectively and promptly to stop a hostile act.
c. When US forces are attacked by unarmed hostile elements, mobs and/or rioters, US forces should use the minimum force necessary under the circumstances and proportional to the threat.
d. Inside designated security zones, once a hostile act or hostile intent is demonstrated, you have the right to use minimum force to prevent armed individuals/crew-served weapons from endangering US/UNOSOM II forces. This includes deadly force.
e. Detention of civilians is authorized for security reasons or in self-defense.
Remember: 1. The United States is not at war. 2. Treat all persons with dignity and respect. 3. Use minimum force to carry out mission. 4. Always be prepared to act in self-defense.

Table 7-1. ROE Used in Operation United Shield

7203. Planning. Commanders and planners thoroughly study each situation to identify the restrictive measures necessary to prevent needless civilian casualties and damage while also protecting the lives of the Marines who will execute the mission.

a. Mission/Command Guidance. The mission statement itself may determine much about the required level of force for a given operation. A humanitarian relief operation may require only individual weapons for security and personal protection. A raid on an insurgent communication center may need more. The requirements for an NEO conducted in a permissive environment would be different from those in a nonpermissive one. The key is to conduct a thorough analysis of the mission and determine the necessary combat power and how it must be applied to accomplish the mission. Many of these missions will be accompanied by a great deal of additional guidance concerning operational restrictions. These restrictions may come from the NCA, State Department, Department of Defense (DoD), or the joint task force (JTF) commander, and other sources. Such guidance must be carefully studied and incorporated into the plan. The commander can impose restrictions on his own forces that are more stringent then those dictated by higher authority. A careful analysis of the ROE will provide key insights into other restraints that the commander may wish to impose on his own forces.

b. Intelligence/Threat. Obtaining detailed and specific intelligence on the target and the area is essential to minimize collateral damage. Information on the physical layout of an area is often as important as information about the enemy. This is especially true for an NEO where avoiding conflict is a primary concern.

c. Location. The geographic area into which operations are to be conducted will normally dictate some restrictions on operations. In an urban environment, where the city is located in a desert, a common restriction is placed on the destruction of water producing facilities. If the operation is conducted where there is a large Hindu population, special protection may be given to cattle. If the operation is on an island nation, restrictions may be placed on destroying port or airfield facilities. In an urban environment with a large amount of shanty towns, restrictions may be placed on the use of incendiary devices.

d. Restrictions on Force Employment. The *right of self-defense is inherent in all military operations.* Marines should not be placed in a situation where they are not provided the means and authority to protect themselves or others. Beyond that, the restrictions placed on the forces conducting a particular operation depend on a variety of factors. These include:

- NCA guidance
- Type of mission
- Nature of the threat
- Political environment
- Host nation attitude.

e. Lethal Fires. Most restrictions in MOOTW are normally concerned with the application of fires. For instance, a NEO in a permissive environment may not require the landing of fire support assets However, a raid may require supporting arms to be landed with detailed instructions on their use. Aviation in MOOTW is commonly restricted in the amount and type of ordnance it can deliver because of the potential for collateral damage. However, in both a permissive and nonpermissive environment, supporting arms still should be planned. The most common types of fire support assets that will be restricted in the urban environment include artillery mortars, tanks, armored vehicles, and fixed-wing attack aircraft.

f. Search and Clear Buildings. Marine forces organize to search and clear a building under restrictive conditions in the same ways as discussed in Appendix C. However, because noncombatants may be occupying rooms within the building, clearing procedures must be modified as follows:

(1) Squad organization remains the same and is reinforced as appropriate (see Appendix A), except assault elements become search elements. Search elements enter and search rooms in the same manner as assault elements enter and clear rooms, as discussed in Appendix A.

(2) ROE must be identified and known by all personnel before entering or clearing a room or building.

(3) PSYOP or CA teams can help remove noncombatants before a search or battle starts.

(4) The building or area of search should be isolated with security forces and the intent to search the building should be communicated via loudspeakers.

(5) Assault forces move to the building and begin systematically to search room by room and floor by floor. Marines should carry ammunition and other equipment as in normal urban operations. However, Marines will be restricted in the use of their weapons and other equipment as dictated by the ROE. If permitted, a riot control agent (such as CS gas) can be used to assist in the clearing of a building. Additionally, shotguns and stun grenades are more appropriate than rifles and machine guns or fragmentation grenades for this type of restricted, close-combat operation.

g. Communications and Information Systems (CIS).

(1) CIS must be able to satisfy the C2 requirements of MOOTW. CIS systems must provide the commander with the ability to rapidly collect, process, analyze, and exchange information in order to keep the potential for violence to a minimum. These systems must make available the information needed, when it is needed, whenever it is needed. When supporting arms are on standby, fire support coordination and air traffic control nets are particularly critical.

(2) CIS officers (G/S-6) must carefully analyze the mission, situation, threat, and specific characteristics of the urban environment to design reliable, flexible, and responsive CIS. These systems should include single and multichannel radios, tactical telephones, and tactical satellite communications. A detailed backup CIS plan should also be prepared. The success of CIS in MOOTW depends on designing, planning, and employing CIS that satisfies the commander's C2 requirements.

Appendix A

Organization for Urban Combat and Fundamental Combat Skills

Successful combat operations in built-up areas depend on training in fundamental skills and proper employment techniques and procedures.

Before conducting urban operations, Marines must be trained in fundamental combat skills to conduct urban operations. Training for this unique environment will reduce casualties because properly prepared Marines will be able to fight more effectively. This appendix supersedes all other fundamental combat skills taught for fighting in urbanized terrain. Urban combat will require Marines to use their best initiative and judgment in applying the techniques and procedures described in this appendix.

Section I

Organization

1. Structure. Basic infantry organization does not change when the battlefield is moved into an urban area. However, infantry units may be task organized or reinforced down to squad level in order for small units to fight and win in this violent, three-dimensional battlespace. Several new terms that address organization and employment have been added to clearly define mission assignments. Figures A-1 through A-3 show platoon through fire team organization breakdown.

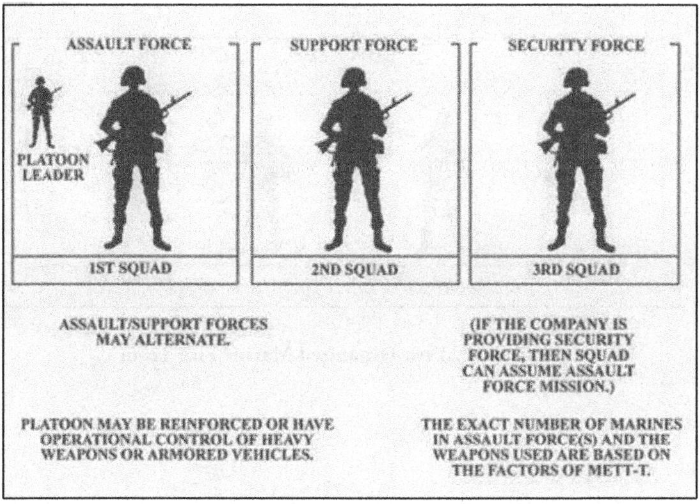

Figure A-1. Task-Organized Marine Platoon

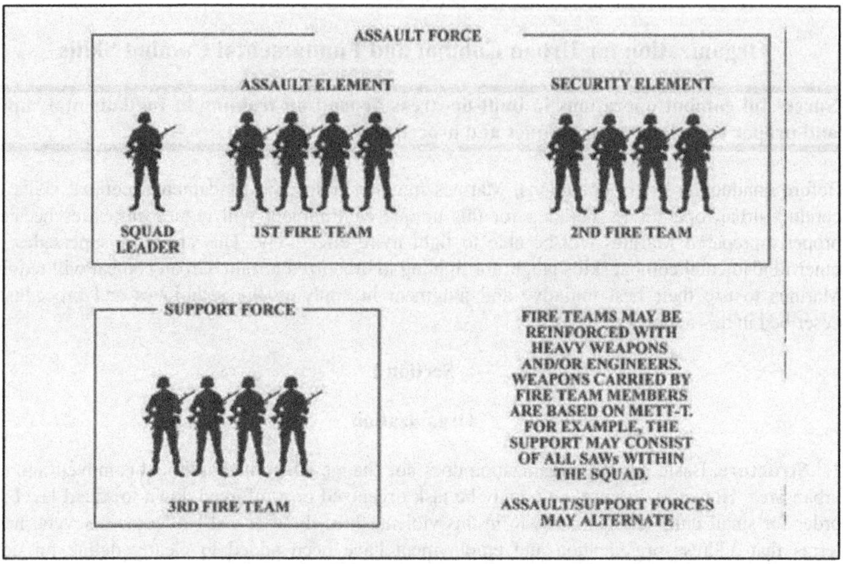

Figure A-2. Task-Organized Marine Squad

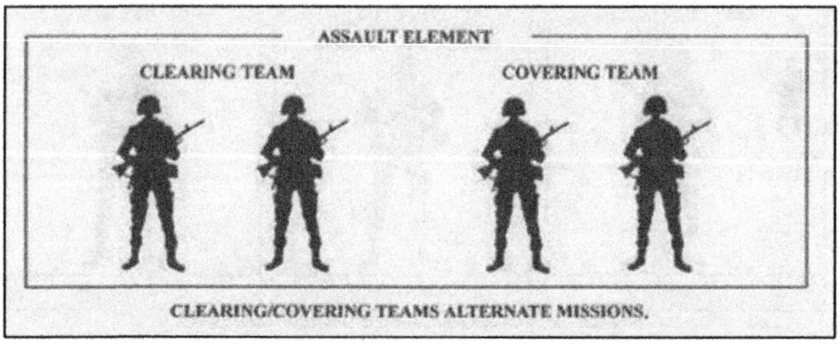

Figure A-3. Task-Organized Marine Fire Team

Section II

Weapons Handling and Firing Techniques

2. Weapons Carries. During MOUT, a Marine often finds himself very unexpectedly engaging targets at close ranges. Weapons carries provide a safe and effective method for handling the service rifle. For additional in-depth discussion of weapons handling and firing techniques, refer to MCWP 3-01X, *Field Firing of the M16A2 Rifle.*

a. Tactical Carry. The tactical carry is used when no immediate threat is present. It permits control of the rifle while moving and still allows quick engagement of the enemy. The buttstock of the rifle is placed alongside the body at approximately hip level, and the barrel is angled upward approximately 45 degrees in the general direction of the enemy (Figure A-4).

Figure A-4. Tactical Carry

b. Alert Carry. The alert carry is used when enemy contact is likely. Engagement of the enemy is faster from the alert carry than from the tactical carry. The buttstock of the rifle is placed in the shoulder with the muzzle angled down approximately 45 degrees and pointed in the likely direction of the enemy (Figure A-5).

In the alert carry, the Marine keeps both eyes open and scans for any threat with the weapon always in line with the Marine's LOS (this is called "guns 'n' eyeballs"). When a threat is spotted, the Marine quickly snaps his weapon into a firing position, looks over the rear sight aperture, and aligns the tip of the front sight post center mass on the target. Once the threat has been eliminated, the Marine returns to the alert carry and continues to scan for targets. Accuracy is the primary goal.

Figure A-5. Alert Carry

c. Ready Carry. The ready carry is employed when contact with the enemy is imminent. It allows for immediate target engagement. The buttstock of the rifle is in the shoulder with the muzzle of the rifle pointed in the direction of the enemy (Figure A-6). As in the alert carry, sight alignment and sight picture are achieved as the shot is fired. Marines should always strive for a clear tip of the front sight post and center mass hold.

Figure A-6. Ready Carry

d. Short Stocking. Rifles may be "short stocked" to increase their maneuverability in an enclosed area and to reduce possible "target indicators" for the enemy (muzzles sticking around corners, for example). The weapon is held in such a manner as to reduce the length of the weapon without sacrificing too much accuracy (Figure A-7). The stock is positioned so that the pistol grip is behind the Marine's head. The Marine may use his index finger or thumb to manipulate the trigger. The handguard of the weapon is placed against the Marine's cheek, and a firm stockweld is attained. The Marine should use the tip of the front sight post. Short stocking may be conducted from either side. However, care should be taken not to mask the ejection port cover. The short stocking technique may be used in the tactical, alert, or ready carry.

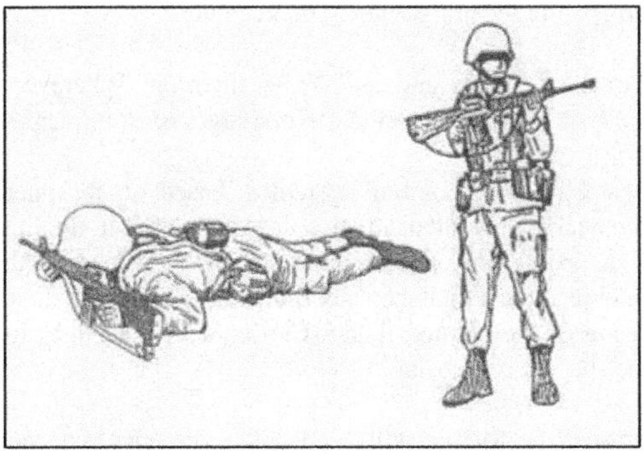

Figure A-7. Short Stocking

e. Weak-Hand Carries. Marines may carry their weapon on the weak-hand side to obtain maximum cover and fire from the weak-hand position to avoid exposing themselves. However, firing from the weak-hand position may reduce a Marine's ability to fire accurately. This may be acceptable when Marines are providing supporting fires, but in close kill-or-be-killed engagements, rapid, accurate firing from the strong-hand position is preferred.

3. Firing Techniques. The speed and uncertainty of combat require Marines to act without hesitation while achieving accurate target engagement. Several firing techniques are listed below. For an in-depth discussion of firing techniques that can be applied to MOUT operations, see MCWP 3-01X.

a. Pieing. Pieing is an effective technique for clearing dead space inside rooms and buildings to gain security of hallways, stairwells, mouseholes, and so on. It is conducted by using the ready carry position or short stocking technique. The weapon is aimed at a sector of a window, doorway, corner, or hallway and slowly moved at different angles, sectoring off the window, doorway, corner, or hallway until each sector is cleared of any threat. As soon as a hostile threat is seen, immediate, accurate fire can be placed on the threat. Figures A-11, A-34, A-35, A-38, A-47, and A-49 show examples of the pieing technique.

b. Aimed Quick Fire. The Marine's initial focus is on the target. As the rifle is brought up, the firing eye looks through or just over the rear sight aperture, and the front sight post is used to aim at the target. The Marine fires two quick shots. Focus remains on the front sight post throughout the aiming process.

.3

Immediately after target engagement, the Marine scans for additional targets. To scan, perform the following steps:

(1) Lower the weapon to look over the sights.

(2) Place the trigger finger straight along the receiver.

(3) Scan the area for targets and assess the situation. Wherever the head moves, the muzzle moves (eyes, muzzle, target). Keep both eyes open to increase the field of view.

c. Pointing Quick Fire. The pointing system is based on the phenomenon that when a person looks at an object and simultaneously points a finger at it, the finger aligns itself on the point of focus of the eyes with no conscious effort on the part of the individual. When a Marine looks at an object and simultaneously brings his rifle to his shoulder, the rifle in effect becomes an extension of the pointed finger. Consequently, it aligns itself naturally with the object on which the shooter is focusing.

When a target appears, the Marine will keep both eyes open, concentrating intensely on a small, specific focal point near the base of the target mass. The rifle is brought simultaneously to the hollow of the shoulder. The head is held high, stock welded to the jaw. The eyes are 2 - 3 inches over the top of the sights, staring intently at the target. As soon as the rifle is brought to the shoulder, two quick shots are fired. Focus on the target is not broken during the interval between initially seeing the target and discharging the rifle.

d. Instinctive Shooting. There may be situations in which a Marine is surprised and may need to react immediately. If possible, the Marine should engage the threat by using the tip of the front sight post. However, *speed* may be more important. The Marine's weapon and body are quickly "pointed," and the target is engaged. It is important that the body be turned with the weapon in order to achieve a natural point of aim. Simply pointing the weapon will usually result in a miss. Once the first two shots have been fired and the Marine regains the initiative, the weapon should be quickly moved to the Marine's shoulder and the tip of the front post used for sighting subsequent shots.

Section III

Movement

Movement techniques within built-up areas are the first fundamental skills that Marines should master. These movement techniques should be practiced until they become habitual. To reduce exposure to enemy fire, Marines should avoid silhouetting themselves, avoid open areas, and select their next covered position before moving.

4. Crossing a Wall. Each Marine should learn the correct method of crossing a wall (Figure A-8). After reconnoitering the other side, the Marine quickly rolls over the wall, keeping a low silhouette. The speed of his movement and a low silhouette deny the enemy a good target.

Figure A-8. Crossing a Wall

5. Observation Around Corners. The area around a corner should be observed before the Marine moves beyond it. The most common mistake the Marine makes at a corner is allowing his weapon to extend beyond the corner before observing, thereby exposing his position. Using the short stocking technique (Paragraph 2.d.) reduces exposure. Additionally, Marines can use a technique called "popping the corner" to reduce exposure time (Figure A-9). The Marine will get into a prone position near the corner of a building or obstacle around which he needs to observe. The weapon is short stocked, and the muzzle is pointed in the direction the Marine is looking. This allows the Marine to engage a target, if necessary, when he observes around a corner. The Marine will crawl to the corner but not expose himself. He will raise his upper body onto his elbows. The Marine will then push his body forward with his feet and legs without moving his elbows. His upper body, with the weapon ready, will move forward. The final position will expose the weapon, the Marine's helmet, and a minimal amount of the Marine's face. The Marine's forearms will come to rest on the deck giving him a low profile, the ability to observe around the corner, and the immediate capability to engage targets with his weapon.

Figure A-9. Correct Technique for Observing Around a Corner

6. Movement Past Windows. Windows present another hazard to Marines. The most common mistake made when passing a window is exposing the head to enemy observation from inside the room. If a Marine shows his head, an enemy gunner inside the building could engage him through the window without exposing himself to return fires.

a. The correct technique for passing a window is for the Marine to stay below the window level. He makes sure he does not silhouette himself in the window; he "hugs" the side of the building. An enemy gunner inside the building would have to expose himself to covering fires to engage the Marine (Figure A-10A).

Figure A-10A. Moving Past Windows

b. The most common mistake Marines make when passing a basement window is not being aware of it. A Marine should not walk or run past a basement window because he presents a good target to an enemy gunner inside the building. When using the correct procedure for negotiating a basement window, the Marine stays close to the wall of the building and steps or jumps past the window without exposing his legs (Figure A10-B).

c. An alternate method is for the pointman to pie off the window (or other opening) as he approaches it. As the pointman moves toward the window, the second Marine (behind the pointman) steps out to provide security to the direct front. The pointman continues to cover the opening by pieing, all the while using the near edge for cover. The pointman covers and sectors off that part of the interior that can be viewed from the outside. The second Marine moves with the pointman and maintains security to the front, eventually moving past the pointman as the pointman clears the inside corners of the opening. Once the opening is cleared, the second man may assume the duties of point, or when subsequent Marines move up to secure the opening, the original pointman may reassume those duties. Figure A-11 illustrates this alternate method.

Figure A-10B. Marine Passing Basement Windows

7. Use of Doorways. Doorways should not be used as entrances or exits because they are normally covered by enemy fire. If a Marine must use a doorway as an exit, he should move quickly through it to his next position, staying as low as possible to avoid silhouetting himself (Figure A-12). Preselection of positions, speed, a low silhouette, and the use of covering fires should be emphasized when exiting doorways.

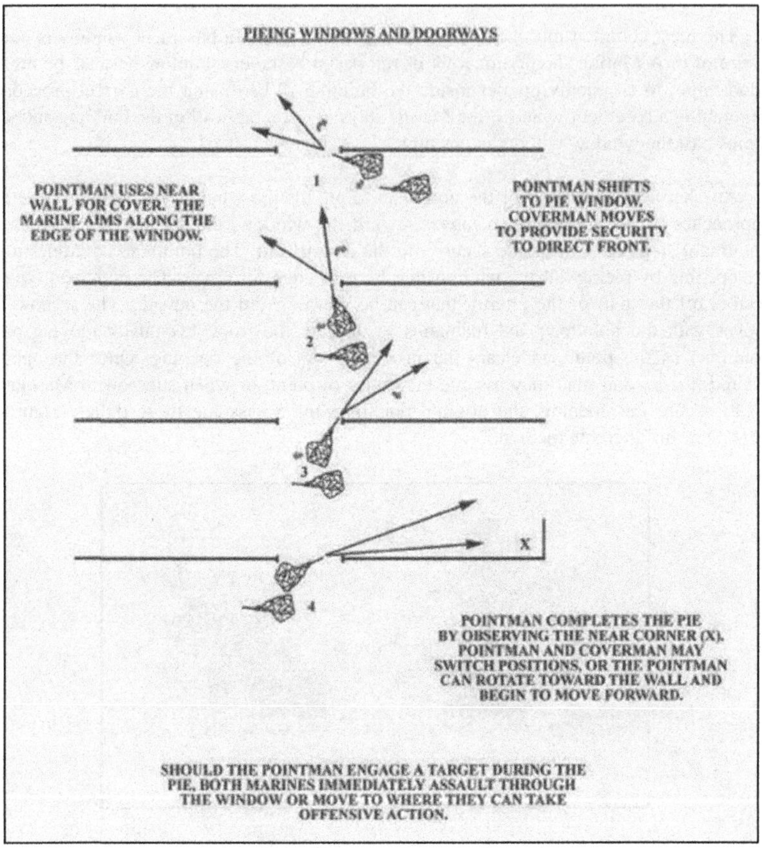

Figure A-11. Alternate Technique for Moving Past Windows

8. Movement Parallel to Buildings. Marines and small units may not always be able to use the inside of buildings as a route of advance. In such cases, they should move on the outside of the buildings (Figure A-13). Smoke, covering fires, and cover and concealment should be used to hide movement. When correctly moving along the outside of a building, the Marine hugs the side of the building, stays in the shadow, presents a low silhouette, and moves rapidly to his next position (Figure A-14). If an enemy gunner inside the building fires on a Marine, he exposes himself to fire from other squad/fire team members who are covering him. Furthermore, an enemy gunner farther down the street may have difficulty detecting and engaging the Marine.

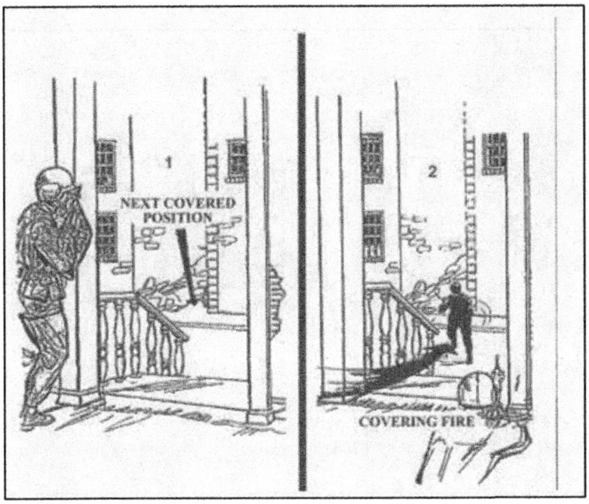

Figure A-12. Exiting a Doorway

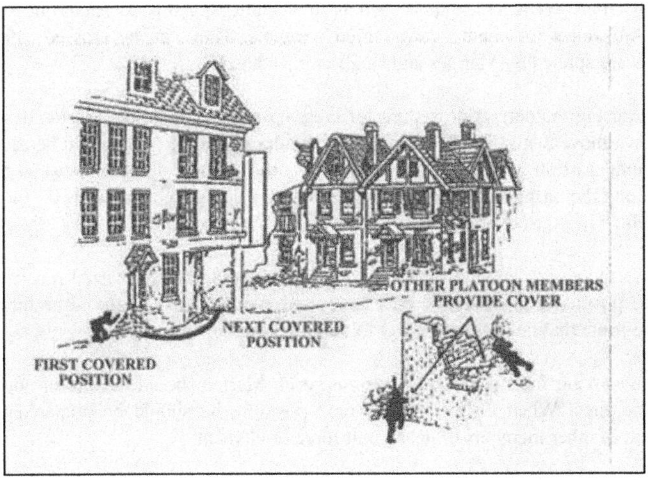

Figure A-13. Moving Outside a Building

Figure A-14. Selection of the Next Position

9. Crossing Open Areas. Open areas such as streets, alleys, and parks should be avoided. They are natural kill zones for enemy crew-served weapons. They can be crossed safely if certain fundamentals are applied by Marines and small-unit leaders.

a. When using the correct procedure for crossing an open area, the Marine develops a plan for his own movement. (Smoke from hand grenades or smoke pots should be used to conceal the movement of all Marines.) The Marine runs the shortest distance between buildings and moves along the far building to the next position. By doing so, he reduces the amount of time during which he is exposed to enemy fire.

b. Before moving to another position, the Marine should make a visual reconnaissance and select the position that offers the best cover and concealment. At the same time, he should select the route that he will take to get to that position.

c. When moving from position to position, each Marine should be careful not to mask his supporting fires. When he reaches his next position, he should be prepared to cover the movement of other members of his assault force or element.

10. Assault Element Employment. Moving as an assault element from building to building or between buildings presents a problem because an assault element presents a large target to enemy fire. When moving from the corner of one building to another, the assault element should move across the open area in groups of two, maintaining their integrity within their respective clearing or covering team (Figure A-15). The assault element leader (fire team leader) is responsible for the secure movement of his unit. Smoke should be used to screen the assault element's movement. The covering team will move first to establish security on the far side. Their movement is protected by the fire of the clearing team. Once the covering team is in position, they provide cover as the clearing team moves across and prepares, then enters, the structure. Moving from the side of one building to the side of another presents a similar problem, and the technique of movement employed is the same (Figure A-16).

There may be situations in which it is best to move in larger groups. An example is that Marines crossing in pairs may key enemy snipers in to the movement, setting them up for good shots at the next pair to cross. Leaders should remain aware of the situation and threat and plan movements accordingly.

Figure A-15. Assault Element Movement

11. Movement Inside a Building. When moving within a building that is under attack (Figure A-17), Marines should avoid silhouetting themselves in doors and windows. If forced to use a hallway (Figure A-18), Marines should move by hugging the wall to avoid presenting targets to the enemy.

Figure A-16. Movement to an Adjacent Building

Figure A-17. Movement Within a Building Under Attack

Figure A-18. Moving Down a Hallway

Section IV

Entry Techniques

Marines should enter a building with minimum exposure. They should select an entry point before moving toward the building; avoid windows and doors; use smoke to conceal their advance to the building; use demolitions, tank rounds, and so on to make new entrances; precede the entry of the clearing team with a grenade; enter immediately after the grenade explodes; and be covered by the covering team.

12. Upper Building Levels. Clearing a building from the top down is the preferred method. Clearing or defending a building is easier from an upper story. Gravity and the building's floor plan become assets when throwing hand grenades and moving from floor to floor.

a. An enemy who is forced to the top of a building may be cornered and fight desperately or escape over the roof, but an enemy who is forced down to the ground level may withdraw from the building, thus exposing himself to friendly fires from the outside.

b. Various means such as ladders, drainpipes, vines, helicopters, or the roofs and windows of adjoining buildings may be used to reach the top floor or roof of a building. In some cases, one Marine can climb onto the shoulders of another and reach high enough to pull himself up. Another method is to attach a grappling hook to the end of a scaling rope so that a Marine can scale a wall, spring from one building to another, or gain entrance through an upstairs window.

13. Use of Ladders. Ladders offer the quickest method to gain access to the upper levels of a building (Figure A-19). Units can get ladders from local civilians or stores or obtain material to build ladders through supply channels. If required, ladders can be built with resources that are available throughout the urban area; for example, lumber can be taken from inside the walls of buildings (Figure A-20). Although ladders may not permit access to the top of some buildings, they may offer security and safety through speed.

14. Use of Grappling Hook. A suitable grappling hook and rope are selected to meet the need. The grappling hook should be sturdy, portable, easily thrown, and equipped with hooks that can hold inside a window. The scaling rope should be 5/8 of an inch to 1 inch in diameter and long enough to reach the objective window. Knots are tied in the rope at 1-foot intervals to make climbing and holding easier. Marines should follow the procedures outlined below.

a. When throwing the grappling hook, the Marine should stand as close to the building as possible (Figure A-21). If the Marine stands close to the building, he is less exposed to enemy fires and he does not have to throw the hook as far.

Figure A-19. Using Ladders To Get to Upper Levels

Figure A-20. Getting Lumber From Inside the Walls

Figure A-21. Grappling Hook Thrown at Close Range

b. Making sure there is enough rope to reach the target, the Marine should hold the hook and a few coils of rope in the throwing hand. The remainder of the rope, in loose coils, should be in the other hand. The rope should be allowed to play out freely. The throw should be a gentle, even, upward lob of the hook, with the other hand releasing the rope as it plays out.

c. Once the grappling hook is inside the window (or on the roof), the Marine should pull on the rope to obtain a good hold before beginning to climb. When using a window, the hook should be pulled to one corner to ensure a good "bite" and to reduce the Marine's exposure to lower windows during the climb.

d. The use of grappling hooks is the least preferred method for gaining entry to upper levels of buildings. Grappling hooks should be used only as a last resort and away from potential enemy positions. This method may potentially be used on adjacent buildings that offer concealed locations and a roof that connects to enemy positions.

15. Scaling Walls. When forced to scale a wall during exposure to enemy fire, all available means of concealment should be used. Smoke and diversionary measures improve the chances of a successful exposed movement. When using smoke for concealment, Marines should plan for wind direction. They should use fire, shouts, and fake movement to distract the enemy.

a. A Marine scaling an outside wall is vulnerable to enemy sniper fire. Marines moving from building to building and climbing buildings should be covered with friendly fire. Areas between buildings offer good fields of fire to the enemy. Properly positioned friendly weapons can suppress and eliminate enemy fire. The M203 grenade launcher is effective in clearing the enemy from rooms inside buildings (Figure A-22).

Figure A-22. Employment of the M203 Grenade Launcher

b. A Marine scaling a wall with a rope should avoid silhouetting himself in windows of uncleared rooms or exposing himself to enemy fires from lower windows. He should climb with his weapon slung over the firing shoulder so that he may quickly bring it to a firing position. He should clear the lower room with a hand grenade before going outside the window. Before doing this, the Marine first loosens the safety pin so that he needs only one hand to throw the grenade. The objective upper story window should not be entered before a hand grenade has been thrown in.

c. The Marine enters the objective window with a low silhouette (Figure A-23). Entry can be head first; however, a preferred method is to hook a leg over the windowsill and enter sideways, straddling the ledge.

16. Rappelling. Rappelling (Figure A-24) is an entry technique that Marines can use to descend from the rooftop of a tall building into a window. (See Army manual TC 21-24 for more information on rappelling.)

17. Entry at Lower Levels. Buildings should be cleared from the top down. However, it may be impossible to enter a building at the top; therefore, entry at the bottom or lower level may be the only course of action. When entering a building at the lower level, Marines avoid entry through windows and doors because both can be easily boobytrapped and are usually covered by enemy fire.

Figure A-23. Entering the Objective Window

Figure A-24. Rappelling

a. Ideally, when entering at lower levels, demolitions, artillery, tank fire, antiarmor weapons fire, or similar means are used to create a new entrance to avoid boobytraps. Quick entry is required to follow through on the effects of the blast and concussion.

b. When the only entry to a building is through a window or door, supporting fire should be directed at that location. (Specific lower level entry techniques are shown in Figure A-25.) If no supporting fire is available, AT4s can be employed instead.

c. Before entering, Marines throw a hand grenade into the new entrance to reinforce the effects of the original blast. When making a new entrance in a building, they consider the effects of the blast on the building and adjacent buildings. If there is the possibility of a fire in an adjacent building, Marines coordinate with adjacent units and obtain permission before starting the operation. In wooden-framed buildings, the blast may cause the building to collapse. In stone, brick, or cement buildings, supporting fires are aimed at the corner of the building or at weak points in the building construction.

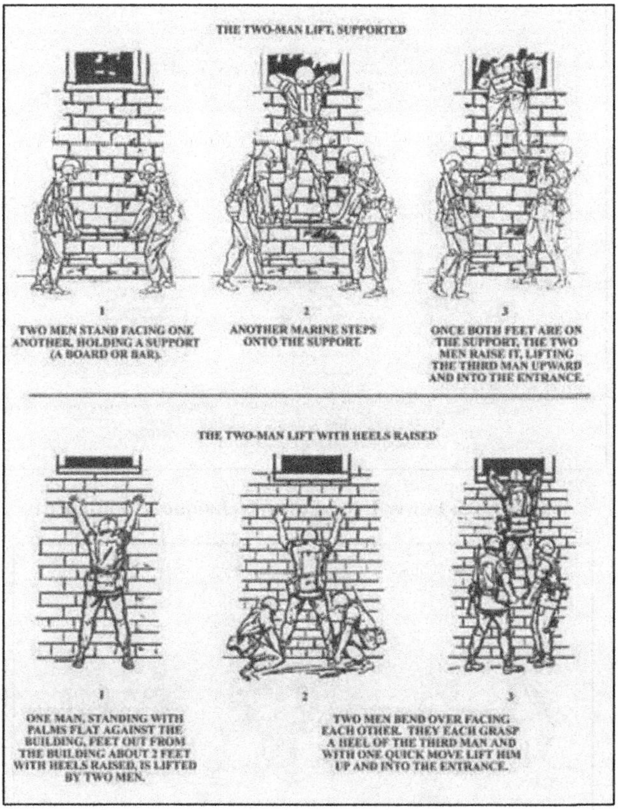

Figure A-25. Lower Level Entry Techniques

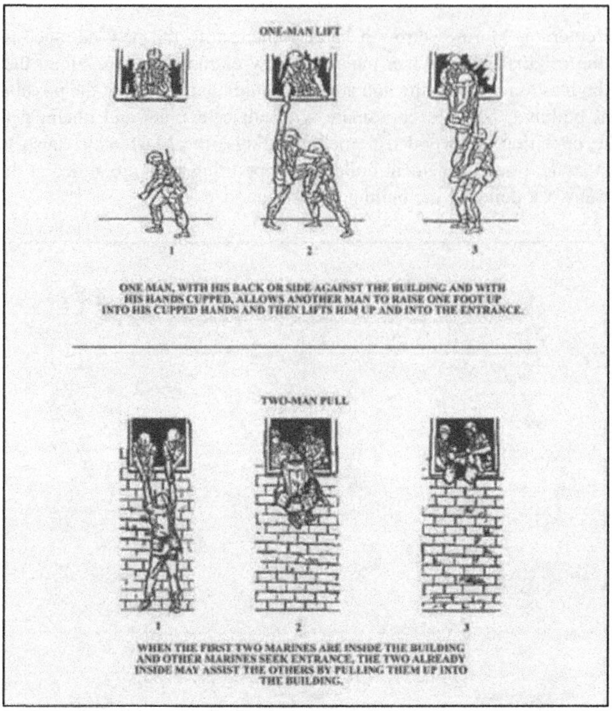

Figure A-25. Lower Level Entry Techniques (Continued)

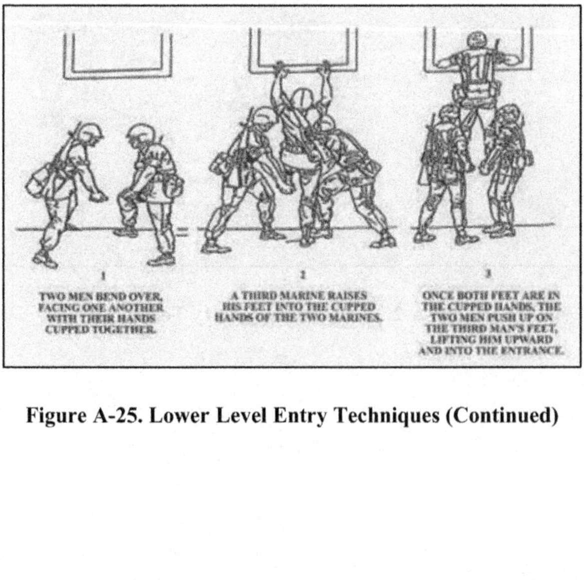

Figure A-25. Lower Level Entry Techniques (Continued)

18. Hand Grenades. Combat in built-up areas (mainly during the attack) requires extensive use of hand grenades. The Marine should throw a grenade before entering rooms or negotiating staircases, mouseholes, and so on. This usually requires the use of both hands and employs both the overhand and underhand methods of throwing.

a. Two techniques may be used in the throwing of a grenade. The *preferred technique* involves throwing the grenade hard enough that it bounces or skips around, making it difficult to pick up. The hard-throw, skip/bounce technique may be used by Marines in training and combat. The *least preferred technique*, which involves cooking-off the grenade for two seconds then throwing it to prevent the enemy from grabbing the grenade and tossing it back, may be used by Marines, as appropriate, during actual combat but not in training.

b. Nonverbal or verbal alerts are used, as appropriate, before throwing a grenade. A visual showing (preferred) of the grenade to be thrown is made to assault element members, and a visual acknowledgment from them is received. A nonverbal alert may ensure that the enemy is surprised when the grenade is thrown. If the situation demands, a voice alert can be used, but the element of surprise may be lost. When or if a voice alert is used, the voice alert is "**FRAG OUT**;" when an enemy grenade has been identified, friendly forces shout, "**GRENADE**."

c. The construction material used in the building being cleared influences the use of grenades. In some situations, concussion grenades may be preferred over fragmentary grenades during offensive operations or when defending from hasty defensive positions. If the walls of a building are made of thin material, such as sheetrock or thin plyboard, the Marine should either lie flat on the floor with his helmet pointing toward the area of detonation or move away from any wall that might be penetrated by grenade fragments.

d. Marines should throw grenades into an opening before entering a building to eliminate any enemy that might be near the entrance (Figure A-26). Using the M203 grenade launcher is the best method for putting a grenade in an upper story window.

e. When a hand grenade must be used, the Marine throwing the grenade should stand close to the building, using it for cover. At the same time, the individual and the rest of the element should have a planned area to which they can move for safety if the grenade does not go through the window, but falls back to the ground.

f. The Marine throwing the grenade should step out far enough to lob the grenade into the upper story opening. The Marine's weapon should be kept in the nonthrowing hand so that it can be used if needed. The weapon should never be laid down outside or inside the building. Once the grenade has been thrown into the opening, assaulting forces should move swiftly to enter the building. This technique should be employed only when the window has already been broken. Otherwise, the chances are high that the thrown grenade will bounce off of the window and fall back onto the ground without going into the room.

Figure A-26. Hand Grenade Thrown Through Window

WARNING

After throwing the grenade, the Marine should immediately announce, "FRAG OUT," to indicate that a grenade has been thrown. He then takes cover because the grenade may bounce back or be thrown back, or the enemy may fire at him.

g. If Marines must enter the building by using the stairs, they first look for boobytraps. Then they throw a grenade through the stairwell door, let it detonate, and move quickly inside. They can use the staircase for cover.

h. The best way to enter a building is to breach the exterior wall. Again, a grenade should be thrown through the hole while using all available cover, such as the lower corner of the building (Figure A-27).

i. Another way to enter a room is to blast mouseholes with demolitions. In moving from room to room through mouseholes, Marines should use grenades to clear the room as they do in moving through open doorways.

j. Although buildings are best cleared from the top down, this is not always possible. While clearing the bottom floor of a building, Marines may encounter stairs that must also be cleared. Once again, grenades play an important role. Before climbing the stairs from the bottom floor, Marines should first inspect for boobytraps then toss a grenade to the head of the stairs, if appropriate (Figure A-28). The use of a grenade in these situations is based on stair/building construction and building layout. *Throwing grenades up staircases is extremely* *dangerous, and extreme caution and common sense should be used.* In most situations, throwing a grenade up a staircase is not recommended. (Again, the grenade is being used in the initial building entry or clearing of the bottom floor of a building having a staircase.) If a grenade is thrown too hard or the staircase is too steep, the grenade can roll back down the staircase. For this reason, if a grenade is thrown, it should be thrown underhand to reduce the risk of it bouncing back and rolling down the stairs. A nonverbal or, as appropriate, voice alert is used when throwing the grenade. Once the first grenade has detonated, another grenade should be thrown over and behind the staircase banister and into the hallway, destroying any enemy hiding to the rear.

Figure A-27. Entering Through a Breached Hole

Figure A-28. Tossing a Grenade Up a Stairway

k. After the stairs have been cleared, assaulting forces move to the top floor and clear it. Upon clearing the top floor, forces move downstairs to clear the center and bottom floors and to continue with the mission.

Note: Because large quantities of hand grenades are used when clearing buildings, a continuous supply must be available to forces having this mission within a built-up area.

Section V

Clearing Techniques

This section will discuss basic clearing techniques; however, the fundamentals may have to be modified depending on the various situations and building/room layouts that Marines may encounter. Caution should be taken when using a stacked type of position as personnel may be concentrating in a danger area where they could be exposed to fires penetrating walls or explosive effects from boobytraps or grenades thrown by the enemy. *Marines should always remain aware of the protective, or lack of protective, capabilities of the materials comprising the walls, floors, ceilings, and doors within the areas they are clearing.*

19. Clearing a Room, Door Closed or Open, From Stacked Positions. The following discussion and series of figures will focus on the two-man clearing team in the conduct of clearing a room. The covering team is positioned by the assault element leader to provide security as the clearing team begins its procedures to enter and clear a room.

a. Shooter Number Two is behind Shooter Number One (stacked position). This positioning is the same if the door is already open or nonexistent (Figure A-29).

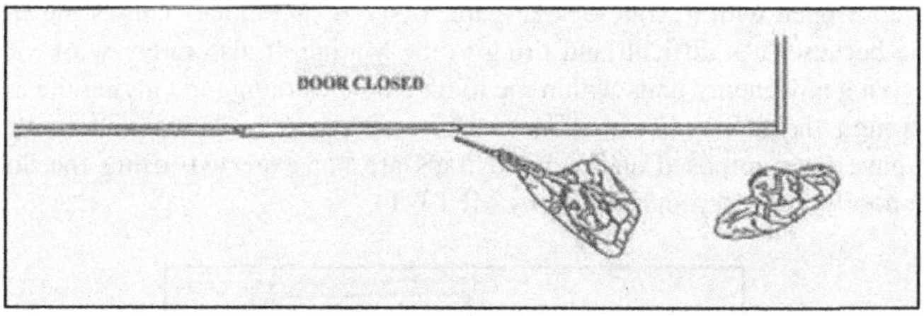

Figure A-29. Clearing a Room, Door Closed

b. Shooter Number One determines the direction in which the door opens then quickly informs Shooter Number Two and forces the door open.

When a door is the only means of entering a room, Marines should beware of fire from enemy soldiers within the room and of boobytraps. When opening a door, Marines should not expose themselves to fires through the door. The two-man clearing team forces the door open by directing a short burst of automatic fire through the door around the latch (Figure A-30) and then kicking it open. Another method of opening the door is to use a shotgun with 12-gauge slugs to shoot the hinges and/or the latch. When the doorknob is locked and/or deadbolt locks are present, shooting the hinges of the door is recommended. The door may then be opened by kicking or using a battering ram.

Figure A-30. Shooting the Door Open

Marines can also use an ax or demolitions if they are available. As a last resort, Marines can kick the door open with a front kick (Figure A-31) or mule kick. This is the least preferred technique because it is difficult and tiring to the Marine. It also rarely works the first time, thereby giving any enemy units within the room ample warning (and giving the enemy time to shoot through the door). In situations where restrictions on firing and limiting structural damage have been imposed and/or boobytraps are not expected, using the door handle to open the door may be appropriate (apply METT-T).

Figure A-31. Opening a Door With a Front Kick

c. Once the door is open, Shooter Number Two immediately throws a hand grenade into the room. Shooter Number One maintains cover for Shooter Number Two as he prepares to throw the grenade into the room. Shooter Number Two then returns to the original stacked position behind Shooter Number One (Figure A-32). If Marines have no grenades, the procedures remain the same except for throwing the grenade. *Note: When the door is already open or nonexistent, the stacked position is always used.*

WARNING

Because fragments from M67 fragmentation grenades may injure Marines outside the room, they should not be used. Marines should use MK3A2 offensive hand grenades instead. Also, the technique of cooking-off hand grenades can be deadly unless properly performed.

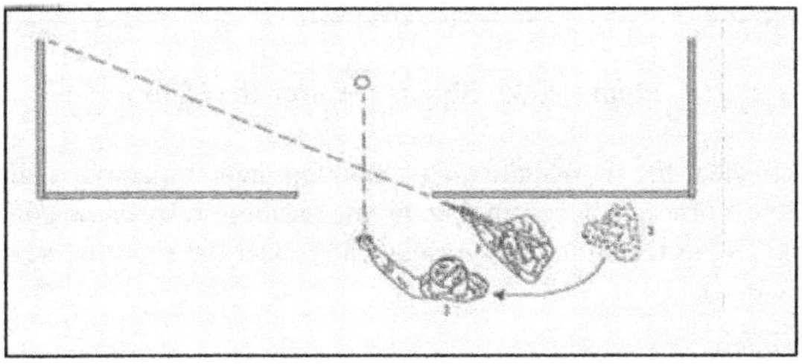

Figure A-32. Throwing a Grenade Into a Room

(1) Throwing Grenades. Two techniques are available for using grenades in room clearing. The *preferred technique* for Marines is to throw a concussion hand grenade into the room so hard that it skips and bounces, making it difficult for the enemy to pick up and throw back. The skip/bounce technique should be used by Marines during training and combat. The *least preferred technique* is to cook-off a concussion hand grenade by removing the grenade's safety pin, releasing the safety lever, counting off two seconds by thousands (one thousand and one, one thousand and two), and then throwing the grenade into the room. Cooking-off the grenade will be used only as appropriate during combat.

(2) Nonverbal and Verbal Alerts. To alert all that a grenade will be thrown, a visual showing of the grenade is made to assault element members, and a visual acknowledgment from them is received. A nonverbal alert may ensure that the enemy is surprised when the grenade is thrown. If the situation demands, a voice alert can be used, but the element of surprise may be lost. When or if a voice alert is used, the voice alert is "**FRAG OUT**;" when an enemy grenade has been identified, friendly forces shout, "**GRENADE**." This allows Marines to distinguish between warnings for outgoing and incoming grenades.

d. After the grenade explodes, Shooter Number One steps across the threshold and clears his immediate area (Figure A-33). He engages targets from the ready carry position. Shooter Number Two follows immediately behind Shooter Number One, buttonhooks, and clears his area.

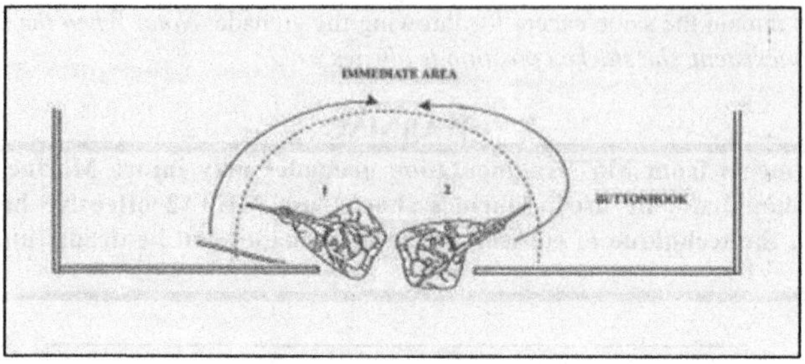

Figure A-33. Shooters Enter the Room

e. Both shooters clear the immediate area and along their respective walls, starting from the nearest respective corner and continuing to the farthest respective corner (Figure A-34). Shooters use the pieing technique to systematically clear the room by sector. (See Paragraph 3.a. of this appendix.)

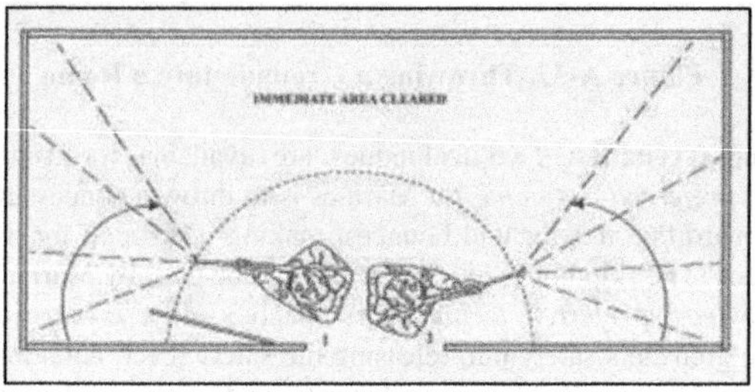

Figure A-34. Clearing a Room

f. Both shooters then establish a dominant position in the room one step away from the wall and two steps into the room and clear the room by sector, pieing to the opposite side of the room (Figure A-35). Meanwhile, the covering team (two-man team) in position outside the room being cleared provides security.

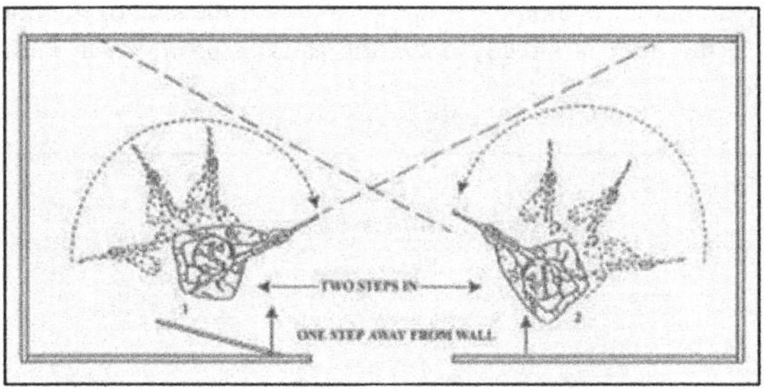

Figure A-35. Clearing a Room (Continued)

g. Another option for entering a room (situation dependent): If Shooter Number One of the clearing team enters a room without Shooter Number Two immediately following him (this could be due to room size, entering the room through a narrow opening, or a number of factors caused by the situation), then Shooter Number One will position himself inside the room to the left or right of the door and quickly scan the entire room. He then gives the command to Shooter Number Two, "**NEXT MAN IN, LEFT (RIGHT)**." Shooter Number Two shouts, "**COMING IN, LEFT (RIGHT)**," enters the room, positions himself up against the wall to the left (right) of the entrance as designated, and scans the room. Once in position, the clearing team leader can call in additional members with the "**NEXT MAN IN**" command, as the situation dictates. It is critical that all assault element members tell each other where they are to avoid fratricide.

h. When the clearing team has cleared the room and is ready to exit the room, it uses the following voice alerts. Once a room has been cleared, the clearing team yells, "**CLEAR**," to inform the covering team. Before leaving the room and rejoining the covering team, the clearing team yells, "**COMING OUT**" and waits to hear the covering team yell, "**ALL CLEAR**." The clearing team then executes its movement out of the room. The assault element then marks the room according to unit SOP.

When moving up or down a staircase, the appropriate team yells, "**COMING UP**" or "**COMING DOWN**." As before, it waits for the response "**ALL CLEAR**" before executing movement and rejoining its element. When exiting a building, the clearing team or assault element yells, "**COMING OUT**" and waits for the support force or covering team to respond with "**ALL CLEAR**."

20. Clearing a Room, Door Closed, Split Positions

a. Before opening a door, Shooter Number Two of the clearing team positions himself opposite Shooter Number One on the other side of the door and away from the wall in a safe position that allows Shooter Number One to shoot the door-opening mechanism (Figure A-36). This positioning allows Shooter Number Two to be in position to immediately move to a wall position opposite Shooter Number One. No matter what method is used to open the door, Shooter Number Two should get in a position on the side of the door opposite Shooter Number One. If the door is already open, the stacked positions in Paragraph 19.a. of this section are used.

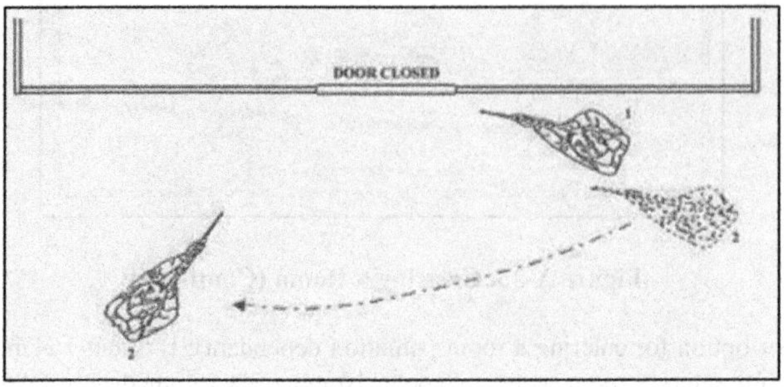

Figure A-36. Positioning To Open a Door

b. After Shooter Number One shoots the door open, Shooter Number Two immediately moves to a kneeling position against the wall (Figure A-37). Shooter Number One kicks the door open from a standing position. If the Marines have grenades, then Shooter Number Two immediately throws a grenade into the room as discussed in Paragraph 19.c. of this section. If the Marines do not have any grenades, the stand/kneel positioning of shooters places the shooters' weapons at different levels, thus allowing each shooter to engage targets inside the room while staying out of the other shooter's line of fire. The shooters scan the room for targets.

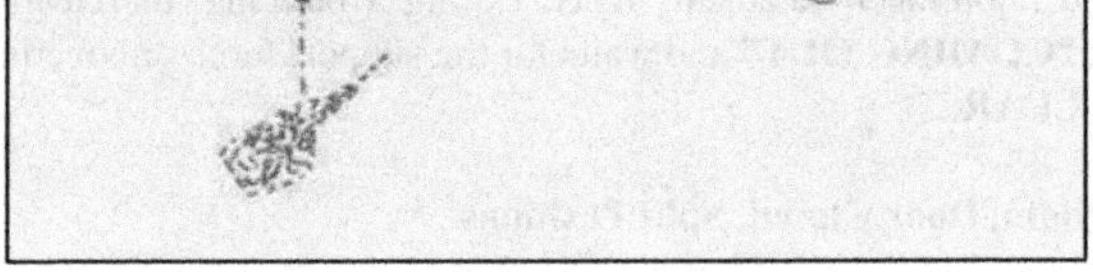

Figure A-37. Positioning To Enter a Room

c. If no grenade was thrown: Once the initial scan of the room is complete, Shooter Number One yells, "**MOVE**," and both shooters pivot from their respective positions toward each other, pieing (see Paragraph 3.a. of this appendix) while maintaining eye-muzzle-target contact. Shooter Number Two is kneeling, and Shooter Number One is still standing (Figure A-38). Both shooters clear their respective sector of fire.

If a grenade was thrown: After the explosion, Shooter Number One yells, "**MOVE**," and both shooters pivot from their respective positions toward each other, immediately enter the room, and buttonhook (Figures A-38, A-39A, and A-39B). Shooter Number Two goes from kneeling to standing as he pivots. Both shooters clear their respective sector of fire.

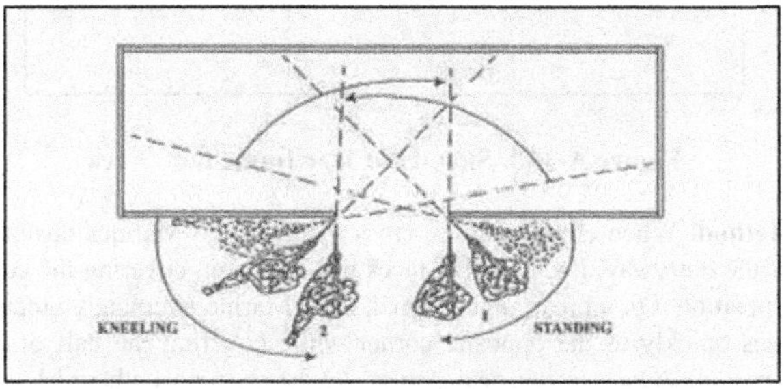

Figure A-38. Shooters Clear Sectors of Fire From Doorway

d. If no grenade was thrown: After scanning is completed, Shooter Number Two stands up and yells, "**READY**," and Shooter Number One Yells, "**MOVE**." Both shooters enter the room together (Figure A-39A).

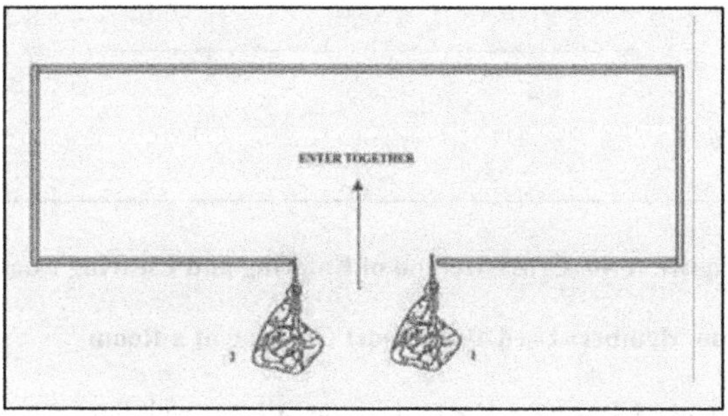

Figure A-39A. Shooters Enter the Room Together

e. Both shooters pass through the doorway together and immediately face respective corners, using the adjacent wall for cover. Both shooters buttonhook their respective areas (Figure A-39B). At this point, clearing the room is the same as discussed in Paragraphs 19.e. - 19.h.

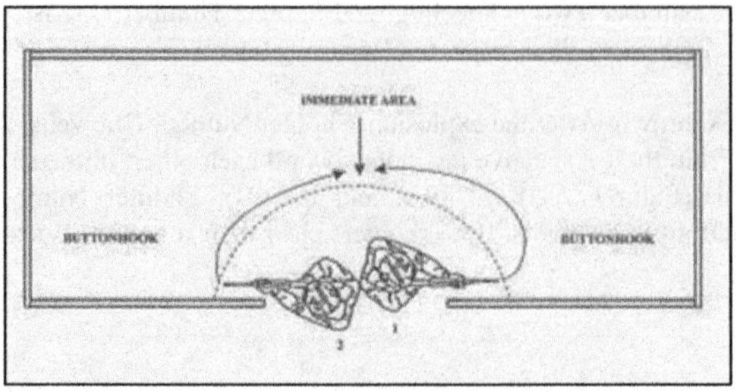

Figure A-39B. Shooters Clear Immediate Area

f. Cross Method. When employing the cross method, two Marines position themselves on either side of the entryway. Each Marine faces into the room covering the corner of the room opposite his position. On a prearranged signal, each Marine alternately enters the room. Each Marine crosses quickly to the opposite corner while covering the half of the room toward which he is moving. Once in the near corner, he assumes an outboard kneeling position to reduce his silhouette and continues to maintain coverage of his half of the room. He may change the position of his weapon to best cover the room. (Figure A-40)

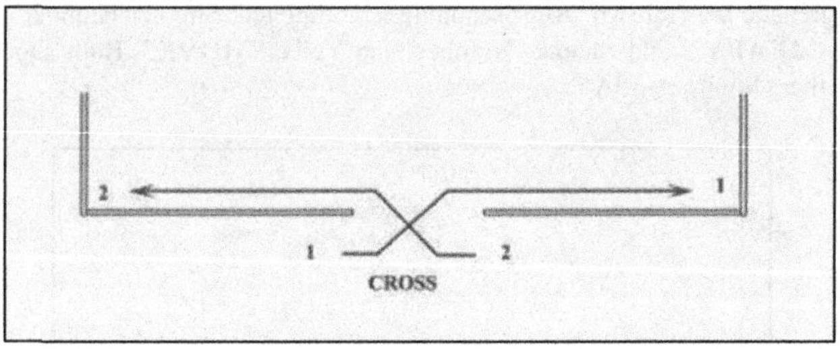

Figure A-40. Cross Method of Entering and Clearing Rooms

21. Covering Team Members Used To Support Clearing of a Room

a. Three Marines Clearing a Room. Clearing a room with three men involves the same procedures used by the aforementioned two-man clearing team, with a third Marine (from the covering team) following immediately (stacked) behind Shooters Number One and Two. Shooter Number Three moves to one side of the door and establishes a center sector of fire coverage (Figure A-41). The remaining covering team Marine provides outside security. If the split position is used, Shooters Number Two and Three are placed on the opposite side of the door.

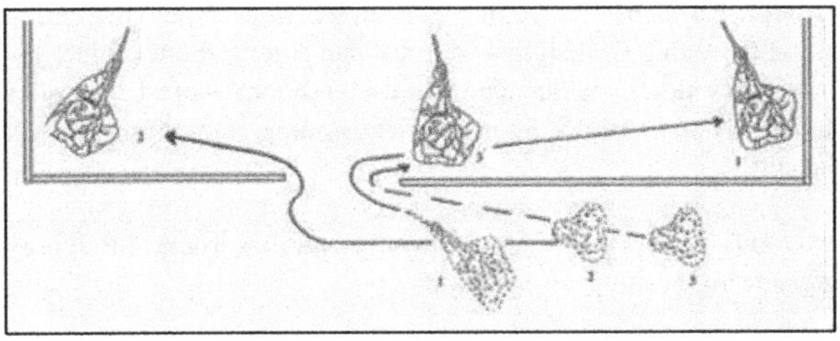

Figure A-41. Three-Marine Clearing Team (Stacked)

b. Four Marines Clearing a Room. When four Marines (entire assault element) are used to clear a room, outside security should be established by another assault element. (If outside security cannot be established, then the two- or three-Marine clearing team should be used.) The two-Marine clearing team procedures are used for initial entry into the room. Shooters Number Three and Four (split position example) stack up opposite Shooters Number One and Two. Shooters Number Three and Four follow Shooter Number Two into the room. Shooter Number Three crosses the threshold and establishes a center sector of fire coverage. Shooter Number Four follows Shooter Number Three and establishes a center sector of fire coverage opposite Shooter Number Three (Figure A-42). If the stacked position is used to enter a room, all shooters stack up on the same side of the door and follow each other in with the same positioning.

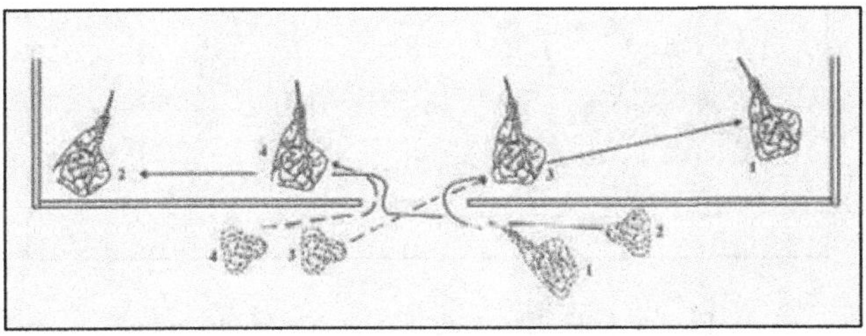

Figure A-42. Four-Marine Clearing Team (Split)

c. Single Marine Clearing a Room. This is the *least preferred* method of clearing a room. However, there may be times when this is the only method available, for example, when gaining access to an upper story window. When this is the case, the Marine gains entry to the room by shooting, kicking, ramming, or breaking a window. He then throws a grenade into the room if one is available and no restrictions apply and takes cover. Immediately after the grenade explodes, he crosses the threshold using the buttonhook technique depicted in earlier figures, scans the room for targets, and engages targets with two shots as they appear.

22. Clearing a Room, Entering Through Mousehole. Mouseholes measure about 2 feet wide and are blown or cut through a wall so that Marines can enter a room (Figure A-43). Mouseholes are safer entrances than doors, which can be easily boobytrapped and should be avoided. Procedures discussed in Paragraph 19 are used. The assault element leader positions the covering team to provide security.

a. In this situation, Shooter Number One would enter the room through a mousehole after throwing a grenade and waiting for the detonation.

b. Shooter Number One will position himself inside the room to the left or right of the mousehole and quickly scan the entire room. He then gives the command to Shooter Number Two, "**NEXT MAN IN, LEFT (RIGHT).**" Shooter Number Two shouts, "**COMING IN, LEFT (RIGHT),**" enters the room, positions himself against the wall to the left (right) of the entrance, and scans the room.

c. The clearing team continues clearing the room as discussed in Paragraph 19.

Figure A-43. Entering Through a Mousehole

23. Clearing an L-Shaped Hallway. After entering a room or building, the clearing team may encounter an L-shaped hallway that should be cleared. The following procedures are used:

a. The clearing team finds an L-shaped hallway.

b. Each shooter takes a dominant position.

c. Shooter Number Two aims his weapon to cover the dead space (Figure A-44).

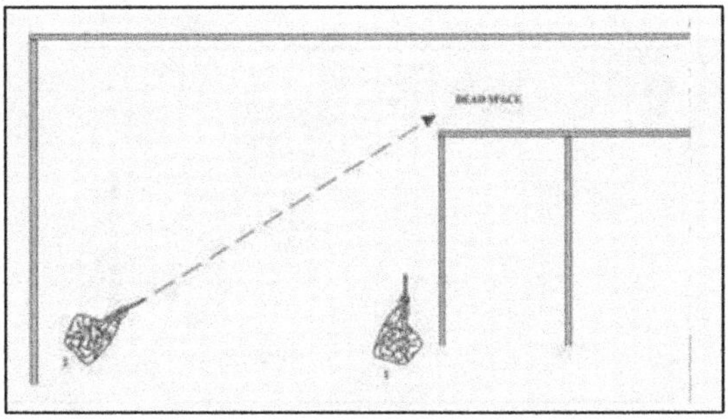

Figure A-44. Positioning To Clear an L-Shaped Hallway

d. Shooter Number Two, with his weapon covering the dead space, pies as far as possible and moves forward to increase his angle of fire further into the dead space. Shooter Number One moves accordingly with Shooter Number Two (Figure A-45).

e. Shooter Number Two stops pieing just before he reaches the position where he cannot cover the dead space. Shooter Number One stops short of the hallway and goes to a kneeling position.

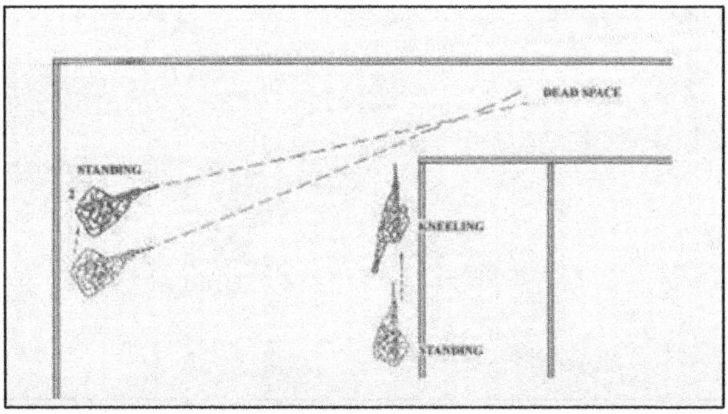

Figure A-45. Clearing an L-Shaped Hallway

f. Shooter Number One says, "**READY!**"

g. Shooter Number Two then says, "**MOVE!**" and both shooters move simultaneously to clear the hallway (Figure A-46). Shooter Number Two remains standing, and Shooter Number One remains kneeling as he pivots around the corner of the wall into the hallway.

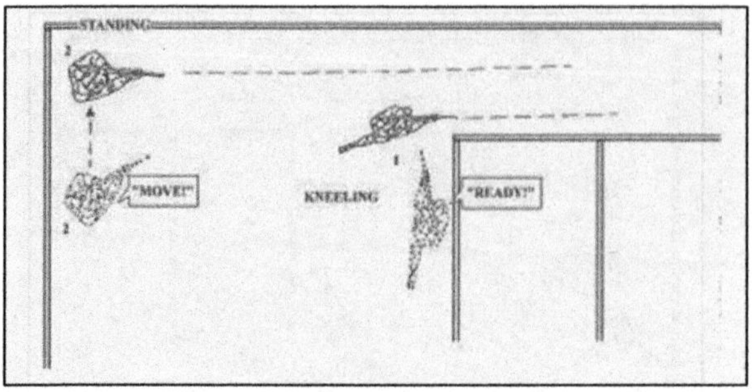

Figure A-46. Clearing an L-Shaped Hallway (Continued)

24. Clearing a T-Shaped Hallway. After entering a room or building, the clearing team may find a T-shaped hallway that should be cleared. The following procedures are used:

a. Both shooters approach the T-shaped intersection together, pieing their individual sectors while maintaining eye-muzzle-target contact (Figure A-47).

b. Both shooters move to dominant positions without entering the intersection.

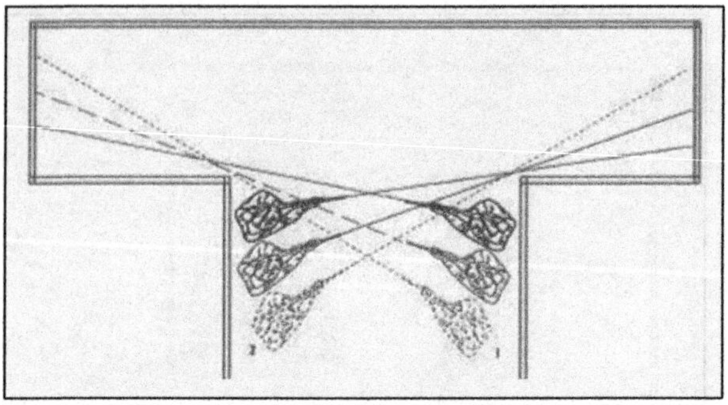

Figure A-47. Positioning To Clear a T-Shaped Hallway

c. Shooter Number One clears his sector of fire and then says, "**READY!**" (Figure A-48).

d. Shooter Number Two clears his sector of fire and then says, "**MOVE!**"

e. Both shooters buttonhook into their next positions.

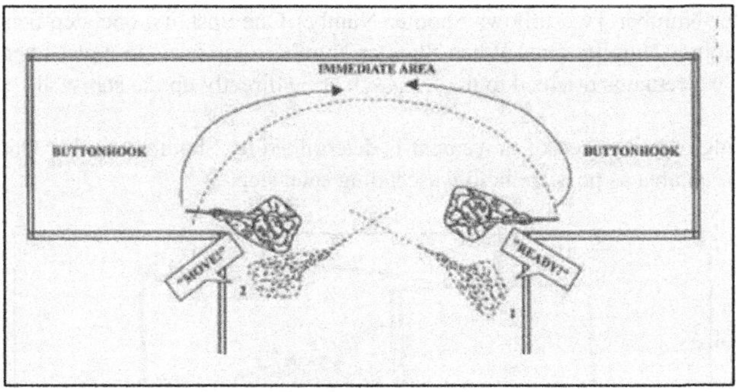

Figure A-48. Clearing a T-Shaped Hallway

f. Both shooters clear their immediate areas along their respective walls, starting from the nearest respective corner and continuing to the farthest respective corner (Figure A-49).

g. Both shooters establish dominant positions that give them control of the hallway and the doorways leading into hallways. The covering team can then be called forward to provide security while the clearing team clears one end of the hallway and associated rooms. The clearing team then clears the other end of the hallway.

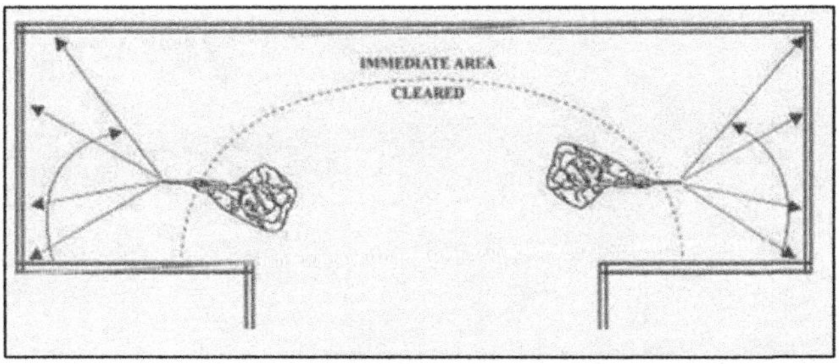

Figure A-49. Clearing a T-Shaped Hallway (Continued)

25. Clearing a Stairwell. After entering a building, the assault element may encounter stairwells. The following procedures are used:

a. Shooter Number One leads upstairs, one step ahead of Shooter Number Two (Figure A-50). Upon reaching the point just before he can be engaged from above, Shooter Number One turns around and covers overhead. From this point, Shooter Number One ascends the stairs, moving backward while covering behind and above.

b. Shooter Number Two follows Shooter Number One upstairs, one step behind and to the side of Shooter Number One. When Shooter Number One turns to cover overhead, Shooter Number Two remains oriented to the front, covering directly up the stairwell.

The clearing team's speed of movement is determined by Shooter Number One. Marines pie as much of an area as possible before ascending each step.

Figure A-50. Clearing a Stairwell

Section VI

Firing Positions

Whether a unit is attacking, defending, or conducting retrograde operations, its success or failure depends on the ability of the individual Marine to place accurate fire on the enemy while providing the least exposure to return fire. Consequently, Marines should immediately seek and properly use firing positions.

26. Hasty Firing Position. A hasty firing position is one that is normally occupied in the attack or the early stages of the defense. It is a position from which a Marine can place fire on the enemy while using available cover for protection from return fire. The Marine may occupy this position voluntarily, or he may be forced to occupy it because of enemy fire. In either case, the position lacks preparation before occupation. Some of the more common hasty firing positions in a built up area involve firing around corners of buildings, firing from behind walls, firing from windows, firing from unprepared loopholes, and firing from the peak of a roof.

a. Corners of Buildings. The corner of a building provides cover for a hasty firing position if used properly.

(1) The shooter should be capable of both right-handed and left-handed firing of his weapon (using the "short stocking" technique) to be effective around corners. Short stocking the weapon will prevent the muzzle from protruding and keep the weapon ready to fire the instant visual contact with the enemy is made. Furthermore, it reduces the Marine's exposure as a target. (Figure A-51)

(2) A common mistake made when firing around corners is firing from the standing position. The shooter exposes himself at the height the enemy would expect a target to appear and risks exposing the entire length of his body as a target for the enemy.

Figure A-51. Firing, Using Either Hand, Around the Corner of a Building

b. Walls. When firing from behind walls, a Marine should fire around cover—not over it (Figure A-52). Marines in this situation should be able to fire from the right or left shoulder. When firing from behind cover, a Marine should remain far enough back from that cover so that the weapon does not extend beyond it. This aids in weapon retention by ensuring that no enemy on the other side of the wall can get control of the barrel.

Figure A-52. Firing Around Cover

c. Windows. In a built-up area, windows provide convenient firing ports. A Marine should avoid firing from the standing position because it exposes most of his body to return fire from the enemy and could silhouette him against a light-colored interior beyond the window. This is an obvious indicator of the firer's position, especially at night when the muzzle flash can easily be observed. When using the proper method of firing from a window (Figure A-53), the Marine is well back into the room to prevent the muzzle flash from being seen, and he is kneeling to limit exposure and avoid silhouetting himself.

Figure A-53. Firing From a Window

d. Loopholes. A Marine may fire through a hole torn in the wall and avoid windows (Figure A-54). He stays well back from the loophole so the muzzle of the weapon does not protrude beyond the wall and the muzzle flash is concealed.

Figure A-54. Firing From a Loophole

e. Roof. The peak of a roof provides a vantage point for snipers that increases their field of vision and the ranges at which they can engage targets (Figure A-55). A chimney, a smokestack, or any other object protruding from the roof of a building can reduce the size of the exposed target and should be used.

Figure A-55. Firing From the Peak of a Roof

f. No Position Available. When a Marine is subjected to enemy fire and none of the positions mentioned are available, he should try to expose as little of himself as possible. When a Marine is in an open area between buildings (a street or alley) with no cover available and he is fired upon by an enemy from one of the buildings to his front, he should lie prone as close as possible to a building on the same side of the open area as the enemy. To engage the Marine, the enemy must then lean out of the window and expose himself to return fire.

g. No Cover Available. When no cover is available, target exposure can be reduced by firing from the prone position, by firing from shadows, and by presenting no silhouette against buildings.

27. Prepared Firing Position. A prepared firing position is one built or improved to allow the shooter to engage a particular area, avenue of approach, or enemy position, while reducing his exposure to return fire. Examples of prepared positions include: barricaded windows, fortified loopholes, sniper positions, antiarmor positions, and machine gun positions.

a. The natural firing port provided by windows can be improved by barricading the window and leaving a small hole for the firer's use (Figure A-56). The barricading may be accomplished by using materials torn from the interior walls of the building or any other available material. When barricading windows, a Marine should avoid:

(1) Barricading only the windows that will be used as firing ports. The enemy will soon determine that the barricaded windows are firing positions.

(2) Neat square or rectangular holes that are easily identified by the enemy. A barricaded window should not have a neat, regular firing port. The window should keep its original shape so that the position of the firer is hard to detect. Firing from the bottom of the window gives the firer the cover advantage of the wall because the firing port is less obvious to the enemy. Sandbags are used to reinforce the wall below the window and to increase protection for the firer. All glass should be removed from the window to prevent injury to the firer. Lace curtains permit the firer to see out and prevent the enemy from seeing in. Wet blankets should be placed under weapons to reduce dust that may create a signature. Wire mesh placed over the window keeps the enemy from throwing in hand grenades.

b. Although windows are usually good firing positions, they do not always allow the firer to engage targets throughout his sector.

(1) To avoid establishing a pattern of always firing from windows, an alternate position is required such as the prepared loophole (Figure A-57). This involves cutting or blowing a small hole in the wall to allow the shooter to observe and engage targets in his sector.

(2) Sandbags are used to reinforce the walls below, around, and above the loophole. If the position is on the second floor or higher, two layers of sandbags are placed on the floor under the firer to protect him from an explosion on a lower floor. A wall of sandbags, rubble, furniture, and so on should be constructed to the rear of the position to protect the firer from explosions in the room.

Figure A-56. Window Firing Position

(3) A table, a bedstead, or other available material provides overhead cover for the position. This prevents injury to the shooter from falling debris or explosions above his position.

(4) The position should be camouflaged by knocking other holes in the wall. This makes it difficult for the enemy to determine which hole the fire is coming from. Siding material should be removed from the building in several places to make loopholes less noticeable.

Figure A-57. Prepared Loopholes

c. A chimney or other protruding structure provides a base from which a sniper position can be prepared. Part of the roofing material is removed to allow the sniper to fire around the chimney. He should stand inside the building on the beams or on a platform with only his head and shoulders above the roof (behind the chimney). Sandbags placed on the sides of the position protect the sniper's flanks.

d. When the roof has no protruding structure to provide protection, the sniper position should be prepared from underneath on the enemy side of the roof (Figure A-58). The position is reinforced with sandbags, and a small piece of roofing material should be removed to allow the sniper to engage targets in his sector. The missing piece of roofing material should be the only sign that a position exists. Other pieces of roofing should be removed to deceive the enemy as to the true sniper position. The sniper should be invisible from outside the building, and the muzzle flash should be hidden from view.

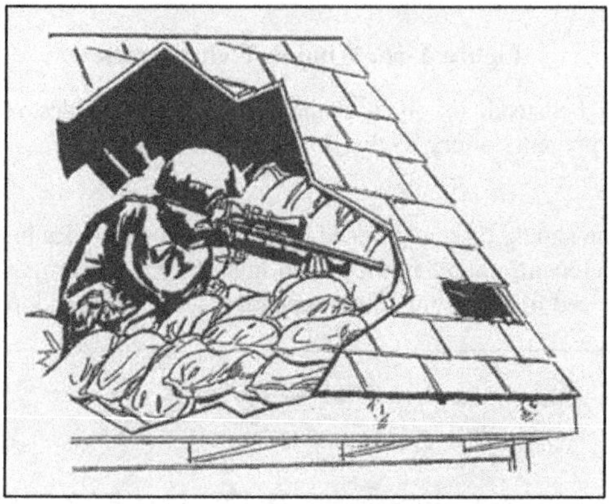

Figure A-58. Sniper Position

e. Some rules and considerations for selecting and occupying individual firing positions are:

(1) Make maximum use of available cover and concealment.

(2) Avoid firing over cover; when possible, fire around it.

(3) Avoid silhouetting against light-colored buildings, the skyline, and so on.

(4) Carefully select a new firing position before leaving an old one.

(5) Avoid setting a pattern; fire from both barricaded and unbarricaded windows.

(6) Keep exposure time to a minimum.

(7) Begin improving a hasty position immediately after occupation.

(8) Use construction material that is readily available in the built-up area for prepared positions.

(9) Remember that positions that provide cover at ground level may not provide cover on higher floors.

f. In attacking a built-up area, recoilless weapon and ATGM crews may be hampered in choosing firing positions by the backblast of their weapons. They may not have enough time to knock out walls in buildings and clear backblast areas. They should select positions that allow the backblast to escape, such as corner windows where the round fired goes out one window and the backblast escapes from another. The corner of a building can be improved with sandbags to create a firing position (Figure A-59).

Figure A-59. Corner Firing Position

g. The rifle squad is often reinforced with attached antitank weapons during an attack on and in defense of a built-up area. Therefore, the rifle squad leader should be able to choose good firing positions for the antitank weapons under his control.

h. Various principles for employing antitank weapons have universal applications, such as making maximum use of available cover, trying to achieve mutual support, and allowing for the backblast when positioning recoilless weapons, TOWs, Dragons, and AT4s.

i. Operating in a built-up area presents new considerations. Marines should select numerous alternate positions, particularly when the structure does not provide cover from small-arms fire. They should position their weapons in the shadows and within the building.

j. Recoilless weapons and ATGMs firing from the top of a building can use the chimney for cover (Figure A-60). The rear of this position should be reinforced with sandbags.

Figure A-60. Antitank Weapon Firing From a Rooftop

k. When selecting firing positions for recoilless weapons and ATGMs, Marines should make maximum use of rubble, corners of buildings, and destroyed vehicles to provide cover for the crew. Recoilless weapons and ATGMs can also be moved along rooftops to obtain a better firing angle from which to engage enemy armor. When buildings are elevated off the ground, positions can be prepared using a building for overhead cover (Figure A-61). The backblast under the building must not damage or collapse the building or injure the crew.

Note: When firing from a slope, ensure that the angle of the launcher relative to the ground or firing platform is not greater than 20 degrees. When firing within a building, ensure that the enclosure is at least 10 feet by 15 feet, is clear of debris and loose objects, and has windows, doors, or holes in the walls to allow the backblast to escape.

l. The machine gun can be emplaced almost anywhere. In the attack, windows and doors offer ready-made firing ports (Figure A-62). For this reason, the enemy normally has windows and doors under observation and fire; they should therefore be avoided. Any openings in walls that were created during the fighting may be used. When other holes are not present, small explosive charges can create loopholes (Figure A-63). Regardless of what openings are used, machine guns should be sited within the building and in the shadows.

Figure A-61. Prepared Positions Using a Building for Overhead Cover

Figure A-62. Emplacement of a Machine Gun in a Doorway

m. Upon occupying a building, Marines board up all windows and doors. By leaving small gaps between the slots, Marines can use windows and doors as good alternate firing positions.

n. Loopholes should be used extensively in the defense. They should not be constructed in any logical pattern, nor should they all be at floor or tabletop level. Varying their height and location makes them hard to pinpoint and identify. Dummy loopholes, knocked-off shingles, or holes that are cut but not intended to be used as firing positions aid in deception. Loopholes located behind shrubbery, under doorjambs, and under the eaves of a building are hard to detect.

Figure A-63. Use of a Loophole With a Machine Gun

o. Increased fields of fire can be obtained by locating the machine gun in the corner of the building or sandbagging it under a building (Figure A-64). Available materials, such as desks, overstuffed chairs, couches, and other items of furniture, should be integrated into the construction of bunkers to add both cover and concealment (Figure A-65).

Figure A-64. Sandbagged Machine Gun Emplacement Under a Building

p. Although grazing fire is desirable when employing the machine gun, it may not always be practical or possible. Where destroyed vehicles, rubble, and other obstructions restrict the fields of grazing fire, the gun can be elevated to a position from which it can deliver plunging fire over obstacles. Firing from loopholes on the second or third story may be necessary. A firing platform can be built under the roof (Figure A-66) and a loophole constructed. Again, the exact location of the position should be concealed by knocking off shingles in isolated patches over the entire roof.

Figure A-65. Corner Machine Gun Bunker

Figure A-66. Firing Platform Built Under a Roof

28. Target Acquisition. Built-up areas provide unique challenges to acquiring targets. Buildings mask movement and the effects of direct and indirect fires. Also, the rubble from destroyed buildings provides concealment and protection for attackers and defenders alike, making target acquisition difficult.

a. The techniques of patrolling and using OPs apply in the city as well as in wooded terrain. These techniques enable units to locate the enemy, to develop targets for direct and indirect fires in the defense, and to find uncovered avenues of approach in the offense.

b. Most weapons and vehicles produce recognizable signatures. These come from design features or from the environment in which the equipment is used. For example, firing a tank main gun in dry, dusty, and debris-covered streets raises a dust cloud; a tank being driven in built-up areas produces more noise than one moving through an open field; people moving through rubble on a street or in the halls of a damaged building create more noise than in a wooded area. Marines should learn to recognize signatures so they can locate and identify targets. Seeing and hearing assist in detecting and identifying signatures that lead to target location, identification, and rapid engagement. Marines should look for targets in areas where they are most likely to be employed.

c. Target acquisition should be continuous, whether halted or moving. Built-up areas provide both the attacker and defender with good cover and concealment but usually favor the defender. This makes rapid and accurate target acquisition extremely important.

d. When a unit is moving and enemy contact is likely, the unit should employ an overwatching element. This principle applies in built-up areas as it does in other kinds of terrain except that in built-up areas the overwatching element should observe both the upper floors of buildings and the street level.

e. Stealth should be used when moving in built-up areas because little distance separates attackers from defenders. Only hand and arm signals should be used until contact is made. The unit should stop periodically to listen and observe to ensure that it is not being followed or that the enemy is not moving parallel to the unit's flank for an ambush. Routes should be carefully chosen so that buildings and piles of rubble can be used to mask the unit's movement.

f. Observation duties should be clearly issued to squad members to ensure all-around security as they move. This security continues at the halt. All of the senses should be used to acquire targets, especially hearing and smelling. Marines soon recognize the sounds of vehicles and people moving through streets that are littered with rubble. The smell of fuel, cologne, or food cooking can disclose enemy positions.

g. OPs are "positions from which military observations are made, or fire directed and adjusted, and which possess appropriate communications" (Joint Pub 1-02). They are positions from which Marines can watch and listen to enemy activity in a specific sector. They warn the unit of enemy approach and should be positioned in the upper floors of buildings to give Marines a better vantage point than they would have at street level.

h. In the defense, a platoon leader positions OPs for local security as ordered by the company commander. The platoon leader selects the general location, but the squad leader establishes the OP (Figure A-67). Normally, there is at least one OP per platoon. An OP consists of two to four men and is within small-arms supporting range of the platoon. Leaders look for positions that allow good observation of the target sector. Ideally, an OP has a field of observation that overlaps those of adjacent OPs. The position selected for the OP should provide cover and concealment for units moving to and from the OP. The upper floors of houses or other buildings should be used. The squad leader should not routinely select obvious positions such as water towers or church steeples because those positions naturally attract the enemy's attention.

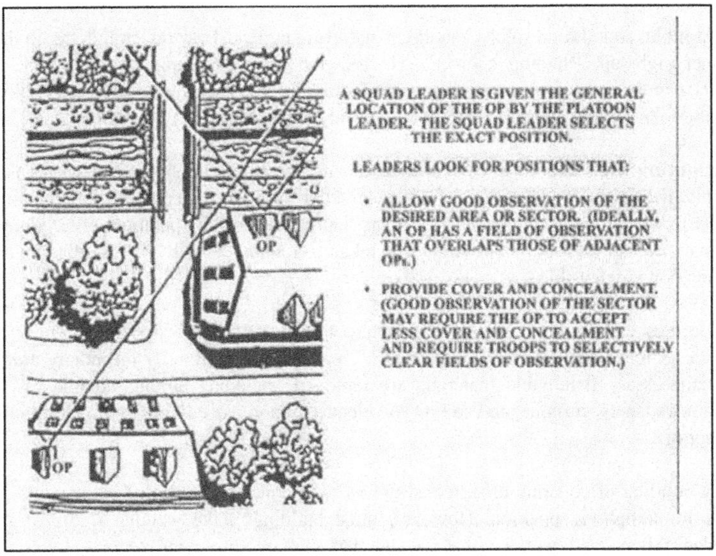

Figure A-67. Selection of an OP Location

i. The Marine should be taught how to scan a target area from OPs or from fighting positions. Use of proper scanning techniques enables observers to quickly locate and identify targets. The Marine searches quickly without optics for obvious targets, using all of his senses to detect target signatures. If no targets are found and time permits, he makes a more detailed search (using binoculars, if available) of the terrain in the assigned sector using the 50-meter method. First, he searches a strip 50 meters deep from right to left. He then searches a strip from left to right that is farther out, overlapping the first strip. This process is continued until the entire sector has been searched. In the city core or core periphery where the observer is faced with multistory buildings, the overlapping sectors may extend up rather than out.

j. Marines who man OPs and other positions should employ target acquisition devices. These devices include binoculars, image intensification devices, and thermal sights. All of these devices can enhance the unit's ability to detect and engage targets, especially at night or during periods of reduced visibility.

k. Target acquisition techniques used at night are similar to those used during the day. At night, whether using daylight optics or the unaided eye, the observer should not look directly at an object but a few degrees off to the side. The side of the eye is more sensitive to dim light. When scanning with off-center vision, the Marine moves his eyes in short, abrupt, irregular movements. At each likely target area, he pauses for a few seconds to detect any motion.

l. At night, sounds and smells can aid in acquiring targets. They transmit better in the cooler, damper night air. Running engines, vehicles, and people moving through rubble- covered streets can be heard for great distances. Odors from diesel fuel, gasoline, cooking food, burning tobacco and aftershave lotion can reveal enemy and friendly locations.

29. Firefighting Planning and Operations. Incendiary ammunition, special weapons, and the ease with which incendiary devices can be constructed from gasoline and other flammables make fire a true threat in operations in built-up areas. During defensive operations, firefighting should be a primary concern. Proper steps should be taken to reduce the risk of a fire that could make a chosen position indefensible.

a. Marines choose or create positions that do not have large openings. These positions provide as much built-in cover as possible to prevent penetration by incendiary ammunition. All unnecessary flammable materials are removed, including ammunition boxes, furniture, rugs, newspapers, curtains, and so on. The electricity and gas coming into the building should be shut off.

b. A building of concrete block construction with concrete floors and a tin roof is an ideal place for a fighting position. However, most buildings have wooden floors or subfloors, wooden rafters, and wooden inner walls that may require improvement. Inner walls are removed and replaced with blankets to resemble walls from the outside. Sand is spread 2 inches deep on floors and in attics to retard fire.

c. All available firefighting gear is prepositioned so that it can be used during combat. For the individual Marine, such gear includes entrenching tools, helmets, sand, and blankets. These items are supplemented with fire extinguishers.

d. Fire is so destructive that it can easily overwhelm personnel regardless of extraordinary precautions. Marines should plan routes of withdrawal so that evacuation can be accomplished from their fighting positions. This allows Marines to exit through areas that are free of combustible material but that still provide cover from enemy direct fire.

e. The confined space and large amounts of combustible material found in built-up areas may influence the enemy to use incendiary devices. First-aid problems that are expected include burns and smoke/flame inhalation. These can easily occur in buildings, thereby rendering the victims combat ineffective. Although there is little defense against flame inhalation and lack of oxygen, smoke inhalation may be greatly reduced by wearing the individual protective mask. Regardless of the fire hazard, defensive planning for combat in built-up areas should include location of corpsmen. Corpsmen must reach victims and their equipment and should stock extra supplies for the treatment of burns and inhalation injuries.

f. Offensive operations also require plans for firefighting because the success of the mission can easily be threatened by fire. Poorly planned use of incendiary munitions can make fires so extensive that they become obstacles to offensive operations. The enemy may use fire to cover his withdrawal and to create obstacles and barriers to the attacker.

g. Every Marine participating in an attack should be ready to deal with the problems of fire. The normal firefighting equipment available includes the entrenching tool, helmet (for carrying sand or water), and blankets (for snuffing out small fires). Fire extinguishers are normally available on vehicles.

30. Employment of Snipers. The value of the sniper to a unit operating in a built-up area depends on several factors. These factors include the type of operation, the level of conflict, and the ROE. Where ROE allow destruction, snipers may not be needed because other weapons systems available to the infantry commander have a greater destructive effect. However, snipers can contribute to the fight. Where the ROE prohibit collateral damage, snipers may be the most valuable tool the commander has. (See MCWP 3-15.3 (under development).)

a. Sniper effectiveness depends in part on the terrain. Control is decentralized by the characteristics of an urban area. To provide timely and effective support, the sniper should have a clear picture of the commander's mission and intent.

b. Where possible, snipers should be positioned in masonry buildings. These buildings should also offer long-range fields of fire and all-around observation. The sniper has an advantage because he does not have to move with or be positioned with lead elements. He may occupy a higher position to the rear or flanks of, and some distance away from, the element that he is supporting. By operating far from the other elements, a sniper avoids decisive engagement but remains close enough to kill distant targets that threaten the unit. Snipers should not be placed in obvious positions such as church steeples and rooftops because the enemy often watches these and targets them for destruction. Also, snipers should not be positioned where there is heavy traffic; these areas also invite enemy observation.

c. Snipers should operate throughout the AO, moving with and supporting the companies as necessary. Some teams may be designated to operate independently of other forces. They may be tasked to search for targets of opportunity, especially for enemy snipers. The sniper team may occupy multiple positions. A single position may not afford adequate observation for the entire team without increasing the risk of detection by the enemy. Separate positions should maintain mutual support. Alternate and supplementary positions should also be established in urban areas.

d. Snipers may be assigned the following tasks:

- Killing enemy snipers (countersniper fire)

- Killing targets of opportunity (These targets should be prioritized by the commander. Types of targets might include enemy snipers, leaders, vehicle commanders, radio men, sappers, and machine gun crews.)
- Denying enemy access to certain areas or avenues of approach (controlling key terrain)
- Providing fire support for covering barricades and other obstacles
- Maintaining surveillance of flank and rear avenues of approach (screening)
- Supporting local counterattacks with precision fire.

Section VII

Advancing/Patrolling Along City Streets

31. Advancing/Patrolling Along City Streets. During MOUT operations, skills for advancing/ patrolling along city streets will be required. The platoon, squad, or fire team mission may be such that building-to-building clearing is neither required nor desired until contact with the enemy has been made. This section will discuss advancing/patrolling along urban streets. The standard tactics, techniques, and procedures for the conduct of a patrol, as outlined in MCWP 3-11.6 *Scouting and Patrolling* (draft), apply except as otherwise discussed within this section.

a. Organization. The platoon's three-squad organization or the squad's three-fire team organization is already structured to conduct an advance/patrol down an urban street. The typical patrol is organized into assault, support, security, and command and control elements. Security is divided into front, flank, and rear security responsibilities. The platoon or squad can move down the street in a double or single column.

b. Formation and Movement

(1) Double Column. The double column is the *preferred method* for moving along an urban street (Figures A-68 and A-69). The double column provides 360 degrees of security and allows for mutual support and interlocking sectors of fire. The double column is flexible but can make control more difficult. There should be at least two squads or fire teams to conduct a double column. The two squads or fire teams move down the sides of the street using the walls of the buildings for cover and concealment. The pointmen for the two columns should always stay abreast of each other and not move in front of or behind the opposite column. The Marines following in trace of the pointmen stagger themselves accordingly. The platoon or squad leaders move where they can best control their units. If a squad is conducting the advance/patrol along the street, the trail fire team is never split.

(a) Sectors of Responsibility. Every Marine moving in a column has a particular sector for whose security he is responsible. The pointmen are responsible for security to the direct front and for pieing off windows and doorways. They are also responsible for staying in line with one another.

The Marines immediately behind the pointmen are designated as the covermen. They are responsible for providing guidance to their respective pointmen from the leader of the unit and for covering the 45-degree oblique angle across from one another. The second coverman is also responsible for covering the direct front when the pointman is pieing and checking for the enemy. The third and fourth covermen in the column are responsible for the far-side flank security; one covers the upper level and one covers the lower level building windows, doors, and stairwells. Marines will alternate providing security on upper level and lower level windows, doors, and stairwells. The fourth Marine also provides rear security if he is the last man in the column or interlocks his sector of fire with that of the lead Marine in the next fire team in the column. Should the second coverman have to take the point while the pointman pies a window or door, the third coverman would automatically shift his sector of fire to the 45-degree oblique angle. If the second coverman has to assist the pointman in pieing a large opening, the third coverman would provide security to the direct front while the fourth coverman would take the 45-degree oblique angle. Marines following in the column would adjust their sectors of fire accordingly.

The two columns provide mutual support for each other and the greatest all-around security. Teamwork and coordination between the two columns are vital.

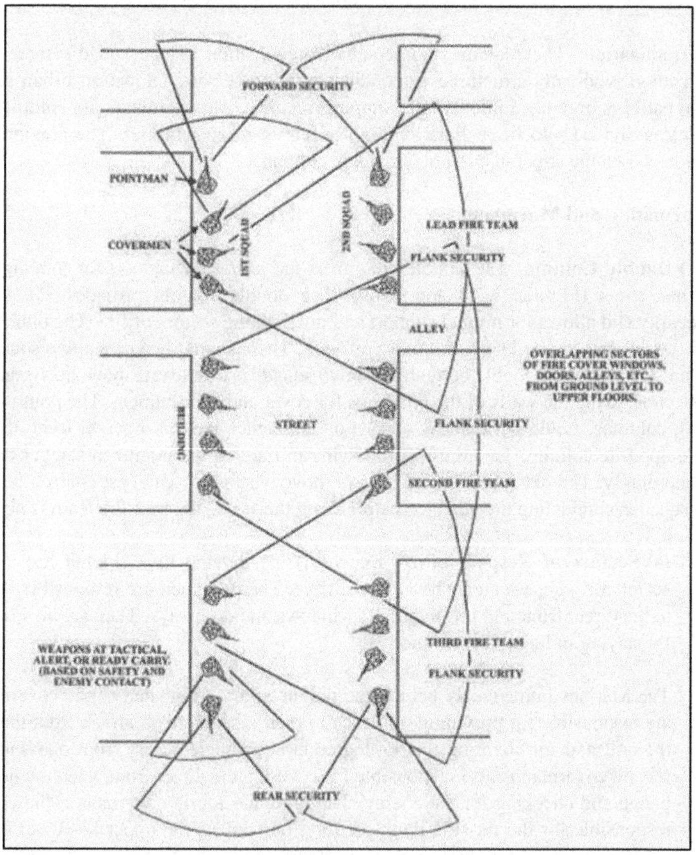

Figure A-68. Advancing/Patrolling Along an Urban Street (Squad, Double Column)

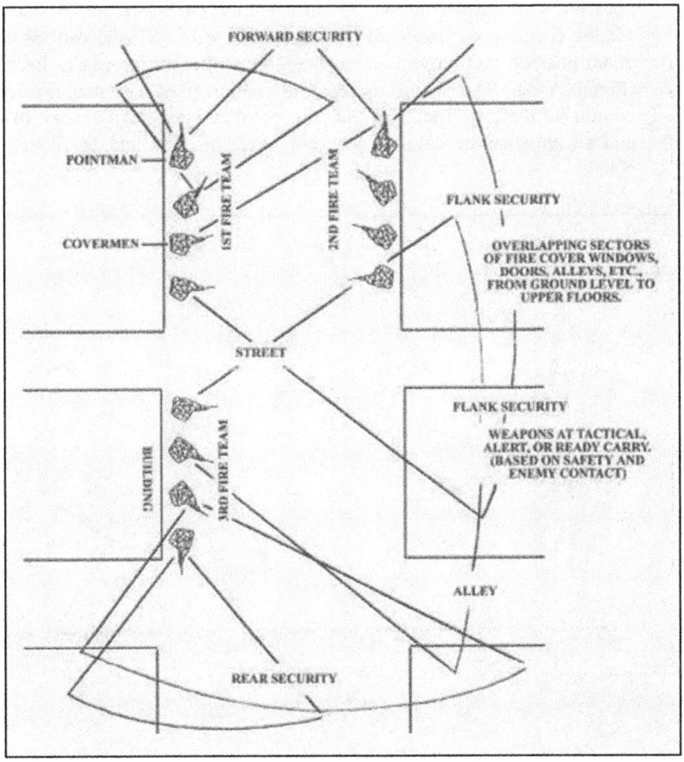

Figure A-69. Advancing/Patrolling Along an Urban Street (Fire Team, Double Column)

(b) Rear Security. The last two Marines in the squad of each column are responsible for the rear security of each of their formations. They will move in bounds in order to allow a Marine to always provide active security to the squad's rear (Figures A-70 and A-71). Two methods of bounding may be used: *alternate* or *successive bounding*.

1 **Alternate Bounding.** In this method, the two rear-security Marines alternate being the last Marine in the squad. The last Marine in the formation will have his weapon at the ready and will be in the kneeling position along the wall of the building. The second to last Marine will move in trace of the unit until he has moved approximately 10 to 15 meters. At this point, he will step out one pace from the wall. He will then assume a good kneeling firing position and announce, "**SET**." The last Marine will rise from his position, turn around, and move next to the wall to catch up with the column. It is important that the Marine moving back

does not step out and mask the other Marine's field of fire. At this point, the Marine covering the rearward movement of the last Marine can assume the rear guard position and move back to the wall as the Marine passes by his position (Figure A-70). The process is repeated continuously. The rear fire team leaders coordinate the movements of the rear guard to keep them abreast of each other and ensure that the column does not move too fast and leave the rear guard behind.

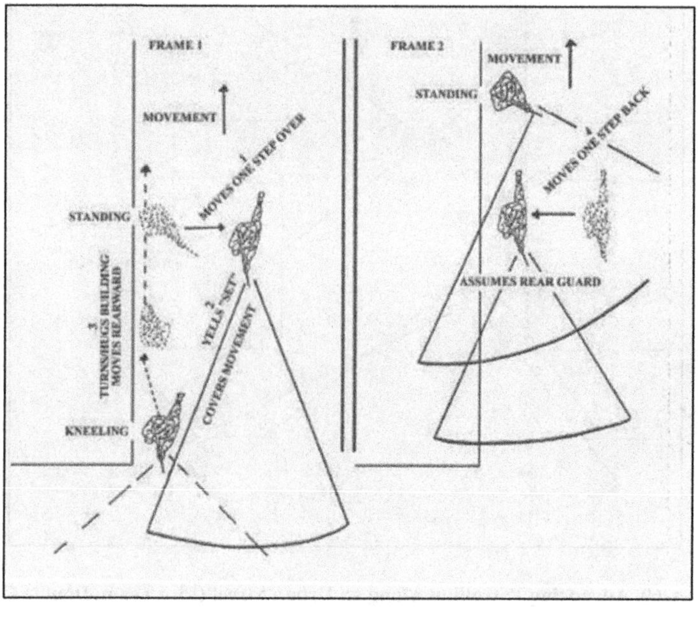

Figure A-70. Rear Security (Alternate Bounds)

2 **Successive Bounding.** In this method, the last Marine in the squad continues to be the rear guard after bounding. The same procedures are employed as in alternate bounding; however, the rear-guard Marine bounds abreast of the Marine providing cover, turns around, drops to a good kneeling position, provides rear-guard security, and announces, "**SET**." The covering Marine moves behind the rear-guard Marine and continues to move 10 to 15 meters before repeating the process (Figure A-71). The rear fire team leaders coordinate the movements of the rear guard as in alternate bounding.

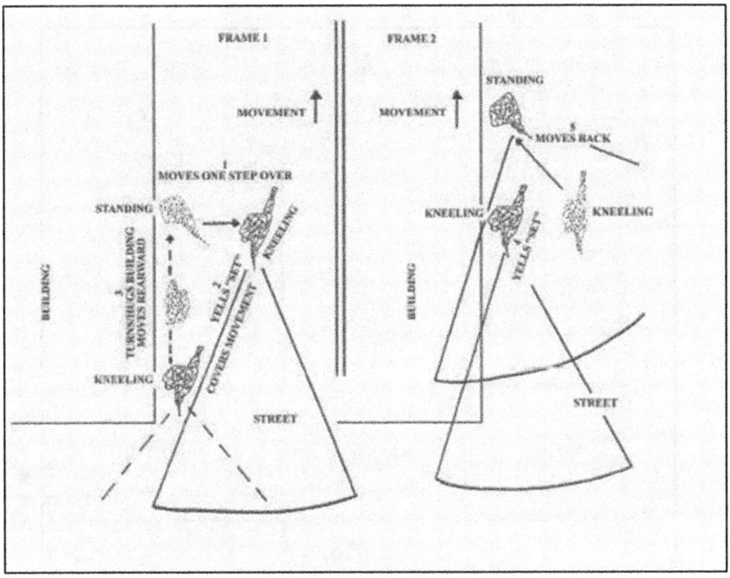

Figure A-71. Rear Security (Successive Bounds)

(2) Single Column. The single column is used to move rapidly along a street or when streets are restricted to a single column. The single column is the least preferred advance/patrol method of moving along an urban street. The same principles apply as in the double column, except the single column does not provide 360-degree security. The single column is normally used when speed is desired and contact with the enemy is considered to be unlikely.

32. Crossing a Street Intersection. Four-way intersections are very dangerous areas that should be crossed quickly. Security at an intersection is extremely important for the protection of crossing units. When moving, the pointmen notify the unit leaders, through the covermen, of an approaching intersection. Crossing the intersection will force the three elements of a unit moving along a city street to rotate positions and assume appropriate responsibilities.

a. First Fire Team To Cross. The lead (1st) fire teams in the two columns will simultaneously set security around the corners of the building. They will provide mutually supporting security to the direct front, security to the building on the far side of the intersection, and intersection security (Figure A-72). The middle (2nd) fire teams provide mutually supporting flank and rear security. On order, the rear (3rd) fire teams move forward, cross the intersection, and assume the lead (1st) fire teams' forward security mission.

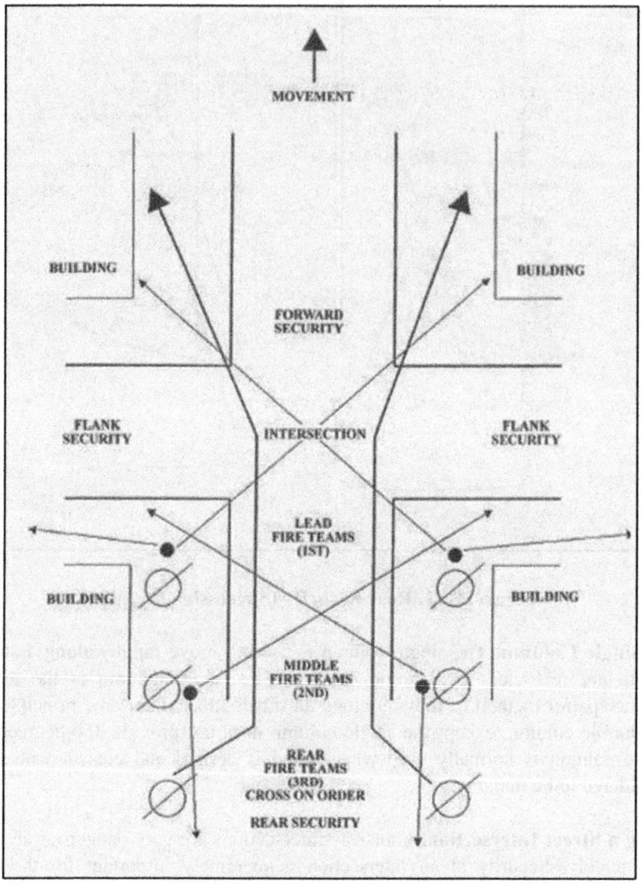

Figure A-72. Crossing an Intersection (Frame 1)

b. Second Fire Team to Cross. The 1st fire teams (now the middle fire teams) continue to provide intersection security and assume rear area security. The 2nd fire teams (now rear fire teams) cross the intersection on order. The second fire teams position themselves to provide rear and intersection security to support the covering of the last fire teams (Figure A-73).

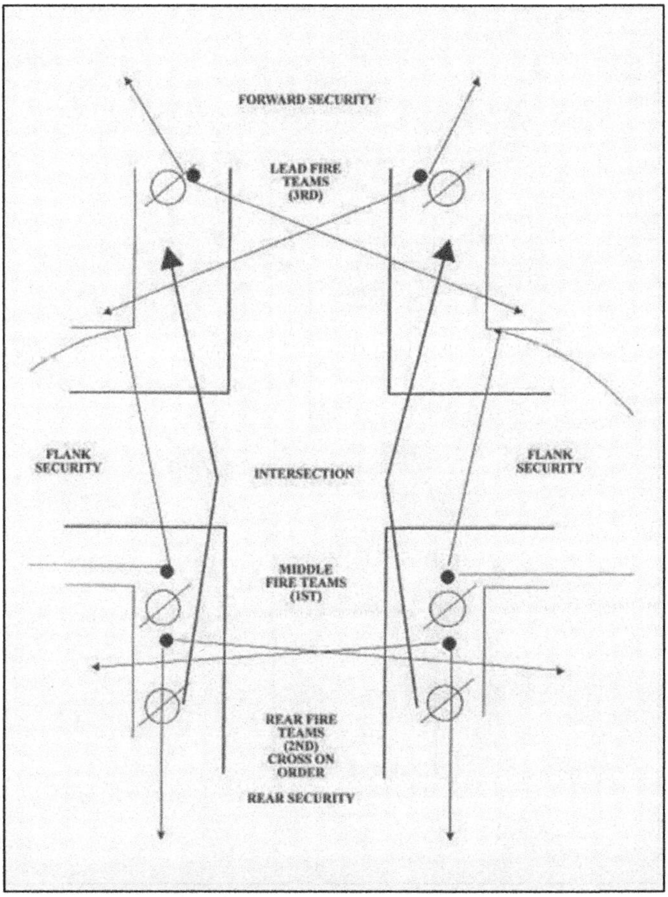

Figure A-73. Crossing an Intersection (Frame 2)

c. **Third Fire Team To Cross.** The 1st fire teams (now the rear fire teams) provide rear security and cross the intersection on order. The 1st fire teams bypass the 2nd fire teams to become the middle fire teams and provide mutually supporting forward and flank security. The 2nd fire teams become the rear security (Figure A-74).

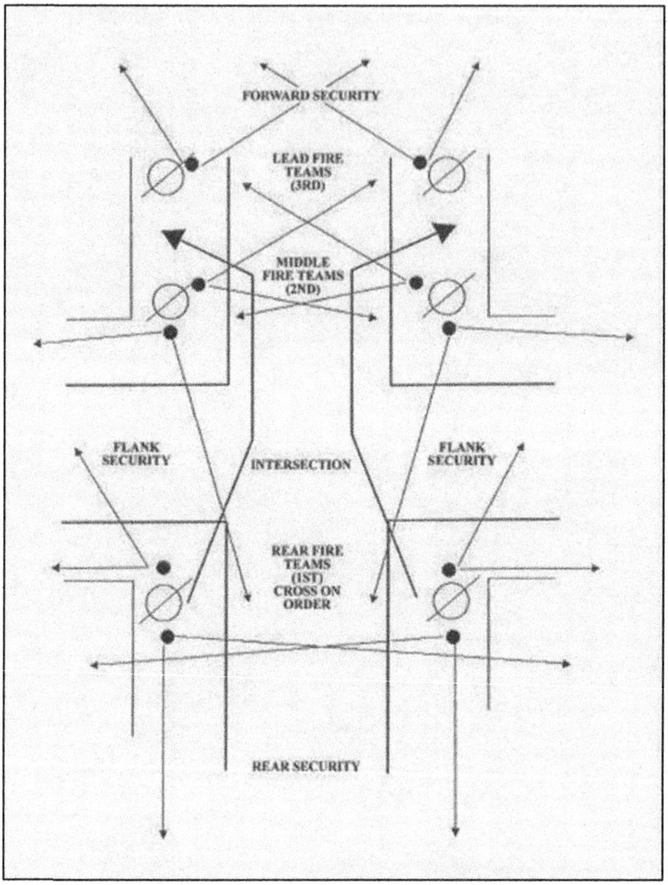

Figure A-74. Crossing an Intersection (Frame 3)

d. The squads and three fire teams have now crossed the intersection, and the fire teams have rotated (Figure A-75). This rotation is convenient because it alternates the lead units, who may quickly become fatigued during MOUT operations. The same procedures are used to cross an intersection with a single column.

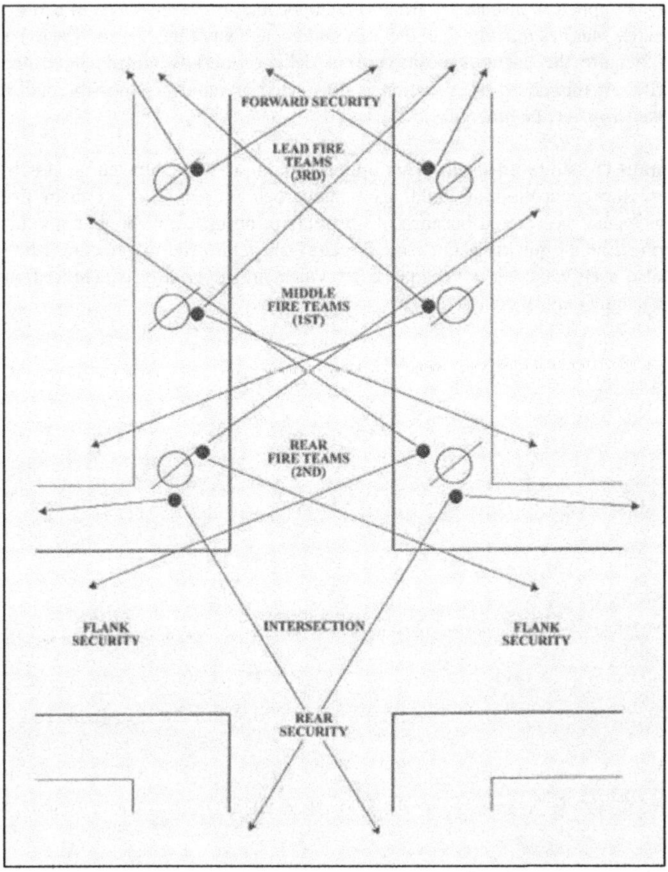

Figure A-75. Crossing an Intersection (Frame 4)

33. Reacting to Enemy Contact. When a platoon or squad is moving along a city street, contact with the enemy could happen at any time, even if the area has been declared secure. If a unit engages or is engaged by the enemy, there are two basic options that Marines may follow. The first and most preferred option is to immediately return fire and conduct a hasty clearing of a structure to seek cover inside of the building to which the unit was adjacent. The second and least preferred method is to remain outside of the building(s) and fight from the street.

a. Hasty Clearing. To conduct a hasty clearing of a structure in order to seek cover, Marines immediately return fire and enter buildings as quickly as possible. The basic principles of entry are applied as discussed earlier in the appendix; however, speed of action is essential. Once inside, Marines quickly scan the area and engage any threat. If no threat is present, the Marines acquire the enemy's positions and deliver carefully aimed shots to achieve fire superiority. At this point, the situation is developed as rapidly as possible and an attack to clear the enemy may be ordered.

b. Remain Outside and Fight. This option should not be conducted unless strict ROE do not allow the occupation of buildings without being fired on from them first or unless buildings cannot be entered because of obstruction, obstacles, or boobytraps. In such cases, the Marines should quickly acquire the target(s) and return fire to gain fire superiority. They should also seek whatever cover may exist. Once fire superiority is achieved, an assault to clear the enemy may be conducted.

Section VIII

Navigation in Built-Up Areas

Navigation in built-up areas present a unique set of challenges. Deep in the city core, the normal terrain features depicted on maps may not apply—buildings become the major terrain features, and units become tied to streets. Fighting in the city destroys buildings, whose rubble then blocks streets. Street and road signs are destroyed during the fighting if they are not switched or removed by the defenders. Operations in subways and sewers present other unique challenges. However, maps and photographs are available to help units overcome these problems. Additionally, the global positioning system (GPS) can help supplement navigational abilities in built-up areas.

34. Military Maps. The military city map is a topographical map of a city that delineates streets and shows street names, important buildings, and other urban elements. The scale of a city map can vary from 1:25,000 to 1:50,000 to 1:100,000 depending on the importance and size of the city, the density of detail, and intelligence information available.

a. Special maps prepared by supporting topographical engineers can assist units in navigating in built-up areas. These maps are designed or modified to provide information that is not covered on a standard military map. This may include maps of road and bridge networks, railroads, built-up areas, and electric power fields. Special maps can be used to supplement military city maps and topographical maps.

b. Once in the built-up area, Marines may use street intersections as reference points as they would use hills and streams in rural terrain. City maps supplement or replace topographical maps as the basis of navigation. These maps enable units moving in the built-up area to know where they are and to move to new locations even though streets have been blocked or a key building has been destroyed.

c. The techniques of compass reading and pace counting can still be used, especially in a blacked-out city where street signs and buildings are not visible. The presence of steel and iron in the MOUT environment may cause inaccurate compass readings. Sewers may be navigated in much the same way. Maps providing the basic layout of the sewer system are maintained by city sewer departments. This information includes directions in which the sewer lines run and distances between manhole covers. Along with basic compass and pace-count techniques, such information enables a unit to move through the city sewers with accuracy.

d. Operations in a built-up area adversely affect the performance of some types of communications-electronic devices such as the GPS, the Position Location Reporting System (PLRS), or other data-distribution systems. These systems function the same as some communications equipment—by LOS. They cannot determine underground locations or positions within a building. These systems should be employed on the tops of buildings, in open areas, and down streets where obstacles will not affect LOS readings.

e. City utility workers are assets to units fighting in built-up areas. They can provide maps of sewers and electrical fields and information about the city. This is especially important with regard to the use of the sewers. Sewers can contain pockets of methane gas that are highly toxic to humans. City sewer workers know the locations of these danger areas and can advise a unit on how to avoid them.

35. Aerial Photographs. Current aerial photographs are also excellent supplements to military city maps and can be substituted for a map. A topographical map or military city map could be obsolete if compiled many years ago. A recent aerial photograph shows changes that have taken place since the map was made. This could include destroyed buildings and streets that have been blocked by rubble as well as enemy defensive preparations. More information can be gained by using aerial photographs and maps together than by using either one alone.

Section IX

Camouflage

To survive and win in combat in built-up areas, a unit should supplement cover and concealment with camouflage. To properly camouflage men, carriers, and equipment, Marines should study the surrounding area and make their fighting positions blend with the local terrain.

36. Application. Only the proper amount of material needed for camouflaging a position should be used. Excess material could reveal the position. Material should be obtained from a wide area around the position. For example, Marines defending a cinder-block building do not strip the front, sides, or rear of the building to camouflage a position.

a. Buildings provide numerous opportunities for establishing concealed positions. Armored vehicles can often find prospective positions under archways or inside small industrial or commercial structures. Thick masonry, stone, or brick walls offer excellent protection from direct fire and may provide concealed routes into or out of the position.

b. After camouflage is completed, the Marine inspects his position from the enemy's viewpoint. He makes routine checks to see whether the camouflage remains natural looking and actually conceals the position. If it does not look natural, the Marine should rearrange or replace it until a natural look is achieved.

c. Positions should be progressively camouflaged as they are prepared. Work should continue until all camouflage is complete. If the enemy has air superiority, work may be possible only at night. Shiny or light-colored objects that attract attention from the air should be hidden.

d. Camouflage face paint is issued in three standard, two-tone sticks. When issue-type face-paint sticks are not available, burnt cork, charcoal, or lampblack can be used to tone down exposed skin. Mud may be used as a last resort because it dries and may peel off, leaving the skin exposed, and may also contain harmful bacteria.

37. Use of Shadows. Buildings throw sharp shadows that can be used to conceal vehicles and equipment (Figure A-76). Marines should avoid movement or positions in areas that are not in shadows. Vehicles may have to be moved periodically as shadows shift during the day. Emplacements inside buildings provide better concealment and cover as well.

38. Color and Texture. Standard camouflage-pattern painting of equipment is not as effective in built-up areas as a solid, dull, dark color hidden in shadows. Because repainting vehicles before entering a built-up area is not always practical, the lighter sand-colored patterns should be subdued with mud or dirt.

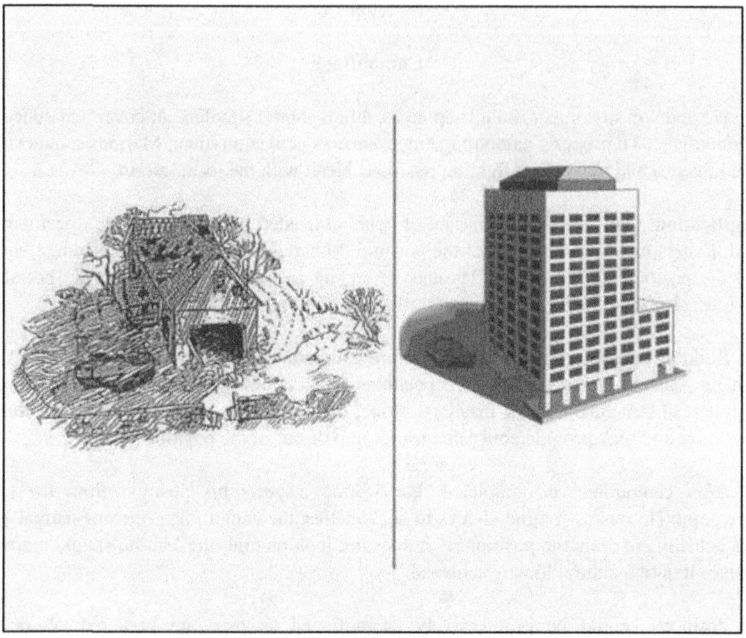

Figure A-76. Use of Shadows for Concealment

a. The need to break up the silhouette of helmets and individual equipment exists in built-up areas the same as it does elsewhere. However, burlap or canvas strips are a more effective camouflage than foliage (Figure A-77). Predominant colors are normally browns, tans, and sometimes grays rather than greens, but each camouflage location should be evaluated to determine ideal colors and patterns.

Figure A-77. Helmet Camouflaged With Burlap Strips

b. CPs and logistics emplacements are easier to camouflage and better protected if located underground. Antennas can be remoted to upper stories or to higher buildings based on remote capabilities. Field telephone wire should be laid in conduits, in sewers, or through buildings.

c. Marines should consider the background to ensure that they are not silhouetted or skylined, but rather blend into their surroundings. To defeat enemy urban camouflage, Marines should be alert for common camouflage errors such as:

- Tracks or other evidence of activity
- Shine or shadows
- Unnatural color or texture
- Muzzle flash, smoke, or dust
- Unnatural sounds and smells
- Movement.

d. Dummy positions can be used effectively to distract the enemy and make him reveal his position by firing on them.

e. Built-up areas afford cover, resources for camouflage, and locations for concealment. Following are the basic rules of cover, camouflage, and concealment to which Marines should adhere:

(1) Use the terrain and alter camouflage habits to suit your surroundings.

(2) Employ deceptive camouflage of buildings.

(3) Continue to improve positions. Reinforce fighting positions with sandbags or other fragment- and blast-absorbent material.

(4) Maintain the natural look of the area.

(5) Keep positions hidden by clearing away minimal debris for fields of fire.

(6) Choose firing ports in inconspicuous spots when available.

Note: Remember that a force that *covers* and *conceals* itself has a significant advantage over a force that does not.

Appendix B

Employment and Effects of Weapons

Success against the enemy in urban operations to a large extent depends on the effective employment of weapons. Marines at all levels must understand the effects that their weapons will produce in urban combat.

1. Effectiveness of Weapons and Demolitions. The characteristics and nature of combat in built-up areas affect the results and employment of weapons. Leaders at all levels must consider the following factors in various combinations when choosing weapons to employ.

a. Hard, smooth, flat surfaces are characteristic of urban targets. Rounds rarely impact perpendicular to these flat surfaces, but rather tend to impact at some angle of obliquity. This reduces the effect of a round and increases the threat of ricochets. The tendency of rounds to strike glancing blows against hard surfaces means that up to 25 percent of impact-fuzed explosive rounds may not detonate when fired into rubbled areas.

b. Engagement ranges in MOUT are close. Studies and historical analyses have shown that only 5 percent of all targets are more than 100 meters away. About 90 percent of all targets are located 50 meters or less from the identifying Marine. Few personnel targets will be visible beyond 50 meters, and most occur at 35 meters or less. Minimum arming ranges and unit safety from backblast or fragmentation effects must be considered.

c. Engagement times are short because of the close fight. Enemy personnel present only fleeting targets. Enemy-held buildings or structures are normally covered by mutually supporting fires and often cannot be engaged with deliberate, well-aimed shots without prolonging exposure to their observation and weapons.

d. Depression and elevation limits for some weapons create dead space. Tall buildings form deep canyons that are often safe from indirect fires. Some weapons, such as the M203, can fire rounds to ricochet behind cover and inflict casualties. Target engagement from oblique angles, both horizontal and vertical, demands superior marksmanship skills.

e. Smoke from burning buildings, dust from explosions, shadows from tall buildings, and the lack of light penetrating inner rooms all combine to reduce visibility and to increase a sense of isolation. Added to this is the masking of fires caused by rubble and manmade structures. Targets, even at close range, tend to be indistinct.

f. Urban fighting often leads to confused melees with several small units attacking on converging axes. The risks from friendly fires, ricochets, and fratricide must be considered during the planning phase of operations, and control measures must be continually adjusted to lower these risks. Marines and leaders must maintain a sense of situational awareness and clearly mark their progress in accordance with unit SOPs to avoid fratricide.

g. The firer and target may each be inside or outside the same or separate buildings. The enclosed nature of combat in built-up areas means that a weapon's signature effect, such as a muzzle blast or backblast, must be considered, as well as the round's impact on the target.

h. Usually the manmade structure that is protecting the enemy must be attacked before enemy personnel inside can be engaged. Therefore, weapons and demolitions can be chosen for employment based on their effects against masonry and concrete rather than against enemy personnel.

i. Modern engineering and design improvements mean that most large buildings constructed since World War II are resilient to the blast effects of bomb and artillery attack. Even though modern buildings may burn easily, they often retain their structural integrity and remain standing. Once high-rise buildings burn out, they are still useful to the military and are almost impossible to damage further. A large structure can take 24 to 48 hours to burn out and become cool enough for people to enter.

j. The most common worldwide building type is the 12- to 24-inch brick building. Table B-1 lists the frequency of occurrence of building types worldwide.

Type of Building	Frequency of Occurrence (Percentage)
30-inch stone	1
8- to 10-inch reinforced concrete	6
12- to 24-inch brick	63
6-inch wood	16
14-inch steel and concrete (heavy clad)	2
7-inch steel and concrete (light clad)	12

Table B-1. Types of Buildings and Frequency of Occurrence

2. M16 Rifle and M249 Squad Automatic Weapon. The M16A2 rifle and the M249 squad automatic weapon are the most common weapons used in combat in built-up areas. The M16A2 rifle and the M249 are used to kill enemy personnel, to suppress enemy fire and observation, and to penetrate light cover. Leaders can use 5.56-mm tracer fire to designate targets for other weapons.

a. Employment. Close combat is the predominant characteristic of urban engagements. Marine riflemen must be able to hit small, fleeting targets from bunker apertures, windows, and loopholes. This requires pinpoint accuracy with weapons fired in the semiautomatic mode. Killing an enemy through an 8-inch loophole at a range of 50 meters is a challenge, but one that may be common in combat in built-up areas.

(1) When fighting and engaging targets inside buildings, the weapons handling and firing techniques discussed in Appendix A are used.

(2) Within built-up areas, burning debris, reduced ambient light, strong shadow patterns of varying density, and smoke all limit the effectiveness of night vision and sighting devices. Aiming stakes and noise/light boobytraps coupled with night vision goggles and firing techniques can be used in the defense. During the offense, night vision goggles and illumination munitions can be used. Any Marine using night vision goggles should be teamed with at least one Marine not wearing them.

b. Weapon Penetration. The penetration that can be achieved with a 5.56-mm round depends on the range to the target and the type of material being fired against. Single 5.56-mm rounds are not effective against structural materials (as opposed to partitions) when fired at close range—the closer the range, the less the penetration.

(1) For the 5.56-mm round, maximum penetration occurs at 200 meters. At ranges of less than 25 meters, penetration is greatly reduced. At 10 meters, penetration by the M16 round is poor as a result of the tremendous stress placed on this high-speed round, which causes it to yaw upon striking a target. Stress causes the projectile to break up, and the resulting fragments are often too small to penetrate.

(2) Even with reduced penetration at short ranges, interior walls made of thin wood paneling, sheetrock, or plaster offer no protection against 5.56-mm rounds. Common office furniture such as desks and chairs cannot stop these rounds, but a layer of books 18 to 24 inches thick can.

(3) Wooden-framed buildings and single cinder-block walls offer little protection from 5.56-mm rounds. When clearing such structures, Marines must ensure that friendly casualties do not result from rounds passing through walls, floors, or ceilings.

(4) Armor-piercing rounds are slightly more effective than ball ammunition in penetrating urban targets at all ranges. They are also more likely to ricochet than ball ammunition, especially when the target presents a high degree of obliquity.

c. Protection. The following common barriers in built-up areas stop a 5.56-mm round fired at less than 50 meters:

- One thickness of sandbags
- A 2-inch concrete wall (unreinforced)
- A 55-gallon drum filled with water or sand
- A small ammunition can filled with sand

- A cinder block filled with sand (block will probably shatter)
- A plate-glass windowpane at a 45-degree angle (glass fragments will be thrown behind the glass)
- A brick veneer
- A car body (5.56-mm round will penetrate but normally not exit).

d. Wall Penetration. Although most structural materials repel single 5.56-mm rounds, continued and concentrated firing can breach some typical urban structures (See Table B-2).

Type	Penetration	Rounds Required
8-inch reinforced concrete	Initial Loophole	35 250
14-inch triple brick	Initial Loophole	90 160
12-inch cinder block with single-brick veneer	Loophole Breach hole	60 250
9-inch double brick	Initial Loophole	70 120
16-inch tree trunk or log wall	Initial[1]	1 to 3
12-inch cinder block (filled w/sand)	Loophole	35
24-inch double sandbag wall	Initial[1]	220
3/8-inch mild steel door	Initial[1]	1

[1]Penetration only, no loophole

Table B-2. Structure Penetration Capabilities of the 5.56-mm Round Against Typical Urban Targets (Range 25 to 100 Meters)

(1) The best method for breaching a masonry wall is by firing short bursts (3 - 5 rounds) in a U-shaped pattern. The distance from the gunner to the wall should be minimized for best results—ranges as close as 25 meters are relatively safe from ricochet. Ballistic eye protection, a protective vest, and a helmet should be worn.

(2) Ball ammunition and armor-piercing rounds produce almost the same results, but armor-piercing rounds are more likely to fly back at the firer. The 5.56-mm round can be used to create either a loophole (about 7 inches in diameter) or a breach hole (large enough for a man to enter). When used against reinforced concrete, the M16 rifle and M249 cannot cut the reinforcing bars.

3. Medium/Heavy Machine Guns and Sniper Rifles (7.62 mm and .50 caliber)

a. Machine Guns. In the urban environment, the .50-caliber machine gun and the 7.62-mm M240G machine gun provide high-volume, long-range, automatic fires for the suppression or destruction of targets. They provide final protective fires along fixed lines and can be used to penetrate light structures; the .50-caliber machine gun is most effective in this role. Tracers from both machine guns are likely to start fires, but the .50-caliber tracer is more apt to do so.

(1) Employment. The primary consideration affecting the employment of machine guns within built-up areas is the limited availability of long-range fields of fire. Although machine guns should be emplaced at the lowest level possible, grazing fire at ground level is often obstructed by rubble.

(a) The .50-caliber machine gun is often employed on its vehicular mount during both offensive and defensive operations. If necessary, it can be mounted on the M3 tripod for use in the ground role or in the upper level of buildings. When mounted on a tripod, the .50-caliber machine gun can be used as an accurate, long-range weapon and can supplement sniper fires.

(b) The M240G machine gun is cumbersome, making it difficult to use inside while clearing a building. However, it is useful outside to suppress and isolate enemy defenders. The M240G can be fired from either the shoulder or the hip to provide a high volume of assault and suppressive fires. The use of the long sling to support the weapon and ammunition is preferred.

(c) Because of their reduced penetration power, M240G machine guns are less effective against masonry targets than .50-caliber machine guns. However, their availability and light weight make them well suited to augment heavy machine gun fire, to be used in areas where .50-caliber machine guns cannot be positioned, or as a substitute when heavy machine guns are not available. The M240G machine gun can be employed on its tripod to deliver accurate fire along fixed lines and can then quickly be converted to bipod fire to cover alternate fields of fire.

(2) Penetration. The ability of the 7.62-mm and .50-caliber rounds to penetrate is also affected by the range to the target and type of material against which the rounds are fired. Close ranges affect penetration of the 7.62-mm round less than the 5.56-mm round; the .50-caliber round's penetration is reduced the least.

(a) At 50 meters, the 7.62-mm ball round cannot penetrate a single layer of sandbags. It can penetrate a single layer at 200 meters, but not a double layer. The armor-piercing round does only slightly better against sandbags. It cannot penetrate a double layer but can penetrate up to 10 inches of sand at 600 meters.

(b) The penetration of the 7.62-mm round is best at 600 meters, but most urban targets are closer. The longest engagement range is usually 200 meters or less. Table B-3 lists the penetration capabilities of a single 7.62-mm (ball) round at closer ranges.

Range (meters)	Pine Board (inches)	Dry, Loose Sand (inches)	Cinder Block (inches)	Concrete (inches)
25	13	5	8	2
100	18	4.5	10	2
200	41	7	8	2

Table B-3. Penetration Capabilities of a Single 7.62-mm (Ball) Round

(c) The .50-caliber round is also optimized for penetration at long ranges (about 800 meters). For hard targets, .50-caliber penetration is affected by obliquity and range. Both armor-piercing and ball ammunition penetrate 14 inches of sand or 28 inches of packed earth at 200 meters if the rounds impact perpendicular to the flat face of the target. Table B-4 depicts the effect of a 25-degree obliquity on a .50-caliber penetration.

Thickness (feet)	100 Meters (rounds)	200 Meters (rounds)
2	300	1,200
3	450	1,800
4	600	2,400

Table B-4. Number of Rounds Needed To Penetrate a Reinforced Concrete Wall at a 25-Degree Obliquity

(3) Protection. Barriers that offer protection against 5.56-mm rounds are also effective against 7.62-mm rounds with some exceptions. The 7.62-mm round can penetrate a windowpane at a 45-degree obliquity, a hollow cinder block, or both sides of a car body. It can also easily penetrate wooden-framed buildings. The .50-caliber round can penetrate all of the commonly found urban barriers except a sand-filled 55-gallon drum.

(4) Wall Penetration. Continued and concentrated machine gun fire can breach most typical urban walls. Such fire cannot breach thick reinforced concrete structures or dense natural stone walls. Internal walls, partitions, plaster, floors, ceilings, common office furniture, home appliances, and bedding can be easily penetrated by both 7.62-mm and .50-caliber rounds (Tables B-5 and B-6).

(a) The M240G machine gun can be hard to hold steady enough to repeatedly hit the same point on a wall. The dust created by the bullet strikes also makes precise aiming difficult. Firing with a tripod is usually more effective than without, especially if sandbags are used to steady the weapon. Short bursts of three to five rounds fired in a U-type pattern are best.

(b) Breaching a brick veneer over cinder block presents a special problem for the M240G machine gun. Rounds penetrate the cinder block but leave a net-like structure of unbroken block. Excessive ammunition is required to destroy the net because most rounds only pass through a previously eroded hole. One or two minutes of work with an entrenching tool, crowbar, or ax can remove this web and allow entry through the breach hole.

(c) The .50-caliber machine gun can be fired accurately from the tripod using the single-shot mode. This is the most efficient method for producing a loophole. Automatic fire in three- to five-round bursts in a U-type pattern is more effective in producing a breach.

Type	Thickness (Inches)	Hole Diameter (Inches)	Rounds Required
Reinforced concrete	8	7	100
Triple brick wall	14	7	170
Concrete block with single brick veneer	12	6 and 24	30 and 200
Cinder block (filled)	12	N/A[1]	18
Double brick wall	9	N/A[1]	45
Double sandbag wall	24	N/A[1]	110
Log wall	16	N/A[1]	1
Mild steel door	3/8	N/A[1]	1

[1]Penetration only, no loophole

Table B-5. Structure Penetrating Capabilities of 7.62-mm Round (NATO) Against Typical Urban Targets (Range 25 Meters)

Type	Thickness (Inches)	Hole Diameter (Inches)	Rounds Required
Reinforced concrete	10	12	50
		24	100
	18	7	140
Triple brick wall	12	8 and 26	15 and 50
Concrete block with single brick veneer	12	10 and 33	25 and 45
Armor plate	1	N/A[1]	1
Double sandbag wall	24	N/A[1]	5
Log wall	16	N/A[1]	1

[1]Penetration only, no loophole

Table B-6. Structure Penetrating Capabilities of .50-caliber Ball Against Typical Urban Targets (Range 35 Meters)

b. Sniper Rifles. The 7.62-mm match ball ammunition fired from the M-40A1 sniper rifle will perform similarly to the 7.62-mm ball round. The M-82A1A special application sniper rifle (SASR) can fire the .50-caliber armor-piercing incendiary (API) ammunition, which will have the same performance as when fired from the M-2 HB. The M-82A1A also fires the .50-caliber RAUFOSS ammunition, which contains a tungsten penetrator and a more powerful explosive charge than the API ammunition. When the RAUFOSS penetrator punches through the metal, the explosive detonates inside, acting as an incendiary round. Although the SASR's exact penetration capabilities have yet to be defined, it has penetrated an inch of steel at 2000 yards. Both sniper rifles are organic to the Marine infantry battalion.

4. Grenade Launchers, 40-mm (M203 and MK19). Both the M203 dual-purpose weapon and the MK19 grenade machine gun fire 40-mm HE and HE dual-purpose (HEDP) ammunition. Ammunition for these weapons is not interchangeable, but the grenade and fuze assembly that actually hits the target is identical. Both weapons provide point and area destructive fires as well as suppression. The MK19 has a much higher rate of fire and a longer range. The M203 is much lighter and more maneuverable.

a. Employment. The main consideration affecting the employment of 40-mm grenades within built-up areas is the typically short engagement range. The 40-mm grenade has a *minimum arming range of 14 to 28 meters*. If the round strikes an object before it is armed, it will not detonate. Both the HE and HEDP rounds have 5-meter burst radii against exposed forces, which means that the *minimum safe firing range for combat is 33 meters*. The 40-mm grenades can be used to suppress the enemy in a building or inflict casualties by firing through apertures or windows. The MK19 can use its high rate of fire to concentrate rounds against light structures. This concentrated fire can create extensive damage. The 40-mm HEDP round can penetrate the armor on the flank, rear, and top of Soviet-made BMPs and BTRs. Marines can use the M203 from upper stories to deliver accurate fire against the top decks of armored vehicles. Multiple hits are normally required to achieve a kill.

b. Weapon Penetration. The 40-mm HEDP grenade has a small shaped charge that penetrates better than the HE round. It also has a thin wire wrapping that bursts into a dense fragmentation pattern, creating casualties out to 5 meters. Because they explode on contact, 40-mm rounds achieve the same penetration regardless of range. Table B-7 lists the penetration capabilities of the HEDP round.

Target	Penetration (Inches)
Sandbags	20
Sand-filled cinder block	16
Pine logs	12
Armor plate	2

Table B-7. Penetration Capabilities of the HEDP Round

(1) If projected into an interior room, the 40-mm HEDP can penetrate all interior partition-type walls. It splinters plywood and plaster walls, making a hole large enough to fire a rifle through. It is better to have HEDP rounds pass into a room and explode on a far wall, even though much of the round's energy is wasted penetrating the back wall (Figure B-1). The fragmentation produced in the room causes more casualties than the high-explosive jet formed by the shaped charge.

(2) The fragments from the HEDP round do not reliably penetrate interior walls. They are also stopped by office furniture, sandbags, helmets, and protective vests (flak jackets). The M203 dual-purpose weapon has the inherent accuracy to place grenades into windows at 125 meters and bunker apertures at 50 meters. These ranges are significantly reduced as the angle of obliquity increases. Combat experience shows that M203 gunners cannot consistently hit windows at 50 meters when forced to aim and fire quickly.

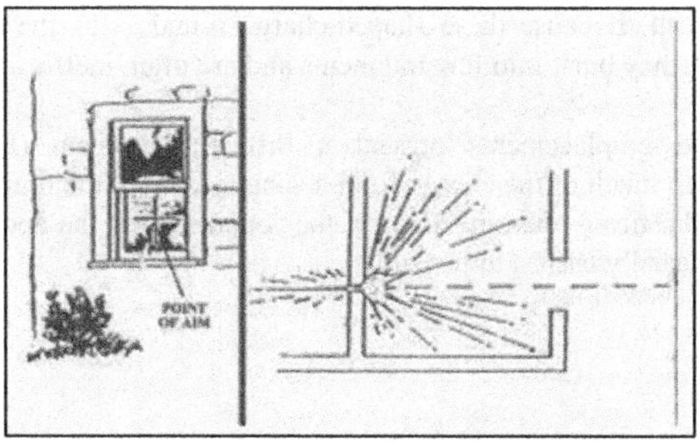

Figure B-1. Aim Point for 40-mm HEDP

c. Wall Penetration. The M203 cannot reasonably deliver the rounds needed to breach a typical exterior wall. The MK19 can concentrate its fire and achieve wall penetration. Firing from a tripod using a locked-down traversing and elevating mechanism is best for this role. Brick, cinder block, and concrete can be breached by using the MK19; individual HEDP rounds can penetrate 6 to 8 inches of brick. The only material that has proven resistant to concentrated 40-mm fire is dense stone such as that used in some European building construction. No precise data exist as to the number of rounds required to produce loopholes or breach holes with the MK19. However, the rounds' explosive effects are dramatic and should exceed the performance of the .50-caliber machine gun.

5. Light and Medium Recoilless Weapons. Light and medium recoilless weapons are used to attack enemy personnel, field fortifications, and LAVs. They have limited capability against main battle tanks, especially those equipped with reactive armor, except when attacking from the top, flanks, or rear. This category of weapons includes the AT4, the M47 Dragon, and the SMAW.

a. Employment. Other than defeating LAVs, the most common task for which light recoilless weapons are used is to neutralize fortified firing positions. Because of the design of the warhead and the narrow blast effect, these weapons are not as effective in this role as heavier weapons such as a tank main gun round. Their light weight allows Marines to carry several AT4s. Light recoilless weapons can be fired from the tops of buildings or from areas with extensive ventilation.

(1) Light and medium recoilless weapons with the exception of the SMAW employ shaped-charge warheads. As a result, the hole they punch in walls is often too small to use as a loophole. The fragmentation and spall that these weapons produce are limited. Normally, shaped-charge warheads do not neutralize enemy forces behind walls unless they are located directly in line with the point of impact.

(2) Against structures, shaped-charge weapons should be aimed about 6 inches below or to the side of a firing aperture (Figure B-2). This enhances the probability of killing the enemy behind the wall. A round that passes through a window wastes much of its energy on the back wall. Because these shaped-charge rounds lack the wire wrapping of the 40-mm HEDP, they burst into few fragments and are often ineffective casualty producers.

(3) Sandbagged emplacements present a different problem (Figure B-3). Because sandbags absorb much of the energy from a shaped charge, the rounds should be aimed at the center of the firing aperture. Even if the round misses the aperture, the bunker wall area near it is usually easiest to penetrate.

Figure B-2. Point of Aim for a Shaped-Charge Weapon Against a Masonry Structure

Figure B-3. Point of Aim for Sandbagged Emplacement

(4) Light and medium recoilless weapons obtain their most effective short-range antiarmor shots by firing from upper stories or from the flanks and rear. When firing at main battle tanks, these weapons should always be employed against these weaker areas in volley or paired firing. They normally require multiple hits to achieve a kill on a tank. Flank, top, and rear shots hit the most vulnerable parts of armored vehicles. Firing from upper stories protects the firer from tank main gun and coaxial machine gun fire because tanks cannot sharply elevate their cannons. The BMP-2 *can* elevate its 30-mm cannon to engage targets in upper stories. The BTR-series armored vehicles *can* also fire into upper stories with their heavy machine gun.

(5) Modern IFVs such as the BMP-2 and the BTR-80 have significantly improved frontal protection against shaped-charge weapons. Many main battle tanks have some form of reactive armor in addition to their thick armor plate. Head-on, ground-level shots against these vehicles have little probability of obtaining a kill. Even without reactive armor, modern main battle tanks are hard to destroy with a light antiarmor weapon.

(6) The easiest technique to improve the probability of hitting and killing an armored vehicle is to increase the firing depression angle. A 45-degree downward firing angle doubles the probability of a first-round hit as compared to a ground-level shot.

b. Backblast. Backblast characteristics must be considered when employing all recoilless weapons. During combat in built-up areas, the backblast area in the open is more hazardous because of the loose rubble and the channeling effect of the narrow streets and alleys. Figure B-4 shows the backblast areas of Marine light recoilless weapons in the open.

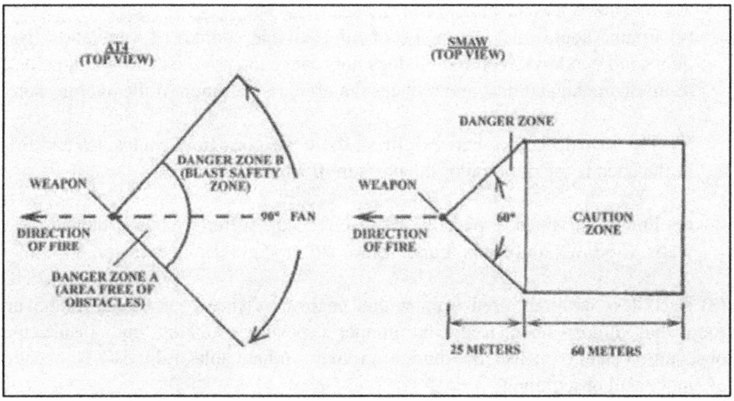

Figure B-4. Backblast Areas of Light Recoilless Weapons in the Open

(1) When firing recoilless weapons in the open, Marines should protect themselves from blast and burn injuries caused by the backblast. All personnel should be out of the danger zone. Anyone not able to vacate the caution zone should be behind cover. Marines in the caution zone should wear helmets, protective vests, and eye protection. The firer and all Marines in the area should wear earplugs.

(2) Extensive testing on the effects of firing recoilless weapons from within enclosures has been conducted since 1948. These tests showed that firing these weapons from enclosures presented no serious hazards, even when the overpressure was enough to produce structural damage to the building. The following were other findings of these tests.

(a) Little hazard exists to the gunnery or crew from any type of flying debris. Loose items were not hurled around the room.

(b) No substantial degradation occurs to the operator's tracking performance as a result of obscuration or blast overpressure.

(c) The most serious hazard that can be expected is hearing loss. This must be evaluated against the advantage gained in combat from firing from cover. To place this hazard in perspective, a gunner wearing earplugs and firing the loudest combination (the Dragon from within a masonry building) is exposed to less noise hazard than if he fired a LAW or AT4 in the open without earplugs.

(d) The safest place for other Marines in the room with the firer is against the wall from which the weapon is fired. Plastic ignition plugs are a hazard to anyone standing directly behind a LAW or TOW when it is fired.

(e) Firers should take advantage of all available sources of ventilation by opening doors and windows. Ventilation does not reduce the noise hazard, but it helps clear the room of smoke and dust and reduces the effective duration of the overpressure.

(f) The only difference between firing these weapons from enclosures and firing them in the open is the duration of the pressure fluctuation.

(g) Frame buildings, especially small ones, can suffer structural damage to the rear walls, windows, and doors. Large rooms suffer slight damage, if any.

(3) Recoilless weapons fired from within enclosures create some obscuration inside the room but almost none from the gunner's position looking out. Inside the room, obscuration can be intense, but the room remains inhabitable. Table B-8 shows the effects of smoke and obscuration.

Building	Weapon	From Gunner's Position Looking Out	Inside the Room	From Outside at a Distance
Masonry	LAW	None	Moderate	Slight smoke
	Dragon	Slight	Moderate	Small flash

Bunker	Dragon TOW	None None	Slight Slight	Moderate flash Moderate smoke
Small frame	LAW Dragon	None None	Moderate Severe	Moderate Moderate
Medium frame	LAW Dragon	None None	Slight Severe	Moderate Slight flash
Large frame	LAW Dragon TOW	None Slight None	Slight Severe Severe	None Slight flash Slight smoke

Table B-8. Smoke and Obscuration

(4) The Dragon causes the most structural damage, but only in frame buildings. There does not seem to be any threat of injury to the gunner because the damage is usually to the walls away from the gunner. The most damage and debris is from flying plaster chips and pieces of wood trim. Large chunks of plasterboard can be dislodged from ceilings. The backblast from a LAW, AT4, Dragon, or TOW rarely displaces furniture. Table B-9 shows the test results of structural damage and debris.

		Damage		
Building	**Weapon**	**Structure**	**Wall Covering**	**Debris Movement**
Masonry	LAW Dragon	None None	Slight Slight	Slight Slight
Bunker	Dragon TOW	None None	None None	None Leaves and dust disturbed
Small frame	LAW Dragon	None Severe	Slight Severe	None None
Medium frame	LAW Dragon	None Slight	None Slight	Slight Lamp and chair overturned
Large frame	LAW Dragon TOW	None Slight Slight	Slight Moderate Severe	Slight None None

Table B-9. Structural Damage and Debris Movement

(5) To fire a LAW, AT4, or SMAW from inside a room, the following safety precautions must be taken (Figure B-5).

(a) The building should be of a sturdy construction.

(b) The ceiling should be at least 7 feet high with loose plaster or ceiling boards removed.

(c) The floor size should be at least 15 feet by 12 feet. (The larger the room, the better.)

(d) At least 20 square feet of ventilation (room openings) should exist to the rear or side of the weapon. An open 7- by 3-foot door would provide minimum ventilation.

(e) All glass should be removed from windows, and small, loose objects should be removed from the room.

(f) Floors should be wet to prevent dust and dirt from blowing around and obscuring the gunner's vision.

(g) All personnel in the room should be forward of the rear of the weapon.

(h) All personnel in the room should wear helmets, protective vests, ballistic eye protection, and earplugs.

(i) If the gunner is firing from the prone position, his lower body must be perpendicular to the bore of the weapon, or the blast could cause injury to his legs.

Figure B-5. Firing an AT4 From Inside a Building

c. Weapon Penetration. The most important tasks to be performed against structures are the neutralization of fortified firing positions, personnel, and weapons behind barriers. Recoilless weapons can be used in this role; none, however, are as effective as heavy, direct-fire weapons or standard demolitions. Each recoilless weapon has different penetrating ability against various targets. Penetration does not always mean the destruction of the integrity of a position. Usually only those enemy soldiers directly in the path of the spall from a high-explosive antitank (HEAT) round become casualties. Other soldiers inside a fortification could be deafened, dazed, or shocked but could eventually return to action.

(1) AT4. The AT4 is heavier than the LAW and has a diameter of 84 millimeters, which gives the warhead much greater penetration. The AT4 can penetrate more than 17.5 inches (450 mm) of armor plate. Its warhead produces highly destructive results behind the armor. Tests against typical urban targets are still ongoing, but the AT4 should penetrate at least as well as the 90-mm recoilless rifle, if not better. The AT4 has a *minimum arming distance of 10 meters*, which allows it to be fired successfully against close targets. Firers should be well covered and protected when firing at close targets.

(2) MK153 83-mm SMAW. The SMAW is a lightweight assault weapon that is easily carried and placed into action by one man. It is used against fortified positions, but it is also effective against LAVs. The SMAW has a 9-mm spotting rifle and a 3.8-power telescope that ensure accuracy over ranges common to combat in built-up areas. The SMAW has excellent incapacitating effects for enemies behind walls and inside bunkers and can arm within 10 meters. Its fuze has the ability to distinguish between armor and soft earth, maximizing its capabilities against buildings, bunkers, or light armor. The warhead detonates immediately against hard targets but delays detonation against soft targets and burrows in to explode inside. The SMAW can destroy most bunkers with a single hit. The 83-mm HE warhead can create an 8-inch mousehole in reinforced concrete but will not cut the steel reinforcing bar. The SMAW is an outstanding urban support weapon that can be decisive during an urban assault. The SMAW allows for quick reduction of obstacles and bunkers as well as creation of entry points. SMAWs fired in a volley can be devastating to a building.

d. Wall Breaching. Wall breaching is a common combat task in built-up areas for which light recoilless weapons can be used. Breaching operations improve mobility by providing access to building interiors without using existing doors or windows. Breaching techniques can also be used to create loopholes for weapons positions or to allow hand grenades to be thrown into defended structures. Breaching holes for unit mobility should be about 24 inches (60 centimeters) in diameter. Loopholes should be about 8 inches (20 centimeters) in diameter (Figure B-6). Only the SMAW provides a one-shot wall breaching ability. However, in breaching walls a number of shots should be planned.

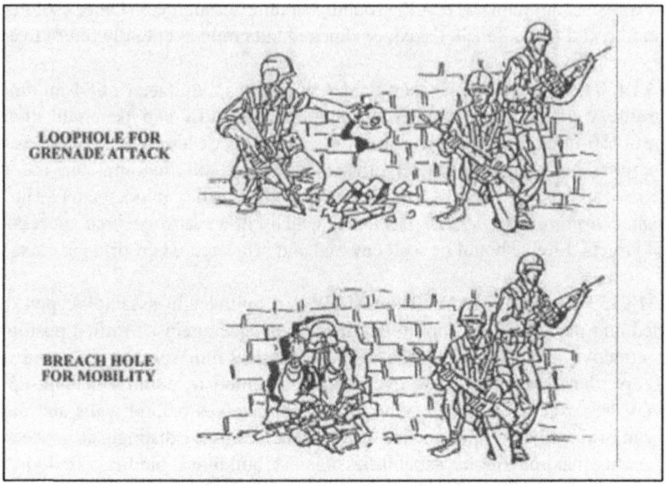

Figure B-6. Tactical Use of Holes in Masonry Walls

(1) Of all the common building materials, heavy stone is the most difficult to penetrate. The AT4 normally will not penetrate a heavy European-style stone wall. Surface cratering is usually the only effect it will have.

(2) Layered brick walls are also difficult to breach with light recoilless weapons. Some brick walls can be penetrated by multiple firings, especially if they are less than three bricks thick. Multiple AT4 rounds fired at the same spot on an 8-inch (double-brick) wall normally produce a loophole. The SMAW produces a hole in brick walls that is often large enough to be a breach hole.

(3) Wooden structural walls offer little resistance to light recoilless weapons. Even heavy timbered walls are penetrated and splintered. Multiple AT4 rounds fired at the same area of a wood-framed wall usually produce a man-sized hole. The SMAW has a devastating effect against wood-framed walls. A single round produces a breach hole as well as significant spall.

(4) Because of its high velocity, the AT4 may penetrate a soft target such as a car body or frame building before exploding.

(5) None of the light recoilless weapons are as effective against structural walls as demolitions or heavier weapons such as tank main guns or field artillery. Of all the light recoilless weapons, the SMAW is the most effective.

6. Antitank Guided Missiles. ATGMs are used mainly to defeat main battle tanks and other armored combat vehicles. They have a moderate capability against bunkers, buildings, and other fortified targets commonly found during combat in built-up areas. This category of weapons includes the TOW and Dragon missiles.

a. Employment. TOWs and Dragons provide overwatch antitank fires during the attack of a built-up area and an extended range capability for the engagement of armor during the defense. Within built-up areas, they are best employed along major thoroughfares and from the upper stories of buildings to attain long-range fields of fire. Their *minimum firing range of 65 meters* could limit firing opportunities in the confines of densely built-up areas.

(1) Obstacles. Rubble or other obstacles could interfere with the flight of missiles when fired from street level. At least 3.5 feet (1 meter) of vertical clearance over such obstacles must be maintained. Figure B-7 shows the most common obstacles to ATGM flight that are found in built-up areas. *Power lines are a special obstacle and present a unique threat to ATGM gunners. If the power in the lines has not been interrupted, the ATGM guidance wires could create a short circuit. This would allow extremely high voltage to pass to the gunner in the brief period before the guidance wires melted. This voltage could either damage the sight and guidance system or injure the gunner. Before any ATGM is fired over a power line, an attempt must be made to determine whether or not the power to that line has been interrupted.*

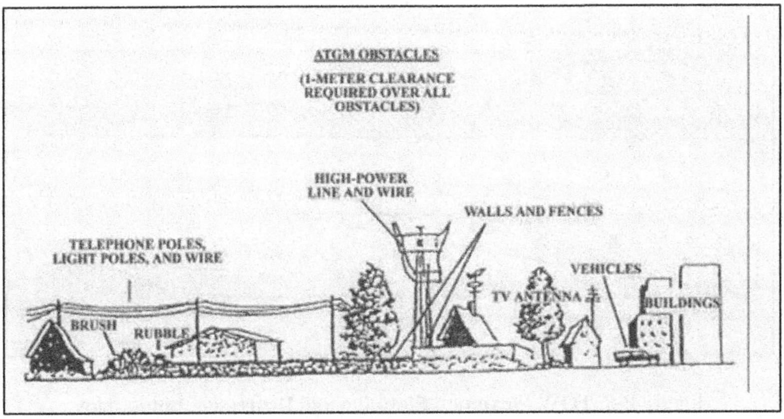

Figure B-7. Common Obstacles to ATGM Flights

(2) Dead Space. Three aspects of dead space that affect ATGM fires are arming distance, maximum depression, and maximum elevation.

(a) Both the Dragon and TOW missiles have a minimum arming distance of 65 meters, which severely limits their use in built-up areas. Few areas in the inner city permit fires much beyond the minimum arming distance. However, ground-level, long-range fires down streets or rail lines and across parks or plazas are possible. ATGMs may be used effectively from upper stories or roofs of buildings to fire into other buildings.

(b) The TOW is limited much more than the Dragon by its maximum depression and elevation. The maximum depression and elevation limits of the TOW mount could result in dead space and preclude the engagement of close targets (Figure B-8). A target located at the minimum arming range (65 meters) cannot be engaged by a TOW crew located any higher than the sixth floor of a building due to maximum depression limits. At 100 meters, the TOW crew can be located as high as the ninth floor and still engage the target.

(3) Backblast. Backblast for ATGMs is of more concern during combat in built-up areas than in open country. Any loose rubble in the caution zone could be picked up and thrown by the backblast, and the channeling effect of walls and narrow streets is even more pronounced due to the greater backblast. If the ATGM backblast strikes a wall at an angle, it can pick up debris or be deflected and cause injury to unprotected personnel (Figure B-9). Both types of ATGM can be fired from inside some buildings. In addition to the helmet and protective vest, eye protection and earplugs should be worn by all personnel in the room.

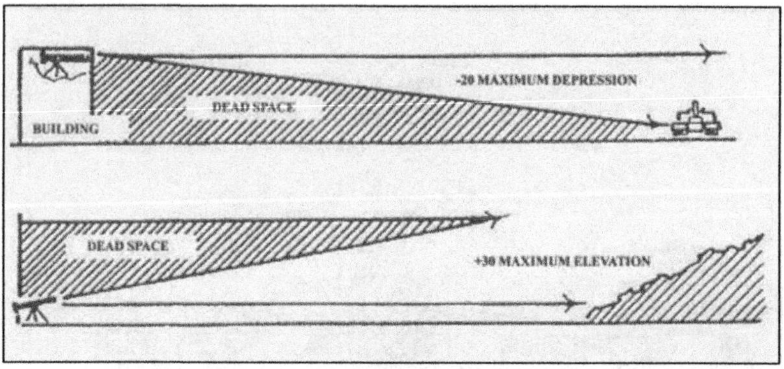

Figure B-8. TOW Maximum Elevation and Depression Limitations

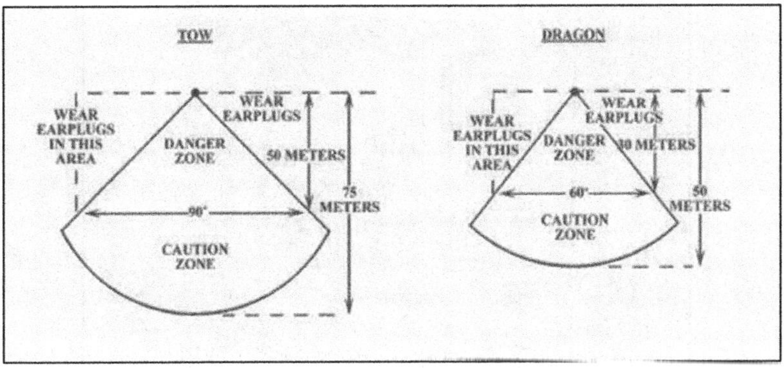

Figure B-9. ATGM Backblast in an Open Street

(a) To fire a TOW from inside a room, the following safety precautions must be taken (Figure B-10):

- The building must be of sturdy construction.
- The ceiling should be at least 7 feet high.
- The floor size of the room should be at least 15 by 15 feet, or larger if possible.
- At least 20 square feet of room ventilation should exist, preferably to the rear of the weapon. An open 7- by 3-foot door is sufficient. Additional ventilation can be created by removing sections of interior partitions.
- All glass must be removed from the windows and all small, loose objects removed from the room.
- All personnel in the room should be forward of the rear of the TOW.
- All personnel in the room should wear ballistic eye protection and earplugs.
- A clearance of 9 inches (23 centimeters) must exist between the launch tube and the aperture from which it is fired. (See FM 23-34, *TOW Weapon System*, for more detailed safety information.)

Figure B-10. TOW Fired From Inside a Room

(b) To fire a Dragon from inside a room, the following safety precautions must be taken.

- The building must be of sturdy construction.
- The ceiling should be at least 7 feet high.
- The floor size should be at least 15 by 15 feet, or larger if possible.
- At least 20 square feet of ventilation should exist (room openings), preferably to the rear of the weapon. An open 7- by 3-foot door would provide minimum ventilation.
- All glass should be removed from windows, and small, loose objects should be removed from the room.
- The room should be clean or the floors must be wet to prevent dust and dirt (kicked up by the backblast) from obscuring the vision of other soldiers in the room.
- All personnel in the room must be forward of the rear of the weapon.
- All personnel in the room must wear ballistic eye protection and earplugs.
- At least a 6-inch clearance must exist between the launch tube and the aperture from which it is fired.

b. Weapon Penetration. ATGMs can penetrate and destroy heavily armored tanks. They have large warheads that contain shaped charges. Because of their size, these warheads can achieve significant penetration against typical urban targets. Penetration, however, does not mean a concurrent destruction of the structural integrity of a position. The shaped-charge warhead produces relatively little spall. Enemy personnel not standing directly behind or near the point of impact of an ATGM may escape injury.

(1) Standard TOW Missiles. The basic TOW missile can penetrate 8 feet of packed earth, 4 feet of reinforced concrete, or 16 inches of steel plate. The improved TOW (ITOW), the TOW 2, and the TOW 2A have all been modified to improve their penetration. They all penetrate better than the basic TOW. All TOW missiles can defeat triple sandbag walls, double layers of earth-filled 55-gallon drums, and 18-inch log walls.

(2) TOW 2B. The TOW 2B uses a different method of defeating enemy armor. It flies over the target and fires an explosively formed penetrator down onto the top armor, which is thinner. *Because of this design feature, the TOW 2B missile cannot be used to attack nonmetallic structural targets. When using the TOW 2B missile against enemy armor, gunners must avoid firing directly over other friendly vehicles, disabled vehicles, or large metal objects such as water or oil tanks.*

(3) Dragon Missile. The Dragon missile can penetrate 8 feet of packed earth, 4 feet of concrete, or 13 inches of steel plate. It can attain effective short-range fire from upper stories or from the rear or flanks of a vehicle. These engagements are targeted against the most vulnerable parts of tanks and can entrap tanks in situations where they are unable to counterfire. Elevated firing positions increase the first-round hit probability. Firing down at an angle of 20 degrees increases the chance of a hit by 67 percent at 200 meters. A 45-degree downward angle doubles the first-round hit probability, compared to a ground-level shot.

c. Breaching Structural Walls. Firing ATGMs is the least efficient means to defeat structures. Because of their small basic load and high cost, ATGMs are better used against tanks or enemy-fortified firing positions. They can be effective against bunkers or other identified enemy firing positions.

7. Flame Weapons. Flame weapons are characterized by both physical and psychological casualty-producing abilities. Flame does not need to be applied with pinpoint accuracy, but it also must not spread to structures needed by friendly forces. Large fires in built-up areas are catastrophic. If they burn out of control, fires can create an impenetrable barrier for hours. The Marine Corps does not possess flame-generating equipment. However, the M34 WP grenade can be used to ignite and destroy flammable objects, especially wooden structures. The Army's M2A1-7 portable flamethrower is stored in war reserve status as a standard "C" item. Its availability is limited; however, Marines could be issued flame-support weapons for urban operations depending on the status of international agreements on the use of flame weapons that are in effect at the time.

a. Employment. Flame weapons used against fortified positions should be aimed directly at the aperture. Even if the round or burst misses, enough flaming material enters the position to cause casualties and disrupt the enemy occupants. The M34 WP grenade is difficult to throw far or into a small opening such as a bunker aperture. However, its effects are dramatic when thrown into a room or building.

b. Effects. The M34 is used to ignite and destroy flammable objects, especially wooden structures. It is also used to create an immediate smoke cloud to conceal movement across a narrow open space such as a street. Its smoke is not toxic but can cause choking in heavy concentrations.

(1) The grenade's explosion, bright flash, smoke, and burning WP particles all combine to make the M34 one of the most effective psychological weapons available. The M34 hand grenade throws WP fragments up to 35 meters from the point of detonation. These fragments can attach to clothing or skin and continue burning. Because of its weight, most infantrymen can throw this grenade only 30 to 40 meters.

(2) The Marine must avoid injury from friendly use of the M34. The M34 can ignite if the WP inside is exposed to the air. Bullets and shell fragments have been known to strike and rupture M34 grenades; therefore, grenades must be protected from enemy fire.

(3) The M34 WP grenade is an effective weapon against enemy armored vehicles when used in the close confines of combat in built-up areas. It can be thrown or dropped from upper stories onto enemy vehicles. The M34 can be combined with flammable liquids, detonating cords, blasting caps, and fuze igniters to create the eagle fireball, a field-expedient antiarmor device. (See FM 21-75, *Combat Skills of the Soldier*, Appendix H.)

(4) The M34 is also excellent as a screening device. A grenade can be thrown from behind cover into an open street or plaza. When it explodes, the enemy's observation is temporarily obscured. Thus, friendly forces can quickly cross the open area—if the enemy fires, it is unaimed and presents less of a danger. If screening smoke is used to cover a squad's movement across short open areas, it will reduce expected casualties from small-arms fire by about 90 percent.

8. Hand Grenades. Hand grenades are used extensively during combat in built-up areas. Smoke grenades are used for screening and signaling. Riot control grenades are used to drive the enemy out of deep fortifications. Fragmentation and concussion grenades are used to clear the enemy out of rooms and basements. Hand grenades are the most-used explosive munition during intense combat in built-up areas. In World War II, it was common for a battalion fighting in a city to use more than 500 hand grenades each day.

a. Employment. Smoke and riot-control grenades have similar employment techniques. Fragmentation grenades are used to produce enemy casualties.

(1) The AN-M8 HC grenade produces a dense white or grey smoke. It burns intensely and cannot be easily extinguished once it ignites. The smoke can be dangerous in heavy concentrations because it makes breathing difficult and causes choking. The M8 grenade is normally used for screening. It produces a slowly building screen of longer duration than the M34 WP grenade without the problem of collateral damage caused by scattered burning particles.

(2) The M18-series smoke grenades produce several different colors of smoke that are used for signaling. Yellow smoke is sometimes difficult to see in built-up areas.

(3) The M7A3 CS riot control grenade can be used to drive enemy troops out of fortifications when civilian casualties or collateral damage constraints are considerations. Built-up areas often create variable and shifting wind patterns. When using CS grenades, Marines must prevent the irritating smoke from affecting friendly forces. The CS grenade burns intensely and can ignite flammable structures. Enemy troops wearing even rudimentary chemical protective masks can withstand intense concentrations of CS gas.

(4) The MK3A2 offensive hand grenade, commonly referred to as the concussion grenade, produces casualties during close combat while minimizing the danger to friendly personnel. For this reason, it is the preferred hand grenade for use during offensive operations in a MOUT environment. The grenade produces severe concussion effects in enclosed areas. It can be used for light blasting and demolitions and for creating breach holes in interior walls. The concussion produced by the MK3A2 is much greater than that of the fragmentation grenade. It is very effective against enemy soldiers within bunkers, buildings, and underground passages.

(5) The M67 fragmentation grenade is the most commonly available grenade during combat in built-up areas. It provides suppression during room-to-room or house-to-house fighting, and it is used while clearing rooms of enemy personnel. The M67 grenade has a 3- to 5-second delay fuze. When used at close ranges, it can be cooked-off for two seconds to deny the enemy the time to throw it back. The fragmentation grenade can be rolled, bounced, or ricocheted into areas that cannot be reached by 40-mm grenade launchers.

b. Effects. Each type of hand grenade has its own specific effect during combat in built-up areas.

(1) Both HC smoke and CS gas *displace* oxygen from enclosed areas. Gas masks will not prevent this from occurring because they only filter air, not provide oxygen. Smoke grenades produce dense clouds of colored or white smoke that remain stationary in the surrounding area. They can also cause fires if used indiscriminately. If trapped and concentrated within a small space, smoke grenades and CS gas *can suffocate Marines.*

(2) The fragmentation grenade has more varied effects in combat in built-up areas. It produces a large amount of small, high-velocity fragments that can penetrate sheetrock partitions and are lethal at short ranges (15 to 20 meters). Fragments lose their velocity quickly and are less effective beyond 25 meters.

The fragments from a fragmentation grenade cannot penetrate a single layer of sandbags, a cinder block, or a brick building, but they can perforate wood-framed and tin buildings if exploded close to their walls.

(3) Inside rooms, fragmentation barriers consisting of common office furniture, mattresses, doors, or books can be effective against the fragmentation grenade. For this reason, a room should never be considered safe just because one or two grenades have been detonated inside. Fragmentation grenades detonated on the floor not only throw fragments laterally, but also send fragments and spall downward to lower floors. Predicting how much spall will occur is difficult because flooring material varies, but wooden floors are usually affected the most.

(4) Some foreign grenades throw fragments that are much larger than those of the U.S.-made M67. Light barriers and interior walls would probably be less effective against these grenades than against the M67. A major problem with the U.S.-made fragmentation grenade is its tendency to bounce back off of hard targets. Grenades are often directed at window openings on the ground floor or second floor. At ranges as close as 20 meters, a thrower's chances of missing a standard 1-meter by 1-meter window are high. The fragmentation grenade normally breaks through standard window glass and enters a room. If the grenade strikes at a sharp angle or the glass is thick plate, the grenade could be deflected without penetrating.

(5) Hand grenades are difficult weapons to use. They involve a high risk of fratricide. Commanders should conduct precombat training with hand grenades as part of normal preparations. Marines must be very careful when throwing hand grenades up stairs to avoid having the grenade roll back down on them.

(6) Once pulled, the pins of both fragmentation and concussion grenades can be replaced if the thrower decides not to use the weapon. This pin replacement must be done carefully (see FM 23-30, *Grenades and Pyrotechnic Signals*) and prior to the spoon being released or detaching.

(7) An estimate of the situation (METT-T) and existing ROE will dictate what type of grenade will be used to clear each room. Because of the high rate of expenditure of grenades, units should use assault packs to carry additional grenades of all types. Additional grenades can also be carried in empty ammunition or canteen pouches.

9. 25-mm Automatic Gun. The 25-mm automatic gun mounted on the LAV-25 is an effective weapon during combat in built-up areas. The primary roles of LAV-25s during combat in built-up areas are to provide suppressive fire and to breach exterior walls and fortifications.

a. Obliquity. The 25-mm gun produces its best urban target results when fired perpendicular to the hard surface (zero obliquity). In combat in built-up areas, however, finding a covered firing position that permits low-obliquity firing is unlikely unless the streets and gaps between buildings are wide. Most shots impact the target at an angle, which normally reduces penetration. With the armor-piercing, discarding, sabot with tracer (APDS-T) round, an angle of obliquity of up to 20 degrees can actually improve breaching. The rounds tend to dislodge more wall material for each shot but do not penetrate as deeply into the structure.

b. Target Types. The 25-mm gun has different effects when fired against different urban targets.

(1) Reinforced Concrete. Reinforced concrete walls, which are 12 to 20 inches thick, present problems for the 25-mm gun when trying to create breach holes. It is relatively easy to penetrate, fracture, and clear away the concrete, but the reinforcing rods remain in place. These create a "jail window" effect by preventing entry but allowing grenades or rifle fire to be placed behind the wall. Steel reinforcing rods are normally 3/4 of an inch thick and 6 to 8 inches apart—there is no quick way of cutting these rods. They can be cut with demolition charges, cutting torches, or special power saws. Firing with either APDS-T or high-explosive, incendiary with tracer (HEI-T) rounds from the 25-mm gun will not always cut these rods.

(2) Brick Walls. Brick walls are more easily defeated by the 25-mm gun regardless of their thickness, and they produce the most spall.

(3) Bunker Walls. The 25-mm gun is devastating when fired against sandbag bunker walls. Obliquity has the least effect on the penetration of bunker walls. Bunkers with earth walls up to 36 inches thick are easily penetrated. At short ranges typical of combat in built-up areas, defeating a bunker should be easy, especially if the 25-mm gun can fire at an aperture.

c. Burst Fire. The 25-mm gun's impact on typical urban targets seems to be magnified if the firing is in short bursts. At close ranges, the gunner might need to shift his point of aim in a spiral pattern to ensure that the second and third bursts enlarge the hole. Even without burst fire, sustained 25-mm gunfire can defeat almost all urban targets.

d. Weapon Penetration. The penetration achieved by the two combat rounds (HEI-T and APDS-T) differs slightly, but both are effective.

(1) APDS-T. The APDS-T round penetrates urban targets by retaining its kinetic energy and blasting a small hole deep into the target. The APDS-T round gives the best effects behind the wall, and the armor-piercing core often breaks into two or three fragments that can create multiple enemy casualties. The APDS-T needs as few as four rounds to achieve lethal results behind walls. Table B-10 lists the number of APDS-T rounds needed to create different-sized holes in common urban walls.

Target	Loophole	Breach Hole
3-inch brick wall at 0-degree obliquity	22 rounds	75 rounds
3-inch brick wall at 45-degree obliquity	22 rounds	35[1] rounds
5-inch brick wall at 0-degree obliquity	32 rounds	50[1] rounds
8-inch reinforced concrete at 0-degree obliquity	22 rounds	75 rounds (Note: Reinforcing rods still in place)
8-inch reinforced concrete at 45-degree obliquity	22 rounds	40[1] rounds (Note: Reinforcing rods still in place)

[1] Obliquity and depth tend to increase the amount of wall material removed.

Table B-10. Breaching Effects of APDS-T Rounds

(a) When firing single rounds, the APDS-T round provides the greatest capability for behind-the-wall incapacitation. The APDS-T round can penetrate more than 16 inches of reinforced concrete with enough energy left to cause enemy casualties. It penetrates through both sides of a wood-framed or brick veneer building. Field fortifications are easily penetrated by APDS-T rounds. Table B-11 lists the number of APDS-T rounds needed to create different-sized holes in commonly found bunkers.

(b) The APDS-T round creates a hazardous situation for exposed personnel because of the pieces of sabot that are thrown off of the round.

Type Bunker	Obliquity	Penetration	Loophole	Small Breach Hole
36-inch sand/timber	0 degree	1 round	25 rounds	40 rounds
36-inch sand/6-inch concrete	0 degree	6 rounds	6 rounds	20 rounds

Table B-11. Number of APDS-T Rounds Needed To Create Different-Sized Holes in Bunkers

WARNING

Personnel not under cover forward of the 25-mm gun's muzzle and within the danger zone could be injured or killed by these sabots, even if the penetrator passes overhead to hit the target.

The danger zone extends at an angle of about 10 degrees below the muzzle level, out to at least 100 meters, and about 17 degrees left and right of the muzzle. Figure B-11 shows the hazard area of the APDS-T round.

Figure B-11. APDS-T Danger Zone

(2) HEI-T. The HEI-T round penetrates urban targets by blasting away chunks of material.

(a) The HEI-T round does not penetrate an urban target as well as the APDS-T, but it does strip away a greater amount of material for each round. The HEI-T does more damage to an urban target when fired in multiple short bursts because the cumulative impact of multiple rounds is greater than the sum of individual rounds. Table B-12 lists the number of HEI-T rounds needed to create different-sized holes.

Target	Loophole	Breach Hole
3-inch brick wall at 0-degree obliquity	10 rounds	20 rounds
3-inch brick wall at 45-degree obliquity	20 rounds	25 rounds
5-inch brick wall at 0-degree obliquity	30 rounds	60 rounds

8-inch reinforced concrete at 0-degree obliquity	15 rounds	25 rounds
8-inch reinforced concrete at 45-degree obliquity	15 rounds	30 rounds

Table B-12. Number of HEI-T Rounds Needed To Create Different-Sized Holes in Bunkers

(b) The HEI-T round does not provide single-round perforation or incapacitating fragments when used against any external masonry structural wall. It can create first-round fragments behind wood-framed and brick veneer walls. HEI-T rounds cannot penetrate a bunker as quickly as APDS-T rounds, but they can create more damage inside the bunker once the external earth has been stripped away. Against a heavy bunker, about 40 rounds of HEI-T ammunition are needed to strip away the external earth shielding and breach the inner lining of concrete or timber. The HEI-T round is also used for suppression against known or suspected firing ports such as doors, windows, and loopholes.

10. Tank Cannon. The powerful, high-velocity cannon mounted on the M1A1 tank provides Marines with a key requirement for victory in built-up areas—heavy direct-fire support. Although the infantry assumes the lead role during combat in built-up areas, tanks and infantry work as a close team. Tanks move down streets after the infantry has cleared them of any suspected ATGM positions and, in turn, support the infantry with fire. The tank is one of the most effective weapons for heavy fire against structures. The primary role of the tank cannon during combat in built-up areas is to provide heavy direct fire against buildings and strongpoints that are identified as targets by the infantry. The wall and fortification breaching effects of the 120-mm tank cannon are major assets to Marines fighting in built-up areas.

a. Obliquity. The tank cannon produces its best urban target effects when fired perpendicular to the hard surface (zero obliquity). During combat in built-up areas, however, finding a covered firing position that permits low-obliquity firing is unlikely. Most shots strike the target at an angle that would normally reduce penetration. With tank cannon APDS rounds, obliquity angles of up to 25 degrees have little effect, but angles greater than 45 degrees greatly reduce penetration.

b. Ammunition. Armor-piercing, fin-stabilized, discarding sabot (APFSDS) rounds are the most commonly carried tank ammunition. These rounds work best against armored vehicles. The 120-mm cannon also carries an effective high-explosive, antitank multipurpose (HEAT-MP) round.

c. Characteristics. The 120-mm tank cannon has two specific characteristics that affect its employment in built-up areas: limited elevation and depression and short arming ranges. In addition, the M1A1 tank has another characteristic that is not involved with its cannon but that affects Marines working with or around the tank—*extremely hot turbine exhaust.*

(1) The M1A1 tank can elevate its cannon 20 degrees and depress it 10 degrees. The lower depression limit creates a 35-foot (10.8-meter) dead space around a tank. On a street 16 meters wide, this dead space extends to the buildings on each side (Figure B-12). Similarly, there is a zone overhead in which the tank cannot fire (Figure B-13). This dead space offers ideal locations for short-range antiarmor weapons and allows hidden enemy gunners to fire at the tank when the tank cannot fire back. It also exposes the tank's most vulnerable areas: the flanks, rear, and top. Infantry must move ahead, alongside, and to the rear of tanks to provide close protection. The extreme heat produced immediately to the rear of the M1A1 tank prevents dismounted infantry from following closely, but protection from small-arms fire and fragments is still provided by the tank's bulk and armor. The M1A1 tank also has a blind spot caused by the zero-degree depression available over part of the back deck. To engage any target in this area, the tank must pivot to convert the rear target to a flank target.

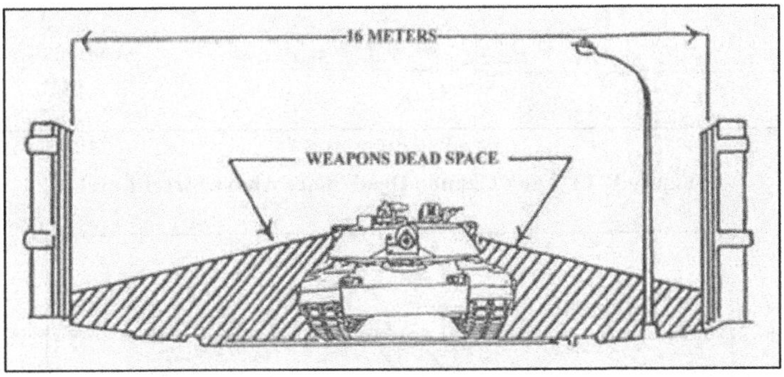

Figure B-12. Tank Cannon Dead Space at Street Level

(2) The 120-mm HEAT-MP round arms at about 36 feet. This arming distance allows the tank to engage targets from short ranges. The armor of the tank protects the crew from both the blowback effects of the round and enemy return fire. The APFSDS round does not need to arm and can, therefore, be fired at almost any range. The discarding portions of the round can be lethal to exposed infantry forward of the tank.

d. Target Effects. HEAT rounds are most effective against masonry walls. The APFSDS round can penetrate deeply into a structure but does not create as large a hole or displace as much spall behind the target. In contrast to lighter HEAT rounds, tank HEAT rounds are large enough to displace enough spall to inflict casualties inside a building. One HEAT round normally creates a breach hole in all but the thickest masonry construction—brick veneer and wood-framed constructions are demolished by a single round. Even the 120-mm HEAT round cannot cut all of the reinforcing rods, which are usually left in place, often hindering entry through the breach hole (Figure B-14). Both HEAT and APFSDS rounds are effective against all field fortifications. Only large earth berms and heavy mass-construction buildings can provide protection against tank fire.

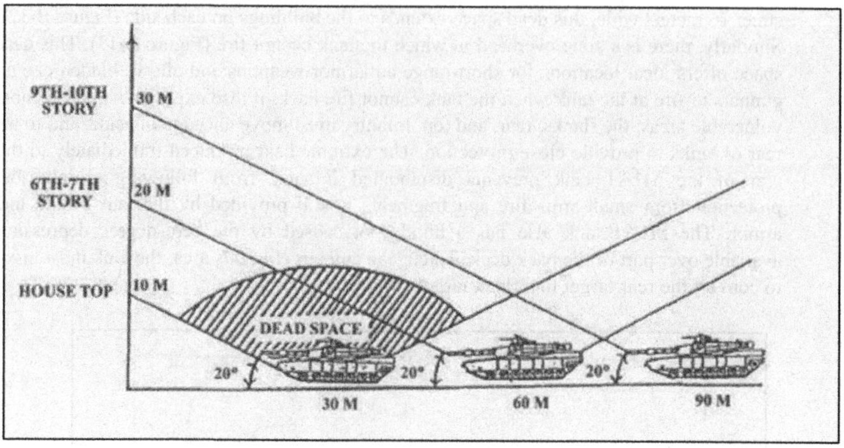

Figure B-13. Tank Cannon Dead Space Above Street Level

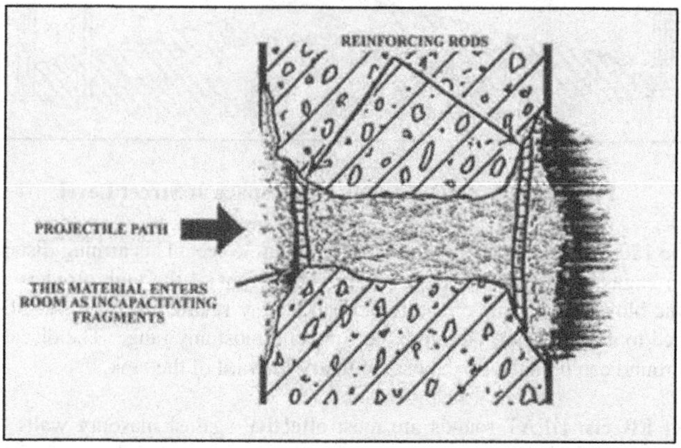

Figure B-14. Tank HEAT Round Effects on Reinforced Concrete Walls

e. Employment. Tank-heavy forces could be at a severe disadvantage during combat in built-up areas, but a few tanks working with the infantry can be most effective, especially if they work well together at the small-unit level. Tank, infantry, and engineer task forces are normally formed to attack a fortified area. The tank platoon or section can work together with rifle squads or platoons.

(1) Tanks need infantry on the ground to provide security in built-up areas and to designate targets. Against targets protected by structures, tanks should be escorted forward to the most covered location that provides a clear shot. On-the-spot instructions by infantry leaders ensure that the tank's fire is accurate and its exposure is limited. The tank commander may have to halt in a covered position, dismount, and reconnoiter his route forward into a firing position.

(2) When the tank main gun fires, it creates a large fireball and smoke cloud. In the confines of a built-up area, dirt and masonry dust are also picked up and added to this cloud. The target is further obscured by the smoke and dust of the explosion. Depending on the local conditions, this obscuration could last as long as two or three minutes. Marines can use this period to reposition or advance unseen by the enemy. Caution must be exercised, however, because the enemy might also move.

(3) A tank cannon creates an overpressure and noise hazard to exposed Marines. All dismounted Marines working near tanks should wear Kevlar helmets and protective vests, as well as ballistic eye protection. If possible, they should also wear earplugs and avoid the tank's frontal 60-degree arc during firing (Figure B-15).

WARNING

- **The overpressure from the tank 120-mm cannon can kill a marine found within a 90-degree arc extending from the muzzle of the gun tube out to 200 meters.**

- **From 200 to 1,000 meters along the line of fire, on a frontage of about 400 meters, dismounted Marines must be aware of the danger from discarding sabot petals, which can kill or seriously injure personnel.**

- **Personnel outside the tank should remain at least 50 meters from the tank in all directions as they may receive damaging effects from firing noise and overpressure.**

- **Personnel must also wear hearing protection when operating within 704 meters of a tank that is firing its main gun.**

TM 9-2350-264-10-2, *Operator's Manual for Operation Under Usual and Unusual Conditions for Tank, Combat, Full-Tracked: 120mm Gun, M1A1 General Abrams*, with Change 4

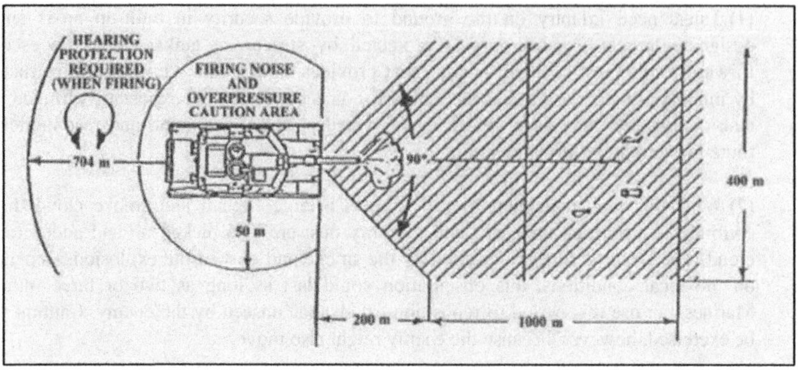

Figure B-15. Danger Areas Around a Tank Firing a 120-mm Gun

(4) Tanks are equipped with powerful thermal sights that can be used to detect enemy personnel and weapons that are hidden in shadows and behind openings. Dust, fires, and thick smoke significantly degrade these sights.

(5) Tanks have turret-mounted grenade launchers that project screening smoke grenades. The grenades use a bursting charge and burning RP particles to create this screen. *Burning particles can easily start uncontrolled fires and are hazardous to dismounted Marines near the tank.* The tank commander and the infantry small-unit leader must coordinate when and under what conditions these launchers can be used. Grenade launchers are a useful feature for protecting the tank but can cause significant problems if unwisely used.

(6) The tank's size and armor can provide dismounted Marines with cover from direct-fire weapons and fragments. With coordination, tanks can provide moving cover for Marines as they advance across small open areas. However, enemy fire that strikes but does not penetrate a tank is a major threat to nearby Marines. Fragmentation generated by antitank rounds and ricochets off of tank armor have historically been a prime cause of casualties while infantry was working with tanks in built-up areas.

11. Artillery. A major source of fire support for infantry forces fighting in built-up areas is field artillery weapons. If the built-up area is near the coast, NGF can be used. Field artillery employment can be in either the indirect- or direct-fire mode.

a. Indirect Fire. Low-angle indirect artillery fire is not effective in attacking targets within walls and masonry structures. It tends to impact on roofs or upper stories rather than on structurally critical wall areas or pillars. High-angle fire can achieve the trajectories necessary to strike targets in urban areas.

.3

Weapons of at least 155 mm are required against thick reinforced concrete, stone, or brick walls. Even with heavy artillery, large expenditures of ammunition are required to knock down buildings of any size. Tall buildings also create areas of indirect-fire dead space, which are areas that cannot be engaged by indirect fire because of a combination of building height and angle of fall of the projectile (Figure B-16). Usually the dead space for low-angle indirect fire is about five times the height of the highest building over which the rounds must pass.

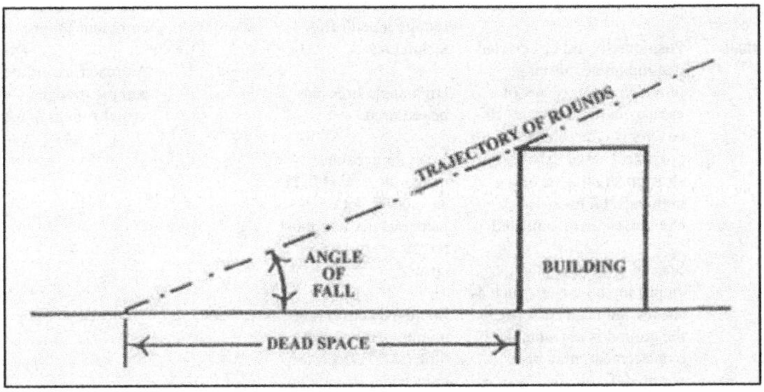

Figure B-16. Indirect-Fire Dead Space (Low Angle)

b. Direct Fire. A likely use of Marine artillery in an urban direct-fire role is to reinforce infantry against tough or important urban targets. Artillery should be used in this role only after an analysis of the need for heavy direct fire and the tradeoff involved in the extreme decentralization of artillery firepower. Artillery has the same need for close security and target designation as tanks.

c. Target Effects. Medium-caliber (155-mm) direct fire has a devastating effect against masonry construction and field fortifications. Smaller artillery (105-mm) pieces in the direct-fire mode are much less destructive than the larger caliber weapons. However, all Marine artillery is towed and, therefore, more difficult to employ in the direct-fire mode.

The 155-mm (M198) towed howitzer is effective because of its rate of fire and penetration. HE rounds can penetrate up to 38 inches of brick and unreinforced concrete. Projectiles can penetrate up to 28 inches of reinforced concrete with considerable damage beyond the wall. HE rounds with concrete-piercing fuzes provide an excellent means of penetrating strong reinforced concrete structures. One round can penetrate up to 46 inches. Five rounds are needed to reliably create a 1.5-meter breach in a wall that is 1 meter thick. About 10 rounds are needed to create the same size breach in a wall that is 1.5 meters thick. Superquick fuzing causes the rubble to be blown into the building, whereas delay fuzing tends to blow the rubble outward into the street. (See Table B-13 for employment considerations for artillery.)

Organization for Combat	Movement/Positioning	Delivery of Fire	Security	Command and Control
Centralized control is required during initial phases; decentralized control is required during later phases to support semi-independent actions of small units.	Movement should occur during night or periods of reduced visibility when possible. There are few displacements, often by platoon or section. Positions should be selected that minimize masking, provide several routes of escape, and afford as much cover and concealment as possible. Use of existing structures (garages, office buildings, highway overpasses) is recommended. Special techniques for emplacing howitzers, such as spades against a curb when the ground is not suitable for emplacement, may be required. Explosives may be required to soften emplacement of howitzers. Reconnaissance, selection, and occupation of position (RSOP) elements should be well armed because they may have to clear areas to be occupied. Extensive route reconnaissance is required. Target acquisition devices are somewhat degraded. Radars should be emplaced to cover likely areas of enemy indirect-fire weapon employment. Radars should not be placed in the midst of an urban area because of masking.	Both direct and indirect fires are delivered for supported units. Destruction of fortifications may require assault fire techniques. High-angle fires may be required. Need for accurate meteorological (MET) and survey data increases because most targets are point targets. Improved conventional munition and variable time (fuze) effects are reduced by structures, although they are effective against personnel on rooftops. HE delay is used for penetration effects. Illumination, chemical incendiary ammunition, and smoke are effective. Ammunition expenditures will be heavy. Lasers and PGMs permit destruction of targets with minimal rubbling of adjacent buildings. Tall buildings may hamper laser use. Batteries must be prepared for hasty survey techniques. Magnetic instruments are impaired.	Positions must be fortified.	Radio communications are impaired by buildings. Wire can usually be run overhead. Make use of civilian communications. A greater use of messengers and prearranged audio and visual signals is required.

Table B-13. Artillery Employment Considerations in Built-Up Areas

12. Mortars. The urban environment greatly restricts low-angle indirect fires because of overhead masking. While all indirect-fire weapons are subject to overhead masking, mortars are less affected than field artillery weapons because of the mortar's higher trajectory. For low-angle artillery fire, dead space is about five times the height of the building behind which the target sits. For mortar fire, dead space is only about one-half the height of the building. Because of these advantages, mortars are even more important than field artillery to the infantry during combat in built-up areas.

a. Employment. Not only can mortars fire into the deep defilade created by tall buildings, but they can also fire out of it. Mortars emplaced behind buildings are difficult for the enemy to accurately locate and even harder to hit with counterfire. Because of their light weight, even heavy mortars can be hand-carried to firing positions that may not be accessible to vehicles.

(1) The 60-mm and 81-mm mortars have limited effects on structural targets. Even with delay fuzes, they seldom penetrate more than the upper stories of light buildings. However, their wide area coverage and multioption fuzes make them useful against an enemy force advancing through streets, through other open areas, or over rubble.

(2) Mortar platoons often operate as separate firing sections during combat in built-up areas. The lack of large open areas can preclude establishing a platoon firing position. Figure B-17 shows how two mortar sections that are separated by only one street can be effective in massing fires and be protected from countermortar fire by employing defilade and dispersion.

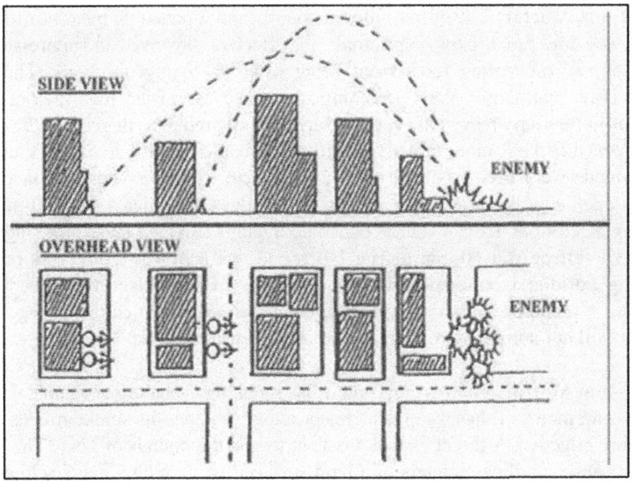

Figure B-17. Split-Section Mortar Operations on Adjacent Streets

(3) All three of the standard mortar projectiles are useful during combat in built-up areas. HE fragmentation rounds are most commonly used. WP is effective for starting fires in buildings and forcing the enemy out of cellars and light-framed buildings, and it is also the most effective mortar round against dug-in enemy tanks. Even near-misses blind and suppress the tank crew, forcing them to button up. Hits are difficult to achieve but are effective.

(4) Because the artificial roughness of urbanized terrain reduces wind speed and increases atmosphere mixing, mortar smoke tends to persist longer and give greater coverage in built-up areas than in open terrain.

(5) Urban masking has an impact on the use of illumination. In built-up areas, it is often necessary to plan illumination behind friendly positions, which places friendly troops in shadows and enemy troops in the light. Illumination rounds are difficult to adjust and are often of limited use because of the deep canyon nature of the urbanized area. Rapidly shifting wind currents in built-up areas also affect mortar illumination, making it less effective.

b. Effects of Mortar Fire. The multioption fuze on newer U.S. mortar rounds makes them effective weapons on urbanized terrain. Delay settings can increase penetration slightly, and proximity bursts can increase the lethal area covered by fragments. Tall buildings can cause proximity-fuzed mortar rounds to detonate prematurely if they pass too closely to the buildings.

(1) 60-mm Mortar. The 60-mm mortar round cannot penetrate most rooftops, even with a delay setting. Small explosive rounds are effective, however, in suppressing snipers on rooftops and preventing roofs from being used by enemy observers. The 60-mm WP round is not normally a good screening round because of its small area of coverage. In combat in built-up areas, however, the tendency of smoke to linger and the small areas to be screened make it more effective. During the battle for Hue in South Vietnam, 60-mm WP rounds were used to create small, short-term smoke screens to conceal movement across open areas such as parks, plazas, and bridges. Fragments from 60-mm HE rounds landing as close as 10 feet cannot penetrate a single sandbag layer or a single-layer brick wall. The effect of a 60-mm mortar HE round that achieves a direct hit on a bunker or fighting position is equivalent to 1 - 2 pounds of TNT. Normally, the blast will not collapse a properly constructed bunker, but it can cause structural damage. The 60-mm mortar will not normally create craters in a hard-surfaced road.

(2) 81-mm Mortar. The 81-mm mortar has much the same effect against urban targets as the 60-mm mortar. It has a slightly greater lethal area, and its smoke rounds (WP and RP) are more effective. A direct hit is equivalent to about 2 pounds of TNT. The 81-mm round cannot cause significant craters in a hard-surfaced road. With a delay setting, the 81-mm round can penetrate the roofs of light buildings.

13. Naval Gunfire. The primary NGF weapon, the 5-inch/54, has a high rate of fire and is roughly equivalent to the 155-mm howitzer in target effect. Target engagement considerations are the same as for artillery; however, the 5-inch/54's flat trajectory is even more affected by terrain masking and a constantly changing gun-target line. NGF can achieve greater target engagement by firing reduced charges at high angles; however, range will be reduced.

14. Aerial Weapons. Both rotary- and fixed-wing aircraft can quickly deliver large volumes of firepower over large built-up areas. Specific targets are hard to distinguish from the air. Good ground-to-air communications are vital in successfully employing aerial firepower. Planners have historically tended to overestimate the effects of HE on defenders. Modern, large buildings are remarkably resistant to damage from bombs and rocket fire.

a. Rotary-Wing Aircraft. Attack helicopters can be used to engage targets in built-up areas. Enemy armored vehicles in small parks, boulevards, or other open areas are good targets for attack helicopters.

(1) The HELLFIRE missile has a larger warhead and greater range than the TOW, but it too is a shaped-charge warhead and is not specifically designed for use against masonry targets. Laser target designation for the HELLFIRE may be difficult as a result of laser reflections off of glass and shiny metal surfaces. The use of attack helicopters to deliver ATGMs against targets in the upper stories of high buildings is sometimes desirable.

(2) The 2.75-inch folding-fin aerial rocket and the 20-mm cannon on Marine attack helicopters are good area weapons to use against enemy forces in the open or under light cover. They are usually ineffective against a large masonry target. The 20-mm cannon produces many ricochets, especially if armor-piercing ammunition is fired into built-up areas.

b. Fixed-Wing Aircraft. Close air support to ground forces fighting in built-up areas is a difficult mission for fixed-wing aircraft. Targets are hard to locate and identify, enemy and friendly forces could be intermingled, and enemy short-range air defense weapons are hard to suppress.

(1) Because enemy and friendly forces may be separated by only one building, accurate delivery of ordnance is critical. Marking panels, lights, electronic beacons, smoke, or some other positive identification of friendly forces is needed.

(2) General-purpose bombs that weigh from 500 to 2,000 pounds are moderately effective in creating casualties among enemy troops located in large buildings. High-dive-angle bomb runs increase accuracy and penetration but also increase the aircraft's exposure to antiaircraft weapons. Low-dive-angle bomb runs using high-drag (retarded) bombs can be used to get bombs into upper stories. Penetration is not good with high-drag bombs. Sometimes aerial bombs pass completely through light-clad buildings and explode on the outside.

(3) Aerial rockets and 20-mm cannons are only moderately effective against enemy soldiers in built-up areas because rockets lack the accuracy to concentrate their effects. The 20-mm cannon rounds penetrate only slightly better than the .50-caliber round, 20-mm armor-piercing rounds can ricochet badly, and tracers can start fires.

(4) The AC-130 aircraft has weapons that can be most effective during combat in built-up areas. This aircraft can deliver accurate fire from a 20-mm Vulcan cannon, 40-mm rapid-fire cannon, and 105-mm howitzer. The 105-mm howitzer round is effective against the roofs and upper floors of buildings. The AC-130 is accurate enough to concentrate its 40-mm cannon and 105-mm howitzer fire onto a single spot to create a rooftop breach that allows fires to be directed deep into the building.

(5) Laser and optically guided munitions can be effective against high-value targets. Specially developed, heavy, laser-guided bombs can penetrate hardened targets. However, problems associated with dense smoke and dust clouds hanging over the built-up area and laser scatter can restrict their use. If the launching aircraft can achieve a successful laser designation and lock-on, these weapons have devastating effects, penetrating deep into reinforced concrete before exploding with great force. If launched without a lock-on, or if the laser spot is lost, these weapons are unpredictable and can travel long distances before they impact.

15. Demolitions. Combat in built-up areas requires the extensive use of demolitions. All Marines, not just engineers, should be trained to employ demolitions. See FM 5-250, *Explosives and Demolitions*, for specific information on the safe use of demolitions.

a. Demolitions. Demolitions come in two types: TNT and C4. Exposed Marines must take cover or move at least 300 meters away from bulk explosives that are being used to breach walls.

(1) TNT comes in 1/4-, 1/2-, and 1-pound blocks. About 5 pounds of TNT are needed to breach a nonreinforced concrete wall that is 12 inches thick if the explosives are laid next to the wall and are not tamped. If the explosives are tamped, about 2 pounds are sufficient.

(2) C4 comes in many different-sized blocks. Normally, it is found in 1 1/4- and 2 1/2-pound blocks within prepackaged satchel charges. About 10 pounds of C4 placed between waist and chest high will blow a hole in the average masonry wall large enough for a man to walk through.

b. Shaped Charges. There are two sizes of shaped charges: a 15-pound M2A3 and a 40-pound M3A3. The M3A3 is the shaped charge that is most likely to be used in built-up areas. It can penetrate 5 feet of reinforced concrete. The hole tapers from 5 inches down to 2 inches. The amount of spall thrown behind the target wall is considerable. There is also a large safety hazard area for friendly forces.

c. Satchel Charges. There are two standard satchel charges: the M183 and the M37. Both come in their own carrying satchel with detonators and blasting cords. Each weighs 20 pounds. The M183 has 16 individual 1 1/4-pound blocks that can be used separately. When used untamped, a satchel breaches a 3-foot-thick concrete wall. Satchel charges are very powerful. Debris is thrown great distances. Friendly forces must move away and take cover before detonation.

d. Cratering Charges. The standard cratering charge is a 43-pound cylinder of ammonium nitrate. This explosive does not have the shattering effect of bulk TNT or C4. It is more useful in deliberate demolitions than in hasty ones.

Appendix C

Attacking and Clearing Buildings

Offensive operations on urbanized terrain will require Marines to attack and clear buildings.

1. Introduction. At the rifle platoon and squad level, the major offensive tasks for combat in built-up areas are attacking and clearing buildings. This involves:

- Suppressing defensive fires
- Advancing Marine assault forces
- Assaulting buildings
- Clearing buildings
- Reorganizing assault forces.

To accomplish these tasks, the assaulting unit task organizes into assault, security, and support forces or elements. (See Appendix A)

2. Requirements. There are five interrelated requirements for attacking a defended building: *fire support, movement, assault, clearing,* and *reorganization.* Proper application and integration of these requirements reduce casualties and hasten accomplishment of the mission. The proper integration of these requirements are determined by METT-T. For example, larger cities may have high-rise apartment buildings and industrial complexes that are separated by parking areas or parks. Increased fire support may be required to suppress and obscure enemy gunners covering the open terrain and spaces between buildings. Conversely, a built-up area with few open areas can provide numerous covered routes that may decrease fire support requirements.

3. Fire Support. Fire support necessary to advance the assault force may include:

- Suppressing and obscuring enemy gunners within the objective building(s) and adjacent structures
- Isolating the objective building(s) with fires to prevent enemy withdrawal, reinforcement, or counterattack
- Breaching walls en route to and in the objective structure
- Destroying enemy positions with direct-fire weapons

- Securing cleared portions of the objective
- Providing replacements for the assault force
- Providing resupply of ammunition and explosives
- Evacuating casualties and prisoners.

a. The size of the support force is determined by the type and size of the objective building(s), whether the adjacent terrain provides open or covered approaches, and the organization and strength of enemy defenses.

b. The support force for a Marine rifle squad may consist of one fire team that could be reinforced with heavy machine guns, antitank weapons, and/or armored vehicles. In situations involving a larger assault force, a rifle platoon or company reinforced with armored vehicles, engineers, and/or artillery may be required to support the movement and assault of an adjacent platoon or company.

c. After seizing objective buildings, the assault force reorganizes and may be required to provide supporting fires for a subsequent assault. Each weapon is assigned a target or area to cover. Individual small-arms weapons place fires on likely enemy weapon positions—loopholes, windows, and roof areas. Snipers are best employed in placing accurate fire through loopholes or engaging long-range targets. M203 fires are directed through windows or loopholes.

d. AT4s and demolitions are employed to breach walls, doors, barricades, and window barriers on the ground level of structures. The tank cannon or LAV-mounted 25-mm gun engages first-floor targets and breaches walls for attacking Marines. Armored vehicle- or wheeled vehicle-mounted machine guns can engage suspected enemy positions on upper floors and in adjacent structures. In addition to destroying or weakening structures, the tank's main gun rounds cause casualties by their explosive effects and by hurling debris throughout the interior of structures.

e. NGF, artillery, and mortars should initially use time fuzes to initially clear exposed enemy personnel, weapons, OPs, and radio sites from rooftops. They should then use delayed fuze action to cause casualties among the defenders inside the structure as a result of shrapnel and falling debris. They may also provide smoke to cover the movement and assault phases of the attack. Artillery can also be used in the direct-fire mode.

f. In support of the movement and assault phases of the attack, buildings can be attacked by fixed-wing aircraft or by attack helicopters.

4. Movement. The assault force minimizes the effects of enemy defensive fires during movement by:

- Using covered routes
- Moving after defensive fires have been suppressed or obscured
- Moving at night or during other periods of reduced visibility
- Selecting routes that will not mask friendly suppressive fires
- Crossing open areas (e.g., streets and spaces between buildings) quickly under the concealment of smoke and suppressive fires
- Moving on rooftops that are not covered by enemy direct fires.

a. The requirement for speed may dictate movement through streets and alleys without clearing all buildings. When moving in this manner, the maneuver force should employ infantry to lead the column, and supported by AAVs, LAVs, wheeled vehicle-mounted machine guns, or tanks following.

b. Marines should move along each side of a street with the leading squad staying abreast of armored vehicles. When not accompanied by armored vehicles, Marines move in a single file along one side of the street under cover of fires from supporting weapons. Individual Marines should be detailed to observe and cover a specific area, such as second-floor windows on the opposite side of the street.

5. Assault. Assault forces quickly and violently execute their assault and subsequent clearing operations. Momentum is maintained to prevent the enemy from organizing a more determined resistance on other floors or in other rooms. Small-unit leaders should keep the assault force moving without allowing the operation to become disorganized.

a. An assault in a built-up area involves the elementary skills of close combat. Leaders must:

- Be trained in the required techniques to defeat the enemy in a face-to-face encounter
- Keep themselves in excellent physical condition
- Have confidence in their abilities and those of their unit.

b. The composition of the assault force will vary depending on METT-T; however, its basic organization at the squad level consists of two assault elements carrying a fighting load of equipment and as much ammunition as possible, especially grenades.

6. Clearing. Entering at the top and fighting downward is the preferred method of clearing a building. Clearing a building is easier from an upper story because gravity and building construction become assets to the assault elements when throwing hand grenades and moving from floor to floor. However, this method is only feasible when access to an upper floor or rooftop can be gained from the windows or roofs of adjoining buildings and assault elements can be transported to the rooftops by helicopter (Figure C-1). Helicopters should land only on those buildings that have special heliports or parking garages on the roofs. When these are unavailable, Marines can rappel or fastrope onto the roof or dismount as the helicopter hovers a few feet above the roof. Assault elements then breach the roof or common walls with explosives and use ropes to enter the lower floors. Stairs are guarded by a covering team or security force when not in use to protect the assault force from a flank attack.

Figure C-1. Helicopters Used To Clear Buildings

a. Although the top-to-bottom method is preferred for clearing a building, assaulting the bottom floor and clearing upward is a common method. When using this technique, the assault force closes on the flank(s) or rear of the building. It then clears each room on the ground floor and then, moving upward, begin a systematic clearance of the remaining floors.

b. Entry can be gained through walls breached by explosives or gunfire. Assault elements should avoid windows and doors because they are usually covered by fire or boobytrapped. If tanks are attached to the company, they can breach the wall by firing their main gun (Figure C-2).

Figure C-2. Main Gun Used To Breach Exterior

c. Shortly before the assault, suppressive fires on the objective should be increased until masked by the advancing assault force. Once masked, fires are shifted to upper windows and continued until the assault force has entered the building. At that time, fires are shifted to adjacent buildings to prevent enemy withdrawal or reinforcement.

d. Hand grenades should be used to clear the space before assault element members enter through a breached wall. Immediately after the explosion, the assault element clearing team enters and uses three round bursts to fire into the interior of the room, concentrating on areas that are possible enemy positions.

e. Once inside the building, the assault force must cover staircases and seize rooms that overlook approaches to the building. These actions are required to isolate enemy forces within the building and to prevent reinforcement. Assault elements should clear each ground floor room and then the basement. Sequence for the assault is that:

(1) The assault force leader determines which room(s) to clear first.

(2) The support force provides suppressive fire while the assault force systematically clears the building. It also provides suppressive fire on adjacent buildings to prevent enemy reinforcements or withdrawal. The support force destroys any enemy trying to exit the building.

(3) The security force isolates the immediate area.

(4) After assault elements establish a foothold in the building, the assault force leader proceeds with the systematic clearing of the building.

f. When clearing a building from the top floor down, assault forces should gain entrance through a common wall or the roof of an adjoining building (Figure C-3). Demolition charges are used to breach the wall and gain entrance to the top floor. Access to lower floors and rooms may be gained by breaching holes in the floor and having the Marines jump or slide down ropes to the lower floors. Stairs can be used if they are first cleared.

g. When using the top-to-bottom method of clearing, requirements remain the same as for other methods. After the floor is breached to gain access to a lower floor, a grenade is allowed to cook-off and is dropped to the lower room. A Marine then sprays the lower room with gunfire using three-round bursts and drops through the mousehole.

Figure C-3. Upper Floors Secured

h. Marines should avoid predictability when entering and clearing rooms. They should vary entry through doors and mouseholes to keep the enemy off guard (Figure C-4). As rooms are cleared, doors should be left open, and a predetermined mark (cloth, tape, spray paint) should be placed on the doorjamb or over the door, as established by unit SOPs.

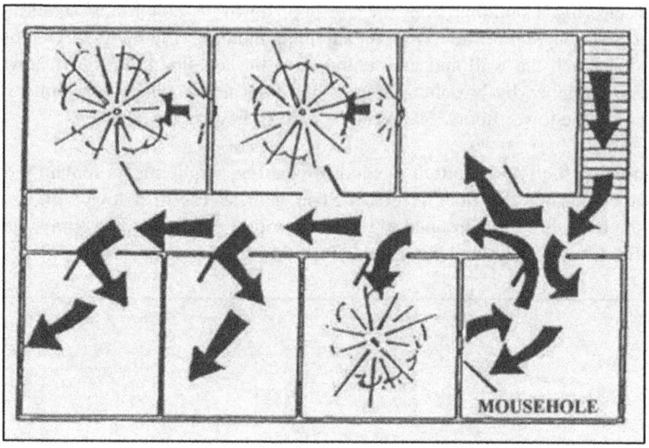

Figure C-4. Varying Techniques for Clearing Rooms

i. If there is a basement, it should be cleared as soon as possible, preferably at the same time as the ground floor. The procedures for clearing a basement are the same as for any room or floor. However, basements often contain entrances to tunnels such as sewers and communications cable tunnels. These should be cleared and secured to prevent the enemy from infiltrating back into cleared areas.

j. Types of buildings that may be cleared include brick buildings and houses, box-wall buildings, heavy-clad framed buildings, and light-clad framed buildings.

(1) Brick Buildings and Houses. The best way to enter a brick building is to blow a breaching hole in the side by using a tank firing HEAT ammunition (Figure C-5). If tanks are not available, a door or window in the rear of the building usually provides better cover and concealment for entry than one in the front. If there is enough cover and concealment, the assault force should enter the rear of the building at an upper level by using a fire escape or grappling hook.

Brick houses usually have similar floor plans on each floor (Figure C-6); therefore, ground floors are cleared the same way as upper floors.

Converely, the floor plans in larger brick buildings typically are different on ground-floor levels than on upper levels (Figure C-7).

Figure C-5. Entering a Brick Building

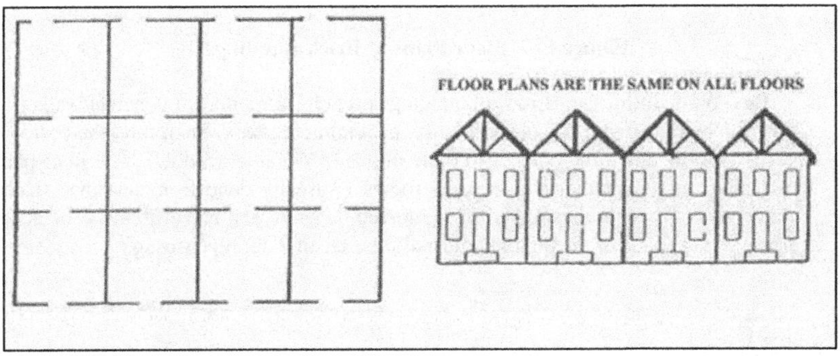

Figure C-6. Brick Houses Have Similar Floor Plans

To clear from building to building, the best method is to move from rooftop to rooftop because the roofs of brick buildings are usually easy to breach. The walls between buildings are at least three bricks thick (total of six bricks between buildings) and require large quantities of demolitions to breach. Walls are normally easier to breach on an upper floor than on a lower floor because the walls are thinner on upper floors. If rooftops are covered by fire and there are not enough demolitions to breach walls between buildings, clearing from rear to rear of buildings is safer than clearing from front to front.

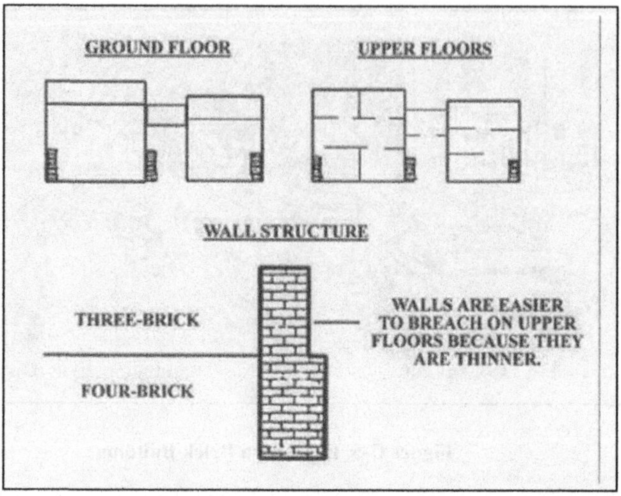

Figure C-7. Floor Plans of Brick Buildings

(2) **Box-Wall Buildings.** Box-wall buildings often have reinforced concrete walls (Figure C-8), which are difficult to breach because of reinforcing bars. Therefore, the best way to enter is to blow down the door or to blow in one of the side windows. The floor plans of these buildings are predictable; clearing rooms is usually done from one main hallway. Interior walls are also constructed of reinforced concrete and are difficult to breach. The stairways at the ends of the building must also be secured during clearing.

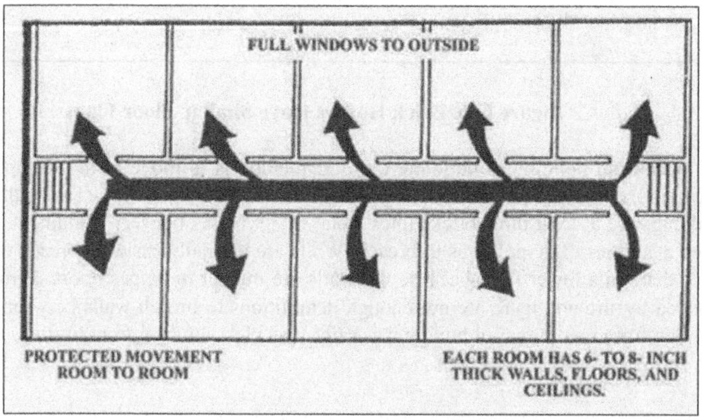

Figure C-8. Box-Wall Buildings

(3) Heavy-Clad Framed Buildings. Tanks can be used to breach a hole in heavy-clad framed buildings. Their floor plans are normally oriented around a stairway or elevator. this must be secured during clearing (Figure C-9). The interior walls of these buildings can be breached, although they may require use of demolitions.

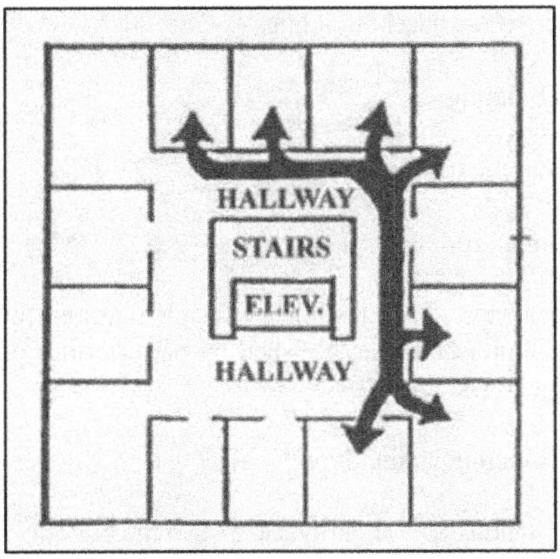

Figure C-9. Heavy-Clad Framed Buildings

(4) Light-Clad Framed Buildings. Light-clad framed buildings (Figure C-10), can be cleared in the same manner as heavy-clad buildings. Secure the central stairway and clear in a circular pattern. Walls are usually thin enough to be breached with an ax.

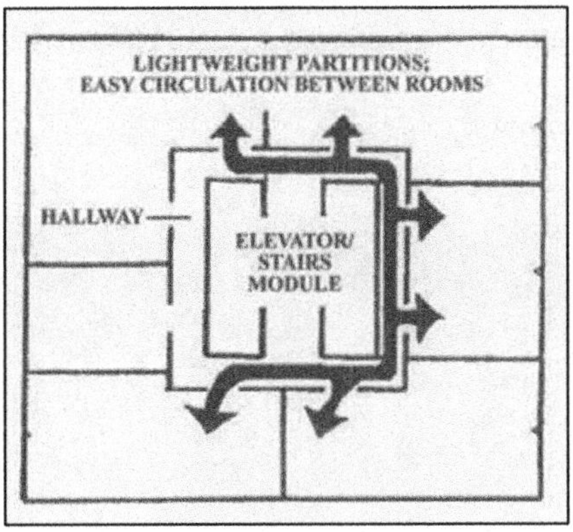

Figure C-10. Light-Clad Framed Buildings

7. Consolidation and Reorganization. Consolidation and reorganization of the assault force in a cleared building must be accomplished quickly to repel enemy counterattacks and to prevent the enemy from infiltrating back into the cleared building. After securing a floor (bottom, middle, or top), selected members of the assault force are assigned to cover potential enemy counterattack routes to the building. Those sentinels alert the assault force and place a heavy volume of fire on enemy forces approaching the building. They should cover:

- Enemy mouseholes between adjacent buildings
- Covered routes to the building
- Underground routes into the basement
- Approaches over adjoining roofs.

As the remainder of the assault force completes search requirements, defensive positions are assigned and tasks are accomplished as established by the priorities of work in the defense. The following actions may then be taken:

- Resupplying and redistributing ammunition
- Marking the building to indicate to friendly forces that the building has been cleared
- Assuming an overwatch mission and supporting an assault on another building
- Treating and evacuating wounded personnel
- Further developing a defensive position if the building is to be occupied for any period.

Appendix D

Fighting Positions

A critical defensive task in combat in built-up areas is the selection and preparation of fighting positions.

1. Considerations. Leaders should consider the following factors when establishing fighting positions:

a. Protection. Leaders should select buildings that provide protection from direct and indirect fires. Reinforced concrete buildings with three or more floors provide suitable protection, while buildings constructed of wood, paneling, or other light materials require reinforcement to gain sufficient protection. One- to two-story buildings without strongly constructed cellars are vulnerable to indirect fires and require construction of overhead protection for each firing position.

b. Dispersion. Normally, a position should not be established in a single building when it is possible to occupy two or more buildings that permit mutually supporting fires. A position in one building without mutual support is vulnerable to bypass, isolation, and subsequent destruction from any direction.

c. Concealment. The fighting position should be concealed. Buildings that are obvious defensive positions should not be selected. The occupation of exposed buildings may be necessary because of requirements for security and fields of fire. Therefore, reinforcement of the structure must be accomplished to provide suitable protection within the building.

d. Fields of Fire. Positions should be mutually supporting and have interlocking fields of fire in all directions. Clearing fields of fire may require the destruction of adjacent buildings by using explosives, engineer equipment, and field expedients. Care should be taken to avoid highlighting the building as a defensive position.

e. Covered Routes. Defensive positions should have at least one covered route that permits resupply, MEDEVAC, reinforcement, or withdrawal from the building. The route can be established by one of the following means:

- Through walls to adjacent buildings
- Through underground systems
- Through communications trenches
- Behind protective buildings.

f. Observation. The building should permit observation of enemy avenues of approach and adjacent defensive sectors.

g. Fire Hazard. Avoid selecting positions in buildings that are a fire hazard. If flammable structures must be occupied, the danger of fire can be reduced by wetting the structure, laying an inch of sand on the floors, and providing fire extinguishers and firefighting equipment. Routes of escape should also be prepared in case of fire.

h. Time. The time available to prepare the defense could be the most critical factor. If enough time is insufficient, buildings that require extensive preparation should not be used. Conversely, buildings located in less desirable areas that require little improvement may be a better choice for a defensive position.

2. Preparation. Preparation of fighting positions depends on proper selection and construction.

a. Selecting Positions. Each weapon should be assigned a primary sector of fire to cover enemy approaches. Alternate positions that overwatch the primary sector should also be selected. These positions are usually located in an adjacent room on the same floor (Figures D-1A, D-1B, and D-1C). Each weapon must be assigned a supplementary position (to engage attacks from other directions) and an FPL.

b. Building Positions. There are many ways to establish a fighting position in a building.

(1) Window Position. Marines should kneel or stand on either side of a window. To fire downward from upper floors, tables or similar objects can be placed against the wall to provide additional elevation, but they must be positioned to prevent the weapon from protruding through the window (Figure D-2). Leaders should inspect positions to determine the width of sector that each position can engage.

(2) Loopholes. To avoid establishing a pattern of always firing from windows, loopholes should be prepared in walls. Marines should avoid firing directly through loopholes to enhance individual protection.

(a) Several loopholes are usually required for each weapon (primary, alternate, and supplementary positions). The number of loopholes should be carefully considered because they can weaken walls and reduce protection. Engineers should be consulted before numerous loopholes are made. Loopholes should be made by punching or drilling holes in walls and should be placed where they are concealed. Blasting loopholes can result in a large hole that can easily be seen by the enemy.

(b) Loopholes should be cone-shaped to obtain a wide arc of fire, to facilitate engagement of high and low targets, and to reduce the size of the exterior aperture (Figure D-3). The edges of a loophole splinter when hit by bullets; therefore, protective linings such as an empty sandbag held in place by wire mesh will reduce spalling effects. When not in use, loopholes should be covered with sandbags to prevent the enemy from firing into or observing through them.

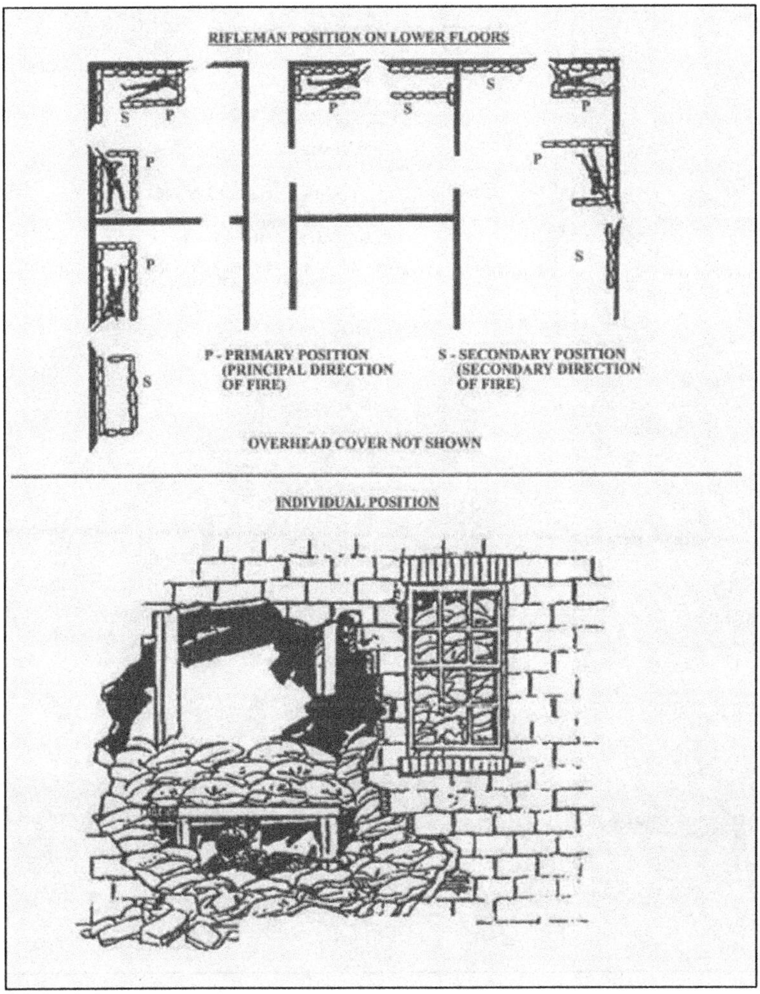

Figure D-1A. Weapon Positions

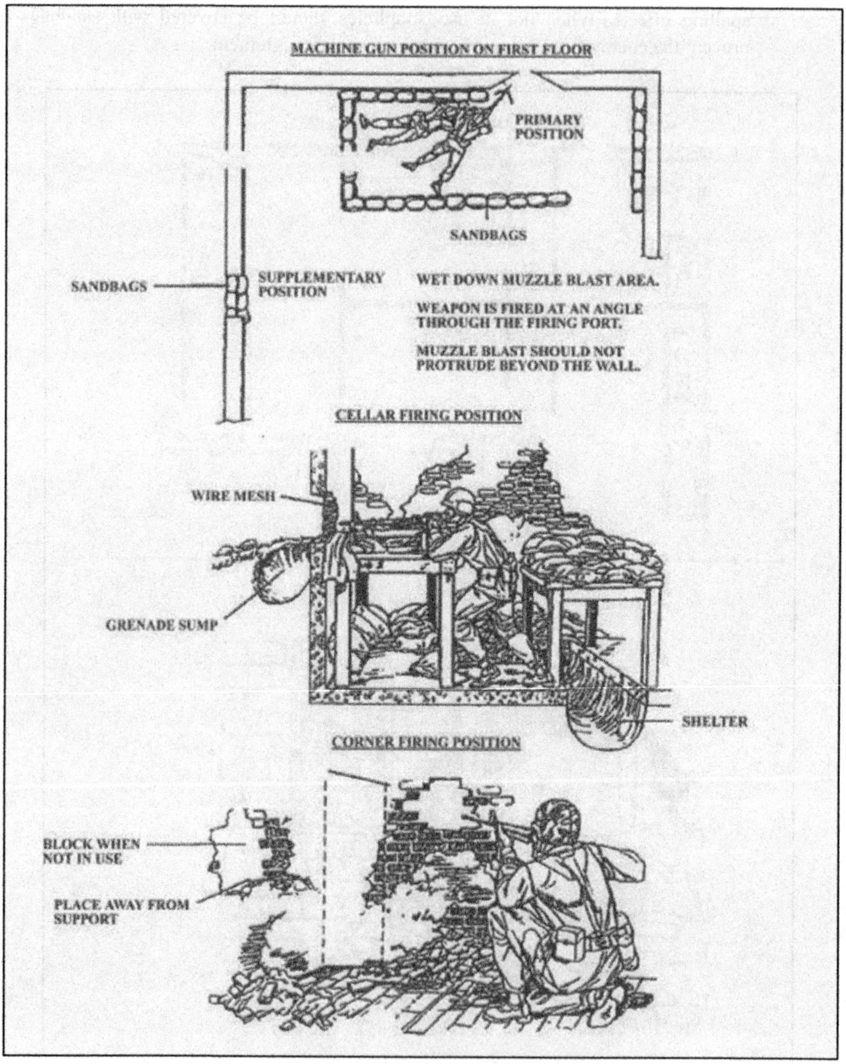

Figure D-1B. Weapon Positions (Continued)

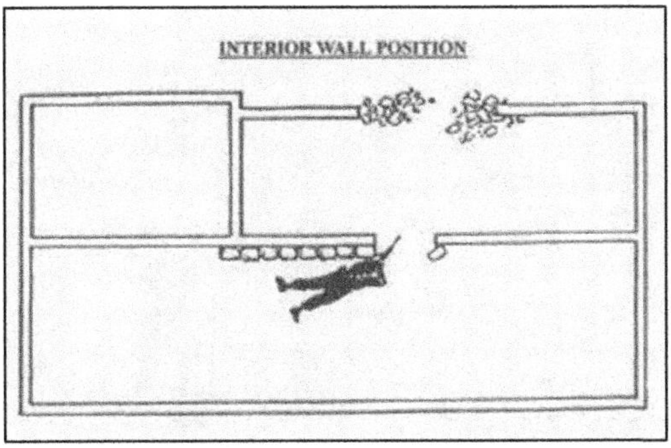

Figure D-1C. Interior Wall Position

Figure D-2. Window Position

(c) Loopholes should also be prepared in interior walls and ceilings of buildings to permit fighting within the position. Interior loopholes should enable overwatch of stairs, halls, and unoccupied rooms and be concealed by pictures, drapes, or furniture. Loopholes in floors permit the defender to engage enemy personnel on lower floors with small-arms fire and grenades.

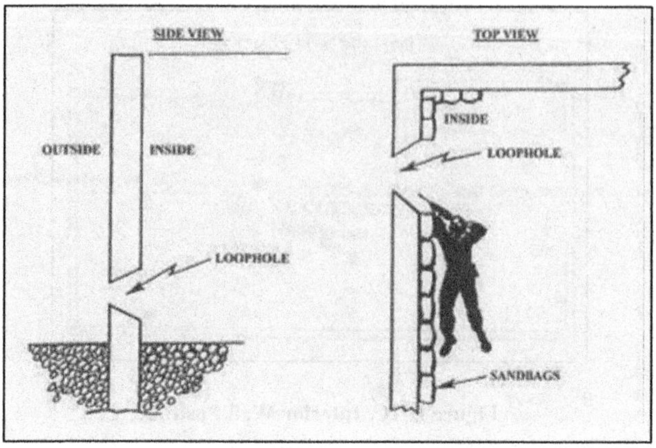

Figure D-3. Cone-Shaped Loopholes

(d) Although walls provide some frontal protection, they should be reinforced with sandbags, furniture filled with dirt, or other expedients. Each position should have overhead and all-around protection (Figure D-4).

Figure D-4. Position With Overhead and All-Around Protection

c. Other Construction Tasks. Other construction tasks may need to be performed. These include work on:

(1) Basements and Ground Floors. Basements require preparation similar to that of the ground floor. Any underground system that is not used by the defender but that could provide enemy access to the position must be blocked.

(a) Doors. Unused doors should be locked, nailed shut, blocked, and reinforced with furniture, sandbags, or other field expedients. Outside doors can be boobytrapped by engineers or other personnel.

(b) Hallways. If not required for the defender's movement, hallways should be blocked with furniture and tactical wire (Figure D-5). If authorized, boobytraps should be employed to impede enemy movement.

(c) Stairs. Defenders should block stairs not used by the defense with furniture and tactical wire (Figure D-5) or remove them. If possible, all stairs should be blocked, and ladders should be used to move from floor to floor and then removed when not being used. Boobytraps should also be employed on stairs.

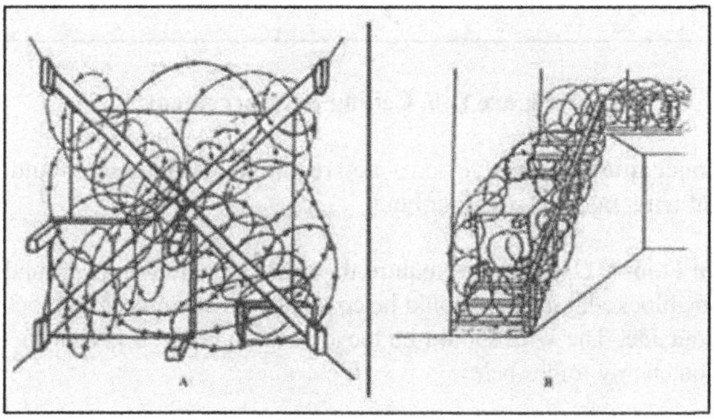

Figure D-5. Blocking Hallways and Stairs

(d) Windows. All glass should be removed. Windows not being used should be blocked with boards or sandbags.

(e) Floors. Fighting positions should be constructed on the ground or basement floors. If there is no basement, fighting positions on the first floor must be constructed to provide additional protection from heavy direct-fire weapons.

(f) Ceilings. Supports that can withstand the weight of rubble from upper floors should be placed under ceilings to provide overhead cover (Figure D-6).

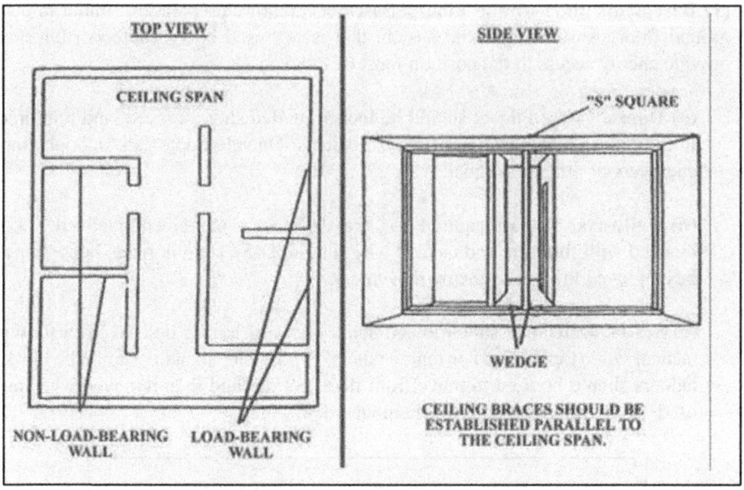

Figure D-6. Ceiling Reinforcement

(g) Unoccupied Rooms. Rooms not required for defense should be blocked with tactical wire, boobytrapped, or both.

(2) Upper Floors. Upper floors require the same preparation as ground floors. Windows need not be blocked, but they should be covered with wire mesh to block grenades thrown from the outside. The wire should be loose at the bottom to permit the defender to drop grenades on enemy forces below.

(3) Interior Routes. Routes are required that permit defending forces to move within the building to engage enemy forces from any direction or reinforce other forces (Figure D-7). Escape routes should also be planned and constructed to permit rapid evacuation of a room or the building. Small holes (called mouseholes) should be made through interior walls to permit movement between rooms. Once the defender has withdrawn to another level, such holes should be clearly marked for both day and night identification. All personnel must be briefed as to where the various routes are located. Rehearsals should be conducted so that everyone becomes familiar with the routes.

(4) Fire Prevention. Buildings that have wooden floors and raftered ceilings require extensive fire prevention measures. The attic and other wooden floors should be covered with about 1 inch of sand or dirt, and buckets of water should be positioned for immediate use. Firefighting materials (dirt, sand, fire extinguishers, and blankets) should be placed on each floor for immediate use. Water basins and bathtubs should be filled as a reserve for firefighting. All electricity and gas should be turned off. Fire breaks can be created by destroying buildings adjacent to the defensive position.

Figure D-7. Movement Between Floors

(5) Communications. Telephone lines should be laid through adjacent buildings or underground systems or buried in shallow trenches. Radio antennas can be concealed by placing them among civilian television antennas, along the sides of chimneys and steeples, or out windows that direct FM communications away from enemy electronic warfare sources and ground observation. Telephone lines within the building should be laid through walls and floors.

(6) Rubbling. Rubbling parts of the building may provide useful cover and concealment for weapons emplacements. Rubbling should be performed only by trained engineers.

(7) Rooftops. Positions in flat-roofed buildings require obstacles that restrict helicopter landings (See Figure F-2). Rooftops that are accessible from adjacent structures should be covered with tactical wire or other expedients and must be guarded. Entrances to buildings from rooftops can be blocked if compatible with the overall defensive plan. Any structure on the outside of a building that could assist in scaling the buildings to gain access to upper floors or to the rooftop should be removed or blocked.

(8) Obstacles. Obstacles should be positioned adjacent to buildings to stop or delay tanks and infantry.

(9) Fields of Fire. Fields of fire should be improved around the defensive position. Selected buildings may be destroyed to enlarge fields of fire. Obstacles to ATGMs, such as telephone or electrical wires, should be cleared. Dead space should be covered with mines and obstacles.

3. Tank and Other Armored Vehicle Positions. Fighting positions for tanks and other armored vehicles are essential to a complete and effective defensive plan in built-up areas.

a. Armored Vehicle Positions. Armored vehicle positions are selected and developed to obtain the best cover, concealment, observation, and fields of fire while retaining the vehicle's ability to move.

(1) If fields of fire are restricted to streets, hull-down positions should be used to gain cover and to fire directly down streets (Figure D-8). From those positions, armored vehicles are protected and can rapidly move to alternate positions. Buildings collapsing from enemy fires are a minimal hazard to armored crews.

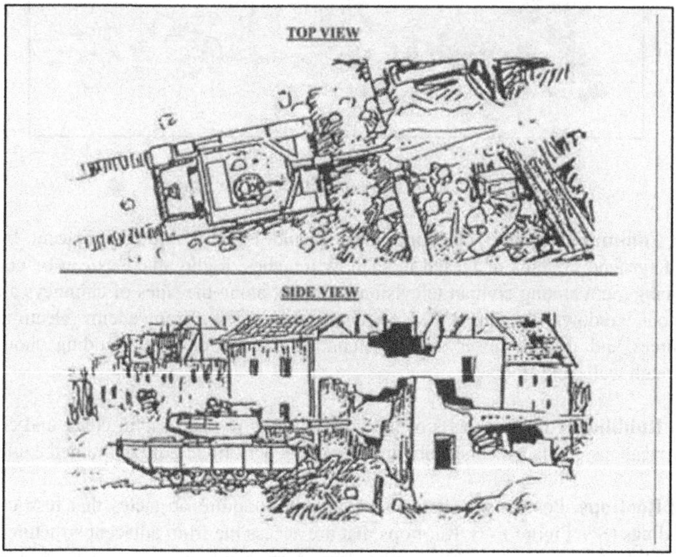

Figure D-8. Hull-Down Position

(2) The hide position (Figure D-9) covers and conceals the vehicle until time to move into position for engagement of targets. Because the crew will not be able to see advancing enemy forces, an observer from the vehicle or a nearby infantry unit must be concealed in an adjacent building to alert the crew. The observer acquires the target and signals the armored vehicle to move to the firing position and fire. After firing, the vehicle moves to an alternate position to avoid compromising one location.

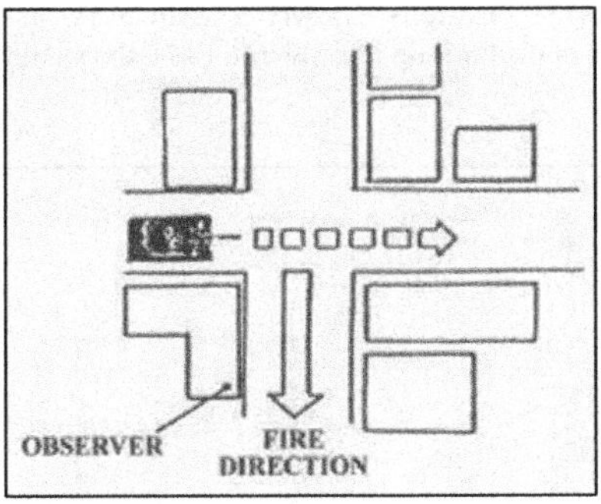

Figure D-9. Hide Position

(3) The building hide position (Figure D-10) conceals the vehicle inside a building. If basement hide positions are inaccessible, engineers can evaluate the building's floor strength and prepare for the vehicle. Once the position is detected, it should be evacuated to avoid enemy fires.

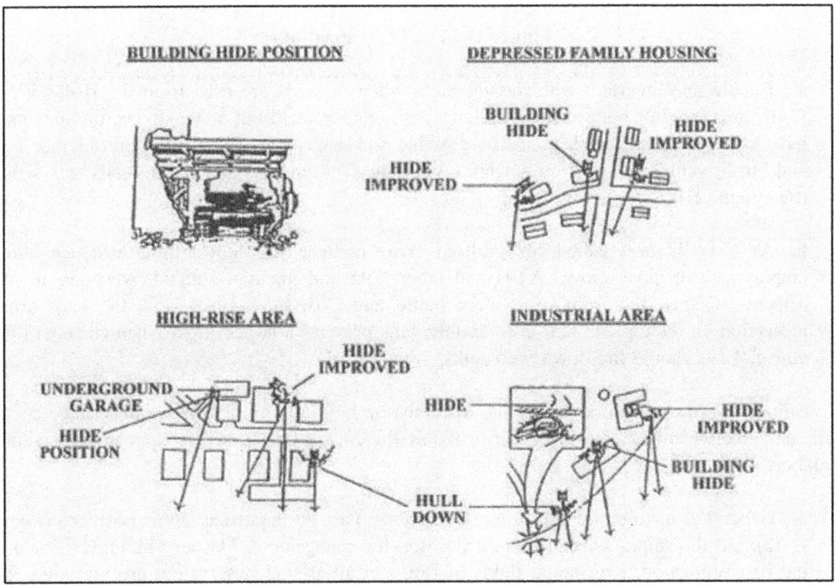

Figure D-10. Building Hide Position

4. Antitank Guided Missile Positions. HMMWVs with ATGMs are employed in areas that maximize their capabilities in the built-up area. Figure D-11 shows mounted and dismounted firing positions.

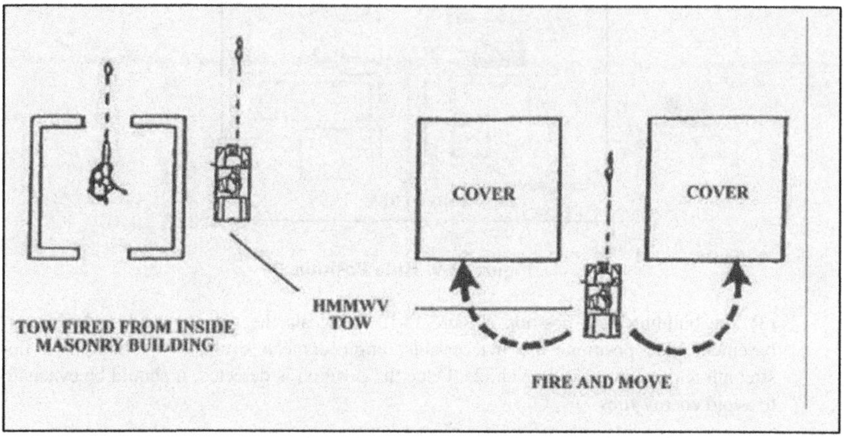

Figure D-11. ATGM Positions

a. Rubble may interfere with missile flight when ATGMs are fired from the HMMWV or from street-level or bottom-floor fighting positions. When firing down streets, missiles must have at least 1 meter of clearance over rubble. Other obstacles to missile flight include trees and brush, vehicles, television antennas, buildings, power lines and wires, walls, and fences (See Figure B-7).

b. An AT4 is best suited for built-up areas because its short arming distance allows employment at close range. AT4s and other light and medium antitank weapons are not effective against the front of modern battle tanks. Because tanks have the least armor protection on the top and rear deck and the tank presents a larger target when engaged from above, AT4s should fire down onto tanks.

5. Sniper Positions. Snipers contribute to combat in built-up areas by firing on selected enemy targets. Effective sniper employment can trouble the enemy far more than its relative personnel numbers reflect.

a. General areas (a building or group of buildings) are designated as sniper positions (Figure D-12), but the sniper selects the best position for engagement. Masonry buildings that offer the best protection, long-range fields of fire, and all-around observation are preferred. The sniper also selects several secondary and supplementary positions to cover his areas of responsibility.

b. Engagement priorities for snipers are determined by the relative importance of the targets to the effective operations of the enemy. Sniper targets may include tank commanders, crewmen of direct fire support weapons, crewmen of crew-served weapons, officers, FOs, and radiotelephone operators.

c. The urban environment often limits snipers to firing down or across streets. Snipers can be employed to cover rooftops, obstacles, dead space, and gaps in final protective fires.

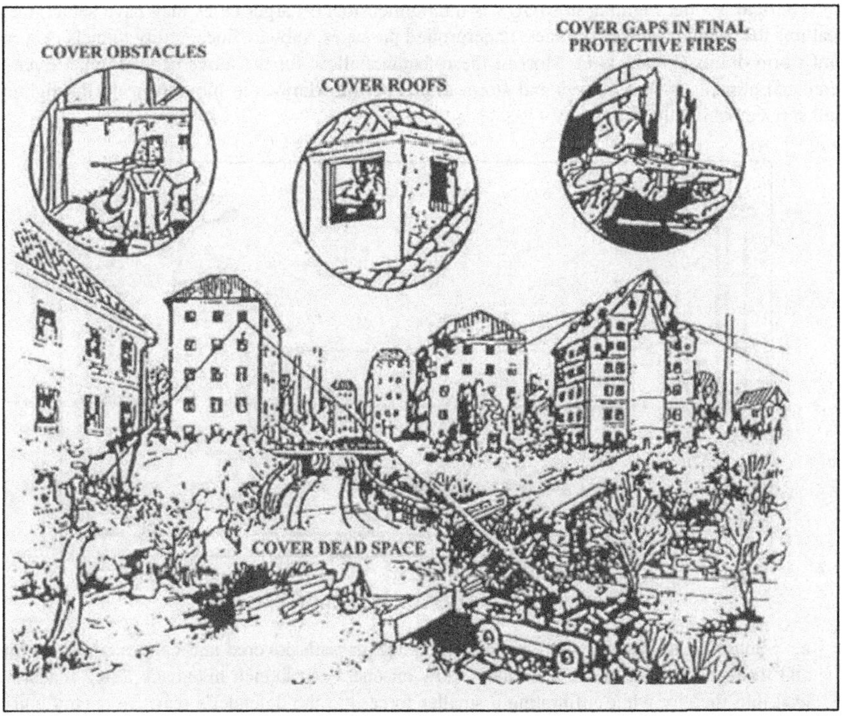

Figure D-12. Sniper Positions

Appendix E

Subterranean Operations

Knowledge of the nature and location of underground facilities is valuable to both the urban attacker and defender.

1. Tactical Value. Fighting in MOUT is multidimensional. Larger cities may have subterranean features that include sunken garages, underground passages, subway lines, utility tunnels, sewers, and storm drains (Figure E-1). Most of these features allow for the movement of units. Even in smaller European towns, sewers and storm drains permit Marines to move beneath the fighting and surface behind the enemy.

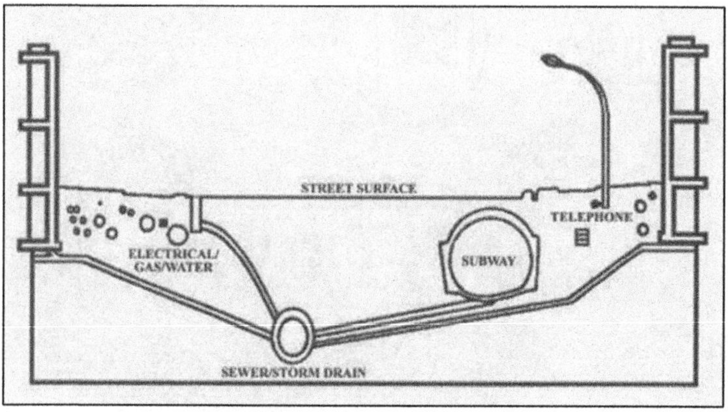

Figure E-1. Potential Underground Passages

a. Subterranean passages may provide the attacker with covered and concealed routes into and through built-up areas. This may allow an enemy to launch his attack along roads that lead into the city while infiltrating a smaller force into the defender's rear. An enemy unit in the defender's rear area may disrupt his defense and obstruct the avenues of withdrawal for his forward defense.

b. An attack along a subterranean avenue of approach could become the main effort. Even if the subterranean effort is not immediately successful, it forces the defender to fight on atleast two levels and to extend resources to more than just street-level fighting.

c. The existence of subterranean passages forces the defender to defend the built-up area above and below ground. While these passages are usually more of a disadvantage to the defender than to the attacker they may offer some advantages in the defense. When thoroughly reconnoitered and controlled by the defender, subterranean passages provide excellent covered and concealed routes to move reinforcements or to launch counterattacks. They also provide ready-made covered lines of communication for the movement of supplies and evacuation of casualties and provide places to cache supplies for forward companies. Subterranean passages also offer the defender a ready-made conduit for communications wire, with protection from armored vehicles and indirect fires.

2. Denial to the Enemy. Subterranean passages are useful to the defender only to the extent that the attacker can be denied their use. The defender has an advantage in that, given the confining, dark environment of these passages, a small group of determined Marines in a prepared position can defeat a numerically superior force.

a. Tunnels afford the attacker little cover and concealment except for the darkness and any manmade barriers. The passageways provide tight fields of fire and amplify the effects of grenades. Obstacles at intersections in tunnels support excellent ambush sites and may turn the subterranean passages into a maze. These obstacles can be created quickly by using chunks of rubble, furniture, and parts of abandoned vehicles interspersed with M18 claymore mines.

b. A thorough reconnaissance of the subterranean or sewer system is required. To be effective, obstacles must be located at critical intersections in the passage network so that they trap attackers in a kill zone but allow defenders freedom of movement (Figure E-2).

3. Subterranean Reconnaissance Techniques. The mission of local reconnaissance for a platoon- or company-sized area of responsibility should be given to a squad-sized element.

a. The squad leader tasks one fire team with providing forward security. One fire team follows in support, and one fire team provides rear security (Figure E-3). The squad leader moves directly behind the pointman and records data collected by the patrol. If available, an engineer should be with the lead fire team. The rear security fire team protects the ground- and subterranean-level point of entry. This fire team is responsible for detecting enemy movement on the patrol's rear and serves as the communications link between the squad leader and higher headquarters. Organization of the patrol may vary based on METT-T.

b. The squad leader should carry a map, compass, street plan, and notebook in which he has detailed the information that he must gather. One Marine is assigned to carry the tools needed to open manhole covers. If the patrol is to move more than 200 to 300 meters, or if the platoon leader directs, the squad may carry a sound-powered telephone (TA-1) and wire dispenser (XM-306A) for communications. Radios are often unreliable in this environment. Marines (especially the pointman, squad leader, engineer, and fire team leaders) should be equipped with night vision devices to maintain surveillance within the sewer. Thermal night devices are also useful as there may not be enough ambient light for accurate visibility with night vision devices.

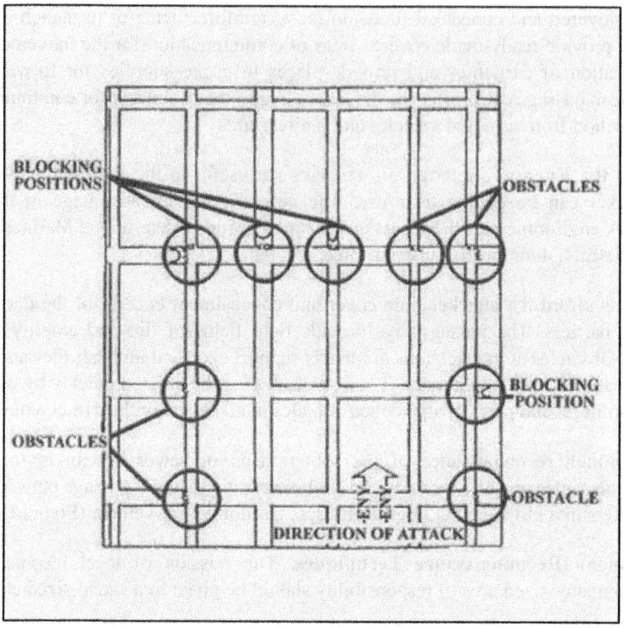

Figure E-2. Defense of a Sewer System

Figure E-3. Organization of the Patrol Unit

c. Marines entering the sewer should carry a sketch of the sewer system that includes magnetic north, azimuths, distances, and manholes. They should also carry protective masks, flashlights, gloves, and chalk (or chemiluminescent lights) for marking features along the route. The patrol should also be equipped with a safety rope to which each man is tied. To improve their footing in slippery sewers and storm drains, the squad members may wrap chicken wire or screen wire around their boots.

d. A constant concern to units conducting a subterranean patrol is chemical defense. Enemy chemical agents used in tunnels may be suddenly encountered in dense concentrations with little dispersion. The pointman should carry an automatic chemical agent alarm system to provide instant warning of the presence of chemical agents. M8 detector paper can also be used to detect chemical agents. At the first indication that harmful gases are present, the squad should mask.

e. In addition to enemy chemical agents, noxious gases from decomposing sewage can also pose a threat. These gases are not detected by chemical agent alarm systems, nor are they completely filtered by the protective mask. Physical signs that indicate their presence in harmful quantities are nausea and dizziness. The squad leader should be constantly alert to these signs and know the shortest route to the surface for fresh air. Some sewer gases deplete oxygen, a situation against which the field protective mask provides no protection.

f. Once the squad is organized and equipped, it moves to the entrance of the tunnel. This may be a manhole. With the manhole cover removed, the squad waits 15 minutes before entry to allow any gases that may be present to dissipate. The pointman then descends into the tunnel to determine whether the air is safe to breathe and whether movement is restricted. The pointman should remain in the tunnel for 10 minutes before the rest of the squad follows. If he becomes ill or is exposed to danger, he can be pulled out by the safety rope.

g. When the squad is moving through the tunnel, the pointman moves about 10 meters in front of the squad leader. Other squad members maintain 5-meter intervals. If water in the tunnel is flowing faster than 2.5 meters per second or the sewer contains slippery obstacles, those intervals should be increased to prevent all patrol members from falling if one man slips. All squad members should stay tied into the safety rope so that they can easily be retrieved from danger. The last Marine in the patrol marks the route with chalk or chemiluminescent lights so other units can find the squad.

h. The squad leader should record the azimuth and pace count of each turn he takes in the tunnel. When he encounters a manhole to the surface, the pointman should open it and determine the location, which the patrol leader then records. The use of recognition signals (Figure E-4) prevents friendly forces from accidentally shooting the pointman as he appears at a manhole.

Figure E-4. Recognition Signals

i. Once the patrol has returned and submitted its report, the platoon leader must decide how, or whether, to use the tunnel. In the offense, a tunnel might provide a covered route to move behind the enemy's defenses. In the defense, a tunnel might provide a covered passage between positions. In any case, the squad members that provided the reconnaissance should act as guides along the route.

j. If the tunnel is to be blocked, the platoon may emplace concertina wire, early warning devices, and antipersonnel mines. A security force should be established at the entrance of the sewer (Figure E-5) to provide warning against any enemy trying to approach the platoon's defense. It should be equipped with command-detonated illumination. The overpressure from grenades, mines, and boobytraps exploding in a sewer or tunnel can have adverse effects on friendly forces, including ruptured eardrums and wounds from flying debris. Also, gases found in sewers can be ignited by the blast effects of these munitions. For these reasons, small-arms weapons should be employed in tunnels and sewers. Friendly personnel should be outside of tunnels or out of range of the effects when mines or demolitions are detonated.

4. Psychological Considerations. Combat operations in subterranean passages are similar to night combat operations. The psychological factors of night operations reduce confidence, cause fear, and increase a feeling of isolation. This feeling of isolation is further magnified by the tight confines of the tunnels. The layout of tunnels could require greater dispersion between positions.

Figure E-5. Security Position Established at the Entrance to a Sewer

a. Leaders must enforce measures to help dispel the feelings of fear and isolation experienced when operating in tunnels. These measures include leadership training, physical and mental fitness, sleep discipline, and stress management.

b. Leaders maintain communication with Marines manning positions in tunnels either through personal visits, messengers, or field telephone. Communication increases situational awareness and provides insight into the mental state of Marines. Marines manning positions below ground should be given as much information as possible on the organization of the tunnels and the importance of the mission. They should be briefed on contingency plans and alternate positions should their primary positions become untenable.

c. Physical and mental fitness can be maintained by periodically rotating Marines out of tunnels so they can stand and walk in fresh air and sunlight. Stress management is also a factor of operations in tunnels. Historically, combat in built-up areas has been one of the most stressful forms of combat. Fighting in continuous darkness and restricted maneuver space causes even more stress to Marines than street fighting.

Appendix F

Obstacles, Mines, and Demolitions

Obstacles are used extensively in combat in built-up areas to allow the defender to canalize the enemy, impede his movement, and disrupt his attack.

Section I

Obstacles

1. Introduction. Obstacles are designed to prevent movement by personnel, to separate infantry from tanks, and to slow or stop vehicles. (See MCRP 3-17A, *MAGTF Breaching Operations.*)

2. Types. Antipersonnel mines, barbed wire, boobytraps, and exploding flame devices are used to construct antipersonnel obstacles (Figure F-1). (See FMFRP 13-5/FM 5-34, *Engineer Field Data*; FM 5-250, *Explosives and Demolitions*; and FM 3-11, *Flame, Riot Control Agent, and Herbicide Operations,* for more detailed information.) These obstacles are used to block the following infantry approaches:

- Streets
- Buildings
- Roofs
- Open spaces
- Dead space
- Underground systems.

a. The approval authority to boobytrap buildings is normally the GCE commander. This authority may be delegated to subordinate commanders.

b. The three types of wire obstacles used in defensive operations are protective, tactical, and supplementary. Wire obstacles should be integrated with other obstacles and covered by fire and observation.

(1) Protective wire is usually located beyond hand-grenade range (40 to 100 meters) from the defensive position.

(2) Tactical wire is positioned to increase the effectiveness of friendly weapons fire. Tactical wire is usually positioned on the friendly side of machine gun FPLs.

(3) Supplementary wire is used to break up the pattern of tactical and protective wire to prevent the enemy from easily locating friendly weapons.

c. Obstacles may be placed in dead spaces to restrict infantry movement in areas that cannot be observed and that are protected from direct fires.

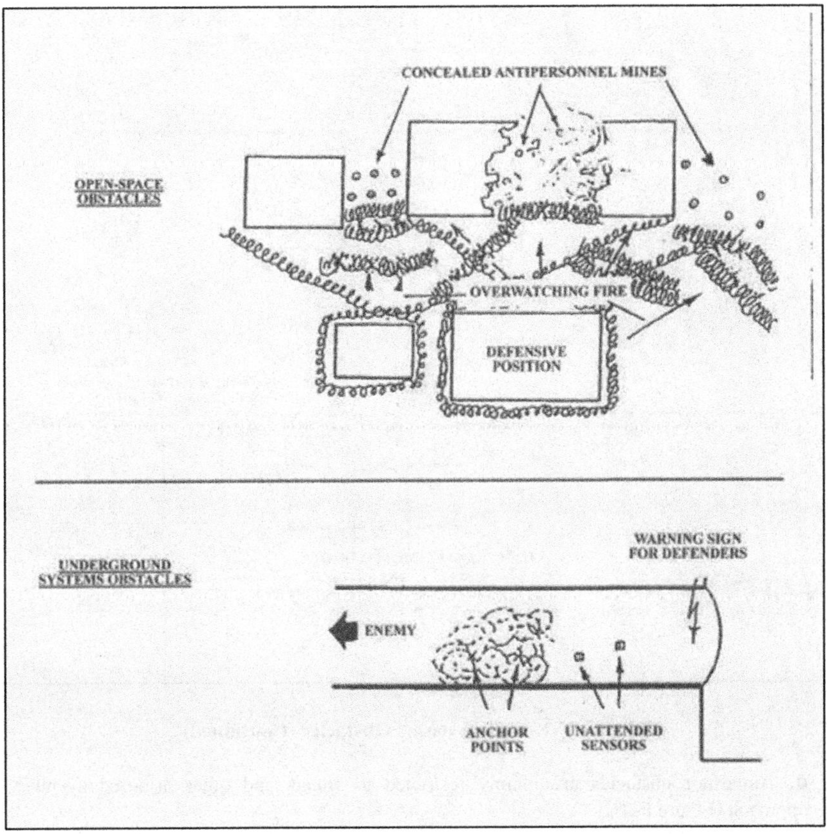

Figure F-1. Antipersonnel Obstacles

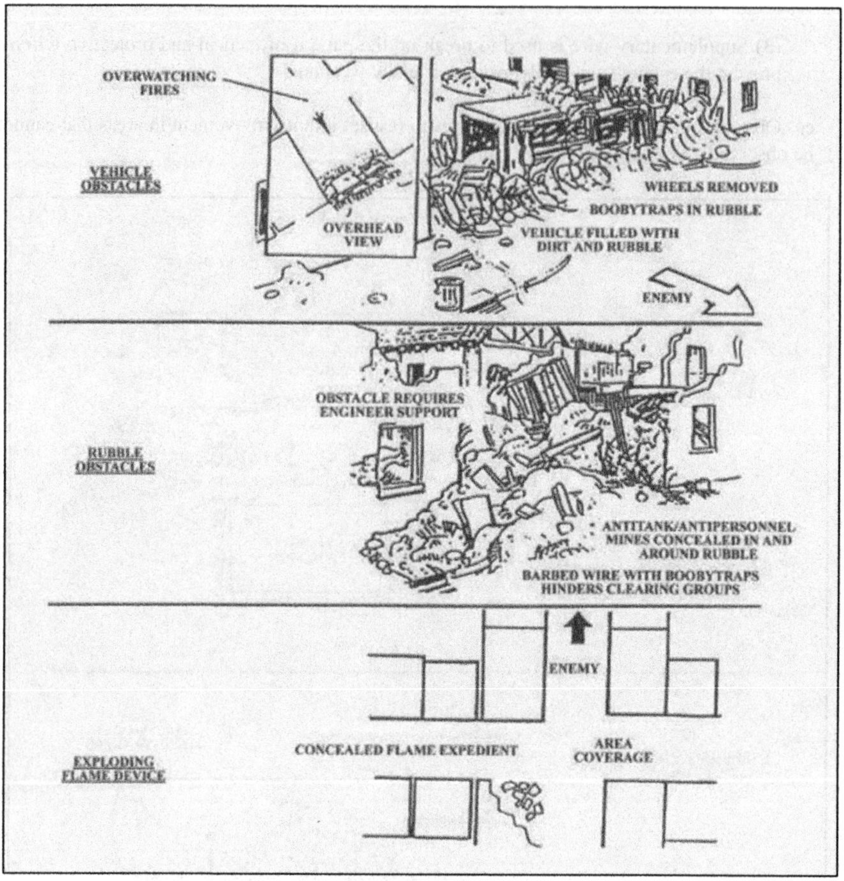

Figure F-1. Antipersonnel Obstacles (Continued)

d. Antiarmor obstacles are mainly restricted to streets and other armored avenues of approach (Figure F-2).

e. Some antitank mines may be sited and employed as antihelicopter mines to cover probable HLZs (Figure F-2).

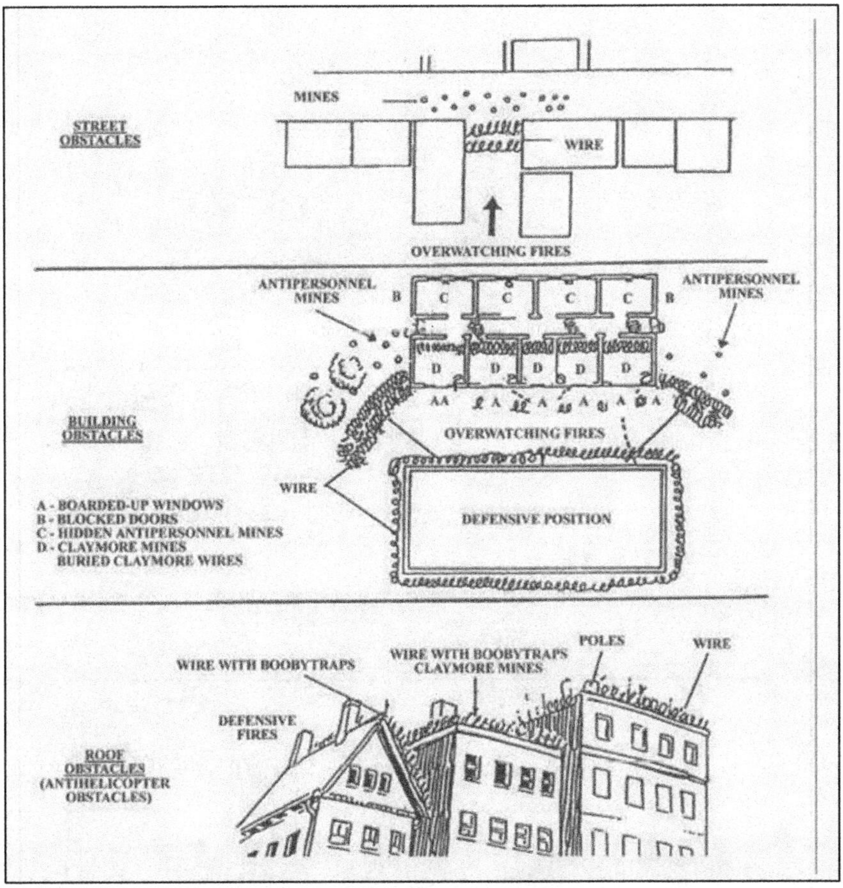

Figure F-2. Antiarmor Obstacles

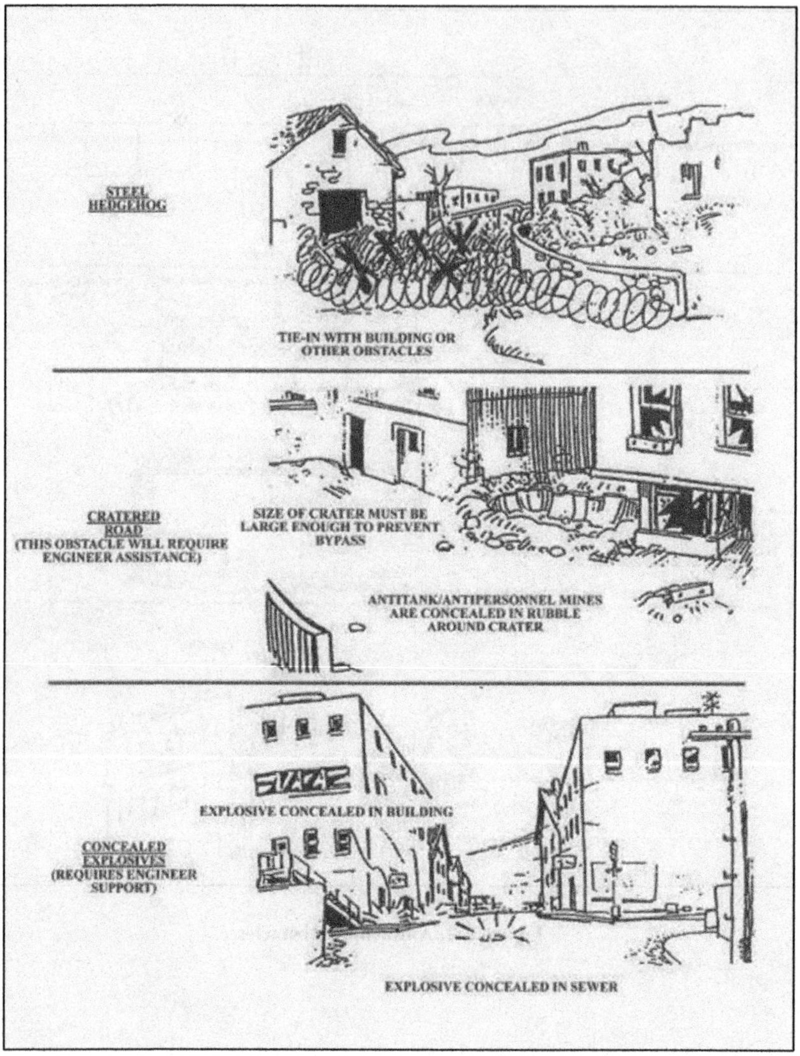

Figure F-2. Antiarmor Obstacles (Continued)

3. Construction of Obstacles. Obstacles are constructed in buildings to deny enemy infantry covered routes and weapons positions that are close to friendly defensive positions. They can be constructed of explosives, by using wire, or by using boobytraps within buildings. The building may be prepared as an explosive or flame trap for command detonation after enemy occupation.

Section II

Mines

4. Introduction. The employment or discovery of mines in built-up areas should be recorded on a building sketch (Figure F-3). (See FM 20-32.) The sketch should include the number of the building (taken from a city map, if available) and all floor plans. It should also include the type of mine and firing device. When possible, mined buildings should be marked on the friendly side (Figure F-4). Clearing areas or buildings that have been mined is extremely difficult and dangerous. Therefore, such areas should be considered "no go" areas for infantry. This factor must be carefully considered when planning and authorizing the placement of mines. (See Table F-1 for notional approving authority for minefields, but with the understanding that information in that table is subject to change with international agreements and ROE.)

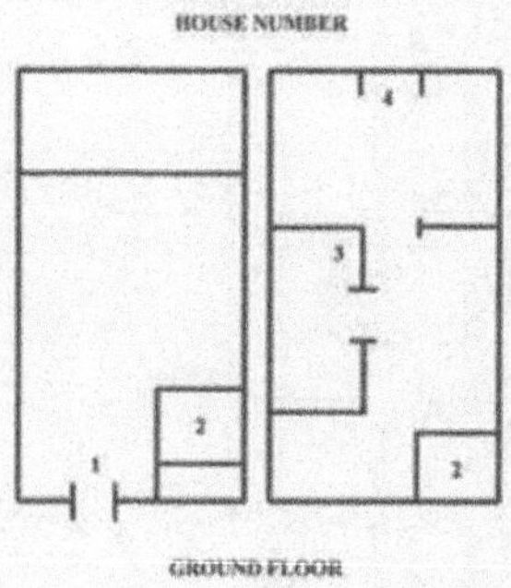

MINE	LOCATION	TYPE	FIRING DEVICE
1	FRONT DOOR	CLAYMORE	COMMAND
2	STAIRS	CLAYMORE	COMMAND
3	CLOSET	CLAYMORE	COMMAND
4	DRESSER	CLAYMORE	COMMAND

Figure F-3. Building Sketch Showing Mines

Figure F-4. Marking Mined Buildings

	Minefields (Type)		FASCAM (Duration)		Other Obstacles	
Level	Tactical	Protective	Long	Short	Tactical	Protective
MEF	X	X	X	X	X	X
Division	X	X	X	X	X	X
Regiment	X	X	X	X	X	X
Battalion	X	X		X	X	X
Company		X				X

Table F-1. Notional Minefield Employment Authority Matrix

5. **Types.** Several types of mines may be encountered or employed in built-up areas.

a. The M14 mine (a plastic type of antipersonnel landmine) cannot currently be used by American forces due to the Leahy Amendment of 1995 and the Certain Conventional Weapons Convention, of which the United States is a signatory nation. It may, however, be encountered in MOUT or be employed by coalition or enemy forces. For this reason, the following information is given as a guide for what U.S. forces may encounter. This type of antipersonnel landmine is normally used in conjunction with metallic antipersonnel, antitank, or chemical mines to confuse and hinder enemy breaching attempts. When encountered, it must be carefully handled because its light weight makes it easy to displace (Figure F-5). However, its size makes it ideal for covering obscure places such as stairs and cellars.

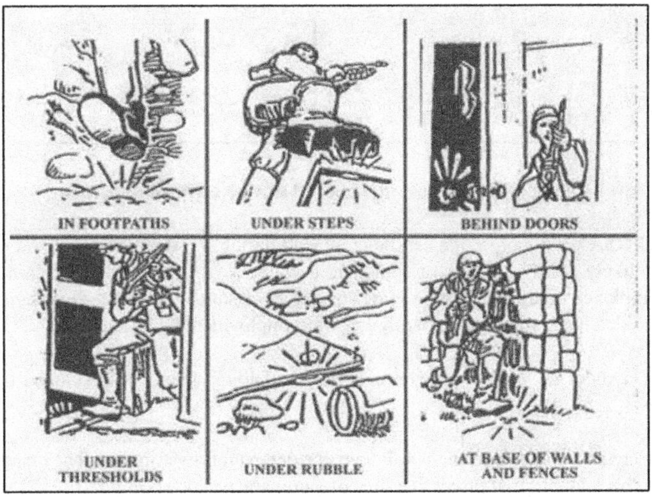

Figure F-5. Typical Employment of the M14 Antipersonnel Mine

b. The M16A2 antipersonnel landmine is subject to the same legal limitations as the M14. It is a boundary type of antipersonnel landmine, which makes it ideal for covering large areas such as rooftops, backyards, parks, and cellars. It can be expediently rigged for command detonation by attaching a rope or piece of communications wire to the pull ring (Figure F-6).

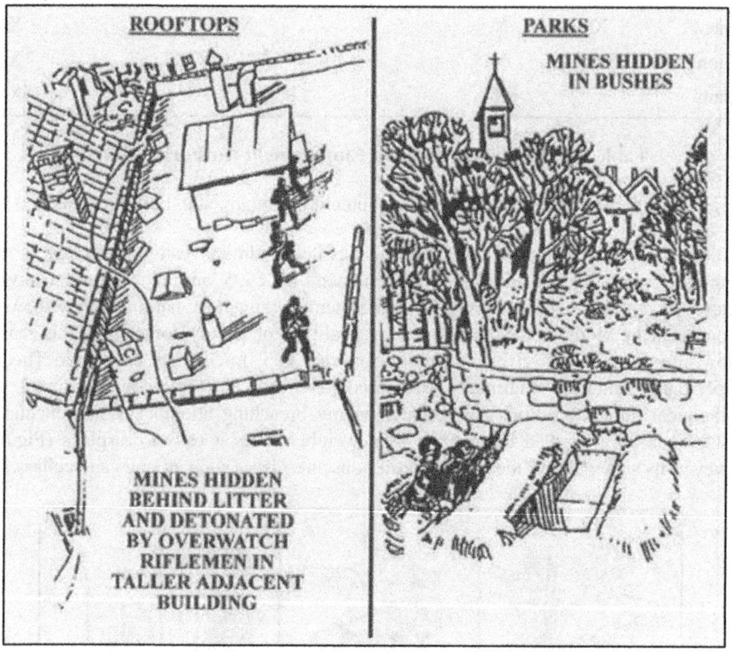

Figure F-6. Emplacements of M16A2 Antipersonnel Mine

c. The M18A1 claymore mine can be employed during the reorganization and consolidation phase on likely enemy avenues of approach. It does not have to be installed in the street, but can be employed on the sides of buildings or any other sturdy structure. The claymore mine, in command-detonated mode, is currently legal for employment by U.S. forces.

(1) Claymore mines can also be used for demolition against thin-skinned buildings and walls.

(2) Claymore mines may be configured for detonation by tripwire. They can help fill the dead space in the final protective fires of automatic weapons (Figure F-7).

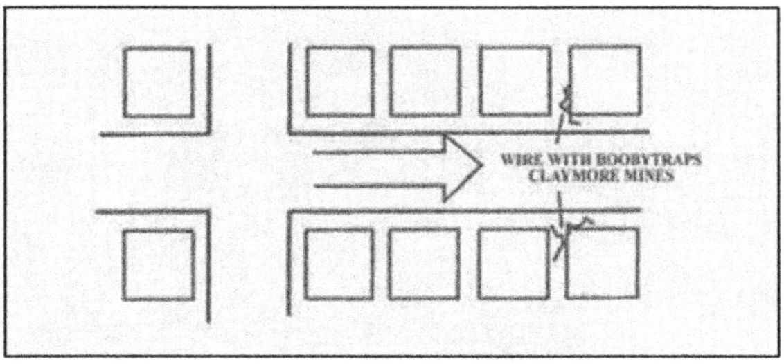

Figure F-7. Claymore Mine Used To Cover Dead Space of Automatic Weapons

(3) Claymore mines can be used in several ways in the offense. For example, if friendly forces are advancing on a city, claymore mines can be used in conjunction with blocking positions to cut off enemy avenues of escape (Figure F-8).

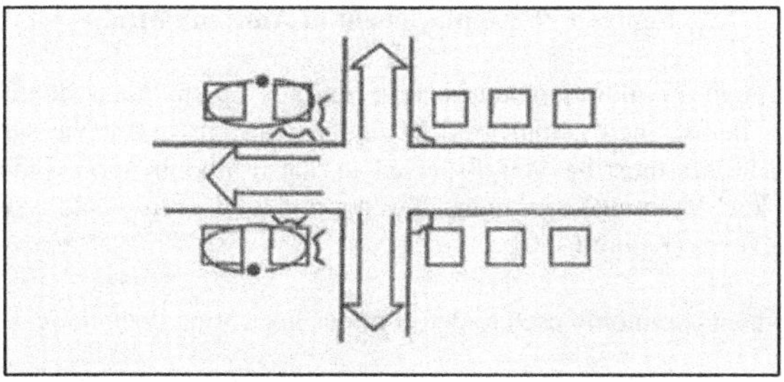

Figure F-8. Claymore Mines Used To Block Enemy Avenues of Escape

d. The M15, M19, and M21 antitank mines are used to stop or turn enemy armor and should be employed (Figure F-9):

- In conjunction with other manmade obstacles and covered by observation and fire
- In streets or alleys to block routes of advance in narrow defiles
- As command-detonated mines with other demolitions.

6. Enemy Mines and Boobytraps. Buildings contain many areas and items that are potential hiding places for boobytraps, for example, doors, windows, telephones, stairs, books, canteens, and so on.

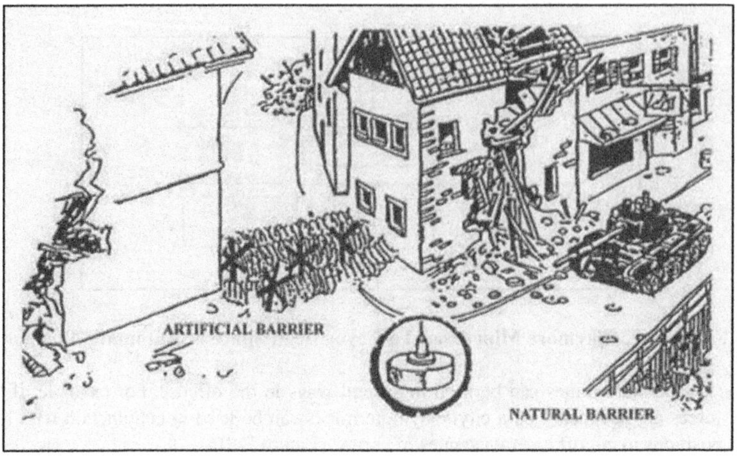

Figure F-9. Emplacement of Antitank Mines

When moving through a building, Marines must not pick up anything—food, valuables, enemy weapons, or other items. Such items could be rigged with pressure devices that explode when they are moved. Marines must be well-dispersed so that if a boobytrap explodes, the number of casualties will be few. Many different types of mines and boobytraps could be encountered during combat in built-up areas (Figure F-10).

a. The equipment commonly used to detect mines in clearing operations includes:

- Mine detectors
- Probes
- Grappling hooks
- Ropes
- Explosives
- Flak vests
- Eye protection
- Engineer tape.

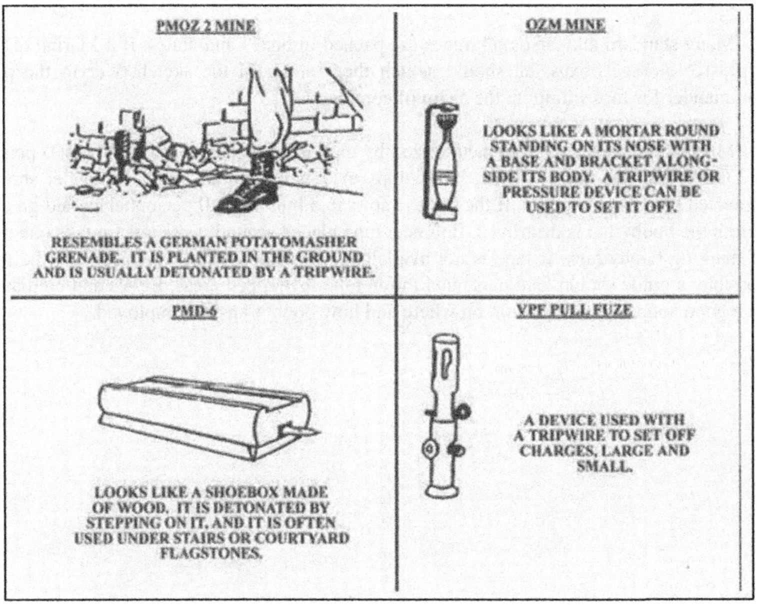

Figure F-10. Common Threat Antipersonnel Mines and Boobytraps

b. To detect tripwires, a Marine can use a 10-foot pole with 5 feet of string tied on one end. A small weight attached to the loose end of the string will snag on the tripwire. This allows the lead Marine to easily detect a tripwire (Figure F-11). Marines can also use a flashlight to cause reflections and shadows to find tripwires. A wand with a dangling thread can be swept over the ground ahead with the thread being watched for any disturbance by a tripwire.

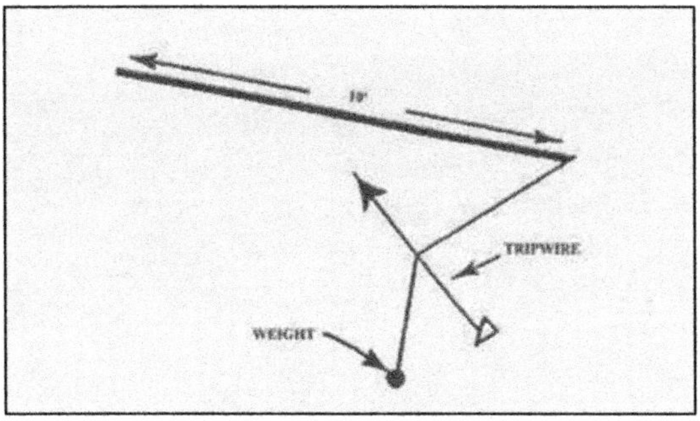

Figure F-11. Tripwire Detection

c. Many standard antipersonnel mines are packed in boxes and crates. If a Marine discovers explosive storage boxes, he should sketch them and turn the sketch over to the platoon commander for forwarding up the chain of command.

d. Most boobytraps should be neutralized by explosive ordnance disposal (EOD) personnel. If EOD teams are not available, boobytraps may be blown in place. Personnel should be protected by adequate cover. If the boobytrap is in a building, all personnel should go outside before the boobytrap is destroyed. Engineer tape placed around the danger area is one method of marking boobytraps. If tape is not available, strips ripped from bedsheets can be used. If possible, a guide should lead personnel through boobytrapped areas. EPWs and civilians may be a good source of information on where and how boobytraps are employed.

Section III

Demolitions

7. Introduction. Demolitions are used more often during combat in built-up areas than during operations in open natural terrain. Demolition operations should be directed by the engineers that support the GCE. However, if engineers are involved in the preparation and execution of the barrier plan, other trained Marines can prepare mouseholes, breach walls, and rubble buildings themselves if assisted and advised by attached engineers.

8. Offensive Use. When assaulting or clearing a built-up area, demolitions may enable the GCE commander to create an avenue of approach through buildings.

a. Every other man in an assault force should carry demolitions, and other selected personnel should carry blasting caps. The same man should *not* carry both the explosives and the blasting caps because of the danger of sympathetic detonation. As the demolitions are expended by the assault force, they should be replaced by explosives carried by the support force.

b. One of the most difficult breaching operations faced by the assault force is the breaching of masonry and reinforced concrete walls. When demolitions must be used, composition C4 is the ideal explosive to use. Normally, building walls are 15 inches thick or less. Assuming that all outer walls are constructed of reinforced concrete, a rule of thumb for breaching is to place 10 pounds of C4 against the target between waist and chest height. When detonated, this normally blows a hole large enough for a man to go through. The amounts of TNT and C4 required to breach concrete are shown in Table F-2.

c. Mouseholes provide the safest method of moving between rooms and floors. They can be created with C4. Because C4 comes packaged with an adhesive backing or can be emplaced using pressure-sensitive tape, it is ideal for this purpose. When using C4 to blow a mousehole in a lath and plaster wall, one block or a strip of blocks should be placed on the wall from neck to knee height. Charges should be primed with detonating cord or electrical blasting caps to obtain simultaneous detonation, which will blow a hole large enough for a man to fit through.

9. Defensive Use. The use of demolitions in defensive operations is similar to their use in offensive operations. When defending a built-up area, demolitions are used to create covered and concealed routes through walls and buildings that can be used for withdrawals, reinforcements, or counterattacks. Demolitions are also used to create obstacles and clear fields of fire.

a. Marines can use demolitions for creating mouseholes and constructing command-detonated mines. Expedient C4 satchel charges can be concealed in firing positions and along movement routes for detonation in the vicinity of the enemy.

Thickness of Material	TNT	C4	Size of Opening
	Reinforced Concrete		
Up to 10 cm (4 inches)	.792 lbs	.591 lbs	10 to 15 cm (4 to 6 inches)
10 to 20 cm (4 to 8 inches)	3.4 lbs	2.54 lbs	15 to 30 cm (6 to 12 inches)
	Nonreinforced Concrete Masonry		
Thickness of Material	**TNT**	**C4**	**Size of Opening**
Up to 35 cm (14 inches)	3.66 lbs	2.73 lbs	35 cm (14 inches)
35 to 45 cm (14 to 18 inches)	7.5 lbs	5.6 lbs	45 cm (18 inches)
45 to 50 cm (18 to 20 inches)	10.2 lbs	7.61 lbs	50 cm (20 inches)

Table F-2. Amounts of TNT and C4 Required To Breach Concrete

b. The engineers should furnish technical assistance for selective rubbling. Normally, buildings can be rubbled by using shaped charges or C4 on the supports and major beams.

c. Charges should be placed directly against the surface to be breached unless a shaped charge is used. Whenever possible, demolitions should be tamped to increase their effectiveness. Tamping materials could be sandbags, rubble, or desks and chairs (Figure F-12).

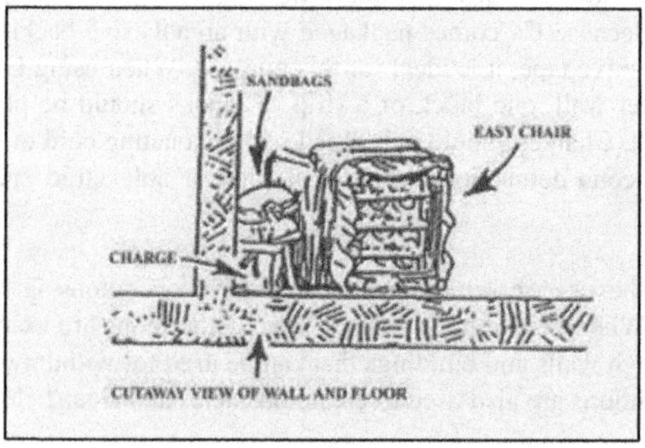

Figure F-12. Chair Used To Tamp Breaching Charge

d. For exterior walls, tamping of breaching charges may be impossible because of enemy fire. Thus, the untamped charge normally requires twice the explosive charge to produce the same effect as a tamped charge (Figure F-13).

METHODS OF PLACEMENT				
THICKNESS OF CONCRETE				
FEET	POUNDS OF TNT	POUNDS OF C4	POUNDS OF TNT	POUNDS OF C4
2	14	11	28	21
2 1/2	27	21	54	41
3	39	30	78	59
3 1/2	62	47	124	93
4	93	70	185	138
4 1/2	132	99	263	196
5	147	106	284	211
5 1/2	189	141	376	282
6	245	186	490	366

Figure F-13. Breaching Reinforced Concrete

e. When enemy fire prevents an approach to the wall, the breaching charge can be attached to a pole and slid into position for untamped detonation at the base of the wall (Figure F-14).

Figure F-14. Charge Placement When Small-Arms Fire Cannot Be Suppressed

f. The internal walls of most buildings function as partitions rather than as load-bearing walls. Therefore, smaller explosive charges can be used to breach them. In the absence of C4 or other military explosives, internal walls can be breached by using a claymore mine (Figure F-15). The claymore should be tamped to increase its effectiveness and to reduce the amount of explosive force directed to the rear.

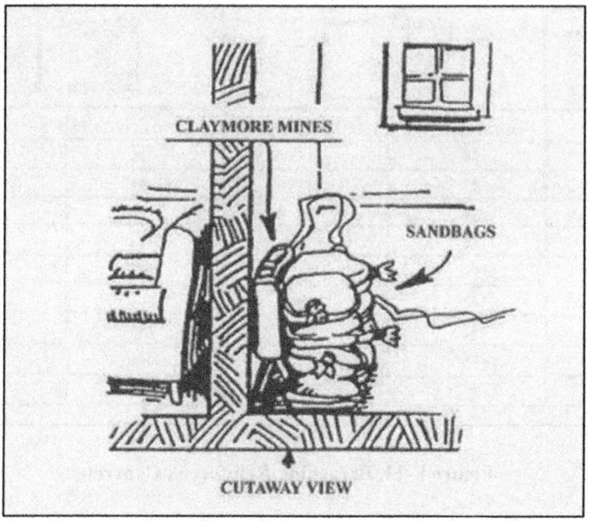

Figure F-15. Tamping a Claymore Mine To Breach Internal Walls

g. The Molotov cocktail (Figure F-16) is an expedient device for disabling both wheeled and tracked vehicles. It is easy to make because the materials are readily available. The results are most effective because of the close engagement in built-up areas. The objective is to ignite a flammable portion of the vehicle such as the fuel or ammunition that it is transporting. The following materials are needed to make a Molotov cocktail:

- Bottle or other glass container
- Gas (60 percent)
- Oil (40 percent)
- Rag for use as a wick.

The gas and oil are mixed thoroughly (60 percent gas to 40 percent oil). The rag is soaked with the mixture, then the mixture is placed into the bottle. The rag is then inserted in the opening of the bottle as a wick. When a target is sighted, the wick is lit and the bottle is thrown hard enough to break.

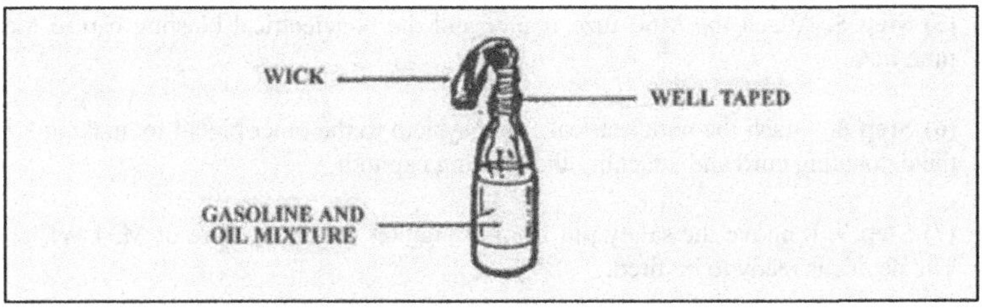

Figure F-16. Molotov Cocktail

WARNING

Ensure that a safe distance is maintained when throwing a Molotov cocktail.

h. The bunker bomb (Figure F-17) is an expedient, explosive flame weapon that is best used against fortified positions or rooms. This expedient munition should be used with a nonelectrical firing system. The following materials are required to make a bunker bomb:

- 1 metal small-arms ammunition container
- 1 gallon of gasoline
- 50 feet of detonating cord
- 1 nonelectrical blasting cap
- 1 M60 fuze igniter
- 7 1/2 feet of M700 time fuze
- 1 M49 trip flare or M34 WP grenade.

(1) Step 1. Fill the ammunition can 3/4 full with fuel and secure the lid.

(2) Step 2. "Hasty whip" the device with 15 turns around the center of the container using 44 feet of detonating cord. Leave 2-foot "pigtails" for attaching the igniter and fuze igniter.

(3) Step 3. Tape the igniter (M49 trip flare or M34 WP grenade) to the container handle.

(4) **Step 4.** Place one detonating cord pigtail end under the igniter spoon handle. Tape it in place.

(5) **Step 5.** Attach the M60 fuze igniter and the nonelectrical blasting cap to the M700 time fuze.

(6) **Step 6.** Attach the nonelectrical firing system to the other pigtail by making a loop in the detonating cord and attaching the blasting cap to it.

(7) **Step 7.** Remove the safety pin from the igniter (M49 trip flare or M34 WP grenade). The device is ready to be fired.

WARNING

Never carry the device by the handle or igniter. Remove the igniter safety pin only when it is time to use the device. Use extreme care when handling or carrying nonelectrical firing systems. Protect blasting caps from shock and extreme heat. Do not allow the time fuze to kink or become knotted. Doing so may disrupt the powder train and may cause a misfire. Prime detonating cord and remove the time fuze igniter safety pin only when it is time to use the device.

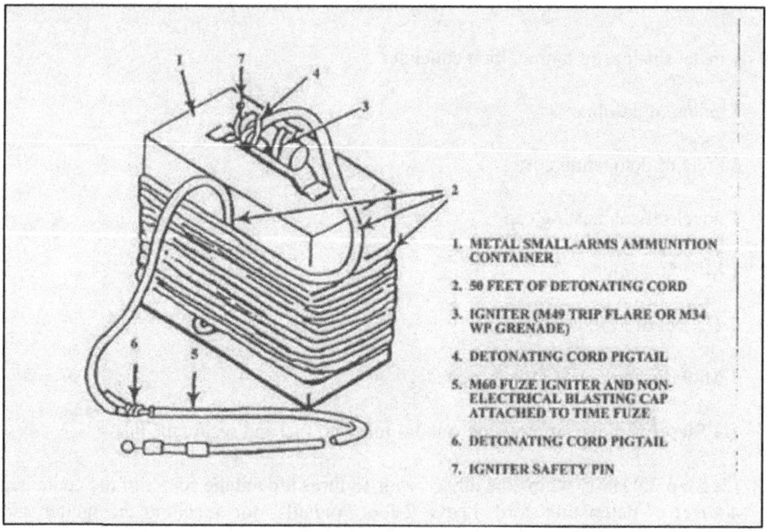

Figure F-17. Bunker Bomb Made From an Ammunition Can

10. Safety. The greatest danger to friendly personnel from demolitions is the debris thrown by the explosion. Leaders must ensure that protective measures are enforced. The safe distance listed in Table F-3 indicates the danger of demolition effects.

Pounds of Explosive	Safe Distance in Meters[1]	Pounds of Explosive	Safe Distance in Meters[1]
1 to 29	300	150	534
30	311	175	560
35	327	200	585
40	342	225	609
45	356	250	630
50	369	275	651
60	392	300	670
70	413	325	688
80	431	350	705
90	449	375	722
100	465	400	737
125	500	425	750

[1]These distances will be modified in combat when troops are in other buildings, around corners, or behind intervening walls.

Table F-3. Minimum Safe Distances for Personnel in the Open

a. The following are the guidelines for using demolitions:

- Keep the blasting machine under the control of a noncommissioned officer.
- Wear helmets and flack jackets at all times while firing explosives.
- Handle misfires with extreme care.
- Clear the room and protect personnel when blowing interior walls.

b. Some charges should be prepared beforehand, without detonators, to save time. Examples include 10- or 20-pound breaching charges of C4 and expedient-shaped charges in No. 10 cans.

- Use C4 to breach hard targets (masonry construction).

- Do not take chances.
- Do not divide responsibility for demolition work.
- Do not mix demolitions and detonators until ready to use.
- Do not carry demolitions and caps together.

Appendix G

Nuclear, Biological, and Chemical Considerations

Current U.S. policy regarding lethal or incapacitating agents is that their use against an armed enemy requires approval at the NCA level. Potential enemies may not operate under the same restrictions. Field commanders must be prepared to assume an adequate NBC defensive posture when engaged in urban fighting.

1. Protection From Nuclear, Biological, and Chemical Weapons. Generally, the lowest floor or basement of a reinforced concrete or steel-formed building offers good protection from nuclear hazards and initial liquid chemical contamination although some chemical agents tend to collect in lower areas. Tanks and LAVs/AAVs also provide protection. (See FMFM 11/MCWP 3-37 series manuals for greater detail.)

a. Biological attacks are difficult to detect and recognize. Biological agents can be disseminated by using aerosols, vectors, and covert methods. Because biological agents can be sprayed or dropped in bomblets, personnel who observe such indicators should promptly report them. Prompt reporting and treatment of the sick speed the employment of medical countermeasures. Although buildings and shelters provide some protection against spraying, they provide little protection against biological agents.

b. Chemical agents cause casualties by being inhaled or absorbed through the skin. They may afford Marines a few seconds to mask. Buildings have a channeling effect and tend to contain the effects of an agent, causing great variations in chemical concentration from room to room or from building to building. Chemical agents usually settle in low places, making sewers and subways hazardous hiding places. A prepared defender should include some collective protective measures in the defensive network. Personnel using fans may be able to put enough overpressure into tunnels to keep some chemical agents from entering. The protective mask and battle dress overgarment provide the best protection against chemical agents.

c. Personal hygiene is a critical defensive measure against infection and disease. Unfortunately, built-up areas are characterized by sophisticated sanitation systems. When those systems are destroyed, the resulting sanitary conditions become much worse than those in areas where sanitation facilities do not exist. Commanders must ensure that personnel employ sanitation measures and that their immunizations are current.

d. Commanders should plan their mission-oriented protective posture (MOPP) with the realization that increased logistics demands in built-up areas also apply to NBC equipment. Protective clothing, detection and decontamination equipment, and sealed containers of food and water must be stockpiled the same as other critical supplies. When operating in protective clothing, commanders must make allowances for the strenuous activities normally associated with combat in built-up areas.

(1) Detection. After an NBC attack, battalions should dispatch their detection and survey teams. Detection in built-up areas is complicated by the numerous rooms within buildings. Teams should conduct tests and surveys of major streets, intersections, and buildings in their area for inclusion in initial NBC reports. A systematic survey of all buildings, rooms, and underground facilities must be accomplished before occupation by unmasked personnel. All data should be forwarded using the appropriate NBC report.

(2) Decontamination. Personnel begin decontamination operations as soon after an NBC attack as the mission allows. Personnel should conduct individual decontamination of themselves and their personal equipment. Unit commanders determine the need for MOPP gear exchange and the requirements for a hasty or deliberate decontamination operation.

(a) Nuclear. Personnel should wear wet-weather gear for certain decontamination operations (hosing down buildings) to prevent radioactive material from touching the skin.

(b) Chemical and Biological. Roads, sidewalks, and other hard surfaces are best decontaminated by weathering, if time permits. Agents can also be covered with several inches of dirt or sand to provide protection. Fragment testing should be conducted periodically to ensure that the agent has not seeped through the covering. For critical sections of roads, a truck-mounted M12A1 (power-driven decontamination apparatus (PDDA)) can be used to spray supertropical bleach (STB) slurry; this aids rapid decontamination. Buildings are difficult to decontaminate, especially wooden ones. Some techniques for their decontamination are scrubbing with STB slurry; washing with hot, soapy water; washing or spraying with a soda solution; and airing.

2. Smoke Operations. The use of smoke is an integral part of either offensive or defensive operations in built-up areas. In the offense, smoke can support the maneuver of combat units and deception operations. Smoke employed in the defense obscures enemy air and ground observation, limiting the accuracy of enemy fires and intelligence collection.

a. Smoke should not be used when it degrades the effectiveness of friendly forces. Likewise, an extremely dense concentration of smoke in a closed area displaces the oxygen, suffocating Marines even when they are wearing protective masks.

b. Smoke pots, generators, or artillery smoke munitions should be used to cover the withdrawal of defending forces or the movement of attacking forces. Artillery-delivered WP can be effective against enemy forces by screening, in addition to causing casualties and fires. Consideration should be given to the incendiary effects of WP, smoke (HC and M825), and illumination munitions on the litter and debris of built-up areas.

c. Smoke grenades can be massed to provide a hasty screen for concealing personnel moving across streets and alleys. Smoke grenades can also be used for signaling; those launched by an M203 can be used to mark targets for fixed and rotary-wing air support.

d. The use of smoke in built-up areas is affected by the complex wind patterns caused by buildings. When covering a built-up area with a smoke haze or blanket, personnel must include all buildings. Failure to obscure tall buildings, towers, and steeples provides enemy observers with reference points for fire placement within the built-up area.

3. Riot Control Agents. When authorized, riot control agents such as CS and CN can be used to drive enemy forces from positions or to deny them areas for occupation. Riot control agents are incapacitating but generally have no lasting effects. They may be appropriate when preventing civilian casualties is a planning consideration. However, riot control agents may not be effective against an enemy that is well trained in chemical defense.

Appendix H

MOUT Under Limited-Visibility Conditions

To be successful, leaders must use limited-visibility conditions to their advantage.

1. Advantages. When fighting in built-up areas during night or periods of limited visibility, attacking or defending forces have several advantages.

a. In many cases, Marines may have a technological advantage in thermal imagery and light intensification over their opponents. This enables Marines to identify, engage, and destroy enemy targets before being detected by the enemy.

b. Possessing the capability to conduct operations as well at night or during periods of reduced visibility as during daylight gives Marines a tremendous advantage over an enemy with reduced capability. It also aids the generation of tempo and allows Marines to gain or retain the initiative.

c. Generally, ranges of direct-fire engagements are greatly reduced in the MOUT environment. During periods of limited visibility, effective unaided target acquisition ranges are further reduced. This enables attacking forces to close to shorter ranges, thus increasing the lethality and accuracy of weapons. Attacking forces can also take advantage of the enemy's reduced ability to see and can more effectively engage the enemy before being detected with thermal imagery or light intensification devices.

d. Helicopterborne assaults are best conducted during periods of limited visibility because the enemy's air defenses are degraded.

e. Attacking during periods of limited visibility gives the attacker a greater chance of achieving surprise.

2. Disadvantages. When fighting in built-up areas during periods of limited visibility, attacking and defending forces also face some disadvantages.

a. Command and control is difficult in any operation in a built-up area; periods of limited visibility increase this difficulty.

b. Marines have an instinctive tendency to move close together during periods of limited visibility. Constant attention must be given to prevent Marines from "bunching up."

c. Marines may become disoriented easily because of the combined effects of low visibility and the characteristics of built-up areas.

d. Target identification becomes more difficult in limited-visibility conditions. Depending on the training of the individual, the Marine may fire at anything seen or may hesitate too long before firing. Improper target identification is one of the leading causes of fratricide, so leaders must pay close attention to individual target engagement.

3. Fratricide Avoidance. The risk of fratricide is much greater during periods of limited visibility. The key to avoiding fratricide is increased situational awareness by leaders and individuals, coupled with realistic training in target identification. (See *Fratricide: Reducing Self-Inflicted Losses*, Newsletter, No. 92-4, April 92, Center for Army Lessons Learned, U.S. Army Combined Arms Command, Fort Leavenworth, Kansas.) Other considerations include:

a. Graphic control measures should be clearly defined and obvious. Examples include distinctive buildings, large boulevards, rivers, and so forth.

b. Leaders must exercise firm control when engaging targets. Movements should also be coordinated and controlled.

c. Cleared rooms and buildings should be distinctly marked to identify cleared areas and friendly forces to base-of-fire elements supporting the maneuver.

d. Visible markers (for example, glint tape or thermal strips) should be attached to individual Marines for rapid identification.

e. Far and near recognition symbols should be coordinated and used properly.

f. Units employing close air support must exercise firm control. Failure to do so may lead to the pilot becoming disoriented and engaging friend and foe alike.

4. Urban Environmental Effects on Night Vision Devices. Built-up areas affect standard night vision devices and sights differently than do open areas. This may cause some confusion because the images Marines receive through their night vision devices are unusual compared to those they may be used to.

a. Most built-up areas have electric power; therefore, street lights and/or building lights may "white out" some light intensification devices unless the power is disrupted or lights are turned off.

b. Fires may be burning in the urban area. Open flames cause problems not only for light intensification devices, but also for thermal devices.

c. Subterranean areas and the interiors of buildings will not have ambient light if the power is off. Passive night vision devices must then use an artificial light source, such as infrared radiation, to provide enough ambient light for the devices to work effectively.

d. The many reflective surfaces found in built-up areas may cause false images, particularly for laser range finders and laser target designators.

e. Thermal imaging devices may not be able to see through large amounts of dust particles suspended in the air.

f. Smoke and dust degrade the effectiveness of night vision devices.

g. Fog degrades long-range target acquisition from thermal sights.

h. Weapons flashes within enclosed areas appear to be much brighter. This causes Marines to lose their night vision and washes out light intensification devices.

5. Considerations. The environment of built-up areas presents special challenges and considerations during periods of limited visibility.

a. The use of glint tape, thermal tape, or chemiluminescent lights is an important consideration. These can be used to mark the FLOT, casualties, cleared buildings and rooms, weapons positions, and individual Marines. Their use must be clearly addressed in the unit's SOP. When markers are used for extended periods, their meanings should change because the enemy may be able to capture or manufacture and use these marking devices to their advantage.

b. The use of tracer and incendiary ammunition may be restricted to prevent fires. The light of fires whites out some night vision devices and may interfere with or confuse thermal devices.

c. The control of power stations may be essential to operations during periods of limited visibility. Control of power stations enables friendly forces to control, to a degree, background illumination. Shutting off the power to the street lights is much easier than shooting out the lights. Commanders must balance the tradeoff between force protection and maintaining law and order after the battle is over. During cold weather, the control of power stations may be critical to the welfare of the civilian population.

d. Identifying friendly units, noncombatant civilians, and enemy troops becomes more difficult during limited-visibility operations.

e. Locating the source of sounds becomes more difficult because of the natural echoing in built-up areas and the tendency of sounds to carry farther at night.

f. Locating boobytraps and obstacles also becomes more difficult at night. Movement rates are normally slower than during periods of normal visibility.

6. Special Equipment. Fighting during periods of limited visibility requires some specialized equipment to maximize advantages.

a. As a rule, thermal imaging devices such as the AN/PAS-7 infrared viewer and the AN/TAS-5 Dragon infrared sight are better for limited-visibility operations than are light intensification devices such as the AN/PVS-7B. Light intensification devices are more easily washed out by background light, weapons flashes in enclosed areas, and fires. Thermal devices, while also affected by fires, are not as easily washed out.

b. The AN/PAQ-4A infrared aiming device is similar to its off-the-shelf laser aiming sight counterparts, except it is not visible to the naked eye. Infrared pen lights can also be attached to weapons and, when used in conjunction with night vision devices, can provide a quick sight picture, illuminate rooms and hallways, identify obstacles and boobytraps, and identify friendly forces.

c. Other night sights for weapons include the AN/TVS 5 crew-served weapon night vision sight, the AN/PVS-4 individual weapons night vision sight, and the AN/UAS-12 night vision sight, which mounts to the modular universal laser equipment (MULE) and the TOW missile system.

d. Trip flares, flares, illumination from mortars, and artillery and spotlights (visible light or infrared radiation) can be used to blind enemy night vision devices or to artificially illuminate the battlefield (Figure H-1). (See FMFM 6-8, *Supporting Arms Observer, Spotter, and Controller*, for more information on illumination from mortars and artillery.)

e. Spare batteries for the night vision devices should be carried to keep the devices operational. Soft, clean rags should be used to clean the lenses.

7. Combat Support. Coordinating employment of forces and fire support in limited-visibility conditions is a major concern to leaders during MOUT. The concentration of forces and fires at the point of decision is facilitated by the technological edge that night vision devices provide.

a. Any degradation of accuracy in artillery fire will likely be a result of the limitations of target acquisition assets. While FOs and FACs may have thermal sights and laser range finders, most Marines on the battlefield do not yet have devices that will enable them to acquire targets accurately. The following are some devices and techniques to improve target acquisition for indirect fires:

(1) If the target is within LOS, tanks and LAVs can rapidly identify the target and provide an accurate range.

(2) Preregistered targets are effective if the target reference point can be observed and the observer has clear communications with the firing unit.

(3) Fixed-wing and rotary-wing aircraft can be used to identify targets and adjust supporting arms.

Figure H-1. Use of Indirect-Fire Illumination During MOUT

8. Combat Service Support. Maneuver unit commanders and their Marines are not the only individuals who must adjust to combat under limited-visibility conditions in MOUT. Logisticians at every level must anticipate requirements for these conditions.

a. Units conducting resupply operations during periods of limited visibility should consider:

- Issuing drivers and vehicle commanders night vision devices so that the vehicles going to and from logistic release points do not require illumination (This also prevents the enemy from acquiring resupply locations by following vehicles with blackout lights on.)

- Maintaining strict noise and light discipline

- Following a clearly marked route to avoid obstacles and prevent the resupply vehicle(s) from becoming disoriented

- Providing radios to resupply vehicles

- Issuing each vehicle a map of the AO (preferably a city map with the street names).

b. Combat units operating for an extended time under conditions of limited visibility should have enough batteries to keep the night vision devices functioning at optimum power and sensitivity.

c. A large operational readiness float of night vision devices should be maintained.

d. Casualty collection during periods of limited visibility is more difficult. Clear methods for marking any casualties must be established before the operation begins.

e. CSS operations located in existing structures should not be visible from a distance. This includes limiting vehicle traffic to an absolute minimum, sealing doors and windows to prevent light leakage, and dispersing assets as much as possible.

9. Operational Considerations. Marine units conduct attacks during periods of limited visibility to gain or sustain the momentum of the attack. Before conducting a limited-visibility attack, the commander must balance the risks and ensure that every Marine understands the mission, intent, and control measures. Rehearsals and strict command and control reduce casualties and greatly enhance the chances for mission accomplishment.

a. To reduce confusion, Marines should clear buildings and rooms using the same techniques they use during the day.

b. Movement rates are slower in the dark. Each Marine must remain alert for mines, boobytraps, and enemy positions.

c. Rifle squads and fire teams should be equipped with a mixture of both thermal imaging and light intensification devices whenever possible. This enables the squads and fire teams to obtain a better picture of the night environment and enables Marines to balance the strengths and weaknesses of each type of night vision device for maximum results.

d. When moving through buildings, assault forces mark cleared rooms and buildings and communicate with the support forces. Marking cleared rooms is especially critical if more than one assault force or element is in the same building. Communications with supporting forces are imperative to avoid fratricide.

e. If flashlights or chemiluminescent lights are used, they should be held away from the head or chest area. This will make it harder for enemy soldiers firing at the light to kill the Marine holding the flashlight or chemiluminescent light.

f. Units must know where everyone is during offensive operations. This not only reduces the risk of fratricide, but also decreases the time spent identifying, locating, and treating casualties. Also, it greatly reduces the chance of Marines becoming disoriented and separated from the unit.

g. Assault units should be aware of adjacent fires that may diminish the effectiveness of night vision devices. Weapons flashes within small rooms cause Marines to lose their night vision and can wash out light intensification devices. Also, enemy soldiers may use flares inside and outside of buildings to deliberately render night vision devices ineffective.

h. Leaders must ensure that all Marines follow the ROE and the laws of land warfare. This is critical if the enemy is intermixed with the local civilian population. Also, leaders must follow all control measures.

i. Enemy forces can be expected to use periods of limited visibility to the same advantages as Marine forces do.

j. Enemy forces may have access to sophisticated night vision devices manufactured in Europe, the United States, Japan, Korea, and the former Soviet Union.

Appendix I

Urban Building Analysis

As in other types of operations, success in urban combat depends largely on the ability to analyze the military aspects of terrain. Marines must be able to recognize certain terrain features when evaluating urbanized terrain.

1. Types of Mass-Construction Buildings. Mass-construction buildings are those in which the outside walls support the weight of the building and its contents. Additional support, especially in wide buildings, comes from using load-bearing interior walls, strongpoints (called pilasters) on the exterior walls, cast-iron interior columns, and arches or braces over the windows and doors (Figure I-1). Modern types of mass-construction buildings are wall and slab structures, such as many modern apartments and hotels, and "tilt-up" structures commonly used for industry or storage. Mass-construction buildings are built in many ways:

- The walls can be built in place by using brick, block, or poured-in-place concrete.
- The walls can be prefabricated and then tilted up, or they can be reinforced-concrete panels.
- The walls can be prefabricated and assembled like boxes.

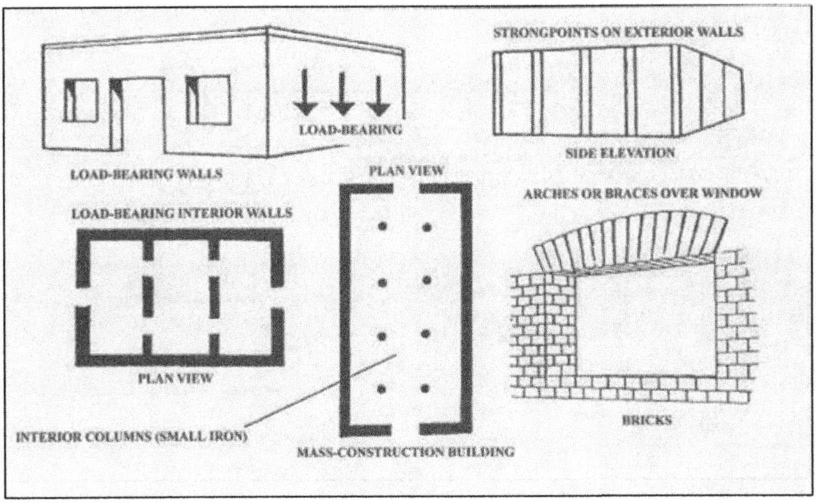

Figure I-1. Mass-Construction Buildings

a. Brick buildings are, in some regions, the most common and most important of the mass-construction buildings. In Europe, brick buildings are commonly covered with a stucco veneer so that bricks do not show (Figure I-2). One of the most common uses of brick buildings is the small store. These buildings are found in all built-up areas but are most common in the core periphery (Figure I-3).

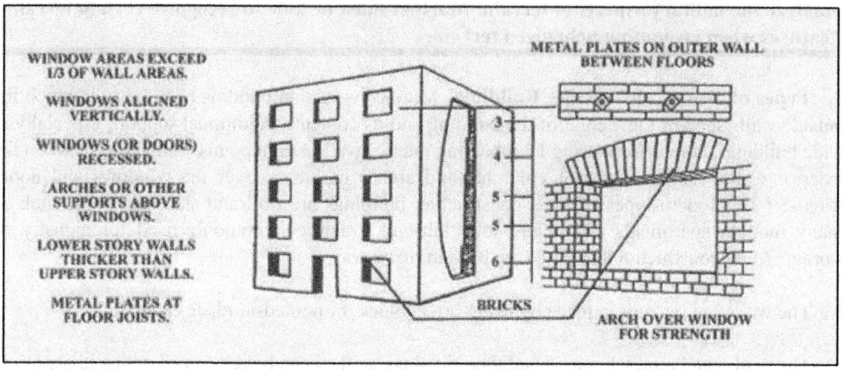

Figure I-2. Brick Buildings

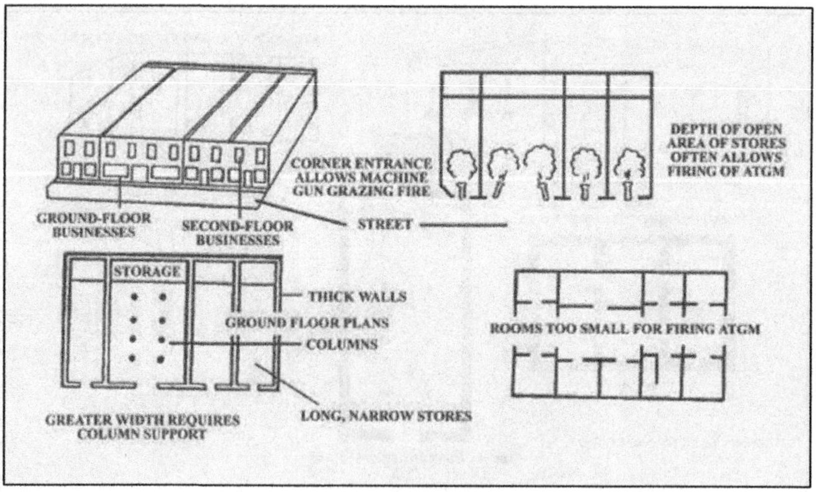

Figure I-3. Brick Store

b. Another common mass-construction building found mainly in industrial areas and along commercial ribbons is the warehouse. It is built of poured-in-place concrete reinforced with steel bars or of prefabricated tilt-up walls. The walls of warehouses provide good cover, although the roofs are generally light and vulnerable. The warehouses' large open bays permit firing of ATGMs or surface-to-air missiles and, because they are normally found in outlying areas, often afford adequate fields of fire for ATGMs. These buildings are built on slabs that can normally support the weight of vehicles and can provide excellent cover and concealment for tanks (Figure I-4).

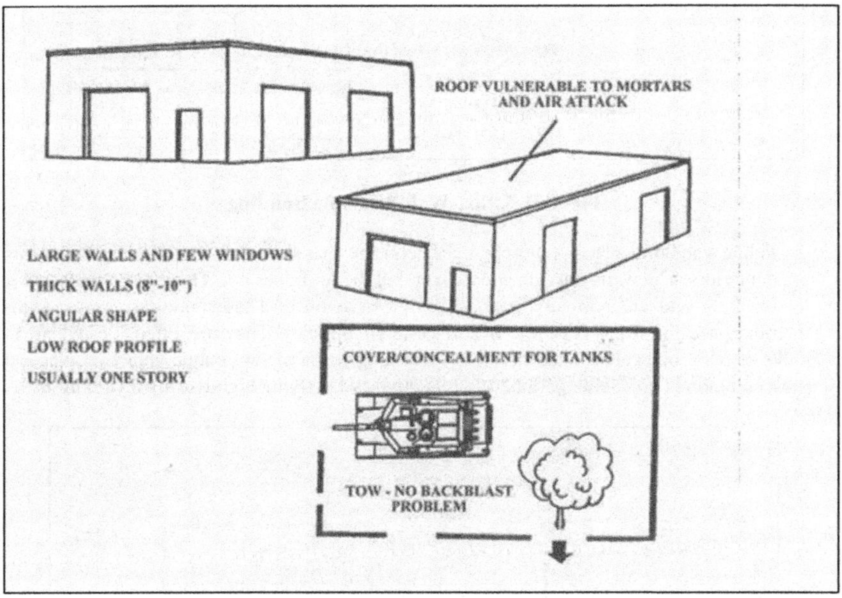

Figure I-4. Warehouse

c. Another type of mass-construction building is the box-wall-principle type. It is made from prefabricated panels of 6- to 8-inch-thick reinforced concrete. The outside wall is often glass. The box-wall-principle building provides good cover, except at the glass wall. The rooms are normally too small for ATGMs to be fired. A good circulation pattern exists from room to room and from floor to floor. These buildings are commonly used as hotels or apartments and are located mainly in residential and outlying areas (Figure I-5).

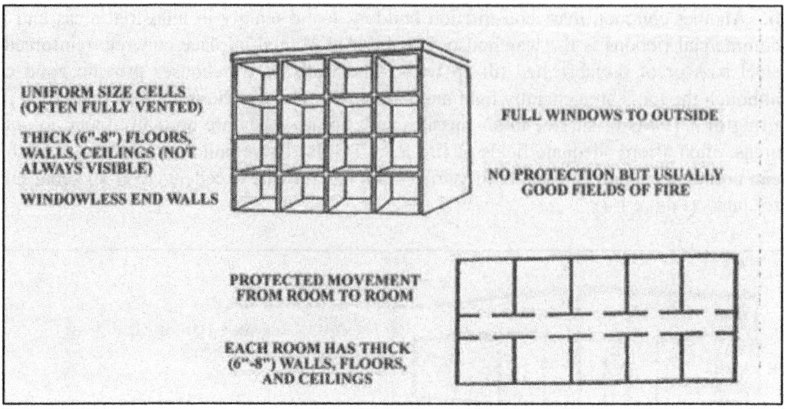

Figure I-5. Box-Wall-Principle Building

d. Public gathering places (churches, theaters) are mass-construction buildings with large, open interiors. The walls provide good cover, but the roof does not. The interior walls are not load-bearing and are normally easy to breach or remove. These buildings have adequate interior space for firing ATGMs or surface-to-air missiles. They are often located next to parks or other open areas and, therefore, have long fields of fire. Public gathering places are most common in core, core periphery, residential, and outlying high-rise areas (Figure I-6).

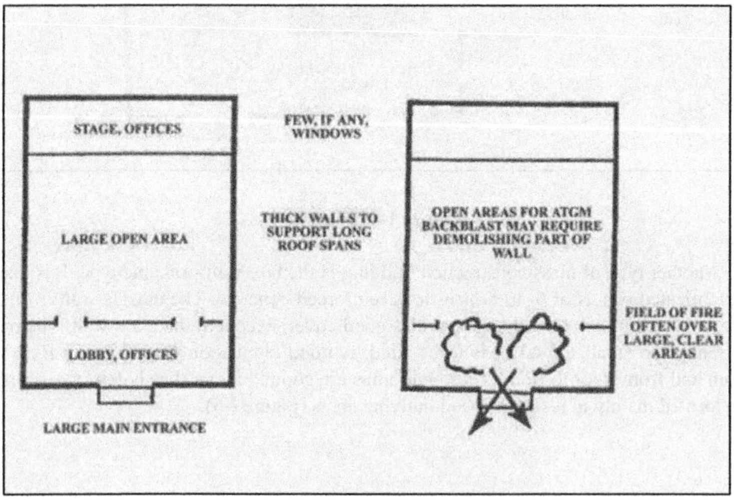

Figure I-6. Public Gathering Places

2. Types of Framed Buildings. Framed buildings are supported by a skeleton of columns and beams and are usually taller than frameless buildings (Figure I-7). The exterior walls are not load-bearing and are referred to as either heavy clad or light clad. Another type of framed building often found in cities is the garage, which generally has no cladding.

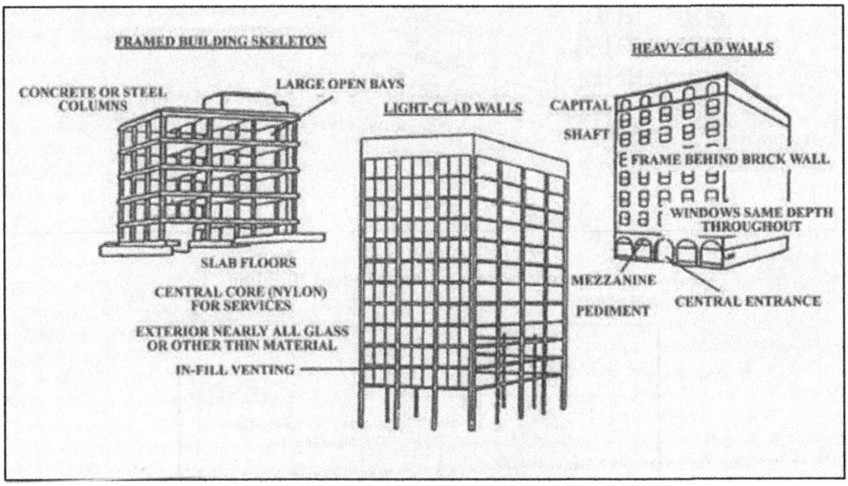

Figure I-7. Framed Buildings

a. Heavy-clad buildings were common when framed buildings were first introduced. Their walls are made of brick and block that are sometimes almost as thick as frameless brick walls, although not as protective. Heavy-clad framed buildings are found in core and core periphery areas. They can be recognized by a classic style or architecture in which each building is designed with three sections: the pediment, shaft, and capital. Unlike the brick building, the walls are the same thickness on all floors, and the windows are set at the same depth throughout. Often the frame members (the columns) can be seen, especially at the ground floor. The cladding, consisting of layers of terra cotta blocks, brick, and stone veneer, does not provide as good a cover as the walls of brick buildings. It protects against small-arms fire and light shrapnel but does not provide much cover against heavy weapons (Figure I-8).

(1) The floor plans of these buildings depend on their functions. Office buildings normally have small offices surrounding an interior hall. These offices have the same dimensions as the distance between columns (some large offices are as large as two times the distance between columns). These rooms are too small to permit firing of ATGMs but do provide some cover for snipers or machine gunners (Figures I-9 and I-10).

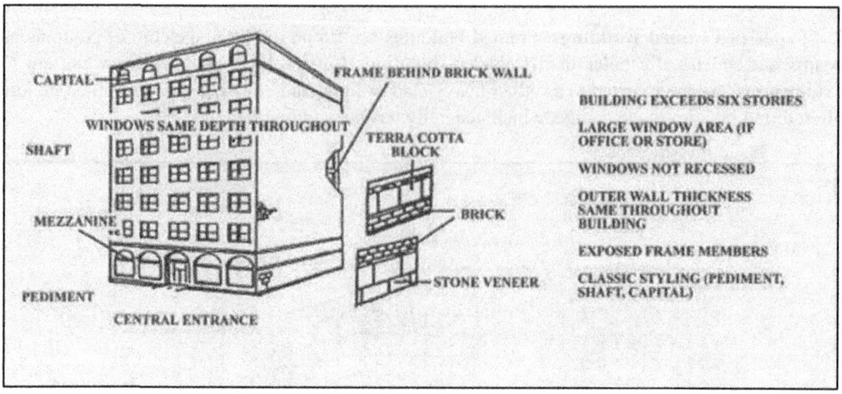

Figure I-8. Heavy-Clad Framed Building

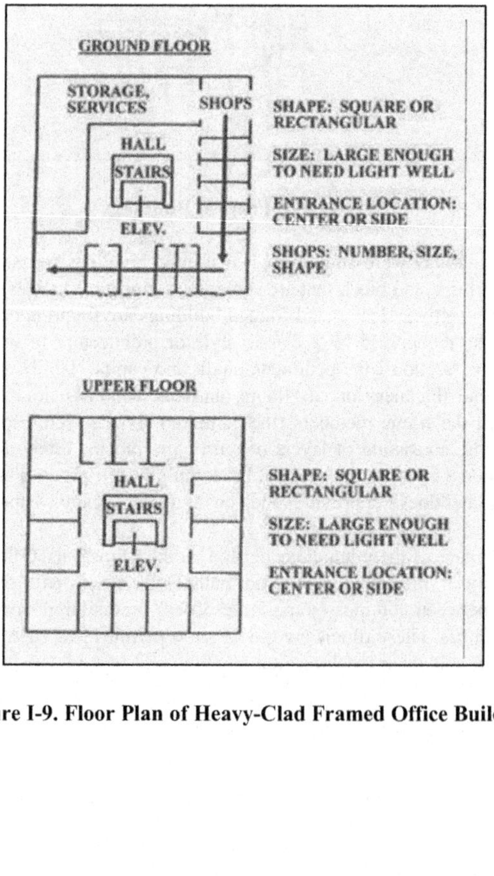

Figure I-9. Floor Plan of Heavy-Clad Framed Office Building

(2) Department stores normally have large, open interiors (Figure I-11). Such areas permit firing of ATGMs or man-portable surface-to-air missiles (if there are adequate fields of fire). Often a mezzanine level with a large backblast area permits firing down onto tanks. Steel fire doors often exist between sections of the store. The steel fire doors are activated by heat. Once closed, they are difficult to breach or force open, but they effectively divide the store into sections (Figure I-12).

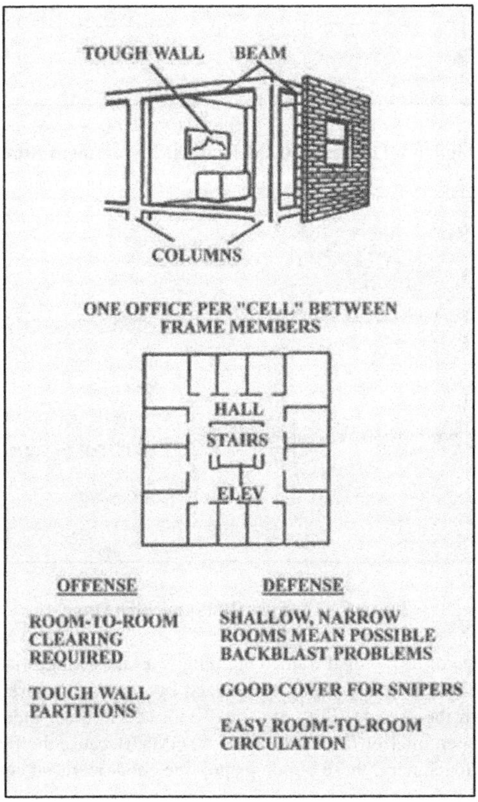

Figure I-10. Heavy-Clad Framed Office

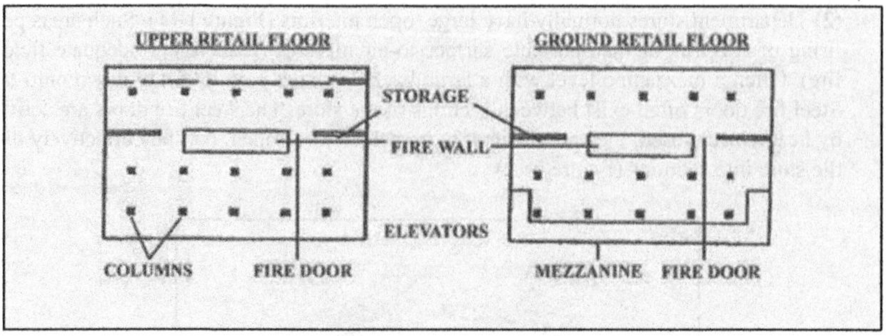

Figure I-11. Heavy-Clad Framed Department Store

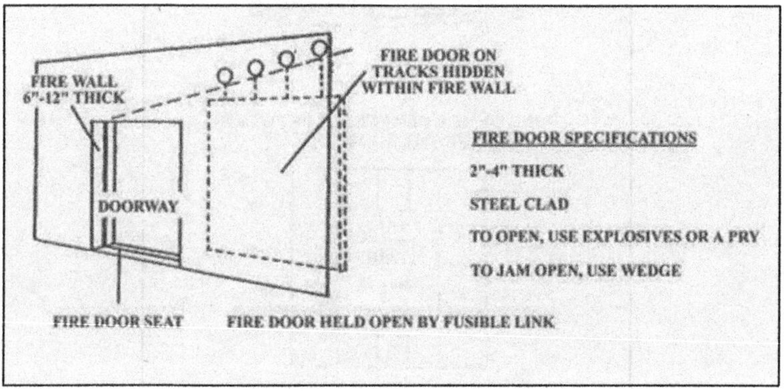

Figure I-12. Fire Wall and Fire Door

(3) Another type of heavy-clad framed building is used as a high-rise factory (Figure I-13). Such buildings are normally easily recognized because the concrete beams and columns are visible from the outside. They are usually located in older industrial areas. The large windows and open interior favor the use of ATGMs. Because the floors are often made to support heavy machinery, these buildings provide good overhead cover.

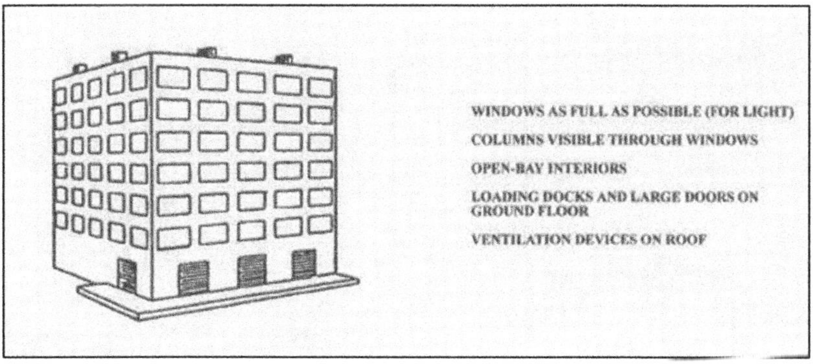

Figure I-13. High-Rise Factory

b. Light-clad buildings are more modern and may be constructed mostly of glass (Figure I-14). Most framed buildings built since World War II are light-clad buildings. They are found in both core and outlying high-rise regions. Their walls consist of a thin layer of brick, lightweight concrete, or glass. Such materials provide minimal protection against any weapon. However, the floors of the buildings are much heavier and provide moderate overhead cover. The rooms in light-clad framed buildings are much bigger than those in heavy-clad buildings. This feature, along with the fact that the buildings usually stand detached from other buildings, favors the employment of ATGMs. The interior partitions are thin, light, and easy to breach (Figure I-15).

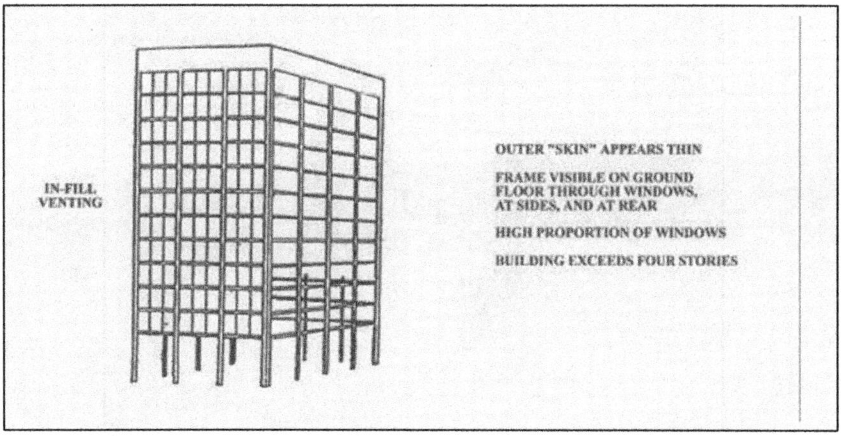

Figure I-14. Light-Clad Framed Building

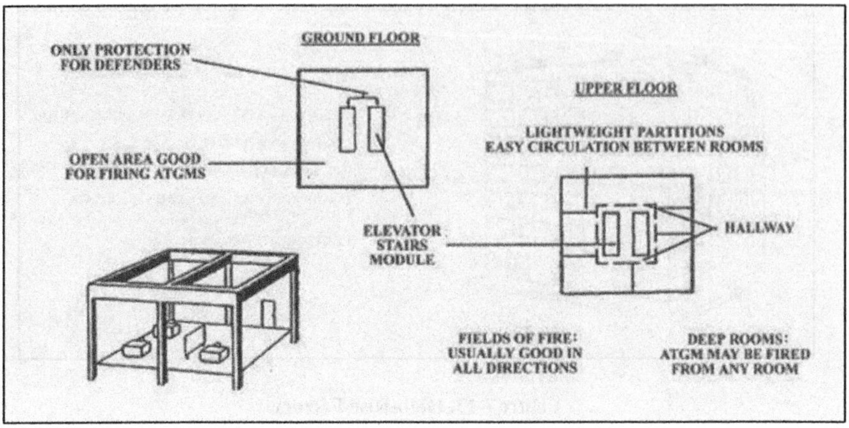

Figure I-15. Light-Clad Framed Room

c. The garage is one of the few buildings in an urban area in which all floors support vehicles. It provides a means to elevate vehicle-mounted TOWs, and the open interiors permit firing of ATGMs. Garages are normally high enough to provide a 360-degree field of fire for antiaircraft weapons. For example, a Stinger could hide under the top floor of the garage, come out to engage an aircraft, and then take cover again (Figure I-16).

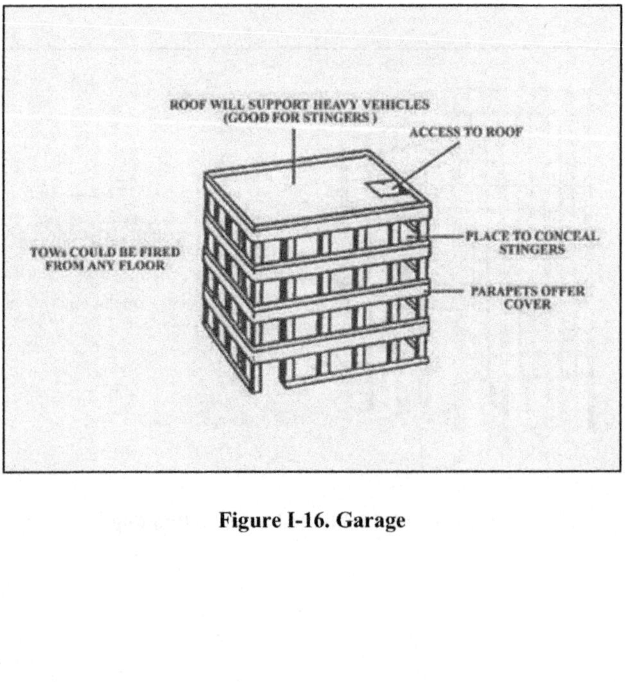

Figure I-16. Garage

3. Floor Plans. Floor plans in buildings follow predictable patterns. One of the factors that determines floor plans is building shape (Figure I-17). The basic principle governing building shape is that rooms normally have access to outside light. This principle helps Marines to analyze and determine the floor plans of large buildings.

4. Residential Areas. The two basic types of houses in the western world are located in and around cities and in rural areas. City-type houses are normally mass-construction brick buildings. Rural-type buildings in the continental United States, South America, and Southeast Asia are commonly made of wood, while in continental Europe, Southwest Asia, and sub-Saharan Africa, where wood is extremely scarce, rural buildings are normally constructed of concrete blocks. Rural-type buildings can also be found within cities (Figure I-18).

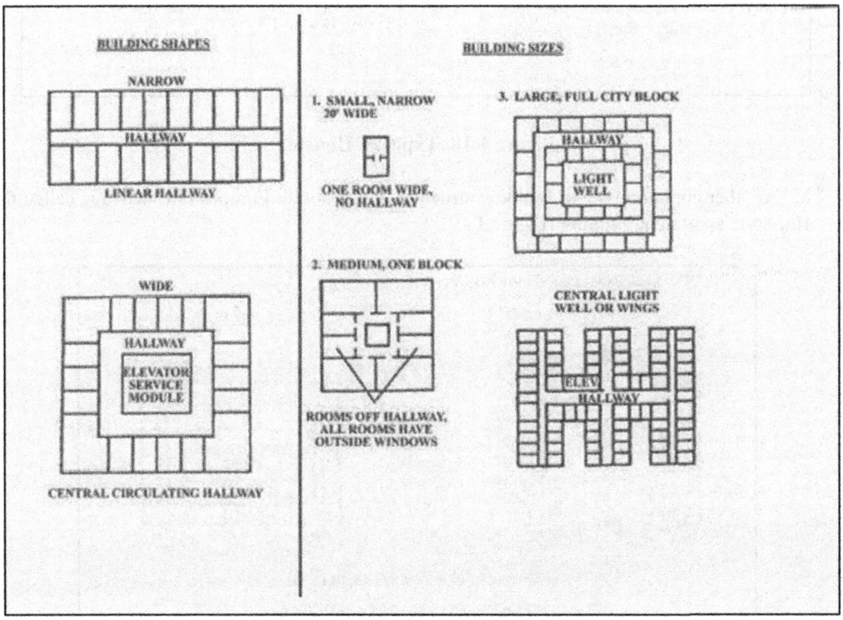

Figure I-17. Building Shapes and Sizes

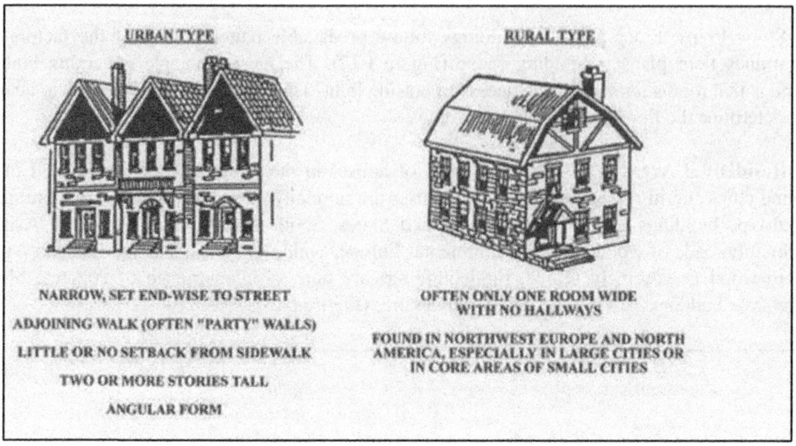

Figure I-18. Types of Housing

a. Another common type of building structure in cities with European influence is called the Hof-style apartment building (Figure I-19).

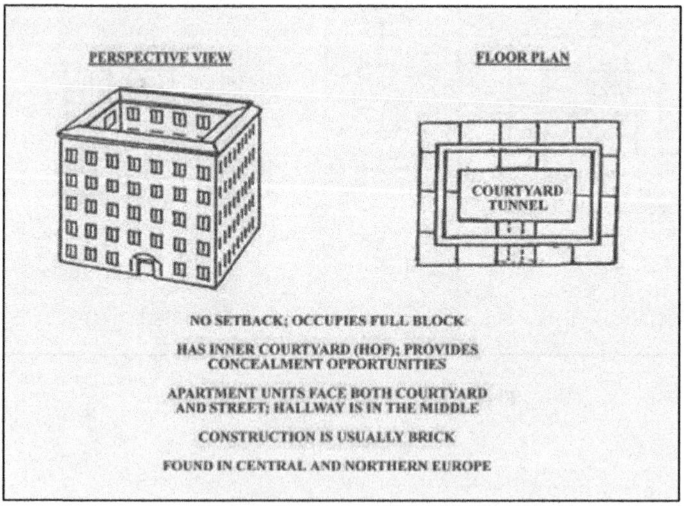

Figure I-19. Hof-Style Apartment Building

b. In the Middle East and tropical regions, the most common type of housing is the enclosed courtyard. Houses are added to one another with little regard to the street pattern. The result is a crooked, narrow maze that is harder to move through or fire in than dense European areas. (Figure I-20).

Figure I-20. Enclosed Courtyard

5. Characteristics of Buildings. Certain characteristics of both mass-construction and framed buildings can be helpful in analyzing a built-up area. Leaders can use Table I-1 to determine how to defend or attack a certain building given the unit's available weapons systems.

Type of Construction	Building Material	Height (Stories)	Average Wall Thickness (cm)
Mass	Stone	1 to 10	75
Mass	Brick	1 to 3	22
Mass	Brick	3 to 6	38
Mass	Concrete block	1 to 5	20
Mass	Concrete wall and slab	1 to 10	22 to 38
Mass	Concrete tilt-ups	1 to 3	18
Framed	Wood	1 to 5	3
Framed	Steel (heavy cladding)	3 to 50	30
Framed	Concrete/steel (light cladding)	3 to 100	2 to 8

Table I-1. Characteristics of Buildings

6. Distribution of Building Types. Certain types of buildings dominate certain parts of a city; this establishes patterns within a city. Analysis of the distribution and nature of these patterns has a direct bearing on military planning and weapon selection (Figure I-21).

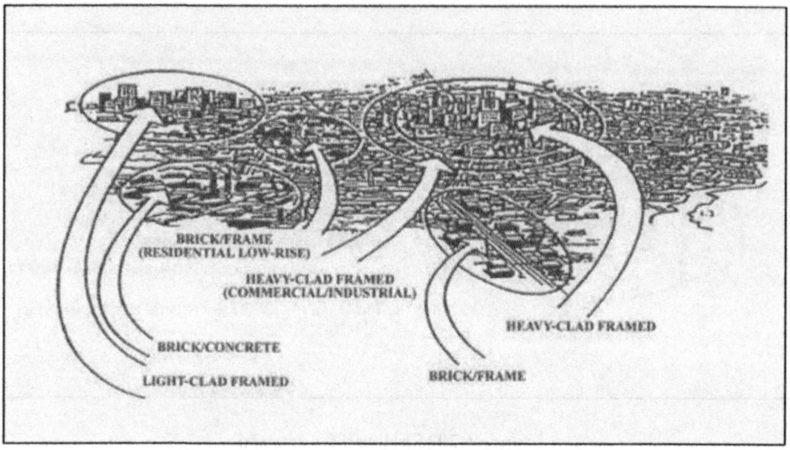

Figure I-21. Distribution of Building Types

a. Mass-construction buildings are the most common structures in built-up areas, forming about two-thirds of all building types. Brick structures account for nearly 60 percent of all buildings, especially in Europe.

b. Steel- and concrete-framed multistory buildings have an importance far beyond their one-third contribution to total ground floor area. They occupy core areas—a city's most valuable land—and as centers of economic and political power, they have a high potential military significance.

c. Open space accounts for about 15 percent of an average city's area. Many open spaces are grass-covered and are used for parks, athletic fields, and golf courses; some are broad, paved areas. The largest open spaces are associated with suburban housing developments where large tracts of land are recreation areas.

d. Streets serving areas consisting of mostly one type of building normally have a common pattern. In downtown areas, for example, high land values result in narrow streets. Street widths are grouped into three major classes: 7 to 15 meters, located in medieval sections of European cities; 15 to 25 meters, located in newer, planned sections of most cities; and 25 to 50 meters, located along broad boulevards or set far apart on large parcels of land. When a street is narrow, observing or firing into windows of a building across the street can be difficult because an observer is forced to look along the building rather than into windows.

When the street is wider, the observer has a better chance to look and fire into the window openings (Figure I-22).

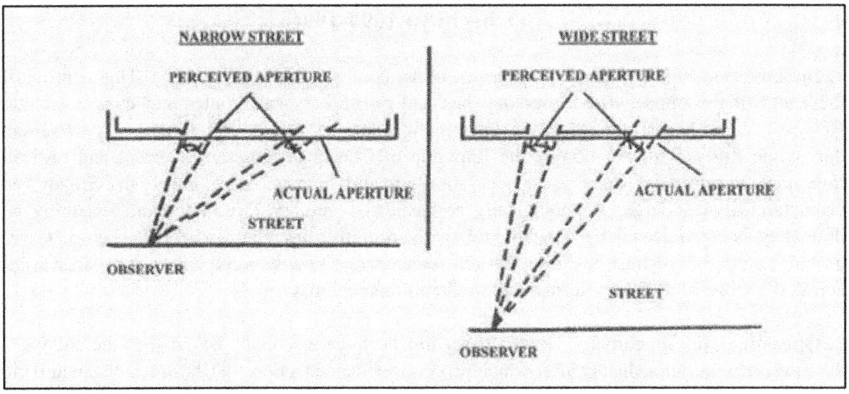

Figure I-22. LOS Distances and Angles of Obliquity

Appendix J

Lessons Learned from Russian Military Operations in Chechnya 1994-1996

1. Introduction. Many historical studies of urban combat have become dated. This is primarily the result of the impact that technology has had on military capabilities and evolving tactics. However, certain tenets of urban combat remain constant. The battles for the city of Grozny during the Russian intervention in the Republic of Chechnya represent a recent and critically important example of large scale operations in urban combat. Combat in Grozny was characterized by a large, technologically sophisticated military force (Russian) engaging and ultimately being defeated by a small, relatively primitive irregular force (Chechen). Grozny provides a number of fresh insights, and reinforcement of time honored tenets of urban warfare, across the scope of activities germane to modern urban combat.

2. Operations in Chechnya. Russia's war in Chechnya essentially began with the collapse of the Soviet Union, immediately after which pro-independence factions in Chechnya began agitating to free the Republic from Russian domination. This eventually led to civil war with pro-Russian factions within the Republic. Russia committed troops to Chechnya on 11 December 1994, for what many in the Russian government and military thought would be a quick campaign to restore Russian primacy. What followed was a bitter, costly, and protracted struggle that ended with the complete withdrawal of Russian military forces from Chechnya.

A primary focus of Russian operations in Chechnya was the capture of the capital city, Grozny. The initial attack on Grozny was conducted by a mechanized force, consisting of 6,000 troops mounted in tanks, BMP infantry fighting vehicles, and BTR armored personnel carriers. The Russians anticipated light opposition, but instead met a determined and heavy resistance from Chechens armed with a large number of antitank weapons. The attack was repulsed with significant Russian losses, both in personnel and armored vehicles (105 of 120 vehicles were lost). A second mechanized attack conducted on New Years Eve, 1994, was also repulsed with the loss of 140 of 200 tanks employed. The Russians captured Grozny some two months later, primarily through the use of excessive, overwhelming firepower, and at great cost to themselves and the local Chechen population.

The capture of Grozny marked the end of the first phase of Russian combat operations in Chechnya. Beginning in March 1995, the Russians became heavily engaged in anti-partisan operations as they tried to gain control of the country. Though the Russians controlled several major Chechen cities and a portion of the countryside by May 1995, they were never able to fully isolate the Republic. Chechen separatist fighters continued to receive a steady flow of arms and supplies from neighboring countries. Consequently, the Chechen fighters were able to maintain the initiative throughout this period, engaging Russian forces whenever and wherever they chose.

Even late in the campaign, Russian intelligence failed to detect Chechen fighters infiltrating Grozny, which enabled the Chechens to launch a major attack against the Russian-controlled city in early August 1996. The Chechens succeeded in capturing the city less than two weeks later.

.3

The recapture of Grozny was a significant loss for the Russians, precipitating a general cease-fire and the withdrawal of Russian forces from the Republic.

Besides the military aspects of combat in Chechnya, other issues such as ethnic, cultural, and religious divides impacted upon the nature of the conflict. The Russians failed to culturally orient their troops, which resulted in many serious cultural mistakes when dealing with the Chechen civilians. Once insulted or mistreated, the Chechen civilians became active participants in the struggle against Russia. The war eventually degenerated into one of abject ferocity and brutality on both sides. The Chechens are reported to have executed Russian prisoners, decapitated bodies, and boobytrapped the dead. Russian wounded and bodies were hung upside down outside of Chechen defensive positions in buildings, causing the Russians to have to shoot through their own people to engage the Chechens. The Russians, initially hesitant to cause civilian casualties, increasingly engaged Chechen units with a growing disregard for the safety of the local populace. As Russian discipline degraded, the incidences of Russian troops engaging in looting, arson, torture, even the summary execution of civilians, rose significantly. The result was the deaths of tens of thousands of civilians.

3. Lessons Learned. The intervention in Chechnya and the battles for Grozny generated many important lessons across the warfighting functions of combat in urbanized terrain. The following information is primarily from the Russian point of view, though some Chechen capabilities are included. The lessons learned are arrayed as warfighting functional considerations.

a. Command and Control.

(1) Russian military commanders lacked clear policy guidance. The Russian military's ability to attain and maintain steady, logical operational progress was hampered, if not negated, by a lack of clear governmental policy. The Russian government was split at the highest levels over the wisdom of committing forces to Chechnya, and the conduct of the intervention.

(2) Poor Russian command and control structure throughout all levels. There was no direct, unified chain of command for the operations in Chechnya. Operational directives often flowed from a number of governmental agencies directly to local unit commanders, often bypassing intermediate military commanders. This resulted in a lack of overall situational awareness, and exceptionally poor coordination among tactical forces.

(3) Russians were unable to contain/prohibit Chechen operations against Russia. When the Russian forces finally began to achieve progress in late May 1995, the Chechens initiated attacks against targets inside Russia. The result was a major propaganda victory for the Chechens.

(4) Excessive senior officer (flag level) interference contributed to operational confusion. There were eight major changes at the senior command level during the course of the intervention. Additionally, there were approximately one hundred general officers on the operational scene, about one general officer per 3,000 troops, which resulted in confusion as they got in each others way.

(5) Poor small unit leadership, particularly at the NCO level, was a primary cause of Russian tactical failures in Grozny. While the Russians recognize that urban combat is small unit intensive, they do not possess a professional NCO corps. Additionally, Russian forces lacked some 12,000 platoon leaders when the intervention began. This had a crippling effect on small unit operations.

(6) Russian inability to isolate Chechnya/Grozny allowed a constant flow of reinforcements and supplies into the country/city. Russian control of any given area was at best, porous. The Chechens continued to receive outside support throughout the campaign, as well as, a large supply of arms and equipment from poorly disciplined Russian troops inside Chechnya.

(7) Both sides used commercial off-the-shelf communication assets. The information war was as vitally important as any other aspect of the campaign. The Chechens had to make do with commercially available communication equipment. They fully exploited the use of cellular telephones, Motorola radios, improvised television stations, video cameras, and the internet. The Russians lacked sufficient military communication equipment, and had to supplement with commercial assets.

(8) Russian tactical communication was very difficult in Grozny. Urban structures in Grozny interfered with Russian military radios, severely hindering tactical communications. The Russians had to establish ground-based and aircraft-based relay stations to overcome the interference.

(9) Boundaries between units were tactical weak points. Units had to think of boundaries in both a horizontal and vertical context. There were many instances when the Chechens held the third floor and above, while the Russians held the first two floors and maybe the roof. This caused a lack of coordination, and mutual support. Additionally, due to poor command and control and coordination between units, boundaries became seams through which the Chechens could pass unopposed.

b. Intelligence.

(1) Lack of high quality intelligence made operations much more difficult and dangerous. Russian intelligence collection and dissemination capabilities were poor throughout the intervention. In March 1996, the Russian Minister of the Interior complained that poor reconnaissance and intelligence allowed Chechen fighters to move in and out of Grozny without opposition, which contributed to the heavy Russian losses.

(2) Directives to minimize civilian casualties ultimately put the Russians at a disadvantage. An advantage accrues to the side with less concern for the safety of the civilian population. Initially, the Russians attempted to limit civilian casualties. The Chechen

.3

fighters took advantage of this reticence, often using civilians, schools, and apartment buildings as a shield. While this may seem to be cold-blooded of the Chechen fighters, what must be understood is that the majority of the populace in Grozny were actually ethnic Russians, and that the ethnic Chechens had no remorse in using them in this manner.

(3) Russian concerns for the safety of the Chechen populace and property declined. As the Russian troops became increasingly frustrated in their attempts to separate Chechen fighters from the local populace, and their casualties began to mount, concerns for civilian safety waned, resulting in greater civilian casualties and further alienating the Chechen people.

(4) Rigorous communications security is essential, even against less sophisticated enemies.Much of the Russian tactical radio traffic was broadcast in the clear, which allowed the Chechens to intercept it. The Chechens were able to redirect Russian airstrikes onto Russian targets, and engage Russian artillery forward observers who broadcast their own position coordinates.

(5) The Russians made extensive use of RPVs for reconnaissance. The Chechen campaign saw the first Russian use of unmanned aerial vehicles for reconnaissance and intelligence gathering. There were primarily used by Russian airborne forces.

c. Maneuver.

(1) Inadequate training in the most basic maneuver and combat skills inhibited Russian operations. Overall, poor Russian combat performance could be traced to a lack of training in fundamental military skills, a situation that was then exacerbated by an additional lack of training in specific urban combat skills. Individual skills training was so inadequate that some units were entirely incapable of undertaking combat operations. The Russian's ability to conduct large scale operations had been seriously eroded by a complete cessation of exercises at the division level in the 2 years prior to the Chechen campaign. Even regimental and battalion level exercises had been curtailed by 75 percent or more. Additionally, The air forces were not any better trained or prepared for their operational role in Chechnya.

(2) Lack of cohesiveness among the Russian forces led to confusion and antagonism. The Russian ground forces consisted primarily of regular army units and Ministry of Interior Internal Troops, which were essentially a police force. With no real unified command and control, the two groups were usually conducting uncoordinated missions, which led to confusion and fratricide.

(3) The nature of cities channels combat operations along narrow lanes of activity. Combat conditions in Grozny were characterized by narrow fields of view, limited fields of fire, and constricted avenues of approach. For the Russians, this meant movement along approaches well known and heavily defended by the Chechens.

(4) Night fighting was the most difficult activity for Russian infantry. The Russian infantry, at all levels, was inadequately trained for night operations, and lacked night vision equipment.

(5) Armored vehicles cannot operate in cities without extensive dismounted infantry support. The Chechens formed antitank hunter-killer teams equipped with RPG-7 and RPG-18 antitank missiles. They engaged Russian tanks with volley fire from above, behind, and the sides. Armor columns not supported by dismounted infantry suffered great losses.

(6) Armored engineer vehicles are indispensable for removing obstacles. The Russians found that armored engineer vehicles were critical to restoring or improving mobility in Grozny. Eventually, the Russians formed special assault groups that included armored engineer vehicles in each group.

(7) The Chechens preferred small unit, hit-and-run ambush tactics. The Chechens relied on small units (15 to 20 troops) to conduct quick, sharp ambushes, from which they would disengage before the Russians could bring significant firepower to bear. Ambushes were often conducted in three tiers, with Chechens in the underground, on the ground floor, and on the roof, with each element having a different task in the ambush. They also used ambush tactics against Russian helicopters.

(8) Urban combat is extremely manpower intensive, and results in significant attrition of personnel and materiel for the attacker. The Russians discovered that a 5 to 1 manpower ratio was often not enough, due to the high casualty rates consistent with urban combat and the requirement to guard virtually every building taken.

(10) Tracked vehicles are preferable to wheeled vehicles in the urban environment. Tracked vehicles provided better mobility than wheeled vehicles due to their significantly greater ability to negotiate obstacles and rubble.

(11) Armored vehicles required additional protection (i.e., reactive armor, sandbags, wire mesh, etc.) to negate attacks from the sides, rear, and top. Russian tank design emphasized greater armor protection on the frontal arc, and less on other areas of the vehicle. This proved inadequate for urban combat, where attacks can come from any direction. The majority of lethal hits came from above, easily penetrating turrets and engine decks, and from the rear.

(12) The spatial qualities and perspectives of urban combat are more "vertical" than non-urban combat. The Chechens regularly fought from basements, rooftops, and upper-floor rooms, making them difficult to locate and engage. Russians soldiers trained in open areas, had to reorient to a confined, highly vertical environment.

(13) Standard Russian military unit configurations were inappropriate for urban combat. Typical Russian army unit organizations, oriented to open spaces and mechanized operations, proved to be inadequate for urban combat. They eventually had to adopt specially configured assault teams that included, among other assets, significantly greater firepower and engineer assets. Essentially, they moved towards a better combined arms team.

(14) The Russians found their helicopters to be far too vulnerable to rooftop snipers and ambushes in the urban setting. While Russian doctrine had long specified the capture of a building from the top-down, it became apparent in Grozny that this tactic was too expensive in terms of helicopter losses. The Chechens made effective use of snipers and RPG-equipped ambush teams to damage or destroy helicopters supporting Russian ground forces.

d. Fires.

(1) Artillery was invaluable in the direct fire mode. The Russians found artillery, to include rocket launchers and mortars, to be very effective for direct fire against point targets , usually at ranges of 150 to 200 meters.

(2) Obscurants (smoke, white phosphorous, tear gas) are especially useful in urban combat. The Russians used significant amounts of smoke and white phosphorus to mask the movement of forces. Every fourth or fifth artillery or mortar round was either smoke or WP. WP also had a toxic effect, easily penetrating Chechen protective masks.

(3) The Chechens effectively used RPGs against helicopters. The Chechens made effect use of ambush teams equipped with RPGs to engage Russian helicopters, either by damaging or destroying the helicopter or simply prohibiting helicopterborne operations.

(4) Air defense guns (i.e., ZSU-23-4) were valuable in suppressing ground targets. The Russians found air defense guns effective against multi-story buildings because they had sufficient elevation to hit targets in the upper stories. Air defense guns became a primary asset in the Russian special assault groups.

(5) The Russians determined that overwhelming firepower can make-up for organizational and tactical deficiencies in the short run if collateral damage is disregarded. Whenever Russian military units proved incapable of effective urban operations, the Russians simply fell back on massive, overwhelming firepower to reduce a strong point. It was through the use of excessive, unrestrained firepower, and a complete lack of regard for collateral damage, that finally enabled the Russians to gain control of Grozny.

(6) The Chechens negated Russian supporting fires by "hugging" Russian units. Whenever possible, Chechen fighters adopted the tactic of staying as physically close to Russian units as possible, so that the Russians could not employ supporting arms without the risk of injuring their own troops.

(7) The Russians found "bunker busting" weapons to be invaluable. The Russians came to rely heavily on "bunker busting" weapons for engaging Chechens, primarily snipers, entrenched in buildings.

(8) Russian helicopters needed stand-off weapons, that would allow them to engage targets while outside the range of Chechen weapons. The Chechens employed 23mm cannons and heavy machine guns against Russian helicopters, causing the Russian's helicopter gunships to fire from ranges of 3,000 meters or more. This required the use of wire-guided munitions and large caliber (240mm) rockets.

(9) The Russians made extensive use of precision-guided weapons. When the weather allowed, the Russians were able to use precision guided weapons, such as laser-guided bombs and missiles. They had great effect against priority targets such as bridges, major road intersections, and buildings.

e. Logistics.

(1) Deficiencies in the Russian logistics system degraded combat effectiveness. Significant inadequacies in the Russian logistics apparatus effected virtually all combat forces and operations. Some soldiers actually entered combat in Grozny without weapons or ammunition. The Russian transportation system failed. Supply officers were unprepared for the high volume demands for ammunition, hand grenades, smoke grenades, demolition charges, and one-shot antitank weapons.

(2) Russian logistics personnel were ill-prepared in basic soldier skills. Russian infantry often had to be used to conduct logistics operations into Grozny, because the logistics personnel lack the basic skills to defend themselves against the Chechens.

(3) The Russian ground forces required special equipment not normally found in unit T/Es. The T/E for the typical Russian infantry unit did not include such items as ropes, grappling hooks, and ladders, all of which proved essential for urban combat.

(4) Armored vehicle recovery is extremely difficult and dangerous. The Russians found the recovery of vehicles to be extremely difficult due to the terrain (i.e., narrow streets, rubble and debris), and Chechen opposition.

(5) Some Russian equipment was modified in the field to counter Chechen tactics and equipment. The Russians had to adopt add-on armor for their vehicles. They fabricated metal screens and wire mesh guards, which helped to defeat RPG and Molotov cocktail attacks.

(6) Medical support was of great importance. Russian casualty evacuation was difficult due to the lack of vehicular mobility, and the vulnerability of helicopters to Chechen fire. Many casualties were psychological, due in great part to the intense, violent nature of close quarter combat. The Russians were ill-prepared for the level of mental health problems they encountered. Additionally, logistics units were often unable to provide fresh drinking water, which caused Russian troops to consume contaminated water that resulted in other health problems.

f. Force Protection.

(1) Operations took much longer than expected, and the cost in terms of casualties was much greater than anticipated. The Russians committed unprepared, poorly trained and poorly equipped forces to Chechnya, based on an unrealistic operational time frame. Ultimately, this cost the Russians significant casualties and the war.

(2) Fratricide was a serious and continuing problem throughout the campaign. The situation facing Russian combat forces was ripe for fratricide. Poorly trained units, operating in a confused and uncertain urban environment, often unable to tell friend from foe, and lacking quality leadership and inter-unit coordination, were often as dangerous to themselves as they were to the Chechens .

(3) Chechen use of adhoc air defense systems eliminated Russian air support, particularly helicopters, in Grozny. The Chechens were able to improvise effective anti-air defenses using a mixture of anti-air and ground (i.e., heavy machine guns, RPGs) weapons, and ambush tactics specifically adapted to the urban environment.

(4) The Chechens made extensive and effective use of snipers. Snipers fired from well inside rooms versus near window openings, as well as, from rooftops and basements. The Russians lacked an effective sniper and counter-sniper capability of their own.

(5) The Chechens improvised crude chemical weapons. The Chechens, lacking access to military chemical weapons, Used off-the-shelf materials and equipment to make chemical mines (i.e., chlorine gas), detonated remotely by radio.

(6) The Chechens made extensive and effective use of boobytraps. It seemed that the Chechens mined and boobytrapped 7-8everything. They had a good understanding of the average Russian soldier's actions and reactions, and boobytrapped accordingly. Additionally, the Russians found it difficult to maintain boobytrap awareness among their troops.

Appendix K

Glossary

I. Acronyms

AAA .... antiaircraft artillery
AAAV .... advanced assault amphibious vehicle
AAV .... assault amphibious vehicle
ACE .... aviation combat element
AO .... area of operations
APC .... armored personnel carrier
APDS-T .... armor-piercing, discarding sabot with tracer
APFSDS .... armor-piercing, fin-stabilized, discarding sabot
API .... armor-piercing incendiary
ARVN .... Army of the Republic of Vietnam
ATF .... amphibious task force
ATGM .... antitank guided missile
BDA .... battle damage assessment
BMNT .... beginning of morning nautical twilight
CA .... civil affairs
CAG .... civil affairs group
CE .... command element
CI .... counterintelligence
COC .... combat operations center
CONAD .... consolidated administration center
CP .... command post
CSS .... combat service support
CSSE .... combat service support element
DOD .... Department of Defense
DS .... direct support
EOD .... explosive ordnance disposal
EPW .... enemy prisoner of war
EW .... electronic warfare
FAC(A) .... forward air controller (airborne)
FASCAM .... family of scatterable mines
FDC .... fire direction center
FEBA .... forward edge of the battle area
FLOT .... forward line of own troops
FM .... U.S. Army field manual
FM .... Frequency modulated
FMFM .... Fleet Marine Force manual
FMFRP .... Fleet Marine Force reference publication
FO .... forward observer
FPL .... final protective line
FRAGO .... fragmentary order
FSC .... fire support coordinator
GCE .... ground combat element
GPS .... global positioning system
GS .... general support
HE .... high explosive
HEAT .... high-explosive antitank
HEAT-MP .... HEAT-multipurpose
HEDP .... high-explosive dual-purpose
HEI-T .... high-explosive incendiary with tracer
HLZ .... helicopter landing zone
HMMWV .... high-mobility, multipurpose wheeled vehicle
HSS .... health services support
HUMINT .... human intelligence
IDF .... Israeli Defense Forces
IFV .... infantry fighting vehicle
IPB .. intelligence preparation of the battlefield/battlespace
ITOW .... improved TOW
JTF .... joint task force
KOCOA .... key terrain, observation and fields of fire, cover and concealment, obstacles, and avenues of approach
LAV .... light armored vehicle
LCAC .... landing craft air cushion
LD .... line of departure
LOS .... line of sight
LZ .... landing zone
MAGTF .... Marine air-ground task force
MCDP .... Marine Corps doctrinal publication
MCRP .... Marine Corps reference publication
MCWP .... Marine Corps warfighting publication
MEDEVAC .... medical evacuation
MEF .... Marine expeditionary force
MET .... meteorological
METT-T ... mission, enemy, terrain and weather, troops and support available-time available
MEU .... Marine expeditionary unit
MOOTW .... military operations other than war
MOPP .... mission-oriented protective posture
MOUT .... military operations on urbanized terrain
MP .... military police
MPF .... maritime prepositioning force
MULE .... modular universal laser equipment
NBC .... nuclear, biological, and chemical
NCA .... National Command Authorities
NCF .... naval construction force
NEF .... naval expeditionary force
NEO .... noncombatant evacuation operation
NGF .... naval gunfire
NSFS .... naval surface fire support
NVA .... North Vietnamese Army
NWP .... naval warfare publication

Appendix J

Glossary

I. Acronyms

OP .... observation post
OPLAN .... operation plan

OPORD
operation order
OPSEC
operations security
PDDA
power-driven decontamination apparatus
PGM
precision-guided munition
PLO
Palestine Liberation Organization
PLRS
position location reporting system
POL
petroleum, oils, and lubricants
PSYOP
psychological operations
ROE
rules of engagement
ROK
Republic of Korea
RP
red phosphorus
RSOP . . . . reconnaissance, selection, and occupation of position
SASR
special application sniper rifle
SMAW
shoulder-launched multipurpose assault weapon
SOP
standing operating procedure
STANAG
NATO standardization agreement
STB
supertropical bleach
TOW
tube-launched, optically tracked, wire-guided missile
UAV
unmanned aerial vehicle
UCMJ
Uniform Code of Military Justice
U.N.
United Nations
VC
Viet Cong
VHF
very high frequency
WP . . . . white phosphorus

II. Definitions

A

assault—1. The climax of an attack, closing with the enemy in hand-to-hand fighting. 2. To make a short, violent, but well-ordered attack against a local objective, such as a gun emplacement, a fort, or a machine gun nest. (Joint Pub 1-02)

assault—The culmination of an attack which closes with the enemy. (MCRP 5-2A)

attack—An offensive action characterized by movement supported by fire with the objective of defeating or destroying the enemy. (FMFRP 0-14)

B

battlespace—All aspects of air, surface, subsurface, land, space, and electromagnetic spectrum which encompass the area of influence and area of interest. (FMFRP 0-14)

block—1. A tactical task assigned to a unit that requires it to deny the enemy access to a given area or to prevent enemy advance in a given direction or an avenue of approach. It may be for a specified time. Units assigned this mission may have to retain terrain and accept decisive engagement. 2. An obstacle effect that integrates fire planning and obstacle effort to stop an attacker on a specific avenue of approach or to prevent an enemy from exiting an engagement area. (MCRP 5-2A)

breach—A tactical task where any means available are employed to break through or secure a passage through an enemy defense, obstacle, minefield, or fortification. (MCRP 5-2A)

built-up area—A concentration of structures, facilities, and populations, such as villages, cities, and towns, that forms the economic and cultural focus for the surrounding area. (MCRP 5-2A, modified)

bypass—A tactical task that involves maneuvering around an obstacle, position, or enemy force to maintain the momentum of advance. Bypassed obstacles and enemy forces are reported to higher headquarters. (MCRP 5-2A)

C

clear—A tactical task to remove all enemy forces and eliminate organized resistance in an assigned zone, area, or location by destroying, capturing, or forcing the withdrawal of enemy forces such that they cannot interfere with the friendly unit's ability to accomplish its mission. (MCRP 5-2A)

combat arms—Units who close with and destroy enemy forces or provide firepower and destructive capabilities on the battlefield. (MCRP 5-2A, modified)

combat operations center (COC)—The primary operational agency required to control the tactical operations of a command that employs ground and aviation combat, combat support, and combat service support elements or portions thereof. The combat operations center continually monitors, records, and supervises operations in the name of the commander and includes the necessary personnel and communications to do the same. (FMFRP 0-14)

combat service support (CSS)—The essential capabilities, functions, activities, and tasks necessary to sustain all elements of operating forces in theater at all levels of war. Within the national and theater logistic systems, it includes but is not limited to that support rendered by service forces in ensuring the aspects of supply, maintenance, transportation, health services, and other services required by aviation and ground combat troops to permit those units to accomplish their missions in combat. Combat service support encompasses those activities at all levels of war that produce sustainment to all operating forces on the battlefield. (Joint Pub 1-02)

combat support (CS)—Fire support and operational assistance provided to combat elements. (Joint Pub 1-02)

combat support (CS)—Units that provide critical combat functions in conjunction with combat arms units to secure victory (MCRP 5-2A, modified).

command post (CP)—A unit's or subunit's headquarters where the commander and the staff perform their activities. In combat, a unit's or subunit's headquarters is often divided into echelons; the echelon in which the unit or subunit commander is located or from which such commander operates is called a command post. (Joint Pub 1-02)

contain—A tactical task to restrict enemy movement. (MCRP 5-2A)

cover—1. Shelter or protection from enemy observation that reduces the effects of enemy direct and indirect fire. 2. A type of security operation that protects the force from surprise, develops the situation, and gives commanders time and space in which to respond to the enemy's actions. (MCRP 5-2A)

D

defend—A combat operation designed to defeat an attacker and prevent him from achieving his objectives. It employs all means and methods available to prevent, resist, or destroy an enemy attack. Forms of defensive operations are area and mobile. Choices of defensive operations are in-depth and forward. The defensive techniques are defend in sector, defend a battle position, and defend a strong point. (MCRP 5-2A)

destroy—1. A tactical task to physically render an enemy force combat-ineffective unless it is reconstituted. 2. To render a target so damaged that it cannot function as intended nor be restored to a usable condition without being entirely rebuilt. (MCRP 5-2A)

disrupt—A tactical task or obstacle effect that integrates fire planning and obstacle effort to break apart an enemy's formation and tempo, interrupt the enemy's timetable, or cause premature commitment of enemy forces, or the piecemealing of his attack. (MCRP 5-2A)

II. Definitions

F

fire support—Assistance to elements of the Marine air-ground task force engaged with the enemy rendered by other firing units, including (but not limited to) artillery, mortars, naval surface fire support, and offensive air support. (FMFRP 0-14)

fix—1. A tactical task in which actions are taken to prevent the enemy from moving any part of his forces either from a specific location or for a specific period of time by holding or surrounding them to prevent their withdrawal for use elsewhere. 2. A tactical obstacle effect that integrates fire planning and obstacle effort to slow an attacker within a specified area—normally an engagement area. (MCRP 5-2A)

H

helicopterborne operation—A military action in which combat forces and their equipment maneuver about the battlefield by helicopters or vertical-landed aircraft. Aviation activities are under the control of the aviation combat element commander who is assigned in direct or general support of one or more combat element(s). (FMFRP 0-14)

hold—1. To maintain or retain possession of by force, as a position or an area. 2. In an attack, to exert sufficient pressure to prevent movement or redisposition of enemy forces. (Joint Pub 1-02)

I

infiltration—The movement through or into an area or territory occupied by either friendly or enemy troops or organizations. The movement is made, either by small groups or by individuals, at extended or irregular intervals. When used in connection with the enemy, it infers that contact is avoided. (Joint Pub 1-02)

isolate—A tactical task given to a unit to seal off (both physically and psychologically) an enemy from his sources of support, to deny an enemy freedom of movement, and prevent an enemy unit from having contact with other enemy forces. An enemy must not be allowed sanctuary within his present position. (MCRP 5-2A)

L

linkup—An operation wherein two friendly ground forces join together in a hostile area. (FMFRP 0-14)

M

Marine air-ground task force (MAGTF)—A task organization of Marine forces (division, aircraft wing, and service support groups) under a single command and structured to accomplish a specific mission. (Joint Pub 1-02)

military operations on urbanized terrain (MOUT)—All military actions that are planned and conducted on a topographical complex and its adjacent natural terrain where manmade construction is the dominant feature. It includes combat in cities, which is that portion of MOUT involving house-to-house and street-by-street fighting in towns and cities (MCRP 5-2A, modified)

military operations other than war (MOOTW)—Operations that encompass the use of military capabilities across the range of military operations short of war. These military actions can be applied to complement any combination of the other instruments of national power and occur before, during, and after war. (Joint Pub 1-02)

N

neutralize—1. To render enemy personnel or material incapable of interfering with a particular operation. 2. To render safe mines, bombs, missiles, and boobytraps. (MCRP 5-2A)

O

occupy—A tactical task in which a force moves onto an objective, key terrain, or other man-made or natural terrain area without opposition, and controls that entire area. (MCRP 5-2A)

P

peace enforcement—Application of military force, or the threat of its use, normally pursuant to international authorization, to compel compliance with resolutions or sanctions designed to maintain or restore peace and order. (Joint Pub 1-02)

peacekeeping—Military operations undertaken with the consent of all major parties to a dispute, designed to monitor and facilitate implementation of an agreement (ceasefire, truce, or other such agreement) and support diplomatic efforts to reach a long-term political settlement. (Joint Pub 1-02)

protect—A tactical task to prevent observation of or engagement or interference with, a force or location. (MCRP 5-2A)

pursuit—An offensive operation designed to catch or cut off a hostile force attempting to escape, with the aim of destroying it. (Joint Pub 1-02)

R

reduce—1. A tactical task to gain control over an enemy position or objective. 2. A task to create lanes through or over an obstacle sufficient to allow the attacking force to accomplish its mission. (MCRP 5-2A)

retain—A tactical task to occupy and hold a terrain feature to ensure it is free of enemy occupation or use. (MCRP 5-2A)

rules of engagement (ROE)—Directives issued by competent military authority which delineate the circumstances and limitations under which United States forces will initiate and/or continue combat engagement with other forces encountered. (Joint Pub 1-02)

S

screen—A task to maintain surveillance, provide early warning to the main body, or impede, destroy, and harass enemy reconnaissance within its capability without becoming decisively engaged. (MCRP 5-2A)

search—A systematic reconnaissance of a defined area, so that all parts of the area have passed within visibility. (Joint Pub 1-02)

secure—A tactical task to gain possession of a position or terrain feature, with or without force, and to deploy in a manner which prevents its destruction or loss to enemy action. The attacking force may or may not have to physically occupy the area. (MCRP 5-2A)

seize—To clear a designated area and obtain control of it. (FMFRP 0-14)

suppression—A tactical task to employ direct or indirect fires, electronic attack, or smoke on enemy personnel, weapons, or equipment to prevent or degrade enemy fires and observation of the friendly forces. (FMFRP 0-14)

T

terrorism—The calculated use of violence or threat of violence to inculcate fear; intended to coerce or to intimidate governments or societies in the pursuit of goals that are generally political, religious, or ideological. (Joint Pub 1-02)

Appendix L

References and Related Publications

Field Manuals (FMs)

3-11	Flame, Riot Control Agent, and Herbicide Operations
5-34	Engineer Field Data
5-102	Countermobility
5-250	Explosives and Demolitions
7-98	Operations in a Low-Intensity Conflict
9-15	Explosive Ordnance Disposal Service and Unit Operations
20-32	Mine/Countermine Operations
21-75	Combat Skills of the Soldier
3-30	Grenades and Pyrotechnic Signals
23-34	TOW Weapon System
27-10	Law of Land Warfare
33-1	Psychological Operations
34-130	Intelligence Preparation of the Battlefield
41-5	Joint Manual for Civil Affairs
41-10	Civil Affairs Operations
90-8	Counterguerrilla Operations
100-20	Military Operations In Low Intensity Conflict

Technical Manual (TM)

9-2350-264-10-2	Operator's Manual for Operation Under Usual and Unusual Conditions for Tank, Combat, Full-Tracked: 120mm Gun, M1A1 General Abrams, with Change 4

Marine Corps Concept Paper

A Concept for Future Military Operations on Urbanized Terrain

Marine Corps Doctrinal Publications (MCDPs)

1	Warfighting
1-3	Tactics

Marine Corps Reference Publications (MCRPs)

5-2A	Operational Terms and Graphics

Marine Corps Warfighting Publications (MCWPs)

3-15.3	Scout Sniping (under development)
3-37.1	Chemical Operations
3-37.4	NBC Reconnaissance

Fleet Marine Force Manuals (FMFMs)

0-9	Field Firing for the M16A2
0-25	The Law of Land Warfare
1-3	Tactics
2-6	MAGTF Rear Area Security
2-11	Antiarmor Operations
3-5	Employment of Military Police in Combat
3-30	Communications
6	Ground Combat Operations
6-7	Scouting and Patrolling for Infantry Units
6-8	Supporting Arms Observer, Spotter, and Controller
6-9	Marine Artillery Support
6-21	Tactical Fundamentals of Helicopterborne Operations
7-14	Combating Terrorism
7-34	MAGTF Civil Affairs
7-34-1	Civil Affairs Guidance for the Small-Unit Leader
7-40	Helicopter Insertion/Extraction
8-2	Counterinsurgency Operations
11-9	NBC Protection
11-10	NBC Decontamination
11-11	Treatment of Chemical Agent Casualties and Conventional Military Chemical Injuries
11-12	Chemical Hazard Assessment Guide
11-13	NBC Handbook
11-17	Chemical and Biological Contamination Avoidance
11-18	Nuclear Contamination Avoidance
11-20	NBC Reconnaissance
13-7	MAGTF Breaching Operations

Fleet Marine Force Reference Publications (FMFRPs)

0-6	Marine Troop Leader's Guide
0-14	Marine Corps Supplement to DOD Dictionary of Military and Associated Terms
4-26	Enemy Prisoners of War and Civilian Internees
7-14A	The Individual's Guide for Understanding and Surviving Terrorism
7-37	Vehicle Bomb Search
12-15	Small Wars Manual

12-18	Mao Tse-Tung on Guerrilla Warfare
12-25	The Guerrilla and How to Fight Him
13-5	Engineer Field Data

Joint Publications

1-02	DOD Dictionary of Military and Associated Terms
3-07	Joint Doctrine for Military Operations Other Than War
3-11	Joint Doctrine for Nuclear, Biological, and Chemical (NBC) Defense
3-15	Joint Doctrine for Barriers, Obstacles, and Mine Warfare
3-53	Doctrine for Joint Psychological Operations
4-06	JTTP for Mortuary Affairs in Joint Operations

Standard NATO Agreements (STANAGs)

2003	Patrol Reports by Army Forces
2014	Operations Orders
2029	Method of Describing Ground Locations, Areas, and Boundaries
2070	Emergency War Burial Procedures
2088	Battlefield Illumination
2101	Establishing Liaison
2868	Land Force Tactical Doctrine (ATP-35(B))

Miscellaneous

Fratricide: Reducing Self-Inflicted Losses, Newsletter No. 92-4, Center for Army Lessons Learned (CALL), U.S. Army Combined Arms Command, Fort Leavenworth, Kansas, April 1992.

HR, 25, Annex to Hague Convention No. IV, *Embodying Regulations Respecting the Laws and Customs of War on Land*, 18 October 1907.

Modern Experience in City Combat, U.S. Army Human Engineering Laboratory, Aberdeen Proving Ground, Maryland, 1987.

*FM 3-19.15 (FM 19-15)

Field Manual
No. 3-19.15

Headquarters
Department of the Army
Washington, DC, 18 April 2005

Contents

Distribution Restriction: Approved for public release; distribution is unlimited.

*This publication supersedes FM 19-15, 25 November 1985.

Page

Preface

Field Manual (FM) 3-19.15 addresses continental United States (CONUS) and outside continental United States (OCONUS) civil disturbance operations. Today, United States (US) forces are deployed on peacekeeping, peace enforcement, and humanitarian assistance operations worldwide. During these operations, US forces are often faced with unruly and violent crowds intent on disrupting peace and the ability of US forces to maintain peace. Worldwide instability coupled with increasing US military participation in peacekeeping and related operations requires that US forces have access to the most current doctrine and tactics, techniques, and procedures (TTP) necessary to quell riots and restore public order.

In addition to covering civil unrest doctrine for OCONUS operations, FM 3-19.15 addresses domestic unrest and the military role in providing assistance to civil authorities requesting it for civil disturbance operations. It provides the commander and his staff guidance for preparing and planning for such operations. The principles of civil disturbance operations, planning and training for such operations, and the TTP employed to control civil disturbances and neutralize special threats are discussed in this manual. It also addresses special planning and preparation that are needed to quell riots in confinement facilities are also discussed.

In the past, commanders were limited to the type of force they could apply to quell a riot. Riot batons, riot control agents, or lethal force were often used. Today, there is a wide array of nonlethal weapons (NLW) available to the commander that extends his use of force along the force continuum. This manual addresses the use of nonlethal (NL) and lethal forces when quelling a riot.

Appendix A complies with current Army directives, which state that the metric system will be incorporated into all new publications.

The proponent for this publication is HQ, TRADOC. Send comments and recommendations on Department of the Army (DA) Form 2028 (*Recommended Changes to Publications and Blank Forms*) directly to Commandant, US Army Military Police School (USAMPS), ATTN: ATSJ-DD, 401 MANSCEN Loop, Fort Leonard Wood, MO 65473-8929.

Unless this publication states otherwise, masculine nouns or pronouns do not refer exclusively to men.

Chapter 1

Operational Threats of the Civil Disturbance Environment

As the Cold War ended with the fall of the Berlin Wall and the collapse of Soviet communism in 1989, the Army was no longer concerned with containing the aggressive Soviet Union. The US Army began to transform because the battlefield on which it would fight had changed. Operations Just Cause, Joint Endeavor, Desert Shield, and Desert Storm have all been recognized as containing nontraditional battlefields. Following these operations, the Army was called upon to conduct peacekeeping, peace enforcement, and humanitarian assistance operations. These operations required US forces to protect refugees and humanitarian workers and keep hostile factions separate to enforce peace.

GENERAL CAUSES FOR CIVIL UNREST

1-1. In these modern times, demonstrations, civil unrest, public disorder, and riots happen for a number of reasons. Some of these reasons are economic hardships, social injustices, ethnic differences (leading to oppression), objections to world organizations or certain governments, political grievances, and terrorist acts. An event can be triggered by a single cause or a combination of causes. For example, operations in the Balkans involving civil unrest and riots were the result of ethnic hatred, a lack of civil authority, food shortages, a revolution, and religious-based fighting factions.

1-2. Demonstrations may range from simple, nonviolent protests that address specific issues, to events that turn into full-scale riots. Gatherings in protest are recognized rights of any person or group, regardless of where US forces may be operating. This fundamental right is protected under the *Constitution of the US*. During peacekeeping or peace enforcement operations, US forces should never violate basic civil or human rights. Most protesters are law-abiding citizens who intend that their protests be nonviolent, but some protest planners insist that the event involve some kind of violence. Often, in the media, protesters can gain sympathy for their cause by prompting authorities to take physical action against them. Violence is often the result of demonstrators beginning to conduct unlawful or criminal acts and authorities (who are responsible for the safety and welfare of all) enforcing the laws of the municipality, state, or nation. The depth of violence is determined by the willingness of demonstrators to display and voice their opinions in support of their cause.

1-3. Commanders must be aware of the possibility that some individuals or groups within an organized demonstration may have the intent to cause disruption, incite violence, destroy property, and provoke the authorities. The tactical situation and actions of the crowd should dictate control and enforcement options. Agitators and criminal infiltrators within the crowd can lead to the eruption of violence. Inciting a crowd to violence or a greater intensity of violence through the use of brutish enforcement tactics should be avoided. Publicity can be detrimental to authorities and beneficial to crowds because it can further their causes.

1-4. Community unrest results in urban conflicts that arise from highly emotional social and economic issues. Economically deprived inner city residents may feel that they are treated unjustly or ignored by people in power and authority. Tensions can build quickly in a community over a variety of issues, such as hunger, poor employment opportunities, inadequate community services, poor housing, and labor issues. Tension in these areas creates the potential for violence. When tensions are high, it takes a small (seemingly minor) incident, rumor, or act of injustice to ignite groups within a crowd to riot and act violently. This is particularly true if community relations with authorities are part of the problem.

1-5. Significant ethnic differences in a community can create an atmosphere of distrust, even hatred. Unrest among ethnic groups competing for jobs, living areas, and sparse essentials can cause an eruption of civil disorder and/ or riots. As emotions run high, violence becomes likely.

1-6. Terrorist organizations may infiltrate groups within a demonstrating crowd. These terrorist groups may intend to embarrass their government or other governments. Terrorist infiltrators can be used to provoke crowds as a diversion, as part of a demonstration, or as cover for terrorist acts.

CROWD DEVELOPMENT

1-7. Crowds are a gathering of a multitude of individuals and small groups that have temporarily assembled in the same place. These small groups are usually comprised of friends, family members, or acquaintances that represent a group belief or cause. Individuals assume a sense of anonymity—they are viewed as just another face in the crowd. People in small groups are known only to companions in their group and to others in the gathering that have come from the same neighborhood or community. Commanders must consider how the individuals assembled and how they are interacting during the gathering process.

GATHERINGS

1-8. The assembly process of a gathering refers to the movement of people from different locations to a common location within a given period. This largely determines who participates. Creating a gathering from a crowd is a process with a beginning, middle, and end. As shown in *Figure 1-1* a crowd has an assembly process that leads to the gathering of the crowd, which is always followed by a dispersal process.

1-9. Gatherings are often assisted by the activities of individuals or groups with a specific agenda, such as yelling catchy slogans and cheers that everyone can easily pick up and join in on. Some groups are so well organized that they can prestage leaders to infiltrate a gathering. This creates unity, even inciting newcomers to join their cause. The first phase of the gathering is the assembly process. It can occur in one of two ways—impromptu or organized.

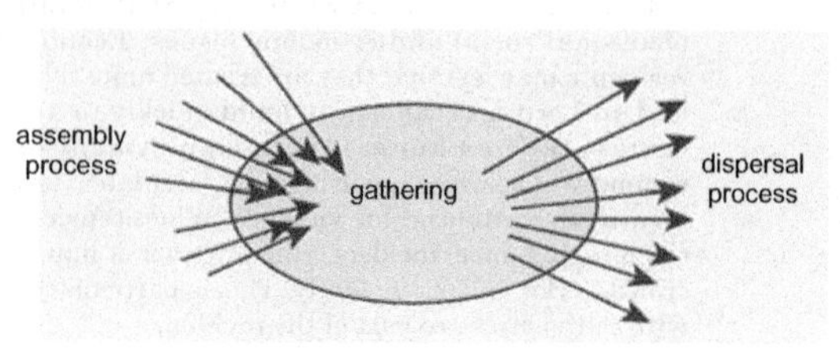

Figure 1-1. Crowd Building

IMPROMPTU GATHERINGS

1-10. Impromptu gatherings usually develop informally and are mostly done by word of mouth (one person to another or one group to another). Participants spread intelligence by telling one another when, where, and what is happening and inviting them to participate. An example of an impromptu gathering would be a gathering at a secured food distribution point after receiving information (by word of mouth) that a large truck carrying much-needed supplies and food is about to arrive. In this case, hunger would be the driving force causing the migration of people to the food distribution facility.

ORGANIZED GATHERINGS

1-11. Like an impromptu gathering, an organized gathering can also involve individuals and groups passing on information to one another. Passing intelligence of police activity and occurrences and on when and where events will take place helps organizers to prestage participants. Rarely is one represented group responsible for pulling together a gathering. Organized gatherings rely heavily on established groups that attract people to gather. Recent examples of these well-organized groups are anarchists, anti-globalization groups, and anti-free-enterprise groups. Groups representing extreme religious faiths and ethnic organizations have been common too. Some gatherings involve groups from rival, or even warring, factions.

1-12. Organized gathering processes rely more on centralized planning and organization. One or more of the groups offer organizers lists of individual names and groups, which they contact as potential participants. Modern technologies are available to these organizers through telephone banks, mailing lists, or e-mail addresses. In preparation for a long-term event, some group organizers have the means to provide logistical support of transportation, food, and water to participants.

> During the preelection period, Bosnian-Serbs (in a bid not to lose any of the territory they had won during the war) began to organize this fiercely contested city. Serbian mobs attacked US forces with bricks, rocks, metal bars, and firebombs. Organization for this altercation started days before, as many witnessed the prestaging of the items to be used in the attack. Wheelbarrows, 55-gallon metal drums, and other large containers were filled with the ammunition that would later be used to attack peacekeepers.
>
> Brcko, Bosnia
> September 1996

CROWD BUILDING

1-13. The second phase of the gathering process is the building of the crowd. Old concepts of crowds based on stereotypes are no longer true. Studies show that—

- Crowds are made of many entities, and all participants are not the same.
- Crowds are not made of isolated individuals but of a minority of individuals and a majority of small groups who may or may not be acquainted with each other.
- Groups and individuals in the crowd are not unanimous in their motivation.
- Groups and individuals in the crowd are usually not anonymous to each other, especially different groups gathered at the same place.
- Individuals within a crowd are not given to unique emotional displays; although known within a group, most individuals would not want to draw attention to themselves.
- Groups within a crowd will often act in unison in an attempt to achieve their agendas.
- Crowds are generally not unique or distinguished by violence or unlawful conduct.

DISPERSAL PROCESS OF A GATHERING

1-14. The last and final phase of any gathering of a crowd is the dispersal process. It involves the movement of people from the common location where they assembled to one or more alternate locations. The dispersal process ends the gathering of a crowd or at least begins its decline. The dispersal can occur on a routine, emergency, or coerced basis.

ROUTINE DISPERSAL

1-15. The routine dispersal may be specified in advance. It can also be included in the assembly instructions given by the organizers of an event.

EMERGENCY DISPERSAL

1-16. An emergency dispersal occurs when people evacuate an area in result of an unexpected crisis, such as a fire, explosion, bomb threat, or terrorist act. Individuals in such an emergency quickly recover from the initial shock. They keep their wits about them and improvise ways of extricating themselves and their companions from the dangerous situation. However, tragedies have occurred in some emergency dispersal situations. Therefore, it is important that forces on the scene are careful not to misinterpret what is occurring.

COERCION DISPERSAL

1-17. Coercion dispersal is caused by the use of force at some level. This is not necessarily the best way to force the dispersal of a crowd. The negotiated management of crowds is the preferred method and has proven to be highly successful in getting crowd organizers to police themselves, especially if the demonstration and/or protest leaders are available and willing to participate.

CROWD DYNAMICS

1-18. Understanding crowds and how individuals and groups form crowds through the gathering process is important because these issues apply to the dynamics of a crowd. Under most circumstances, gathered crowds are orderly and present little or no problems for authorities. Crowds consist of people who, although very motivated and passionate, are also subject to their own need for creature comforts. Rainy, cold, and nasty weather has a way of disheartening all but the few highly motivated and disciplined individuals. If problems exist, they usually fall into the following three categories:

- **Public disorder.** Public disorder is a basic breach of civic order. Individuals or small groups assembling have a tendency to disrupt the normal flow of things around them.
- **Public disturbance.** Public disturbance is designed to cause turmoil on top of the disruption. Individuals and groups assembling into a crowd begin chanting, yelling, singing, and voicing individual or collective opinions.
- **Riot.** A riot is a disturbance that turns violent. Assembled crowds become a mob that violently expresses itself by destroying property, assaulting others, and creating an extremely volatile environment.

1-19. Being part of a crowd of people has certain effects on different people. Each individual in a crowd is susceptible to behaving in a way that is contrary to their normal behavior. Some reasons for these behaviors are as follows:

- Crowds provide individuals with a sense of anonymity. With so many others, an individual realizes that he is just another face in the crowd, giving a sense of invulnerability.
- Crowd and individual behaviors are impersonal by nature. The "them-against-us" attitude affords those within the crowd the ability to freely (without hesitation or reservation) be verbally abusive, throw objects, or attack anyone who gets in the way.

- Crowds provide individuals with the idea that their moral responsibilities have shifted from themselves as a person to the crowd as a whole. Large numbers of people discourage individual behavior, and the urge to imitate others within the crowd is strong. Individuals look to others around them for cues of what to do next, disregarding their own background and training. Often, it is only the strong, well-disciplined person who can resist the prevailing behavior of a crowd.
- Crowd behavior influences the actions of both the disorderly individuals of the crowd and the authorities tasked to control them.

1-20. Individuals within a crowd are at times driven by deep-felt emotions. Emotional contagion is the most dramatic psychological factor of crowd dynamics. It provides the crowd with a temporary bond of psychological unity. Lasting long enough, this unity can push a simple organized crowd into a mob. Normal law and authority are rejected en masse under these conditions, increasing the potential for violence and panic to erupt.

1-21. Panic can erupt quickly, especially when crowds turn into mobs. Individuals within the mob can easily sense that their safety and well-being are at risk, putting them in a "fight-or-flight" mentality. Adding to the panic and confusion is the use of riot control agents (RCAs) by authorities in an attempt to gain control. Individuals in a mob, during the heat of confrontation, may attempt to leave and find that there are no escape routes and that roads are blocked. This can often lead to violent, physical attacks.

CROWD TYPES

1-22. Gaining an understanding of crowds and the dynamics that cause individuals or groups to join together in a crowd should be every leader's responsibility. Active, expressive, acquisitive, and hostile are not acceptable terms used to describe a crowd. They are the motives inferred from the actions the crowd takes. No gathering of a crowd is persistently or exclusively active, expressive, acquisitive, or hostile. For the sake of description, crowds can be identified as casual, sighting, agitated, or mob-like.

CASUAL CROWDS

1-23. Casual crowds are identified as individuals or small groups with nothing in common to bind them together. If they have an agenda, it is their own. They arrive separately and leave separately. Casual crowds are made up of individuals or small groups occupying the same common place, such as a shopping mall where these individuals or small groups meet.

SIGHTING CROWDS

1-24. Sighting crowds are similar to casual crowds with one additional element—an event. People migrate as a crowd to sporting events, are attracted to fires and accidents, and attend music concerts. Individuals or small groups gather at these events for the same purpose. It is the event and/or one's curiosity that compels a crowd to come together.

AGITATED CROWDS

1-25. Agitated crowds add responses that are based on the elements (people, space, and event). Individuals with strong emotional feelings within a crowd can quickly spread and infect the rest of the crowd. As more people within the crowd become emotionally involved, a sense of unity can develop, causing changes in the overall demeanor of the crowd. Yelling, screaming, crying, and profane name-calling are all associated with an agitated crowd.

MOB-LIKE CROWDS

1-26. Mobs have all the elements found in the first three types of crowds, with the addition of aggressive, physical, and sometimes violent actions. Under these conditions, individuals within a crowd will often say and do things they usually would not. Extreme acts of violence and property damage are often a part of mob activities. Mobs consist of (or involve) the elements of people and groups being mixed together and becoming fluid.

CROWD TACTICS

1-27. During a public disorder or disturbance, individuals and small groups within a crowd use any number of tactics to resist authority and disrupt and add turmoil in order to achieve their goals. These tactics can be unplanned or planned and violent or nonviolent. The more organized and purposeful a crowd becomes, the more likely a tactic will be used.

NONVIOLENT BEHAVIOR

1-28. As explained earlier in this chapter, most gatherings of individuals and small groups into a crowd do not involves violent behavior. A public disorder or disturbance usually involves some harmless name-calling, demonstrations to express views, corporate yelling and chanting, and even singing and dancing. In some instances, there may be property damage and the erection of makeshift barricades. All four types of crowds can be nonviolent, but mobs and agitated crowds have the greatest tendency to turn violent. Nonviolent actions of a crowd are disruptive because they are in direct conflict with what authorities want them to do, such as refusing to leave when directed, locking arms, and sitting in front of or around areas and buildings that the authorities are attempting to clear.

VIOLENT BEHAVIOR

1-29. A crowd that becomes a mob can be very violent and destructive. Although some nonviolent activities occur, violent crowds strike out physically at bystanders or others in the crowd, destroying both private and government property, setting fires, and employing bombs (in extreme cases). The only limitations for violent crowds are their own imaginations, the training of their leaders, and the materials readily available.

1-30. Violent crowds may erect barricades and physical barriers to impede the movement of authorities and prevent them from entering certain areas and/or buildings and to hide their activities. Common materials used for barricades include vehicles, felled trees, furniture, felled light poles, fencing materials, or anything else on hand and readily available.

1-31. Setting fires and preventing firefighters access to the structure by blocking it is another violent crowd tactic. Often, fires are set to create confusion in authorities or as a diversion to another activity.

RIOTS

1-32. A riot is one or more groups or individuals who are part of a larger crowd that involves threats of violence against persons or property. In some cases, a crowd will continue to gather until it evolves into a riot.

1-33. Riots vary considerably in both targets and players. A communal riot, for instance, deals with deep-seated ethnic, religious, and language differences. Commodity riots involve an attack on property by acts of vandalism, looting, and arson. Protest riots, such as the riot around the World Trade Organization (WTO) Assembly in Seattle, Washington (30 November 1999), illustrated individuals and groups aggressively and sometimes violently acting out or voicing their opposition to the assembly. The Democratic Convention in Chicago, Illinois, in 1968 reflects a riot that directly targeted police and authority in general. Celebration riots occur across the US as a result of home team victories in sporting events, among other reasons. Celebrating crowds look to make the moment more memorable through raucous acts that demonstrate their joy or happiness, for example, the riots that took place in Chicago in 1992 as a result of the Chicago Bulls winning the National Basketball Association (NBA) Championship.

1-34. Commanders and leaders must be aware that highly organized groups of protesters and/or demonstrators have developed tactics to disrupt the control force. These crowd tactics were published in handbooks for communist organizers during the Cold War, and today the Internet contains sites that are devoted to sharing these tactics with the general public and various protest groups. Although these tactics are somewhat outdated, commanders and leaders must be aware of them and their potential consequences. Many of these groups, such as anarchists, often mock the tactics employed against large crowds; for example, they may identify the police as "Darth Vader Cops" when the control force response is in full riot control gear. Commanders must be aware that well-organized crowds may attempt to engage the control force, surround it, and overpower it by sheer numbers. To avoid this, commanders must do a detailed terrain analysis (intelligence preparation of the battlefield [IPB]) of the area to include all approach and exit routes in the area. The control force flanks have to be protected to avoid being enveloped by the mob. To avoid being enveloped, the on-site commander should keep the crowd at a comfortable distance from the control force formation, which is usually the lethal zone of his NL munitions. Establishing predetermined rally points for the control force is critical in the event that this type of tactic is used by the crowd.

TACTICS USED TO DEFEAT AUTHORITIES

1-35. Organized mobs will try to defeat the control force by employing several different types of tactics. These tactics include the following:

- Constructing barricades.
- Using Molotov cocktails, smoke grenades, rocks, and slingshots.
- Feinting and flanking actions.

BARRICADES

1-36. By constructing barricades, the rioters are trying to protect themselves from assault by the control force. If the control force does advance toward the barricade, rioters hope that the barricade will split the control force and leave it vulnerable to actions by the mob.

MOLOTOV COCKTAILS, SMOKE GRENADES, ROCKS, AND SLINGSHOTS

1-37. Angry mobs will often attempt to disrupt the control force by throwing rocks and other projectiles. As the situation deteriorates, the mob may escalate the violence by using a battery of slingshots that will pellet the control force with a barrage of projectiles ranging from small rocks to marbles. They may also use smoke grenades (homemade or store-bought) to mask their movements. Finally, Molotov cocktails are used against personnel, employed vehicles, and portions of the control force. The mob will attempt to gain a position above an armored vehicle to enable them to drop a Molotov cocktail into an open hatch.

FEINTING AND FLANKING ACTIONS

1-38. Organized mobs may attempt to disrupt the movement of the control force by feinting an assault. When the control force moves to blunt the assault, the mob will assault the exposed flanks in an attempt to split the control force and envelop a portion of it.

WEAPONS

1-39. Mobs will often use various types of weapons against authorities to achieve their agendas and goals. It is extremely important that leaders train soldiers to recognize possible threats of the various weapons used by crowds or mobs, enabling the soldiers to react with the appropriate minimum force.

VERBAL ABUSE

1-40. In almost every instance of a civil disturbance or riot, verbal abuse will be an aggressive tool. Obscene language, racial remarks, taunts, ridicules, and jeers should always be expected. It is apparent that the purpose for using verbal abuse is to anger, demoralize, and provoke a physical response. Undisciplined, untrained soldiers who face such an attack could cause the situation to escalate. Just one provoked action of a soldier could be interpreted as an act of brutality by the media.

PERCEIVED INNOCENT PEOPLE OR THE WEAK

1-41. Women, children, and the elderly have often been used as pawns in civil disturbance operations. On many occasions, the innocent and the weak are moved to the front of the crowd and used as a barrier. With the innocent and the weak directly facing authorities, aggressive and violence-prone individuals or groups behind them attempt to provoke authorities to react. Often, this tactic is to exploit the situation in the eyes of the media for propaganda value. Despite the perception of sometimes being weaker, women, children, and the elderly can be just as committed to the insurgency as those behind them.

BARRICADES

1-42. Barricades are made of anything large or heavy enough to impede or prevent the movement of authorities. Trees, vehicles, fires, and furniture are all examples of items used to erect a barricade.

THROWN OBJECTS

1-43. In many cases, a crowd will throw anything in their reach that can be picked or picked up. Rotten fruit and/or vegetables, rocks, bricks, bottles, sticks, and pieces of lumber are most often used because they are readily available. More sophisticated types of thrown objects are improvised bombs, such as petrol bombs or tennis balls with nails sticking out of them. Never discount that these individuals and groups could acquire hand-thrown explosives, such as sticks of dynamite or grenades.

VEHICLES AND OTHER OBJECTS

1-44. Vehicles and other objects can be placed on higher ground (atop a hill or in a building) to disrupt civil disturbance operations. For example, a vehicle or tire can be set on fire and rolled down a hill, carts or barrels filled with hazardous or flammable materials can be pushed down a hill, or burning furniture and petrol bombs can be thrown from two-story or higher buildings or toward a formation of soldiers.

FIRE

1-45. Rioters have set buildings, vehicles, and other structures on fire to block the advances of authorities and to create confusion and diversions. If property damage is the goal of a crowd, fire is an effective tool. A sniper, for example, may set a building on fire to divert attention or provide maximum target possibilities. Other tactics used, particularly in the Balkans, were to flood the lower floors or basements of structures with flammable liquids or gas and then ignite it in an opportune moment.

FIREARMS

1-46. Sniping or massing fire from within the crowd, in buildings, or other covered positions can (at times) be effective tools for terrorists using a crowd to cover their acts. Firearms are anything that fires a projectile toward a target of opportunity.

EXPLOSIVES

1-47. A bomber's imagination is the only limiting factor in the use of explosives. Explosives can be used as simple diversions to block the advances of formations or provide an escape for rioters. The worst use of explosives is to purposely kill, injure, or demoralize authorities that are protecting and assuring the safety of everyone.

Chapter 2

Control Force Operations

Information is the key to developing civil disturbance plans. One must know who the demonstrators are; when, where, and why they are demonstrating; what their capabilities are; and what their possible course of action (COA) is. A commander's need for current, valid information cannot be overemphasized. The commander must learn as much as he can about the participants (their motivations, strategies, tactics, targets, and dedication). The more knowledge a commander has about the participants, the better equipped he is to counter their actions. He needs sound information to decide how to best use his available resources.

CIVIL DISTURBANCE PLANNING CONSIDERATIONS

2-1. The side that possesses better information and uses that information to gain understanding has a major advantage over its opponent. Forces that have this advantage can use it to bring about changes in attitudes, decisions, and actions in assembled demonstrators. In planning for crowd control or civil disturbance operations, planners must decide what data is needed to develop threat assessments. This decision is based on a multitude of information sources.

STANDARDIZED POSTURES IN COMPANY TACTICAL STANDING OPERATING PROCEDURES

2-2. Within squads, platoons, and companies, equipment may be increased or decreased, as needed. Some examples are as follows:

- Choose the M9 pistol for extraction and apprehension teams. The use of a long weapon (for example, an M16 with an M203, an M4 with an M203, or a 12-gauge shotgun) with NL munitions capability is also recommended, especially for overwatch personnel.
- Add nonstandard weapons such as shotguns for greater NL capabilities.

NOTE: The shotgun is used to protect the M203 gunner as he reloads.

- Add NL munitions to existing organic weapons systems, such as the M203.
- Arm soldiers in the front line of the formation with their standard weapon. If the weapon is a long weapon, it should be carried across the back from left to right or vice versa, with the butt up and muzzle down. Ensure that the weapon is cleared and that the magazine is in the appropriate ammunition pouch.

- Balance the mix of weapons and munitions according to the mission, enemy, terrain, troops, time available, and civilian considerations (METT-TC).
- Establish command relations and determine who has or gives the authority to fire NL munitions.
- Maximize the distance and barriers between the crowd and control formations. Use NL munitions to create a standoff distance.
- Maintain a lethal overwatch of the control force at all times.
- Maintain a reserve force at all times to reinforce the control force, as needed.
- Create NLW range cards for static positions.
- Consider environmental conditions and their effect on the performance of NL munitions. Proper storage, periodic inspections, and the rotation of NL munitions are critical to maintaining the effectiveness and viability of NL munitions.

WEAPON AND AMMUNITION CONFIGURATIONS

2-3. Individuals designated as NL shooters must have the means to transition to lethal rounds, if required. Lethal rounds are carried separately from NL rounds so that the shooter will not confuse them in the heat of confrontation. However, in the heat of confrontation where a lethal option is necessary, NL rounds can be used lethally by adjusting the point of aim to a vulnerable part of the body and closing in on the distance to the target.

2-4. Squad leaders should designate NL shooters within their squad. Ideally, the squad should not change its organization to accommodate NLW.

2-5. Commanders should not dispatch NL patrols. They should plan a combat or security patrol with NL capabilities. Soldiers are never sent in harm's way without lethal protection. NLW are only considered additional tools for the mission and not a mission in itself. There is no such thing as a NL mission.

2-6. Soldiers manning a static position should have NL capabilities depending on METT-TC. Shotguns and the M203 work well at static positions.

2-7. Recovery, apprehension, and/or extraction teams should be established before deployment. Team members should be equipped with personal protective equipment (PPE), an M9, NL munitions and weapons, and Flex-Cufs®. They should have some training in open-hand control, pain compliance, and handcuffs and/or Flex-Cufs.

2-8. Crowd control formations (see *Chapter 6*) should be well trained and well rehearsed. Rapid, coordinated movements of a well-trained and well-rehearsed control force can often be a strong enough deterrent. A lethal overwatch marksman always covers the control force.

2-9. During a NL engagement, the use of designated marksmen (DM) provides confidence and safety to those facing a riot. The DM in an overwatch position scans the crowd to identify threats and designate personnel for recovery and lethal rounds firing (if a lethal threat is presented). They are ideally suited for flank security and countersniper operations.

NONLEHTAL PLANNING

2-10. The fundamental principles of mission planning are not changed by the incorporation of NLW into a plan. NLW are intended to be operational-enhancement tools. The first concern of a commander is the successful completion of the mission. NL munitions provide a commander with a wider range of response options, but these munitions are not a replacement for lethal capabilities. Commanders must always be capable of answering a lethal attack with a lethal response.

2-11. When a commander commits his soldiers and equipment to a crowd control situation, he commits his forces with the additional tools of NL capabilities. This is an addition to the force continuum that the force commander now has available to him. Commanders that properly employ NL munitions and weapons have a tactical advantage over those who rely completely on lethal means.

2-12. When developing a tactical standing operating procedure (TSOP) (see *Chapter 6*), units should consider the following:

- Crowd control formations.
- Extraction teams.
- Apprehension teams.
- Lethal overwatch marksmen and/or observer teams.
- Reserve and/or security forces.

MILITARY WORKING DOG TEAMS

2-13. Military working dog (MWD) teams may be employed with a control force formation as a method of increasing crowd apprehension about approaching or engaging the formation. The teams should be in the rear of the formation in plain sight of the crowd, but in front of the command element and the M33A1 squad RCA disperser. The MWD teams work back and forth behind the formation as an intimidation measure. The presence of the MWD may produce a profound psychological effect on the crowd. These teams may also be used to help control individuals who have been captured by the recovery and apprehension teams.

CAUTION

Do not unleash an MWD on a crowd.

COMBAT CAMERAMEN

2-14. Video and still cameramen should make a photographic record of the individuals in the crowd who are leaders and instigators. Events must be documented to hold personnel, factions, gangs, or groups accountable for acts that violate law, destroy property, or cause physical harm. Electronically recording events aids in the prosecution of such cases and eliminates the sense of anonymity that people in large crowds often feel.

INFORMATION OPERATIONS

2-15. Information superiority helps forces to anticipate problems and requirements. It allows commanders to control situations earlier and with less force, creating the conditions necessary to achieve the optimal end state. Public affairs, psychological operations (PSYOP), and civil military operations are activities that will allow the commander to control situations earlier and with less force.

2-16. Information is available from a multitude of sources. A diversity of sources is the best approach because it prevents biased perspectives. Primary sources are as follows:

- Open sources.
  - Libraries.
  - Newspapers and news periodicals.
  - Radio and television.
  - Internet.
- Law enforcement sources.
  - Local law enforcement agencies.
  - National law enforcement agencies.
- Military sources.
  - Department of Defense (DOD) intelligence community (most restrictive source).
  - Local military intelligence (MI) field offices.

2-17. Police intelligence operations (PIO) is one of the five military police functions. This function is a process of actively and passively collecting information that is of a police, criminal, or combat nature. As military police perform the other four functions of maneuver and mobility support (MS), area security (AS), law and order (L&O), and internment and resettlement (I/R), they are gathering information that supports, enhances, and contributes to the commander's protection program, situational awareness, and battlefield visualization by portraying relevant threat information that may affect operational and tactical environments.

2-18. IPB is a continuous process for analyzing the threat and the environment of a specific geographic area. During the IPB process, the Intelligence Officer (US Army) (S2) or Assistant Chief of Staff, G2 (Intelligence) (G2) uses all available databases, intelligence sources and products, and related MI discipline to analyze the threat and the environment. The PIO function supports this process by providing the S2 with collected police, criminal, and combat information that can directly and significantly contribute to the success of the MI effort. In addition to combat information, the PIO function provides additional information on possible criminal threats and COAs. This is intended to support the S2 IPB process and can be used by the commander to upgrade force protection. (See *FM 19-10*.)

THREAT ANALYSIS

2-19. Crowd control requires its own thought process. Emphasis should be on prevention rather than confrontation. In combat, military forces are taught to fight and eliminate threats. In crowd control, military forces must deal with noncombatants that have internationally recognized rights. These rights must be respected while maintaining public order. This is an issue that law enforcement agencies (LEAs) have been struggling with for years.

2-20. Dealing with crowd control incidents is a matter of using some basic guidelines. It also includes asking the right questions in a logical manner so that key issues are not omitted. The commander must use these guidelines in context with METT-TC and the location of the incident (CONUS or OCONUS).

PREINCIDENT PLANNING

2-21. This phase of planning begins before the incident and is initiated at the operational level with guidance from the strategic level. The planning includes guidance on crowd control and addresses responsibilities, training, organization, operating procedures, use of force (CONUS), and/or rules of engagement (OCONUS). The most difficult and productive decisions are those made in the preincident planning process.

Avoid Confrontation

2-22. Crowd situations are highly unpredictable, but one thing seems certain—confrontation will likely cause crowd resistance. When pushed, people tend to resist opposition to the realization of their purposes.

Focus on Prevention

2-23. Planning should key in on the prevention of unfavorable outcomes. Experience has shown some LEAs attempting to help crowds accomplish their goals within the law that have been beneficial and even led to conceding some violations for the purpose of avoiding confrontation. However, LEAs maintain a law enforcement presence, which signifies social restraint. LEAs also decide when and where they will not compromise and the amount of force to use.

Define Goals

2-24. When defining a goal, deciding what must be accomplished is the first step. Defining the goal is fundamental. However, actually working toward and accomplishing the goal is easier said than done, as the process can drive the situation. Commanders and leaders must be aware of this. The military force must focus on what they are trying to accomplish.

BACKGROUND INFORMATION AND INTELLIGENCE

2-25. Seek to know as much as possible about social protest groups (within the limits of the law, see *Appendix B*) before an incident. This will provide insight into the organization and its functions and provide a warning as to what to expect. When you know as much about these organizations as possible, then it is possible to anticipate their next move. See *Appendix C*.

2-26. Continually assess what is to be accomplished against what others are trying to accomplish. Some groups may not have goals that conflict with those of US forces, but their actions may. Other groups may have goals and activities that do conflict. In either case, the assessment of group goals compared to the goals of US forces helps to understand and avoid potential conflicts.

CIVIL DISTURBANCE OPERATIONS PREPARATION

2-27. Counterdemonstration workgroups are a mixture of traditional and nontraditional staff proponents brought together to coordinate the resources to be employed by task force (TF) units before a potential civil disturbance. Meetings of the counterdemonstration workgroup are held on a regular basis (generally weekly) and are chaired by the Assistant Chief of Staff G3 (Operations and Plans) (G3). The following staff proponents may be involved, and others may be involved as needed:

- Public affairs office (PAO).
- Joint military commission (JMC), if organized.
- Provost marshal office (PMO).
- PSYOP.
- G2, Assistant Chief of Staff G5 (Civil Affairs) (G5), Assistant Chief of Staff G6 (Signal) (G6).
- Civil affairs (CA).
- Army airspace command and control (A2C2).
- Chaplain.
- Engineer.
- Surgeon.
- Fire support element (FSE).

RELATIONSHIPS DEVELOPMENT

2-28. Working relationships between commanders and protest group leaders are increasingly seen as the best means for preventing bad outcomes in crowd situations. This is called the negotiated management model of crowd control. LEAs in large metropolitan cities in the US and Europe practice it.

2-29. Open dialog helps develop working relationships between commanders and protest group leaders, providing an opportunity to communicate clearly. Talking allows group leaders to tell authorities and tactical commanders what they want to accomplish. It also allows authorities and commanders to tell group leaders what they are prepared to do and how they might respond to certain crowd behaviors. Such communications can do much to resolve issues and prevent violence.

2-30. Commanders may not be able to talk to all leaders before a crowd assembles. Some groups do not have recognized leaders and are ad hoc organizations. Other groups may have several leaders, but only some of those leaders will negotiate. In such cases, meetings with some leaders may tell you who the other leaders are and provide critical information. Negotiations may also encourage more moderate leaders to do things that will support the commander.

2-31. Initiate communication with group leaders to work out issues before an event. Commanders should make a concentrated effort to maintain a working relationship with the leaders of protest groups.

2-32. Communication also means persuasion. Commanders should make a concerted effort to win over demonstration leaders. Commanders may tell group leaders that they want to help them complete their mission in a safe manner. They should encourage demonstrators to protest in an acceptable way and, if possible, offer favors to get them to do just that.

NOTE: Commanders must only offer what they are willing and able to deliver.

2-33. Several rules apply in negotiations, and credibility is key. Only communicate necessary information and those actions that authorities and commanders intend to do. Negotiations are made from a position of strength, and by negotiating an agreement may be reached. However, contingencies should be developed in the event that the agreement is violated.

2-34. Working relationships between commanders and protest group leaders often result in protest groups policing themselves. This is one of the basic premises of negotiated management. Protesters are more likely to listen to their own leaders, as opposed to listening to US forces. Protest organizers are likely to buy into a cooperative effort and agree to ground rules set during preprotest negotiations. Many protest leaders seek to stay on the right side of the law.

ACCEPTABILITY

2-35. Deployed US forces will find themselves engaged in crowd control operations under difficult circumstances. The host nation (HN) may have groups that do not accept the presence of US forces, which will make imposing order and the protection of citizens difficult at best. US forces will be under intense media and political scrutiny (an environment much like what LEAs operate in on a day-to-day basis).

2-36. Winning in this environment is not like winning in combat. US forces may appear to be invincible and formidable, but they risk being portrayed as oppressors. Thus, US forces can lose by appearing to win. Groups that perceive themselves as oppressed will readily seek victimhood in an effort to gain the support of public opinion. Winning in this environment is about seizing and holding the moral high ground. US forces must maintain the authority and legitimacy of what they are doing.

2-37. Projecting a favorable image will require outreach to local leaders and citizens. It will also require developing a relationship with the media. For example, the Los Angeles Sheriffs Department often invites the media to accompany their tactical commander during crowd control situations. This shows that they have nothing to hide. It also provides an opportunity for individuals to see the commander's side of an event.

2-38. In this environment, commanders must consider how actions will play among several audiences—local, allied, US, and international. The media effect is inescapable. In some cases, it will be difficult to accommodate all of these audiences. However, the most important audience will be the US public, which is key to the continued support of US forces.

CROWD ASSESSMENT QUESTIONS

2-39. Assessing crowds requires its own set of questions. These questions should be answered before a crowd assembles or as quickly as possible if a gathering occurs without notification. In some cases, all the questions may not be answered. These questions represent a logical way to think about crowds. The questions are as follows:

- **Who are they? What is the overarching identity of the crowd? Are they strikers, ethnic factions, or social protesters? Do they identify themselves as strikers, ethnic groups, religious factions, or protesters against some perceived social injustice?** Understanding who they are will indicate what they may do. It may be possible to determine identities and goals from advance assembling instructions, leaflets distributed to bystanders, placards and banners, and chants and songs.
- **What are their goals?** What the group wants to accomplish by assembling could determine the extent to which they can be accommodated once they have assembled. They may only seek recognition for their cause (being seen and heard). If so, this goal is usually easy to accommodate. However, some groups may have more demanding goals, for example the demonstrators in Seattle who sought to stop the WTO. Goals that cannot be accommodated make confrontation very likely.
- **What is the composition of the crowd and are there any known factions?** Seattle demonstrations against the WTO were comprised of groups that were protesting environmental issues, wages, and child labor laws. Differing goals and the resulting friction were evident between such organizations as the Ruckus Society and the American Federation of Labor-Congress of Industrial Organizations (AFL-CIO). Factions within a crowd represent threats and opportunities.
- **What are they capable of doing?** Protest groups often claim that they will assemble large numbers of people to produce some disruptive action. However, there is often a difference between the claim and reality. An organization may claim that it can mass a 100,000 people, but in reality can only get 40,000. Organizers exaggerate for the following reasons: they want to boost the morale of their own people and they want the media to report that they have strength in numbers. Studying the past activities of a group may provide indications of what they are capable of doing in the future.
- **What are their traditional behaviors or cultural repertoires?** What people do during protests is not universal. It varies with the group and the culture. Social protest organizations and striking unions will carry placards and banners. Other groups will protest in a more quiet way, like the 5,000 women in Sarajevo whose standard Sunday behavior was to sit and block traffic. They were protesting the loss of male relatives in the Bosnian war. Understanding the goals of

the protesters can be helpful in deciding how to respond to their behaviors.

- **When and where will they assemble?** Every protest organizer has a time and place for assembling and perhaps a destination for the crowd to move toward. If the organizer attempts to mobilize large numbers of participants, the time and place for assembling and dispersing must be made known in the instructions. This information may be stated in the mobilization instructions or disseminated by an informal network, such as word of mouth. Such impromptu networks in densely populated areas can enable rapid assembling.
- **Where will they go?** Many crowds have destinations. Organizations may march a specified distance to ensure that their cause gets sufficient attention. Commanders need to know the route to minimize disruptions to the rest of the community. In Los Angeles, LEAs attempt to reroute traffic and prevent congestion caused by protest marches. It is also necessary to provide security along the route to prevent counterdemonstrators from confronting the marchers. This will prevent an even greater problem for the community.
- **What are the possible targets of violence?** Riots, in particular, may focus on target facilities. In the 1992 Los Angeles riots, gun stores were major targets. In developing countries, targets may be more basic. In Somalia, throngs of people stormed food supply facilities. In Haiti, riots broke out over trash dumps where US forces had discarded the remnants of meals, ready to eat (MREs).
- **What is the worst-case scenario?** The worst-case scenario must be recognized for the sake of avoidance. This may be when a peaceful crowd degenerates into a violent riot.
- **When and where will they disperse?** Crowds have a life cycle that includes how they disperse (see *Chapter 1*). Commanders must consider this. It is essential that there be one or more avenues that individuals can use to disperse. These dispersal routes should be clearly marked, visible, and open-ended. There may be a need for US forces to assist the crowd with dispersal. For example, in Los Angeles a 70,000-man march went from east to central Los Angeles. Once at their destination, they engaged in peaceful activities and then dispersed. The LEAs provided transportation back to the assembly area (AA); otherwise, the protesters would have returned to the AA in a disorganized and unsupervised manner.
- **Are there plans for subsequent gatherings?** A crowd may disperse for a short time so participants can take care of personal needs. This could be days or only a few hours.

LEGAL CONSIDERATIONS

2-40. The legal authority in which a peace operation is conducted defines the parameters of the operation. All commanders, leaders, and soldiers must know the legal authority that regulates their operations.

MANDATES

2-41. There are two types of mandates—United Nations (UN) and non-UN. Mandates are the authority under which an operation is conducted. UN mandates are created from resolutions by the UN Security Council (UNSC) or the UN General Assembly (UNGA) that authorize and define an operation. Non-UN mandates are created from treaties, accords, resolutions, or agreements evolving from other international or regional organizations.

STATUS OF FORCES AGREEMENT AND STATUS OF MISSION AGREEMENT

2-42. A status of forces agreement (SOFA) and a status of mission agreement (SOMA) are key documents that define the legal authority and responsibilities of a force and force personnel participating in an operation. A SOFA or SOMA may be a treaty or memorandum of understanding. It is an agreement negotiated between the UN and the host countries, which details the rights, privileges, immunities, and nature of services to be provided to the force and its personnel and their responsibilities and obligations.

INTERNATIONAL LAW

2-43. Legally, international law affects most stability operations and support operations (SOSO). These laws consist mainly of international agreements, treaties, and customary international law, which include the agreements and laws known as the law of war. International agreements prescribe the rights, duties, powers, and privileges of nations relative to particular undertakings. International agreements will affect US participation in SOSO in the—

- Right of US forces to enter a foreign country.
- Status of US forces while in the foreign country.
- Construction and operation of US bases of operations.
- Aircraft overflights and landing rights.
- Processing of claims for any damage to persons and property of the foreign country.

2-44. These agreements or customary international laws govern all aspects of operations carried out. In the absence of a viable HN government or proper international agreement, many aspects of the operations are controlled by domestic (US) or customary international law.

UNITED STATES LAW

2-45. Operations in peace and conflict must comply with US law, whether as a statute, executive order, regulation, or other directive from a branch or agency of the federal government. The *Uniform Code of Military Justice (UCMJ)* applies to questions of military justice. Various statutes, *Executive Order 12333; Section 3, Code of Federal Regulations (CFR);* and DOD and service regulations govern all intelligence-gathering activities. The staff judge advocate (SJA) must actively advise and participate in all activities from initial planning to redeployment.

HOST NATION LAW

2-46. If a viable HN government exists, whether at the national or the local level, all laws of the HN will apply to US forces in that country unless an international agreement provides otherwise. The types of laws that may inhibit US operations are in the fields of immigration, labor, currency exchange, the procurement of goods and services, customs and taxes, and criminal and civil liabilities. A SOFA usually covers these areas before an operation begins. Therefore, planners must understand the law to assess it and determine if it will adversely affect the operation. Assistance may be available from the local US embassy or the command judge advocate; if not, the command may have to rely on other sources for guidance. If local law hinders the operation, the commander must inform the US embassy and request that it negotiate a solution.

CIVIL AUTHORITIES

2-47. There will be times that US forces will deploy to nations where the infrastructure of civil authorities has survived the conflict and is functional. Local authorities and police are responsible for ensuring the safety of their citizens, including the UN High Commissioner for Refugees (UNHCR). US forces assist in this by providing general military security and by facilitating negotiations. In countries where the civil authority infrastructure has collapsed, the UN mandate may require the coalition of forces to help ensure the safety of citizens until their own civil authorities can be rebuilt.

RULES OF ENGAGEMENT

2-48. The rules of engagement (ROE) are the commander's rules for the use of force (RUF). Commanders are responsible for interpreting, drafting, disseminating, and training on the ROE. Ensure that the SJA assists throughout the ROE development process. This will ensure that the ROE will not improperly constrain actions, but are still consistent with domestic and international law and policies and orders of the chain of command.

2-49. Clearly stated ROE are published before Army forces are committed. The highest military authority, with input from subordinate commanders, will continually evaluate the ROE and modify them as appropriate. When the use of NLW has been authorized in the ROE, the decision to use NLW should be delegated to the lowest possible level, preferably the platoon or squad level. This requires that all personnel, not just leaders, have a clear understanding of the ROE and the commander's intent (see *FM 3-22.40*).

2-50. In peace operations, the use of force is restrained while the diplomatic solution is sought through negotiation or mediation. Accordingly, ROE are more restrictive in peace operations, making them more politically sensitive. In today's world, peace operations are usually conducted by a coalition of forces operating under the purview of the UN charter and customary international law. Therefore, the UN may mandate certain restraints. Using overwhelming force could compromise diplomatic efforts to reach a peaceful settlement. Even a single, relatively small engagement could jeopardize the legitimacy of the peacekeeping force and upset negotiations or mediations. Commanders must beware that in any confrontation ROE decisions made by soldiers can have strategic political implications on current and future operations. Therefore, it is vital that leaders and soldiers hold a common understanding of when, how, and to what degree force is to be used. This requires training.

MULTINATIONAL OPERATIONS

2-51. Multinational operations include military forces from different nations. These operations are usually interagency operations. Conducting operations with foreign military partners, like operations with civilian partners, is uncommon to many soldiers, so a clear understanding of this different environment is necessary.

2-52. The majority of US peace operations will be part of a UN peace operation. Their multinational character merits particular attention because national interests and organizational influence may compete with doctrine and efficiency. Commanders can expect contributing nations to adhere to national policies and priorities, which at times complicate the multinational effort.

2-53. When working within a multinational force and with allies, do the following:

- Train coalition partners and allies on tactics and techniques.
- Remember that teamwork and trust are essential.
- Achieve unity of effort.
- Consider equipment and communications capability and interface.
- Strive to eliminate sources of confusion and misunderstanding.
- Remember that military doctrine varies from country to country.
- Remember that national culture influences the way units operate.
- Remember that language barriers represent a significant challenge.

SCALABLE EFFECTS

2-54. Scalable effects concepts are a measured approach in response to a crowd gathering. By recognizing a use-of-force policy, soldiers must be taught and understand that they use the minimum force necessary. Without the appearance of a graduated response, the gathering crowd may consider actions as excessive, causing a possible escalation of hostilities or violence. Do the following to aid in the scalable effects process:

- Try to persuade the crowd to quietly disperse by talking with leaders.
- Use translators as necessary.
- Let the first approach be by local authorities (the mayor or police).
- Pass out handbills requesting that the crowd return home.
- Use video and still cameras to photograph individuals and events for later use at trials.
- Give warnings before moving to the next level of force.

2-55. Ideally, the force should be positioned out of sight of the crowd. Have the troops move into position with few shouted commands. For maximum effect on the crowd, have them form their formation decisively and professionally. Do the following to aid in the formation process:

- Display force in a graduated manner (for example, a UH 60, hardened vehicles, and soldiers with charged weapons).
- Exploit the psychological effects of show of force.
- Demonstrate force (do not use on unarmed civilians).
- Escalate the PSYOP message.
- Highlight the target pointer.
- Demonstrate sniper precision strike capability.
- Employ riot control measures.
- Use NL munitions.
- Use RCAs (if approved).
- Move through the crowd using riot control formations and movement techniques.
- Use the graduated response matrix (GRM)

2-56. With the increased participation of US forces in contingency operations around the world, it accentuates the need to establish procedures for applying graduated military responses to situations that threaten these missions. Numerous GRMs and similar products exist throughout the military. These products graphically portray available responses in a graduated manner. The intent of these products is to give the on-scene commanders a list of options with which to control or defuse a situation before it gets out of hand. Most threats can be eliminated without loss of life or collateral damage by effectively applying the resources available.

PLAN A GRADUATED RESPONSE MATRIX

2-57. The planning for and development of a GRM begins with the mission analysis portion of the military decision-making process (MDMP). Some missions require US forces to enforce treaties or accords to protect the lives of civilians in uncertain or hostile environments (such as noncombatant evacuation operations [NEO]) or to provide large-scale humanitarian assistance. These missions require some sort of graduated response criteria to maintain order and prevent uncertain environments from becoming hostile. The development of a GRM requires the following seven steps:

- Identify the need for a GRM.
- Establish a team to develop the GRM.
- Develop targets.
- Coordinate staff functions.
- War-game.
- Receive command approval.
- Operate deliberately.

Identify the Need for a Graduated Response Matrix

2-58. Not all missions require a GRM. The decision to use a GRM requires careful consideration. Once it is agreed that a GRM is necessary, it requires guidance from the commander regarding the response options available.

Determining the appropriate responses is based on the facts, assumptions, and limitations identified during mission analysis.

2-59. Planners (staff) must agree on the intent of the GRM. The GRM can be used as a training and rehearsal tool. It provides leaders with the most likely vignettes that can be incorporated into a COA analysis, predeployment training, and rehearsals. The GRM can also be used as a handy reference during situations that require graduated responses.

Establish a Team to Develop the Graduated Response Matrix

2-60. Establish a GRM development team with a broad range of skills, knowledge, and professionalism. The following list of individuals makes up a typical GRM team:

- Provost marshal or senior military police advisor.
- FSE.
- Brigade legal officer.
- PSYOP representative.
- Land information warfare or information operations officer.

2-61. Since the GRM is designed to give commanders graduated options for dealing with both hostile and nonhostile threats to the mission, this team composition allows for target selection, application of the ROE, and attack methods using both NL and lethal means.

Develop Targets

2-62. The FSE, in conjunction with the S2 section, develops targets for both lethal and NL attacks. In the case of stability operations, these targets are usually not the conventional specific point or piece of equipment on the ground. They are more situational than specific. The GRM identifies situations or acts that subordinate elements could face during the mission. The sample GRM in *Figure 2-1* shows three possible situations or acts that on-scene commanders could expect to encounter. From the targeting standpoint, these are groups of more specific targets.

2-63. During mission analysis, the fire support officer (FSO) identifies both lethal and NL assets available to the unit. A tactical PSYOP team (TPT) attached to the unit is an example of a NL attack asset that should not be overlooked. The following are examples of what the FSO should look for:

- RCAs.
- TPTs.
- Electronic warfare assets.
- CA teams.
- Information operations (IO) teams.
- Artillery smoke projectiles.
- Aircraft (AH-64s, OH-58Ds, and AC-130s).
- Mortars.

2-64. The lethal assets described could be used in a NL show of force or demonstration to defuse a situation before it requires lethal force. The critical element of this mission analysis is not to focus solely on lethal assets. In

<table>
<tr><th>Crowd Action</th><th colspan="4">Nonlethal Response</th></tr>
<tr><th></th><th>Command Presence/PSYOP</th><th>Show of Force</th><th>Demonstration of Force</th><th>Riot Control Means (if approved)</th></tr>
<tr><td>Unarmed Civilians</td><td rowspan="3">- Ensure that the on-site commander of the TPT directs the broadcast of the dispersal proclamation and/ or passes handbills.
- Escalate the tone of the dispersal proclamation from information to a warning of force.</td><td rowspan="3">- Display force along with escalating the dispersal proclamation.
- Display force in a graduated manner, such as a helicopter hovering over a crowd or soldiers with charged weapons.
- Exploit the psychological effect of a show of force.</td><td>- Do not demonstrate force toward unarmed civilians.</td><td rowspan="2">- Employ RCAs at the point of penetration.
- Use PSYOP to exploit the psychological effect.
- Move through the crowd using riot control formations and movement techniques.</td></tr>
<tr><td>Armed Civilians (knives, clubs)</td><td rowspan="2">- Display force along with escalating the dispersal proclamation.
- Highlight the target pointer.
- Demonstrate sniper precision strike capabilities.</td></tr>
<tr><td>Armed Crowds/ Military (firearms)</td><td>- Do not use RCAs—they may escalate the situation.</td></tr>
<tr><td colspan="5">Hostile intent/hostile act occurs by armed threat.</td></tr>
<tr><th>Crowd Action</th><th colspan="4">Lethal Response</th></tr>
<tr><th></th><th>Sniper Response</th><th>Small Arms Direct Fire</th><th>Aerial Fires</th><th>Indirect Fires</th></tr>
<tr><td>Unarmed Civilians</td><td rowspan="3">- Ensure that target leaders or troublemakers are targeted.
- Use the minimum response necessary.
- Exploit the psychological effect of an attack.</td><td rowspan="3">- Determine that sniper attack is insufficient.
- Consider demonstrating capabilities.
- Exploit the psychological effect of a lethal response.
- Escalate gradually, starting with a small caliber, single round and work up to a large caliber, automatic.</td><td>- Determine that small arms direct fire is ineffective.
- Use the minimum response necessary.
- Use a minimal precision strike initially; use subsequent fires based on the situation.
- Exploit the psychological impact of each strike.</td><td>- Determine that air assets are unavailable or ineffective.
- Use the minimum response necessary.
- Ensure that the response is directed by the on-site commander.</td></tr>
<tr><td>Armed Civilians (knives, clubs)</td><td colspan="2" rowspan="2">- CAS/indirect fires must be authorized by the MACOM commander.
- Consider requesting permission for use when—
• All lesser means have been ineffective.
• There are physical eyes on target.
• Proximity to civilians has been considered.
• Risk to friendly forces/evacuees outweighs the risk of collateral damage.</td></tr>
<tr><td>Armed Civilians/ Military (firearms)</td></tr>
</table>

Figure 2-1. Sample GRM Card

stability operations, the Army wants to prevent acts of hostility first and then be prepared, if necessary, to apply lethal force.

2-65. Graduated responses can range from command presence through the show of force, a demonstration, the use of RCAs, and other techniques (such as the application of lethal force using snipers, small arms, AC-130s, and indirect fires).

Coordinate Staff Functions

2-66. To coordinate staff functions the rest of the GRM team assembles to complete the escalation sequence for each response. PYSOP and legal representatives are critical attendees during the escalation sequencing process. In the area of psychological operations, the TPT must exploit the effects of all responses.

2-67. The legal officer evaluates each escalation of force option and graduated response to ensure that it is consistent with the ROE. The GRM is designed to recommend applications of force consistent with the ROE, yet not limit the leader or individual soldier's right of self-defense. A GRM should clearly show that if a hostile act occurs, lethal options would be first and foremost.

2-68. In the case of lethal responses, the commander's guidance must again be applied. For example, lethal responses are allowed only in self-defense.

War-Game

2-69. Once the types of escalations for each potential graduated response are determined and annotated, the GRM must then be war-gamed. The staff must walk through each act or situation from the on-scene commander's standpoint.

Receive Command Approval

2-70. Once the GRM has been war-gamed, it must be submitted to the commander for approval. This is the final check to ensure that the GRM team has applied the commander's guidance correctly and met his intent.

Operate Deliberately

2-71. Operating deliberately affords commanders ample opportunity to plan and prepare for possible civil disorder situations. Through the effective gathering of information and a working cooperation with local government and police officials, commanders can often be made aware of dates, times, locations, and what groups may assemble before the operation. The purpose is to gather as much information as possible who is involved, where they are assembling, what incident promoted the activity, and what seems to be the prevailing attitude of the assembling crowd.

2-72. Civil disturbance operations are dynamic, ever-changing environments requiring effective communications both up and down the chains of command. They require advance preparation and planning using the established troop-leading procedures (TLP) outlined in *FM 3-19.4*. The eight-step process of TLP aids commanders and their subordinate leaders in planning and preparing for an operation. Most steps can be accomplished concurrently, but rarely is there enough time to go through each step in detail. Despite this, leaders must follow these steps to ensure that nothing is left out of the planning process. As outlined in *FM 3-19.4*, the following are the eight steps of TLP:

- Receive and analyze the mission.
- Issue a warning order.
- Make a tentative plan.
- Initiate movement.
- Conduct a reconnaissance.
- Complete the plan.
- Issue the order.
- Supervise, refine, and rehearse.

CONDUCT HASTY OPERATIONS

2-73. Conducting hasty operations seems to be the norm rather than the exception. Unlike deliberate operations, hasty civil disorder operations are reactionary in nature with little or no time for planning. In these situations, the event of a crowd gathering is already underway. There is very little, if any, advanced warning of the unfolding situation, and commanders are usually put in the position of sending their soldiers into an already volatile, and perhaps hostile, environment.

2-74. Commanders should be in immediate and constant communication with local civil and police authorities. The purpose is to gather as much information as possible about who is involved, where they are assembling, what incident promoted the activity, and what seems to be the prevailing attitude of the assembling crowd.

2-75. It is essential that commanders operating in environments where civil disturbance operations are likely ensure that their soldiers receive training at all levels. Commanders and subordinate leaders need to instinctively rely on the same TLP that they would in a deliberate operation. See *FM 3-19.4*.

CONDUCT TACTICAL OPTIONS

2-76. In the peacekeeping and peace enforcement environment, the use of conventional firearms or the threat of their use may not be the solution to a situation where US forces must separate two belligerent, hostile ethnic groups or prevent a similar group from entering an area that is off limits to them.

2-77. Crowd control options are often combined. Commanders choose their options based on an evaluation of the particular crowd. Commanders select any combination of control techniques and force options they think will influence the particular situation (METT-TC). Commanders must always try to choose the response that can be expected to reduce the intensity of the situation.

Monitoring

2-78. Monitoring an assembled crowd consists of gathering necessary intelligence and watching them to determine progress and development. It is this gathered information that helps commanders select the appropriate actions. Gathered information can often be used to help defuse a situation through persuasion.

2-79. Monitoring is continuous. Without timely information, responses could cause the situation to escalate beyond what it should have or cause a response to be inadequate based on the situation.

2-80. Commanders may task teams with specific missions to monitor crowd activity and note any new developments. Monitoring done by these observation teams gives the commander up-to-the-minute information so he can gauge crowd activity and intent in relation to the overall situation.

2-81. Observation teams monitor crowd activities to gather information. They observe and report on crowd size, location, and mood and on the developing situation. An observation team may consist of a marksman, a radio operator, and an observer equipped with binoculars. They may be posted strategically on rooftops and other high terrain that overlooks the crowd. Sometimes observers use helicopter-mounted observation devices. This also affords security for the operational forces. Commanders must know where their observation teams are located so they are not mistaken for teams of snipers.

2-82. The timely flow of information may allow the commander to influence the outcome of the situation with simple negotiations. Monitoring is appropriate when more decisive action is not feasible because of crowd size or when the intensity of a situation might escalate. It is particularly useful in large, nonviolent demonstrations. Monitoring can serve as an interim measure until more control forces arrive. It includes establishing communication with crowd leaders to convey official interest and intent to the crowd. Monitoring also includes efforts to gain the cooperation of crowd leaders.

2-83. Communication with crowd leaders and participants can help a commander control a situation without more severe measures. If communications exist with crowd leaders, the authorities may be able to divert the leaders or the crowd from their stated or apparent goal. Pressure can be put on leaders to channel the crowd into an area that minimizes disruption to the community and aids crowd control operations. March routes and demonstration areas can be limited to those that will help contain the crowd and reduce their potential for disrupting the community. Pressure can be positive (offering concessions) or negative (with deterrents).

2-84. If the commander can gain the cooperation of crowd leaders, it can decrease the potential for disorder, and if crowd leaders seek cooperation from authorities, officials should try to be accommodating. Crowd leaders can be placed in liaison positions between the crowd and the control force. Crowd leaders can be made responsible for managing the crowd by policing their own activities.

2-85. Taking still pictures or videotaping the faces of individuals within a crowd may prevent or reduce unlawful and violent acts. This could lessen their sense of anonymity. If needed, photographs or videotapes can be used as evidence for prosecution. To be effective, crowd members must see their presence being recorded. The photographer or cameraman should be in uniform to let the crowd know who he or she is. Recorders must be close enough to the crowd to be seen easily, but not close enough to be endangered.

Blocking

2-86. It is not uncommon for protests and demonstrations against the US or its forces to occur outside the main entrances to military installations or US Embassies. Because of the sensitive nature of equipment and information that can be found at these locations, blocking unauthorized entry is critical.

2-87. Blocking is the physical denial of advance upon a facility or an area by a crowd. Commanders may have to task their forces to block a crowd. Blocking physically denies crowd advance. Crowd control formations, particularly line formations, along with barricades can be used to block advancing crowds. Barricades of vehicles, concertina wire, and water-filled barrels are used to block or channel the movement of the crowd.

Dispersing

2-88. Dispersing is taking deliberate actions to fragment an assembled crowd in order to prevent the destruction of property or prevent injury. It is extremely effective against smaller crowds in congested urban environments. However, this may increase and spread lawless activity rather than reduce it. Thus, the commander's forces must control dispersal routes and the areas in which the dispersal will occur. Forces must protect the facilities that could be considered likely targets of opportunity for dispersing individuals or groups. Dispersal of the crowd may require apprehension of small groups still active in the area. Proclamations, shows of force, crowd control formations, and RCAs can disperse crowds.

2-89. Avoid (if possible) the dispersion of crowds into wide-open areas because it gives the crowd the opportunity to grow rapidly in size. The direction of crowd movement is difficult to anticipate and channel.

2-90. Once the crowd has started to disperse, it may occasionally be necessary for the forces to advance, keeping the crowd on the move and in the right direction. A crowd on the move should not be hurried to avoid panic. At no time should the crowd be cornered in a position where there is the perception of no escape. This invokes the "fight-or-flight" syndrome, possibly escalating violent activity.

2-91. Issuing a proclamation (see *Figure 2-2,* page 2-20) can help disperse a crowd. Proclamations officially establish the illegal nature of crowd actions and put the populace on notice that the situation demands special measures. Proclamations prepare the people for the presence of military authority and it tends to inspire respect. Proclamations support law-abiding elements and psychologically bolster military forces trying to restore order. They also demonstrate the gravity of the situation to all concerned and are excellent ways to make a commander's intentions known to a crowd. It is also a good way to reduce crowd size before direct action is taken. Commanders can make a verbal proclamation similar to the following statement: "*In the name of the President of the United States, I command that you disperse and retire peaceably to your homes.*" Such a proclamation may even make direct action unnecessary.

I. DISPERSAL

Unlawful Gathering

"Attention! Attention! This area must be cleared at once! Further unlawful behavior will not be tolerated. Clear this area at once or the necessary force to do so will be used."*

Demonstration

"Disperse and retire peaceably! Disperse and retire peaceably! Attention all demonstrators!"

"The demonstration in which you are participating ends at ______. The permit that was agreed to by the leaders of the demonstration expires at that time. All demonstrators must depart from the ______ by _____. All persons who wish to leave voluntarily may board the buses. These buses will go to the _____. Those who wish to take buses should move to ______. Those demonstrators who do not leave voluntarily by ______ will be arrested and taken to a federal detention center. All demonstrators are urged to abide by the permit."*

Warning of Looting

"Return to your homes! Someone may be looting them at this moment! During a disturbance, criminal activity is at its peak. Your family or your property may be in danger."*

II. EMPLOYMENT OF TROOPS

Employment

"Attention! Attention! Troops are present in this area. They are preparing to advance. Order must and will be maintained. Disperse peaceably and leave this area. To avoid possible injury, leave at once."

"Disperse now and avoid possible injury! Disperse now and avoid possible injury!"*

(Repeat until troops are committed.)

When Troops Are Committed

"Troops are advancing now. They will not stop until this crowd is dispersed and order is restored. To avoid injuries, leave the area at once. Return to your homes as peaceful citizens. Troops have their orders, and they will not stop until the crowd is dispersed. Do not get hurt. Leave this area."*

III. PRESENCE OF CHILDREN

(Used in conjunction with other announcements.)

"Attention! Attention! Do not attempt to cause further disorder. Disperse now in an orderly manner and avoid possible injury to children. Return at once to your homes."*

IV. ADDITIONAL INSTRUCTIONS

*Indicate the method, streets, and direction that the crowd should use when dispersing.

Figure 2-2. Sample Proclamation

2-92. In making a proclamation, a commander may consider imposing a time limit. However, the situation may change, and not imposing a time limit leaves the commander free to choose other actions when he wants. A proclamation must be specific in its instruction.

2-93. If a time limit is stated, it must be a reasonable length of time for the crowd to comply with the instructions. When drafting a proclamation, the commander must consult closely with the SJA. He must use the simplest language possible to maximize the effects of the proclamation. If proclamations must be translated to a local language, the translation must be made with great care.

2-94. The commander at the scene may direct that a proclamation be issued over public address (PA) systems. The force of the words used in the proclamation must be gauged to the composition of the crowd. If the crowd consists of normally law-abiding citizens who are presently assembled to show disagreement with an existing situation, the proclamation requires less force. On the other hand, if the crowd consists of militant rioters, the proclamation requires more force. The text may take a number of forms, depending on the situation.

2-95. A show of force is often a useful measure for dispersing a crowd. When troops arrive, the psychological impact of their arrival can be used. Soldiers can dismount from the helicopters, buses, or trucks in plain sight of the crowd, but they must be far enough away to prevent a provoked attack of thrown objects. The first echelon to dismount from the vehicle secures the surrounding area.

2-96. When small groups are scattered throughout a large disturbance area, a show of force can be made by marching troops conducting motor marches, conducting saturation patrolling, and setting up static posts. Sometimes marching well-equipped, highly-disciplined soldiers in view of a crowd may be all that is needed to convince them to disperse and retire peaceably. On the other hand, a show of force may attract people to an event. It may also provoke a nonviolent crowd into a violent confrontation. Intelligence can help best in preparing a show-of-force response.

Containing

2-97. Containment is the process of limiting a crowd to the area they are presently occupying. It is a suitable option when the crowd must be prevented from spreading to surrounding areas and communities. Additionally, it prevents those outside the already assembled crowd from joining the gathering. Containment is useful when apprehensions become necessary for preventing those in the crowd from escaping. Crowd control formations, perimeter patrols, and barriers are effective methods to accomplish containment. In all instances, caution must be used to avoid the "fight-or-flight" syndrome common to people feeling trapped with no escape.

2-98. Armored vehicles, armored security vehicles (ASVs), and up-armored high-mobility multipurpose wheeled vehicles (HMMWVs) are adaptable to roadblock operations as they can serve as barriers. They also provide added protection for the soldiers inside and those outside manning the roadblocks because they provide an easily accessible barrier for them to crouch behind.

2-99. Military vehicles traveling at close intervals in a column formation next to a crowd are largely a psychological barrier (see *Chapter 6*). They can be used to contain a large, fast-moving crowd. The moving cordon creates a temporary obstacle between the crowd and the line beyond which they will not be allowed to cross. A well-trained driver in a mobile cordon can do a better job than dismounted soldiers. In order to execute safe, quick reversals of direction for mobile cordons on narrow roads, the soldiers executing this formation must be highly-trained vehicle operators.

2-100. By-the-number commands issued over vehicle radios are the most common method for coordinating a cordon movement. Vehicles equipped with PA systems can also prove effective to issue both commands to troops and directives to the crowd. The cordon speed is no more than 5 miles per hour with an interval of at least 20 feet. Blue and red lights, sirens, blinkers, and horns may or may not be used. Armored vehicles can also serve as mobile command posts. When used properly, vehicles provide security, communications, and mobility.

DECIDE TO CONTAIN OR DISPERSE

2-101. Current crowd control doctrine places emphasis on crowd dispersal. Forced dispersal may result in a crowd breaking up into multiple groups that scatter over a large area. This may pose even greater public order problems and may pose a continued threat to control forces. A crowd is often better controlled by means of containment (confining its activities to a given area). A crowd has a limited duration, and its numbers are likely to diminish as individual needs take precedence over those of the crowd.

2-102. During the G8 summit protest in Ottawa, Canada, in June 2002, several thousand people showed up to demonstrate against the evils of world corporate leaders. Canadian Police determined that their best response to this event was a "soft response." Throughout the two-day event, hundreds of officers in standard patrol uniforms operated in four-officer teams, prestaged throughout the city. As the demonstrators began to assemble and eventually took to marching, officers provided escorts both in the front and rear of the enormous crowd along with officers marching to the flanks of the demonstration. This "soft approach" frustrated demonstrators, especially those who had it within their agenda to cause property damage and personal injury. Even more frustrating to the demonstrators was the heavy downpour of rain they endured on both days. Their organizations quickly disbanded out of concern for their own creature comforts. In the end, 10,000 people participated in the protest, with only three arrests made in two days.

USE PATROLS

2-103. Alert, aggressive patrolling of the disturbance area deters the gathering of crowds. The use of saturation patrolling is most effective for this purpose. It allows information to be collected, and it creates the psychological impression of the control force being everywhere at once.

2-104. Standard military transport vehicles provide mobility and communications for area coverage. Soldiers must be deployed with enough vehicles to provide the flexibility to handle all situations in the disturbance area. The vehicles can be modified with sandbags, wire screens, or similar materials to protect against sniper fire and thrown objects. If sniper fire is expected or has been encountered, then up-armored HMMWVs or light armored vehicles should be used for patrolling.

2-105. Armored vehicles patrolling an area of violence provide an added psychological effect and allow troops to maneuver close to snipers to make an apprehension. They may also be used as rescue vehicles to extract hostages or people surrounded by a hostile crowd.

2-106. Varying patrol routes and times for mounted and dismounted patrols keep lawbreakers from being able to pick a safe place or time to act. Patrols are particularly useful in preventing overt arson and violence. Patrol members can also spot and promptly report fires. Whenever possible, military patrols are integrated with civil police patrols. Joint patrols conserve military forces and aid civilian military communications. They also help troops become familiar with an area quickly. In addition, the civil police are available to make apprehensions, if needed. Patrol members must practice proper standards of conduct and fair treatment of civilians at all times. They are performing an important community relation and control function.

2-107. Motorized patrols, because of their speed and mobility, provide timely reconnaissance and broad area coverage. Motorized patrols are in radio contact with TF headquarters (HQ). They can make periodic contact with foot patrols and stationary posts, while patrolling their own areas. Moreover, motorized patrols can respond quickly to calls for help from other patrols and guard posts. Motorized patrols should have at least three vehicles with three men in each. Strength in numbers is necessary for protection. When planning patrol routes, avoid areas where the patrols can be isolated or placed in jeopardy. In addition, motorized patrols are equipped with fire extinguishers to put out small fires, thus reducing the burden on the fire department.

2-108. Besides foot and motorized patrols, air patrols provide a third dimension in directing the overall control effort in the disturbance area. They can perform reconnaissance and surveillance and provide near real time formation over the disturbance area. They are an excellent means of providing timely information to the commander concerning the demonstration area and its surrounding perimeter. They can monitor the actions of rioters, the extent of damage, the status of access routes, locations and conditions of road barriers, and other important conditions.

USE DELAYING TACTICS

2-109. AOs that units are assigned to patrol are often larger than the unit has resources to manage effectively. In addition, units must manage taskings, contingency plans, and other commitments, which will quickly drain available manpower. This problem will manifest itself when factions threaten to riot or gather in large numbers to demonstrate. In order for the commander to stall for time while he shifts unit assets, he may want to delay the arrival of buses and vehicles full of demonstrators. This technique is based on the assumption that demonstrators are moving to the demonstration site by vehicle. Key and secondary routes into the demonstration site must be analyzed by the S2 when he does the IPB.

NOTE: There should be an overwatch when this technique is attempted.

2-110. As demonstrator buses and/or vehicles move toward the demonstration site, soldiers in two HMMWVs block the road just ahead of the approaching vehicles. Once the demonstrators dismount their vehicles and begin to close in on the soldiers, the soldiers leapfrog (see *Figure 2-3*) 1 kilometer farther down the road to repeat the process again. This is repeated as often as necessary. This process, if repeated enough times, should discourage the demonstrators enough that they give up and go home. It should also provide enough time to shift assets to the demonstration site.

2-111. Demonstrators are just as capable of blocking secondary and key routes leading into a demonstration or riot site to prevent relief forces from assisting with riot control as US forces are capable of preventing demonstrators access to the site. One must remember that demonstrators are often well organized and have handheld radios and other items. A great example of this was the 1999 demonstration against the WTO in Seattle, Washington. Police had erected temporary portable fences in the downtown areas to keep demonstrators from disrupting the delegates. Determined to be disruptive, the demonstrators relocated the temporary fences by not allowing police to enter the area. Police were forced to back patrol cars over their own fences to gain access.

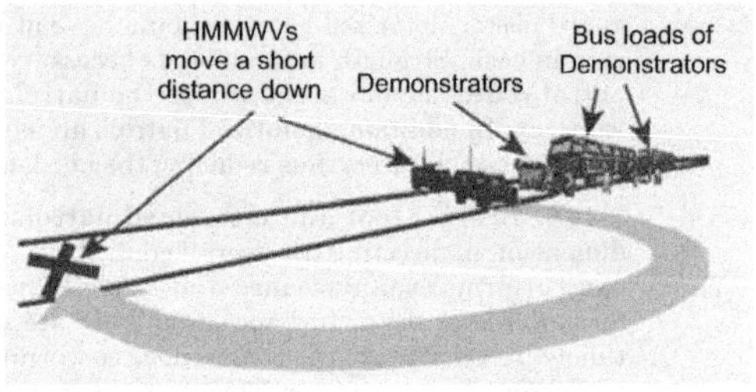

Figure 2-3. Delaying Tactics

Chapter 3

Apprehension, Search, and Detention

During the course of a civil disturbance, some members of the crowd may take part in unlawful activities, such as looting and assault. This will require the civil authorities or attached elements to apprehend, search, and detain people who are participating in the demonstration. Circumstances of the unfolding situation may require US military forces to be called upon to search, take custody of, and detain people who are participating in some form of protest and violating the law. Search, apprehension, and detention operations are conducted to halt these violations and to deter future violations.

LEGAL CONSIDERATIONS

3-1. All apprehensions are made by the civil police force, unless it is not possible for them to do so. Individuals must be known to have committed an illegal act, or there must be probable cause to believe that an individual committed such an act to be apprehended. If it becomes necessary for a military control force to apprehend or temporarily detain such violators, control force members often do so with the approval of the civil authorities. This ensures that all searches, apprehensions, and detentions (if necessary) are conducted within the parameters of the law.

3-2. Commanders operating overseas, similar to Bosnia and Kosovo, may find themselves without an established government or LEAs to work with. Commanders on the ground are then thrust into positions of authority, making them responsible for the actions, conduct, searches, apprehensions, and detentions performed by their troops.

3-3. If US military forces are called upon by the local authorities, certain policies have to be observed during search, apprehension, and detention operations. First, treat all people firmly but with reasonable courtesy and dignity. Remember that participating in a legal demonstration to express views is a right of all people, not just Americans in the US. The action, attitude, and behavior of US soldiers performing these operations are very important.

3-4. Our forces should never be seen as a military policing force on our own soil or as an occupying force in another country. Treating people with contempt, hostility, or excessive force increases the likelihood of resistance and violence. Searching people, placing them under apprehension, and detaining them without probable and just cause or without concern for their constitutional rights create problems and hinder due process. Numerous cases exist where authorities violated an individual's rights, resulting in the prosecution of police authorities or civil suits being waged against the officers and their leaders. This includes the military personnel assisting those officials.

3-5. It is very important that civil law enforcement (if available) be present and supervise all activities. Military personnel conducting a search or making an apprehension must carry out all the procedures carefully within the parameters of their training and the authority given to them. For each search and apprehension performed, the apprehension element must document the specific conduct of the person that violated the law.

3-6. Apprehending officers and supporting US servicemen must provide careful descriptive data for each subject. This description must be sufficient so that, at a later time, the suspected person can be clearly identified as the subject in question. Names and addresses of witnesses to the violation must be obtained. This information can be recorded on *Department of Defense (DD) Form 2708 (Receipt for Inmate or Detained Person).* Apprehension teams must retain and tag each item of physical evidence, such as weapons or stolen goods, that supports the apprehension on *DA Form 4137 (Evidence/Property Custody Document).* The team must give the owner a receipt for all property collected as evidence and retained.

TYPES OF SUBJECTS

3-7. There are two types of subjects (cooperative and uncooperative); they are as follows:

- **Cooperative.** Cooperative subjects actively and willingly comply with instructions. Cooperative subjects are usually compliant and offer no resistance to verbal directions or physical actions. However, if they provide any resistance to authority, it is usually passive. Some examples of this resistance are as follows:
  - Yelling.
  - Screaming.
  - Tightening the body.
  - Pulling away.
  - Letting the body go limp.
- **Uncooperative.** Uncooperative subjects will display preattack cues. Uncooperative subjects will actively resist being apprehended. Often, this resistance comes in the form of physical actions, such as—
  - Assaulting verbally.
  - Refusing to be handcuffed.
  - Refusing to be searched.
  - Attempting to push away, forcing a pursuit.
  - Wrestling with the apprehension team.
  - Striking out at the apprehension team with fists, hands and elbows, and feet.
  - Striking out at the apprehension team with weapons, such as knives, clubs, bats, bricks, rocks, or firearms.

APPREHENSION

3-8. US military forces will be in a supporting role to the local civilian authorities or acting as the control force in civil disturbance operations. When making apprehensions, it is best to use apprehension teams. These teams provide quick, organized responses to developing situations. Teams can be organized at the team, squad, or platoon level depending on the number of apprehensions expected. The intelligence information gathered helps to prepare for events where apprehensions may be necessary.

3-9. Each team (if possible) consists of a civilian police or apprehension officer, security element, and recorder. Team apprehension officers make the actual apprehension with the help of the security elements. If there is not a civilian police officer available to act as the apprehension officer, then it may be an officer or noncommissioned officer (NCO).

3-10. Ensuring that the apprehension process is properly conducted and documented, determine who is to be apprehended. Then inform the subject of why he is under apprehension and supervise his handling.

3-11. Security elements help the apprehension officer handle subjects. Security elements move, restrain, and search subjects under the supervision of the apprehension officer. The primary responsibility of the security elements is to watch the crowd and act as a blocking force, keeping the crowd from interfering with the apprehension. Recorders document the event, preferably by video, and complete *DD Form 2708*. If possible, recorders also photograph the apprehension officer with each subject. This aids in the identification process because it links the apprehension officer to the subject and provides the apprehension officer with documentation of the apprehension for use in court.

3-12. Electronically capturing the events as they unfold is essential in all civil disturbance operations. If personnel and equipment are available, videotape the events as they happen or, at a minimum, take still photographs to provide pictorial documentation that may be used later in court. Recording the scene before, during, and after the apprehension provides necessary evidence and can also help eliminate additional hostile and illegal activities. Having control force members take videos or pictures can help control crowd reaction. It reduces the feelings of anonymity that a crowd often enjoys and makes crowd members less prone to unruly or illegal acts.

3-13. Verbal commands given to the subject should be in a voice and manner appropriate for the purpose and the situation. Commands should be simple, concise, and in an understood language. Verbal commands should be convincing and convey an expectation that the instructions will be obeyed. With desperate, frightened people, sympathy combined with an attitude of firm assurance on the soldier's part may avoid additional violent behavior or trouble. An attitude of forceful authority may be needed with an uncooperative person. Commands given by soldiers must be done in a manner and context so that the subject has no confusion as to what the soldier is ordering them to do.

3-14. A subject will need to be transported from the area. In CONUS operations, civil authorities will usually be responsible for providing the transportation of a subject. Exceptions to this may be if the demonstration occurs on an installation. OCONUS operations may require US forces to provide vehicles, such as trucks, buses, or sedans, that will be used to transport a subject. Whenever possible, vehicles used to transport a subject should be modified with barriers to separate them from the drivers. If a barrier cannot be provided, a guard is to be placed in the vehicle. Passenger compartments are checked for items that could be used as weapons. When guards are required, the guard always sits with his weapon away from the prisoner. Prisoners must be placed where they can be best controlled, but they are not fastened to the vehicle with hand irons or Flex-Cufs. When seat belts are available they must be fastened. See *Soldier Training Publications (STPs) 19-95C1-SM and 19-95C24-SM-TG.*

3-15. Female escorts should always be used to transport a female subject. If female escorts are not available, measures must be taken to avoid false charges of molestation or sexual misconduct. Net control stations should be notified of the departure and arrival times and vehicle mileage readings (before and after). At least two male escorts must be present at all times when a female escort is not available. These procedures also apply when the male and female roles are reversed.

FLEX-CUFS

3-16. The two main positions used to place Flex-Cufs on a subject are the standing-supported and the prone.

3-17. When using the standing-supported position, the search man—

- Escorts the subject to a wall or other sturdy vertical surface (see *Figure 3-1*).

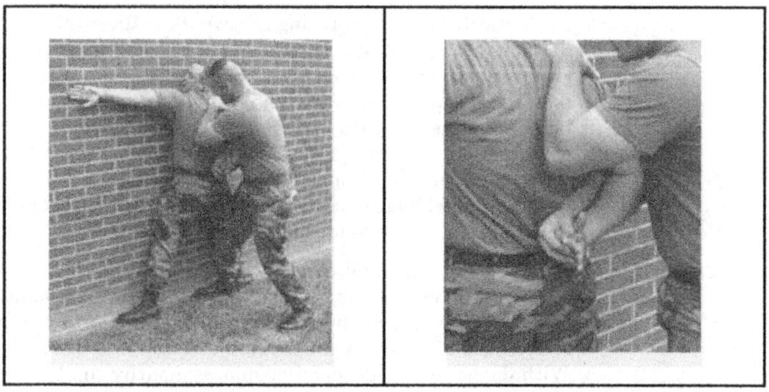

Figure 3-1. Standing-Supported Position

- Instructs the subject (verbally) to do the following:
  - Turn his head away.
  - Place his free arm (palm out, thumb down, and elbow locked) against the surface of the wall.

    - Spread his feet until instructed to stop.
    - Place his heels against the wall.
- Uses the leg closest to the controlled arm to pin the subject's leg (above the knee) to the wall.
- Reinforces verbal commands with pain compliance techniques if the subject fails to follow instructions.
- Maintains a double 90° hold on the controlled arm with his inside hand. With his free hand, the search man grasps the middle portion of the Flex-Cuf, keeping one cuff up and one cuff down.
- Places the first cuff (bottom cuff) on the subject's controlled arm, maintaining wrist manipulation (see *Figure 3-2*).
- Reaches through the free cuff with his outside hand and instructs the subject to place his free hand in the center of his back. With his outside hand, the search man takes the subject's free hand and completes the restraint by securing the cuff. He ensures that there is at least one-finger distance between the cuff and the subject's wrist (see *Figure 3-3,* page 3-6).

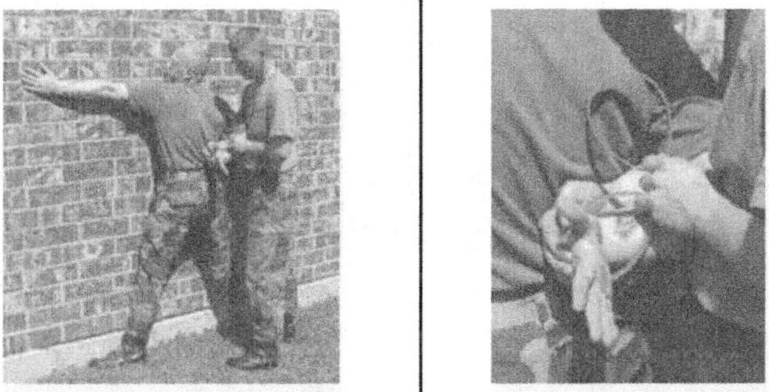

Figure 3-2. Cuffing the First Hand From the Standing Supported Position

3-18. When using the prone position, the search man—

- Grasps the Flex-Cuf with his outside hand, maintains tension on the subject's arm, and places the first cuff on the subject's controlled wrist (see *Figure 3-4,* page 3-6). Next, the search man secures the cuff, keeping at least a one-finger distance between the cuff and the wrist.
- Reaches through the free cuff with his outside hand and instructs the subject to place his free hand in the center of his back (see *Figure 3-5,* page 3-7).
- Reinforces verbal commands with pain compliance techniques if the subject fails to follow instructions. The search man bends the subject's controlled arm (using the natural break of the elbow as a hinge) toward the hand in the middle of the subject's back. With the outside hand, the search man takes the subject's free hand and completes the restraint by securing the cuff.

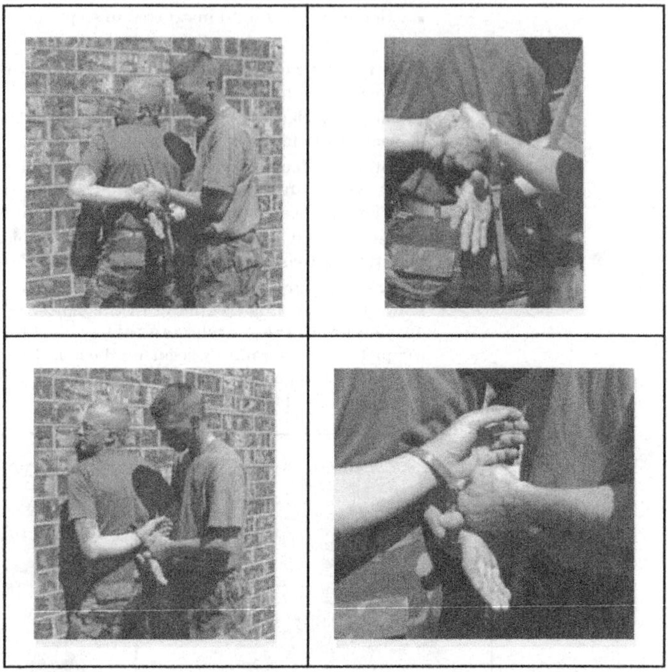

Figure 3-3. Cuffing the Second Hand From the Standing Supported Position

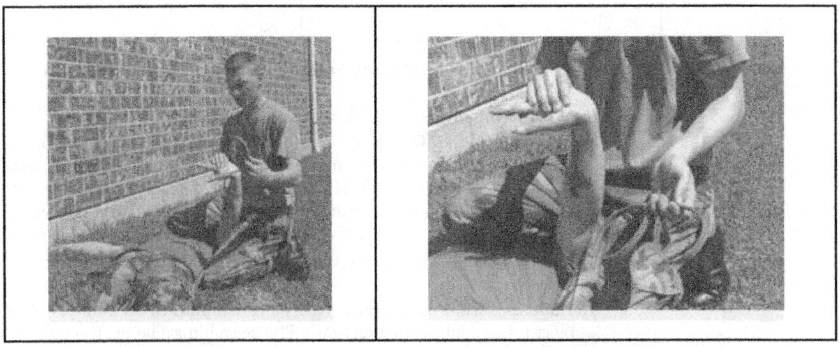

Figure 3-4. Cuffing the First Hand From the Prone Position

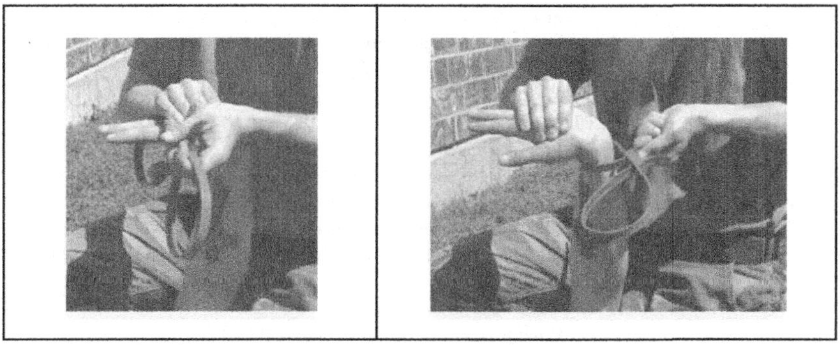

Figure 3-4. Cuffing the First Hand From the Prone Position (continued)

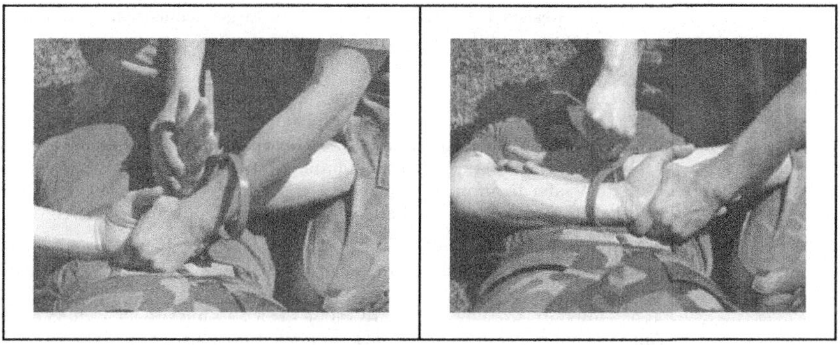

Figure 3-5. Cuffing the Second Hand From the Prone Position

SEARCHES

3-19. Initial contact with the subject is the most dangerous time during any apprehension; therefore, initial contact must be a swift and decisive team effort. Do not give the subject time to think or react. Distract his attention when possible. Apprehension teams are comprised of a search man, threat cover man, and security man (if necessary). Their responsibilities are as follows:

- The search man—
  - Issues verbal commands.
  - Conducts searches.
  - Handles radio communication.
  - Holsters his weapon before approaching a subject and informs the threat cover man that he is holstering his weapon.
  - Performs control procedures and handcuffs the subject.

   - Moves the subject to a safe area.
   - Removes weapons (if any) and sensitive evidence from the subject.
   - Disengages after the subject is restrained.
   - Controls and escorts the subject to a detention area.
- The threat cover man—
   - Provides cover for the search man.
   - Discourages escape and resistance.
   - Discourages hostile interference.
   - Alerts the search man of threats.
   - Resists distractions and maintains target acquisition or scanning.
   - Assesses threats constantly and prepares for the unexpected.
- The security man (if necessary)—
   - Provides cover for the search man.
   - Discourages hostile interference.
   - Alerts the search man of threats.
   - Resists distractions and maintains target acquisition or scanning.
   - Assesses threats constantly and prepares for the unexpected.

Search Principles

3-20. A subject should always be put at a disadvantage and searched immediately, but circumstances may dictate postponing the search until a safer place can be found. At a minimum, a quick pat down can be done en route to a safer location. The longer the apprehension takes, the greater the risk of drawing a hostile crowd and violent acts toward the control force.

3-21. Crowds often sympathize with a subject and try to help him or escalate the activities of the crowd toward violence. A subject's actions may also incite the crowd to violence. The longer a subject is the focal point of an apprehension, the easier it is to stir a crowd to hostile or violent acts. Apprehension teams should choose to move the subject out of the crowd's view. For the safety of the team, always apply Flex-Cufs or handcuffs and escort the subject to an area where a search can be done safely.

3-22. Search team members never operate alone; searches are conducted by a minimum of two team members. Initial steps must be taken to put the subject at a physical disadvantage. Before conducting a search (regardless of rank, sex, position, age, or physical condition), a team member applies the handcuffs or Flex-Cufs. Verbal commands must be short, clear, and distinct.

3-23. A search is conducted on everyone that has been apprehended to ensure that the team and those being apprehended are safe. The search man must abide by the following procedures when conducting a search:

- Search in a clear, secure area (when possible).
- Be ready to counter any resistance or an assault by the subject. Consider the subject armed until proven (by search) otherwise.
- Employ the appropriate level of force immediately, based on the actions of the subject.

- Do not step over the subject when he is at a disadvantage on the ground; go around him and remain out of sight.
- Do a 360° search of the subject's waistline for weapons. Search obvious and inconspicuous areas. Inconspicuous areas such as the groin, armpits, chest, and small of the back must be searched on both men and women; however, do not linger in these areas.

NOTE: Conduct same-gender searches when possible. If mixed gender searches are necessary for speed or security, conduct them in a respectful manner and avoid any action that could be interpreted as sexual in nature. To prevent allegations of sexual misconduct, the apprehension officer shows careful control of the soldiers who perform mixed-gender searches.

- Maintain security.
- Brief subjects quickly on what to do and who to follow.
- Remember that when conducting a search, a subject's body can be divided for a more efficient and thorough search. For example, search from head to toe or from side to side.

NOTE: The search man announces the items found and secures weapons and contraband.

- Military forces (when supporting civil authorities) may perform searches of people and property that are not usually subject to military law during a civil disturbance.
- Remember that it is always better for civilian police to search civilians and civilian property (when possible).
- Remember that if military forces are called upon to support police it is usually a last resort and there are not enough civilian officers available to accomplish the mission.
- Military personnel may search people incident to an apprehension.

3-24. The prescribed methods of searching someone are the frisk, the standing, and the prone. The method of search used largely depends on the situation and the mission. Searches are made not only of the subject and his personal effects such as purses, bags, and wallets, but also of the immediate surrounding area. This prevents the subject being searched from grabbing a weapon or destroying evidence. It may also be necessary for military forces to search private property, including vehicles, if there is—

- Reasonable belief that a person has committed or is committing a violent crime and is hiding in a building or a vehicle.
- Reasonable belief that a vehicle contains weapons or instruments of violence.
- Probable cause for searching a building or a vehicle.
- Probable cause to believe that (unless immediate action is taken) evidence of a crime will be destroyed before a warrant can be obtained.

NOTE: Search teams should always consist of two personnel—one to conduct the search, and the other to provide security and overwatch for the soldier conducting the search. Before being searched, a subject should always be restrained and put at a disadvantage by

securing his hands with handcuffs or Flex-Cufs. Ask the subject if there is anything in his pockets that could hurt the searcher, such as knives or needles.

Frisk Search

3-25. The frisk search (also known as the stop and frisk, pat down, and terry stop) is a search incident to a lawful stop. Subjects are not under apprehension at this point, but if it were an apprehension situation, handcuffs or Flex-Cufs would be applied first. Frisk searches are used when there is reasonable belief that a subject may be armed with anything that could be considered a safety threat.

3-26. Legal and regulatory circumstances for conducting a frisk search are covered in the *Manual for Courts-Martial (MCM*). When a lawful stop is performed, the person stopped may be frisked for weapons when that person is reasonably believed to be armed and dangerous.

3-27. The primary purpose of a frisk search is safety. Frisk-search techniques are to be used only on a compliant subject. Frisk searches only cover the outer garments. Do not search the pockets of the individual unless a weapon is suspected. A frisk search begins when the search man informs the subject of the intent to search.

- During the preparation phase, the search man—
  - Tells the subject that he must perform a search for weapons for the subject's safety and his own.
  - Tells the subject to turn and face away from him and instructs the subject to keep his hands out to his sides where they are visible.

NOTE: The cover man providing security moves around the subject and positions himself out of reach and in front of the subject.

  - Instructs the subject to spread his feet until instructed to stop (more than shoulder width) and point his toes outward compromising his balance (see *Figure 3-6*).

NOTE: Do not kick at the subject's feet to spread them apart; it is unnecessary and unprofessional.

  - Tells the subject to bring both of his hands back to the small of his back with palms out, thumbs up, and the back of his hands together.
  - Instructs the subject not to move (see *Figure 3-7*).
- During the performance phase the search man—
  - Approaches the subject from behind (if possible) at a 45° angle, off center. The search man approaches with caution to ensure that his feet are to the outside of the subject's stance. See *Figure 3-8,* page 3-12.
  - Stays just inside arms reach and grips the subject's fingers with his support hand. The search man firmly grips at least one finger on each of the subject's hands (both thumbs, both index fingers, or an index and middle finger). An overhand or an underhand grip can be used effectively.

Figure 3-6. Approaching the Subject to Perform a Frisk Search

Figure 3-7. Giving Instructions to Subject Before Frisk Search

NOTE: The primary purpose of gripping the fingers is to control the subject's hands while performing the frisk search (see *Figure 3-9*, page 3-12). If the subject is going to resist or become aggressive, he will attempt to free his hands first. If the subject attempts to pull his fingers from the search man's grip, the search man must respond instantly by increasing his grip. If the grip on the fingers is lost, the search man should immediately push away from the subject and create a safe distance. This will allow the search team to respond with the necessary force to regain control of the situation.

Figure 3-8. Performing a Frisk Search

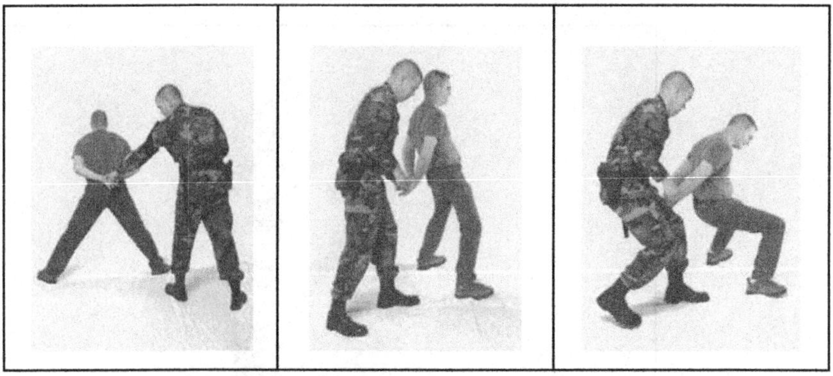

Figure 3-9. Gripping the Subject's Fingers During a Frisk Search

- Ensures that the subject always looks away from him.
- Places his strong hand on the outside of the subject's elbow (similar to the handcuffed escort position).

NOTE: The search man searches one side (the front, back, or side) at a time from head to foot.

- Divides the subject (visually) straight down his middle and does not cross the line while conducting the search.

NOTE: The search man is just feeling or searching for possible weapons, so an open hand should be used.

- Works at arm's reach. It is the sense of touch not the sense of sight that is being used. While performing the search, search men should stay behind and to the side of the subject. This provides the search man with a tactical advantage and prevention from being struck by the subject's elbows or knees if he should begin to resist.

3-28. When searching below the subject's knees, the search man takes a half step back and to the outside. He squats down, but does not drop his knees to the ground.

3-29. If no weapons are found and the subject is to be detained, the hands are secured with Flex-Cufs and the subject is escorted to the holding area. If the subject is to be released, collect information (name and address) and escort him away from the crowd.

Standing Search

3-30. Safety is the primary concern when dealing with a subject under apprehension. Therefore, the search man must protect himself by restraining subjects under apprehension before conducting the standing search. All subjects under apprehension have the propensity to turn violent or hostile. Putting a subject at a physical disadvantage provides safety for the subject and the search man. It limits the subject's ability to hide, destroy, or remove any evidence he may have on him.

3-31. Apply Flex-Cufs to the subject quickly during the apprehension. It is impossible to predict what a subject will do when put under apprehension. A subject may initially be cooperative and compliant, but as the subject becomes aware of his situation he may be overcome with emotion and begin to struggle or pull away.

3-32. There is no legal or regulatory prohibition against opposite-gender searches. Common sense prevails and whenever possible, searches should be done by soldiers of the same gender to avoid obvious questions and complaints. Once a subject is put at a physical disadvantage by cuffing and is more easily controlled, it affords the necessary time to get someone of the same gender there to conduct the search.

- During the preparation phase, the search man—

NOTE: Standing searches always begin with the subject in the handcuffed escort position (see *Figure 3-10,* page 3-14).

  - Takes his outside hand (farthest from subject) and places it on the outside of the subject's elbow while his inside hand (closest to subject) has a reverse-handshake grip of the subject's hand. At this point, both hands are on the same arm (see *Figure 3-10,* page 3-14).
  - Ensures that his position is on the same side of the subject's body being controlled (behind and to the side). When changing sides, change hand position.
  - Instructs the subject to spread his feet until instructed to stop (usually more than shoulder width) and point his toes outward, compromising the subject's balance.

Figure 3-10. Handcuffed Escort Position

NOTE: Do not kick at a subject's feet to spread them; it is unnecessary and unprofessional. If the subject becomes noncompliant or aggressive, pain compliance is usually the best tool to use, but only use enough force to gain compliance.

  - Uses his outside hand to conduct the search. His inside hand is always used to maintain control during the search.
- During the performance phase, the search man—

NOTE: Only one side (the front, back, and side) of the subject is searched at a time from head to toe (see *Figure 3-11*).

  - Divides the subject (visually) (straight down the middle using the spine as the centerline). He does not cross this line while conducting the search. Searches must be systematic, from head to foot. If there is an interruption, it must be started all over again.
  - Searches with an open hand, using a rub down or squeezing method. Do not only pat down like in the frisk search.
  - Works at arms reach. It is the sense of touch, not the sense of sight that is being used. To prevent the possibility of being struck by an elbow or knee, the search man stays behind and to the side of the subject.
  - Takes a half step back and to the outside when searching below the subject's knees. The search man squats down, but does not drop his knees to the ground. Maintaining positive control is essential throughout the search, so never the reverse handshake is never released.
  - Change sides when satisfied that all items have been removed from the side in which the search started. Without losing positive control, the search man conducts the search on the other side of the subject.

Figure 3-11. Performing the Standing Search

NOTE: Changing sides is accomplished by moving the outside hand (the searching hand) to the subject's uncontrolled handcuffed hand located on the outside. The outside hand and forearm will cross over the top of the reverse-handshake grip of the inside hand (see *Figure 3-12,* page 16).

- Crosses over the reverse-handshake grip with the outside hand and grips the subject's uncontrolled handcuffed hand in a reverse-handshake grip.
- Releases (quickly) the first reverse-handshake grip moving that hand to the subject's elbow as he sidesteps to the side yet to be searched. The search man repeats the standing search technique used on the first side of the subject.
- Gives strong verbal commands and uses only the force necessary to control the subject if he resists or becomes aggressive at any time during the search.
- Removes all items found during the search to be evaluated as potential evidence or a weapon. Items not considered a weapon or of evidentiary value will be returned to the subject at a later time.

Prone Search

3-33. Safety is the primary concern when dealing with a person under apprehension. Therefore, the search man must protect himself by restraining subjects under apprehension before conducting this search. Remember that all subjects under apprehension have the propensity to turn violent or hostile. That is the reason a prone search would be necessary. After performing an arm bar takedown on a noncompliant subject, it is especially important for the search man to put him at a physical disadvantage to provide for the subject's safety and his own. By doing so, the subject's ability to hide, destroy, or remove any evidence is limited.

Figure 3-12. Performing the Standing Search (Changing Sides)

3-34. The safest way to handcuff or Flex-Cuf a noncompliant subject is in the prone position, when the search man has greater control of the subject. It is usually the safest position in which to conduct a search

3-35. Handcuff the subject quickly during the apprehension. It is impossible to predict what a subject will do when put under apprehension. Often a subject may initially be cooperative and compliant; however, as the he becomes aware of his situation, he may become overcome with emotion and begin to struggle or pull away.

3-36. There is no legal or regulatory prohibition against opposite-gender searches. Common sense prevails and whenever possible, searches should be done by the same gender to avoid obvious questions and complaints. Once a subject is at a physical disadvantage by cuffing and is more easily controlled, it affords the necessary time to get someone of the same gender there to conduct the search.

NOTE: The subject is put into a physical disadvantage when he is taken facedown to the ground using an arm bar takedown and cuffed.

- During the preparation phase, the search man—
  - Ensures that the Flex-Cufs are not too tight.
  - Rotates his entire body around the subject's shoulder and center at the outside of the subject's hip with his shin over the subject's shoulder and one knee on the ground. At least one knee should remain over the subject's back (near side), with the other on the ground.
  - Pins the subject down by pressing his knee on the side of the subject's back or down on his closest elbow if resistance or aggression is encountered. He simultaneously grips the subject's closest handcuffed hand, grabs it with a reverse handshake, and applies pain for compliance.

- Uses strong, forceful verbal commands, such as "Stop resisting, stop moving around."
- Ensures that the subject's feet and ankles are crossed.

> **CAUTION**
> Stay alert. If the subject uncrosses his feet and ankles, it is usually a sign of preattack.

- Removes all items found during the search so they can be evaluated as potential evidence. Items not considered a weapon or of evidentiary value will be returned to the subject at a later time.
- During the performance phase, the search man—

NOTE: The search man must be systematic from head to foot while conducting the prone search. If interrupted during the search, it is important to start from the beginning to ensure that nothing is missed. It is better to go over the same area two or more times than to miss something that can be used as a weapon.

NOTE: The search man searches only one side (the front, back, or side) of the subject at a time until that side is complete.

- Divides the subject (visually) straight down the middle, using the spine as the centerline.

NOTE: He does not cross this line while conducting the search.

- Searches with an open hand, using a rub down or squeezing method for best results. He does not just pat down the subject like in the frisk search.
- Works at an arm's reach, searching only the area of the subject's body that is not in contact with the ground. The sense of touch, not the sense of sight, is being used.
- Stops the search as the subject's knees are reached. He taps one of the subject's legs and firmly commands him to raise his foot up and back, allowing for an inspection of the bottom of his foot. He then searches from the knee to the ankle area. This is performed on both the subject's legs and feet.

NOTE: At this point, the entire back side of has been searched. What remains to be searched is the front side of the subject (the chest and front of the hips, waist, groin, and thighs).

- Pivots back and faces the subject. Using the hand closest to the subject's head, the search man reaches across the subject's back and firmly grips the shoulder, rolling the subject up on the his side toward him to expose the areas not yet searched.
- Rolls the subject back down on his chest once all areas have been searched.

3-37. The search man must be completely satisfied that he has found everything. He must be especially sensitive to areas where weapons are often concealed like the waistband, the small of the back, ankles, pockets, shoes, the tongue of the shoes, and the groin area.

NOTE: Before removing subjects from the area, a brief summary must accompany them. The summary must include the subject's name, what he is suspected of doing, the time and date of the incident, and who saw him do the action.

DETENTION

3-38. Civil authorities must provide adequate detention facilities for all subjects. Authorities must be prepared to detain large numbers of people. They may choose to expand existing detention facilities or set up temporary facilities to accommodate the extra load. If possible, large-scale arrests are delayed until sufficient detention facilities have been set up.

3-39. If US military forces are committed to support local authorities, commanders should coordinate with them to ensure that adequate detention facilities are available and to learn their locations and capacities. If there are more detainees than civil detention facilities can handle, civil authorities may ask the military to provide support by setting up and operating temporary facilities. Army correctional facilities cannot be used to detain civilians. A temporary military detention facility can be set up if—

- Federal troops have been employed according to provisions.
- The TF commander has verified that available civilian detention facilities can no longer accommodate the number of prisoners awaiting arraignment and trial by civil courts.
- The Army Chief of Staff has granted prior approval.

3-40. Use of the temporary facility ends as soon as civil authorities can take custody of the detainees. Military forces are responsible for the custody, health, comfort, and sustenance of all detainees in their facilities until custody is transferred to civil authorities. Temporary facilities cannot be used to confine people arraigned or convicted in civil courts.

3-41. Women are detained in their own facility. Temporary facilities are supervised and controlled by officers and NCOs trained and experienced in military correctional operations. Females guard females. If two female guards are unavailable, use two male guards or one male and one female guard (see *FM 3-19.40).*

3-42. The same operational procedures that apply to the management of an installation confinement facility (see *FM 3-19.40*) apply to the management of temporary detention facilities, except for training, employment, and administrative discipline. Guards and support personnel under direct supervision and control of military officers and NCOs need not be trained or experienced in military correction operations. They must be specifically instructed and closely supervised in the proper use of force, custodial procedures, and the completion of military and civilian forms and reports that may be used. Soldiers that may be tasked to operate a temporary detention facility should be familiar with the forms and reports used for civilian apprehensions too.

3-43. Temporary facilities are set up on the nearest military installation or on suitable property under federal control. Ideally, the facility should be close enough to the disturbance area to minimize transportation and escort needs. However, it should be far enough away not to endanger those being detained. Whenever possible, existing structures are adapted for this use, but construction may be needed to provide segregation for ensuring effective control and administration.

3-44. The basic structure must include the following:

- Search areas.
- Holding areas for incoming men and women.
- A processing area.
- Holding areas for men and women who have committed misdemeanors.
- Holding areas for men and women who have committed felonies or are violent.
- A holding area for detainees' property and evidence obtained during a search.
- A holding area for administrative support and records.
- A medical station.
- Latrines.

For more information on these structures see *FM 3-19.40.*

3-45. Facility personnel must ensure that proper sanitation is maintained. When large numbers of people are detained or processed through a facility, sanitation becomes a problem. Medical personnel must conduct regular health inspections to detect unsanitary practices and conditions.

3-46. Facilities must be organized for a smooth flow of traffic. Processing stations must be set up so there is a linear or circular sequence of movement. These stations are set up out of sight of the holding areas. If possible, they are separated by a door to reduce noise. To reduce distractions, each station may be partitioned. Detainees may be more cooperative if they are out of sight of each other.

3-47. Injured people are given prompt medical treatment and transportation to medical facilities when necessary. A medical aid station for screening detainees and treating minor injuries is set up inside or next to the detention facility. Treatment areas, however, must be out of sight of the processing and holding areas. If possible, access to the medical facility bypasses the holding and processing areas. Facility personnel may also consider setting up a separate holding area for injured detainees.

3-48. Detention facility operation plans must contain emergency procedures. As a minimum, the following areas must be addressed:

- Fire evacuation.
- Disturbance control.
- Facility security.

NOTE: Depending on the situation, other emergency procedures may be needed.

3-49. Upon arrival at the detention facility, the detainee is logged in and searched, regardless of whether a complete search has already been conducted by the apprehending team in the field. Separate search areas are set up for men and women. Weapons, contraband, flame-producing devices, suspected evidence, money, and high-value items are confiscated. Medications are also confiscated. Medical personnel screen detainees that have had medications confiscated. Receipts are provided for any property or evidence that is confiscated. Confiscated items are tagged, and the items are stored in a controlled-property area.

3-50. As a detainee is brought to the facility, a file is initiated. Assigned detainee case numbers are used on all paperwork, such as logs, evidence tags, reports, and visual documentation. All paperwork (including photographs) that began outside the detention facility is marked with the case number. Facility personnel may also use hospital identification tags. Using indelible ink, the case number is written and attached to the tag on the detainee's wrist. Different colors may be used to identify different subject classifications, such as misdemeanors, felonies, or violent offenders. If opposing factions are involved, procedures are established to ensure that members of opposing factions are not processed together or detained in the same holding area.

3-51. After a detainee has been searched and classified, he is taken to a processing station where his paperwork is processed. If subjects passively resist by going limp, they may be moved by wheelchair. This reduces the number of escorts needed and the fatigue to facility personnel. Paperwork is reviewed to ensure that all information is complete, including charges, witnesses, and reasons for the apprehension.

3-52. Equipment such as height charts, scales, fingerprinting equipment, and cameras must be available for completing the police report and identifying the subject. Detainees may use aliases and not carry identification, so pay special attention to obtaining as much information as possible about the detainee's physical characteristics.

3-53. After processing is complete, the detainee is placed in the appropriate holding area. Paperwork for the detainee is forwarded to the administrative section. Files are reviewed for completeness and to determine the disposition of the detainee. Information from the file may be placed in a computer to find repeat subjects. Computers may be used to file criminal information only.

3-54. Custody transfers and release procedures must be coordinated with civil authorities and the appropriate legal counsel. Every effort must be made to arraign subjects quickly. The purpose of a detention facility is not to keep people off the streets, but to aid in processing subjects through the legal system. To speed up the release process, planners should consider issuing citations or subpoenas for minor offenses.

3-55. Military force members may have to respond to writs of habeas corpus. Writs are court orders addressed to a prisoner's custodian. Such a writ directs the custodian to bring the prisoner to court to determine the legality of the prisoner's apprehension and detention. Custodians must ensure that all documentation concerning the case is brought to court with the detainee.

3-56. Military personnel must obey writs issued by federal courts. For writs issued by a state court, the custodian or his legal advisor should respectfully reply that US authority is holding the prisoner. The Office of the SJA can answer any questions and explain the correct procedures.

Chapter 4

Riot Shield and Riot Baton Techniques

A soldier may be required to be part of a riot control formation where his primary weapon is a riot baton and his only protection a riot shield. A riot baton in the hands of a well-trained soldier is an invaluable weapon in a crowd control situation and may also be the most appropriate weapon. A riot baton is a very versatile weapon that can be used as an offensive and/or defensive weapon. The riot shield offers protection from the top of the head to just below the knees.

OVERVIEW

4-1. Use of the riot baton and riot shield is based on METT-TC and the commander's choice-of-force option. Formations committed to a riot with riot batons must have lethal overwatch, lethal weapons, and NLW employed within the formation and reserve forces with additional NL and lethal capabilities. These forces are positioned to rapidly reinforce the control force formation.

4-2. During the termination phase of the disturbance (when the violence has subsided and is expected to remain that way), soldiers are placed in a defensive posture. They may be equipped with a riot shield and a riot baton to perform their tasks. The riot shield is primarily a defensive weapon, which may be used in an offensive mode. It is primarily used as an offensive weapon when the formation is in close contact with rioters. The riot baton can also be used when formations are on the offensive and in contact with the crowd. However, crowd control elements employed with riot shields and riot batons must have lethal weapons and NLW available.

WARNING

Riot control formations are never employed with riot shields and riot batons only. There must always be lethal overwatch elements in position and lethal and NL capabilities within the riot control formation with a ready reserve force armed with lethal and NL capabilities.

RIOT SHIELD

4-3. In the hands of a properly trained soldier, the riot shield has proven to be the first line of defense for a formation. The riot shield may be used as an offensive weapon when in contact with an aggressive crowd.

4-4. The riot shield may be held with one arm. It is secured to the weak side arm with a self-sticking strap system and a handle. This system is designed to make the riot shield an extension of the soldier's arm. The riot shield is held at a slightly inward angle. This allows for debris thrown at the soldier to fall harmlessly to the ground (see *Figure 4-1*).

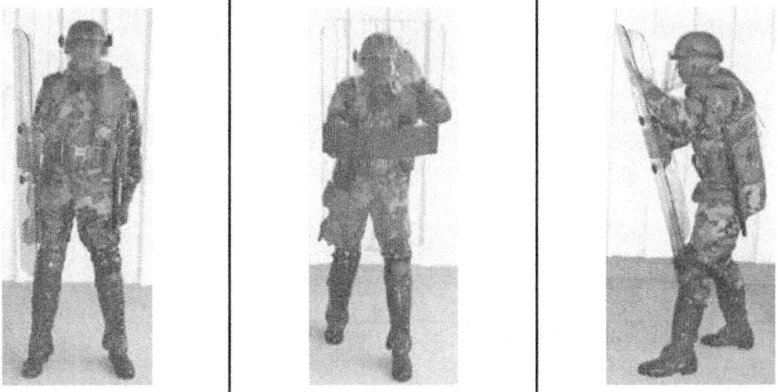

At the Ready | On Guard | On-Guard Profile

Figure 4-1. Riot Shield Positions

RIOT SHIELD CARRIES

4-5. The two basic riot shield carries are as follows:

- **At the ready.** The at-the-ready position is used while in a column formation or on the move toward hostilities. This position is designed for rapid movement. It should not be used when in contact with a crowd.
- **On guard.** The on-guard position is used when the riot control formation is in contact with the crowd or contact is eminent. It is used primarily in the following formations:
  - Line.
  - Wedge.
  - Echelon.
  - Diamond.
  - Circle.

DEFENSIVE AND RETENTION TECHNIQUES

4-6. The primary purpose of the riot shield is for defense. However, each riot shield holder must be proficient in riot shield retention techniques.

DEFENSE

4-7. The first line of defense for the formation is the riot shield. It is designed to withstand strikes from objects thrown by the crowd. In addition to protecting the riot shield bearer, it also protects other members of the formation from thrown objects.

RETENTION

4-8. When the control force is in contact with the crowd, the crowd will attempt to strip the riot shield from the bearer. When a subject grabs the top of the riot shield (see *Figure 4-2*), the riot shield holder slaps the riot shield with his strong hand near the subject's hands and gives the following commands (while continuing to slap the riot shield until the subject releases it):

- "Get back."
- "Get away."
- "Stop."

Figure 4-2. Riot Shield Retention Technique (Top Attempt)

4-9. If the riot shield is grabbed from the bottom (see *Figure 4-3,* page 4-4), then the riot shield holder forcefully drops to one knee, pinning the subject's fingers and/or hands to the ground. This must be done quickly and with force.

RIOT BATON

4-10. A riot baton in the hands of a properly trained soldier is a formidable weapon. However, it must be used in conjunction with other measures to be most effective. Soldiers must be trained with the riot baton to the point that its various techniques are automatic to them. This training must also include

Figure 4-3. Riot Shield Retention Technique (Bottom Attempt)

learning the vulnerable points of the human body so they can avoid areas that may cause permanent injury or death when struck.

WARNING
The riot baton is never raised above the head to strike a subject in a club fashion. Not only is it likely to cause permanent injury; but it also gives an unfavorable image of the control force. The soldier is also vulnerable to an attack on his rib cage when his arm is raised.

TARGET AREAS OF THE HUMAN BODY

4-11. The target areas of the human body are divided into three areas that represent the primary-, secondary-, and final-target areas. These areas are color coded as green (shown as gray in *Figure 4-4*), yellow (shown as white in *Figure 4-4*), and red (shown as patterned gray in *Figure 4-4*).

Primary-Target Areas

4-12. The primary-target areas are those areas which, when struck or restrained, cause the least amount of trauma to the body. These include the meaty or muscular areas of the body, such as the—

- Foot, shin, and instep.
- Inside and outside of the thigh.
- Lower abdominal region.
- Fore- and upper-arms.

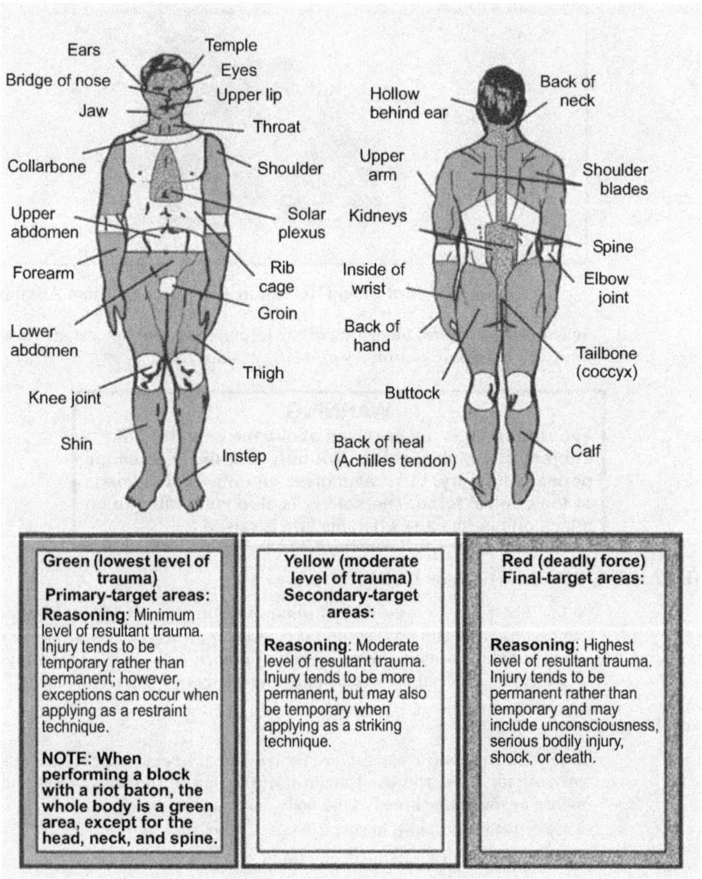

Figure 4-4. Escalation of Trauma Chart

Secondary-Target Areas

4-13. The secondary-target areas are those areas of the body which, when struck or restrained, cause a moderate level of trauma to the body. This type of trauma tends to be longer lasting because the time to heal is longer than with other parts of the body. If these areas are struck with a riot baton, serious injury may occur. However, some of these areas are used for control or restraining holds that tend to lessen the threat of injury. Some secondary targets are the—

- Collar bone and shoulder blades.
- Elbows and knees.
- Upper abdominal region.

Final-Target Areas

4-14. The final-target areas are those areas which, when struck with a riot baton, cause serious trauma that tends to be fatal or critical. However, there are some areas that (when used for restraining only), will not cause death or serious injury. Some final-target areas are the—

- Front and back of the head (eyes, ears, nose, upper lip, and hollow behind the ear).
- Neck and throat.
- Upper chest.
- Xiphoid process.
- Spinal column.
- Groin area.

JOINTS OF THE HUMAN BODY

4-15. The joints of the human body are often described as hinges and, when manipulated in a certain fashion, allow for better control of the subject being restrained, especially if he is aggressive. When applied correctly, minimal pressure is needed to control the subject. When a subject attempts to resist, he creates additional pain for himself. By anatomically locking out these hinges, the subject will be forced to comply. When restraining a subject, emphasis must be placed on locking out two of the three joints of the upper body or two of the three joints of the lower body. The joints are as follows:

- **Shoulder.** Apply pressure to the shoulder to prevent the subject from freely moving the upper portion of his body.
- **Elbow.** Apply pressure on the elbow or just above the elbow on the ulner nerve.
- **Wrist.** Apply pressure by pancaking the hand to twist or lock out the wrist. Apply pressure at the break of the wrist and push the subject's thumb toward a stationary object (the wall or the ground) to lock the hand out.
- **Hip.** Apply pressure to the hip to knock the subject off balance or alter his center of gravity.

- **Knee.** Apply pressure to the knee by pushing outward on it to cause the subject to be placed off balance, or use the knee as a hinge for pain compliance.
- **Ankle.** Apply pressure to the ankle by clasping and turning it in a counterclockwise manner to lock the hinge out.

TYPES OF RIOT BATONS

4-16. The Army is currently fielding two types of riot batons, wooden and expandable. The most common riot baton is the 36-inch hickory riot baton with thong. There is also the 24- to 36-inch expandable riot baton, which has been added to the nonlethal capabilities set (NLCS). Regardless of which riot baton the soldiers are equipped with, all techniques for blocking and striking are identical. Similar to the riot shield, the riot baton offers soldiers both offensive and defensive capabilities.

4-17. Soldiers must be properly trained in all blocking and striking techniques. Improper use of the riot baton by an untrained soldier has the potential for creating a greater problem than what already exists.

NOMENCLATURE OF THE RIOT BATON

4-18. The riot baton is divided into four basic parts: the long end, long portion, grip portion, and grip end (see *Figure 4-5*). Although the expandable and wooden riot batons differ, their nomenclature is quite similar.

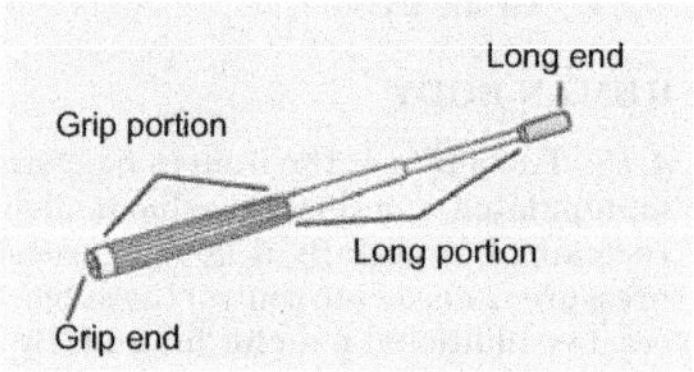

Figure 4-5. The Riot Baton

TYPES OF RIOT BATON CARRIES AND POSITIONS

4-19. As with any other weapon that the soldier will use, there are certain carries and positions from which the riot baton may be used offensively or defensively.

Two-Hand Carry

4-20. The two-hand carry is the most common carry for the riot baton. It is intended for executing all blocks and returning with strikes. When the expandable riot baton is in the collapsed position (at 24 inches), the long end of the riot baton is angled downward approximately 45°. When it is extended to 36 inches the long end is angled upward approximately 45°.

4-21. The wooden riot baton is secured with the soldier's hand by means of a leather thong. He hooks the thumb of his strong hand (palm down) in the thong, pulls the thong across the back of his hand, and grips the riot baton. This provides him with a very secure grip (see *Figure 4-6*).

4-22. The two-hand carry is intended for executing all blocks and returning with strikes. The soldier grips the riot baton with both hands. His strong hand grips the short end of the riot baton, and his weak hand grips the long portion of the riot baton, 2 to 4 inches from the end. His palm is down with his strong hand pulled in close against his strong side hip.

4-23. Wherever the long portion of the riot baton is pointed is where the soldier will strike or jab. The riot baton should be kept at a 45° angle to the ground. All blocks and two-hand jabs and strikes are executed from the two-hand carry.

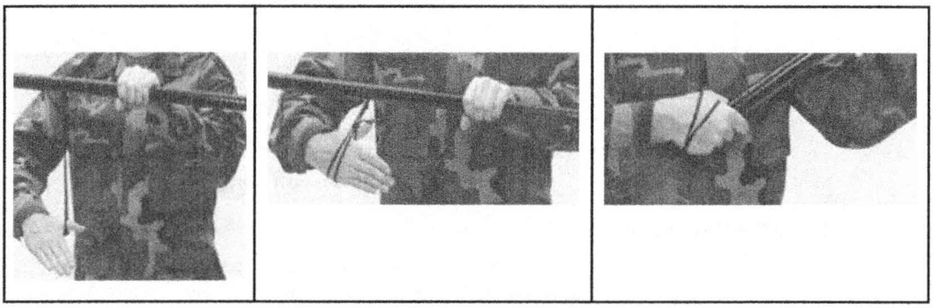

Figure 4-6. Two-Hand Carry

Outside-Arm Carry (Expandable Riot Baton Only)

4-24. The outside-arm carry (see *Figure 4-7*) is used with the expandable riot baton in the collapsed (24-inch) position only. Assuming a wide-based defensive stance (knees slightly bent and feet shoulder width apart), a soldier draws the grip portion of the riot baton with his strong hand. Without striking out or widely swinging the riot baton, he brings the long portion up against the outer part of his strong arm (the tricep area). The soldier's weak hand should be raised in a position to protect and block strikes from an aggressor. It is from this carrying position that a soldier can effectively use both the front and rear strikes. Front and rear strikes should never be performed with an expandable riot baton in the expanded position or with a 36-inch wooden riot baton.

CAUTION

The expandable and 36-inch wooden riot batons provide a greater risk of injury to the soldier's wrist.

Figure 4-7. Outside-Arm Carry

Vertical High-Profile Carry

4-25. Assuming a wide-based defensive stance, the soldier draws the riot baton with his strong hand by the grip portion. Once the riot baton has been drawn from the carrier, the soldier swings the riot baton in a downward motion, fully extending the arm and holding the riot baton down and parallel to the strong side leg. The soldier's weak hand is raised in a position to protect and block strikes from an aggressor (see *Figure 4-8*, page 4-10). Although this carry can be used with both riot baton systems, it is best suited for the expandable riot baton collapsed to 24 inches. The vertical high-profile carry clearly shows the soldier with a drawn riot baton in an effective defensive posture and ready for attack. A drawn riot baton in the hands of a trained soldier may offer just the kind of deterrence needed to stop the activity of an aggressor.

Vertical Low-Profile Carry

4-26. Assuming a wide-based defensive stance, the soldier draws the riot baton with his strong hand by the grip portion. Once the riot baton has been drawn from the carrier, the soldier swings the riot baton in a downward motion, fully extending his arm and holding the riot baton down and parallel to his strong side leg. His weak side hand is raised in a position to protect and block strikes from an aggressor (see *Figure 4-9*, page 4-10). Although this carry can be used with both types of riot batons, it is best suited for the expandable riot baton collapsed to 24 inches. This position portrays the soldier in an effective defensive position ready for an attack, but with a possible hidden weapon. The major difference between the vertical low-profile and high-profile carry is that in the low-profile carry the soldier hides the riot baton behind his strong side leg, allowing for some element of surprise to a potential aggressor. A drawn riot baton in the hands of a trained soldier may offer the deterrence needed to stop the activity of an aggressor.

Figure 4-8. Vertical High-Profile Carry

Figure 4-9. Vertical Low-Profile Carry

RIOT BATON BLOCKING TECHNIQUES

4-27. There are five basic types of riot baton blocking techniques that are executed from the two-hand carry position.

High Block

4-28. The high block is very effective in blocking a downward, vertical strike directed at the top of the head and shoulders. This type of downward, vertical strike may be from a club, pipe, or similar object. The high block is a two-step movement. To perform a high block the soldier performs the following steps:

Step 1. Raises the riot baton straight across his body until it is parallel to the ground (see *Figure 4-10*).

Step 2. Brings the riot baton straight-up in front of his body until the long portion of the riot baton is in front of his forehead (approximately 2 inches from his head). The riot baton should be in a horizontal position above his head and slightly angled down toward his weak hand, as it absorbs the shock from the strike (see *Figure 4-10*). The fingers of his weak hand should be open and behind the long portion of the riot baton when blocking the object to protect his fingers from being smashed by the strike.

Figure 4-10. High Block (Steps 1 and 2)

Low Block

4-29. A low block is very effective in blocking an upward vertical strike directed at the groin, lower abdomen, chest, or chin. This upward vertical strike may be from an individual's foot, knee, or fist (for example, an upper cut to the chin). To perform a low block, the soldier performs the following steps:

Step 1. Raises the riot baton straight across his body until it is parallel to the ground (see *Figure 4-11*, page 4-12).

Step 2. Brings the riot baton straight down in front of his body, bending at his knees and keeping his body upright until the long portion of the riot baton is just below his knees (see *Figure 4-11*). The riot baton should be horizontal and parallel to the ground as it absorbs the shock from an upward strike. His weak hand fingers should be open and behind the long portion of the riot baton when blocking the object to protect his fingers from being smashed by the strike.

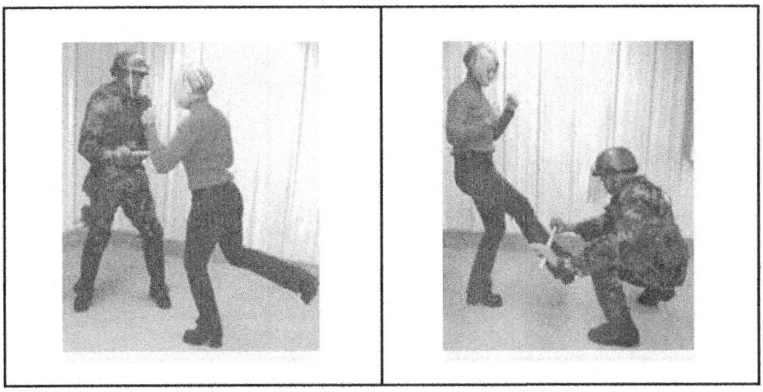

Figure 4-11. Low Block (Steps 1 and 2)

Strong Side Block

4-30. A strong side block is very effective in blocking a horizontal strike directed at the strong side of the head, neck, chest (or flank), or hip area. The horizontal strike may be from an individual's foot, fist, elbow, knee, or weapon. To perform a strong side block, the soldier performs the following steps:

Step 1. Raises the riot baton straight across his body in a vertical position where the grip portion is in his strong hand and the long portion is in his weak hand with the fingers straight up (see *Figure 4-12*).

Step 2. Turns his body laterally by rotating his hips toward his strong side and moving the riot baton across his body, keeping the riot baton vertical (see *Figure 4-12*).

4-31. The strong side block provides protection for a soldier from the front to the rear of his body. A riot baton held in the vertical position absorbs the shock of a strike coming from the strong side. The fingers of his weak hand should be open and behind the long portion of the riot baton, when blocking the object to protect his fingers from being smashed by the strike.

Weak Side Block

4-32. A weak side block is very effective in blocking a horizontal strike directed at the weak side of the head, neck, chest, or hip area. The horizontal strike may be from an individual's foot, fist, elbow, or knee. To perform a weak side block, the soldier performs the following steps:

Step 1. Moves the riot baton in a vertical position toward his weak side, ensuring that the grip portion is pointing straight down and the long portion is pointing straight up.

Figure 4-12. Strong Side Block (Steps 1 and 2)

Step 2. Keeps the fingers of his weak hand open and behind the long portion of the riot baton when blocking the object to protect his fingers from being smashed by the strike. At the moment of contact with the opposing strike, the blocking surface of the long portion should be at a right angle to the opposing object.

Step 3. Assess the situation following a weak side block, and take appropriate follow-up action, as needed.

Middle Block

4-33. A middle block is very effective in blocking a front horizontal strike directed at the face, throat, chest, abdomen, or hip. The horizontal strike may be from an individual trying to tackle near the waist or shove or punch at the face. To perform a middle block, the soldier performs the following steps:

Step 1. Moves the riot baton toward the middle of his body at a 45° angle, keeping the long portion pointed up and slightly forward of the grip portion (see *Figure 4-13*, page 4-14).

Step 2. Keeps the fingers of his weak hand open and behind the long portion of the riot baton when blocking the object to protect his fingers from being smashed by the strike. However, his weak hand may have to grip the long portion when pushing an individual away or forcing an individual down on the ground if the individual is trying to tackle.

RIOT BATON STRIKING TECHNIQUES

4-34. Seven strikes can be executed using the riot baton. Two of these strikes come from the outside-arm or vertical carry. The other five strikes come from the two-hand carry.

Figure 4-13. Middle Block

One-Hand Forward Strike

4-35. The one-hand forward strike is a very quick and effective offensive strike. It is usually employed as a countermeasure to an attack from the front that has been blocked effectively.

4-36. Target areas can be as high as the outside upper arm, down to the outer thigh region. An effective one-hand forward strike must be aggressively quick, with the purpose of distracting, disabling, or altering behavior. Therefore, those employing this technique must be sensitive to what part of the riot baton they use to strike the individual. Strike the target with the last 2 to 4 inches of the long end for optimal results.

4-37. A one-hand forward strike may be employed in one of the following three ways. When selecting available targets, use the principles in *Figure 4-4,* page 4-5.

- A horizontal manner.
- A downward diagonal manner to destabilize an individual.
- A vertical manner in front of the body to clear an aggressor's hands.

NOTE: When the one-hand forward strike is used in a vertical manner, keep the long portion straight up, not angled forward. By keeping the long portion up rather than angled forward, the chance of striking the individual in the face or head is significantly reduced. For the counterstrike to be effective, follow through with the technique when the baton makes contact with the aggressor and/or his object to destabilize the him. A pattern of movement such as a

forward shuffle, forward pivot, strong side step, or rear pivot may enhance this technique.

WARNING

Do not execute the one-hand forward strike with the wooden or expandable riot baton in the extended position. It has been shown to cause injury to the wrist of the user.

4-38. When using the vertical or outside-arm carry position (see *Figures 4-7,* page 4-9 and *4-8,* page 4-10), the soldier quickly moves the riot baton across his body using the strength and power of his hips, moving from his strong side to his weak side (see *Figure 4-14,* page 4-16). Ending this strike leaves the riot baton positioned under the weak side armpit in preparation for the one-hand reverse strike (see *Figure 4-14*). When delivering this strong side strike technique (in a horizontal or downward diagonal manner), ensure that the palm is facing up. In a vertical delivery, the palm will be toward the chest rather than up. Pause following the one-hand forward strike and assess the situation. Take appropriate follow-up action, as needed. If the situation does not call for additional strikes, avoid striking again.

One-Hand Reverse Strike

4-39. The starting position for the one-hand reverse strike is under the weak side armpit (see *Figure 4-14*). The one-hand reverse strike is used as a follow-up strike and is quick and effective.

4-40. Target areas can be as high as the outside upper arm, down to the outer thigh region. An effective one-hand reverse strike must be aggressively quick, with the purpose of distracting, disabling, or altering behavior. Therefore, those employing this technique must be sensitive to what part of the riot baton they use to strike the individual. Strike the target with the last 2 to 4 inches of the riot baton for optimal results.

4-41. A soldier may employee a reverse strike in one of the following three ways:

- A horizontal manner.
- A downward diagonal manner to destabilize an individual.
- A vertical manner.

4-42. In front of the body to clear an individual's hands. When the one-hand reverse strike is used in a vertical manner, keep the long portion straight up, not angled forward. By keeping the long portion up rather than angled forward, the chance of striking the individual in the face or head is significantly reduced. When selecting available targets, use the principles in *Figure 4-4,* page 4-5.

4-43. The one-hand reverse strike is considered a follow-up strike to the one-hand forward strike. As a follow-up strike, it is important that soldiers are taught the discipline of assessing the target. An aggressor may not require another strike, based on his condition and actions after enduring the one-hand forward strike.

Forward Strike | Target Assessment | Reverse Strike

Figure 4-14. One-Hand Forward and Reverse Strikes

4-44. When using the one-hand reverse strike, the solder quickly moves the straight baton across his body, using the strength and power of his hips (from weak side to strong side). When the strike is completed, the soldier's arm will be back in the outside-arm carry position (see *Figure 4-7,* page 4-9). The soldier ensures that his palm is facing down when delivering the counterstrike in a horizontal or diagonal manner. This will prevent injury to the user's wrist. In a vertical delivery, the palm will be toward the chest rather than up. Following the one-hand reverse strike, assess the situation and take appropriate follow-up action, as needed. If the situation does not call for additional strikes, the soldier should avoid striking again.

4-45. When the one-hand reverse strike is used in a vertical manner, keep the long portion straight up, not angled forward. By keeping the long portion up rather than angled forward, the chance of striking an individual in the face or head is significantly reduced. For the counterstrike to be effective, follow through with the technique when the baton makes contact with the aggressor and/or his object to destabilize him. A pattern of movement such as the forward shuffle, forward pivot, strong side step, or rear pivot may enhance this technique.

WARNING

This strike will not be executed with a 36-inch wooden or expandable riot baton in the extended position. It has been shown to cause injury to the wrist of the user.

Two-Hand Strong Side Horizontal Strike

4-46. Starting from the two-hand carry position, the two-hand strong side horizontal strike can be aggressively quick. The purpose of this strike is to create distance, distract, disable, or alter behavior. Strong side horizontal strikes are quick and effective offensive strikes, usually employed as a countermeasure to close frontal attacks that have been effectively blocked. When selecting available targets, use the principles in *Figure 4-4,* page 4-5.

4-47. The two-hand carry position enables soldiers to use riot batons at the 24- or 36-inch length. To effectively employ this strike, use the strength and power of the hips to thrust the riot baton in a horizontal manner. Simultaneously, pull back with the weak hand as the strong hand drives the grip end toward the target, striking a rib or the abdominal region of the aggressor (see *Figure 4-15,* page 4-18). Following the two-hand strong side horizontal strike, assess the target before following up with another strike. An aggressor may not require another strike, based on his condition and actions after enduring the one-hand forward strike.

NOTE: A pattern of movement such as a forward shuffle or forward pivot may enhance this technique.

Two-Hand Weak Side Horizontal Strike

4-48. An effective two-hand weak side horizontal strike must be aggressively quick to create distance, distract, disable, or alter behavior. The weak side horizontal strike is a quick and effective offensive strike usually employed as a follow-up strike to the strong side horizontal strike. It is a countermeasure designed to close frontal attacks that have been effectively blocked. When selecting available targets, use the principles in *Figure 4-4.* Maintaining a strong defensive stance after employing a strong side horizontal strike puts the soldier in position to execute a two-hand weak side horizontal strike.

4-49. Use of the two-hand carry position enables soldiers to use riot batons at 24 or 36 inches. To effectively employ this strike, use the strength and power of the hips to thrust the riot baton in a horizontal manner. Simultaneously, pull back with the strong hand as the weak hand drives the grip forward toward the aggressor, striking a rib or the abdominal region of the target (see *Figure 4-15*). Following the two-hand weak side horizontal strike, it is important to assess the aggressor. An aggressor may not require another strike, based on his condition and actions after enduring the one-hand forward strike.

NOTE: A pattern of movement such as a forward shuffle or forward pivot may enhance this technique.

Two-Hand Front Jab

4-50. The two-hand front jab starts from the two-hand carry position (see *Figure 4-16,* page 4-18). It is aggressively quick and creates distance, distracts, disables, and/or alters behavior. Front jabs are effective offensive strikes, usually employed as a countermeasure to charging and overpowering frontal attacks. When selecting available targets, use the principles in *Figure 4-4.*

Figure 4-15. Two-Hand, Strong Side, and Weak Side Horizontal Strikes

4-51. Use of the two-hand carry position enables soldiers to use riot batons at 24 or 36 inches. Use the strength and power of the hips to effectively employ this strike in a slightly downward or horizontal manner. Simultaneously, use both arms to shoot the long end of the riot baton straight out from the body to the target area. Immediately, pull back the riot baton to the two-hand carry position (see *Figure 4-16*).

Figure 4-16. Front Jab

4-52. Forward jabs drive the long end of the riot baton toward the aggressor, striking a rib, a hip flexor, or the abdominal region (see *Figure 4-4*, page 4-5). Following the two-hand front jab, it is important that soldiers assess the aggressor before following through with another strike, as the aggressor may not require another strike.

NOTE: A pattern of movement such as a forward shuffle will enhance the power of this technique.

Two-Hand Rear Jab

4-53. Starting from the two-hand carry position, the two-hand rear jab is aggressively quick and creates distance, distracts, disables, or alters behavior. Rear jabs are quick and effective offensive strikes usually employed as a countermeasure to charging and overpowering attacks from the rear. When selecting available targets, use the principles in *Figure 4-4,* page 4-5.

4-54. Use of the two-hand carry position enables soldiers to employ riot batons at 24 or 36 inches. To effectively employ this strike, use the strength and power of the hips to thrust the riot baton in a slightly downward or horizontal manner. Simultaneously, turn the head to the rear to see and assess the target. Use both arms to shoot the riot baton straight back from the body to the targeted area and immediately pull back the riot baton to the two-hand carry position (see *Figure 4-17*).

NOTE: A pattern of movement such as a rear shuffle will enhance the power of this technique.

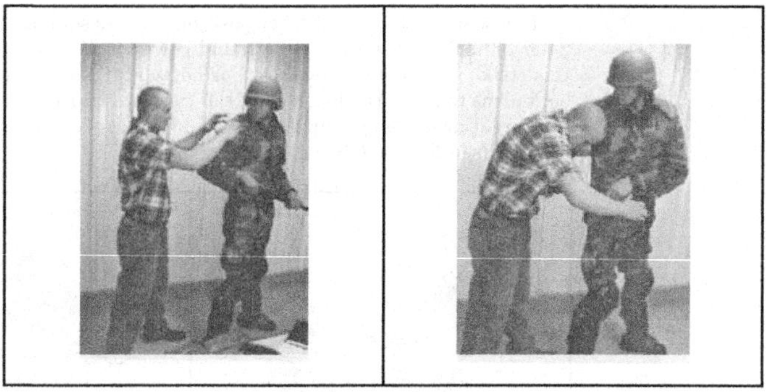

Figure 4-17. Rear Jab

Two-Hand Middle Strike

4-55. The two-hand middle strike starts from the two-hand carry position. It is aggressively quick and creates distance, distracts, disables, or alters behavior. It is proven to be an effective follow-up technique to a two-hand weak or strong side block or to destabilize an individual by pushing him back and away. The two-hand middle strike is used as a two-count movement. When selecting available targets, use the principles in *Figure 4-4.*

4-56. To perform a two-hand middle strike (see *Figure 4-18,* page 4-20) the soldier does the following:

Step 1. Brings the riot baton up to chest level and slightly off parallel to the ground.

Step 2. Steps forward aggressively, simultaneously thrusting the long portion of the riot baton forward in a horizontal manner. Both arms are fully extended at the end of the movement with a quick snappy return to the step 1 position. Use the entire body to generate the power.

NOTE: It is important that soldiers assess the aggressor before applying a follow-up strike. If the soldier steps forward as he employs the middle strike, it will afford greater impact and power.

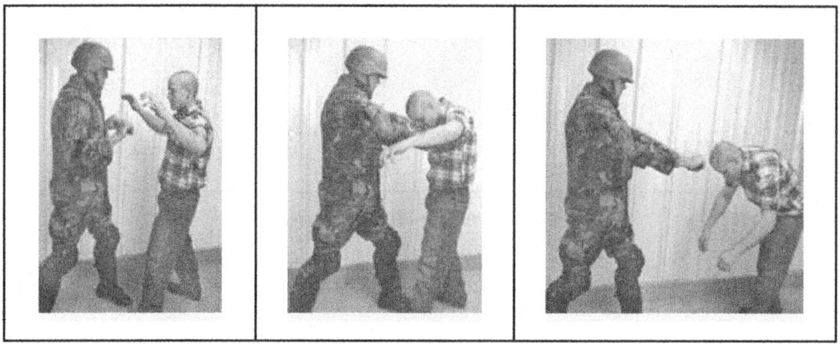

Figure 4-18. Middle Strike

Riot Baton Retention

4-57. A common technique an aggressor may use is grabbing for a soldier's riot baton or, in some cases, even taking the riot baton and using it against the soldier. This causes a disruption or distraction.

4-58. Maintaining physical control of the riot baton is essential for the safety of the soldier and the control force as a whole. Aggressors will generally attempt to grab the riot baton where they can get the strongest hold. Soldiers must resist the instinct of getting into a tug-of-war match over the riot baton, which often proves to be ineffective.

4-59. Using the momentum of the aggressor to pull the riot baton away, the soldier simply steps into or closer to the aggressor with the weak side foot. With the momentum created, the aggressor's hold on the riot baton is lessened. The soldier then rapidly uses the tracing-C technique to recover the riot baton.

4-60. A soldier using the tracing-C technique does the following:

Step 1. Pushes up on the riot baton.

Step 2. Pulls the riot baton around in an upward swing as if tracing a C in the air from bottom to top with the long end.

NOTE: This movement seems to wrap the aggressor's arms around themselves (see *Figure 4-19*).

Step 3. Drives (immediately with a quick and forceful movement) the long end of the riot baton straight down, as if slicing the C in half (see *Figure 4-19*).

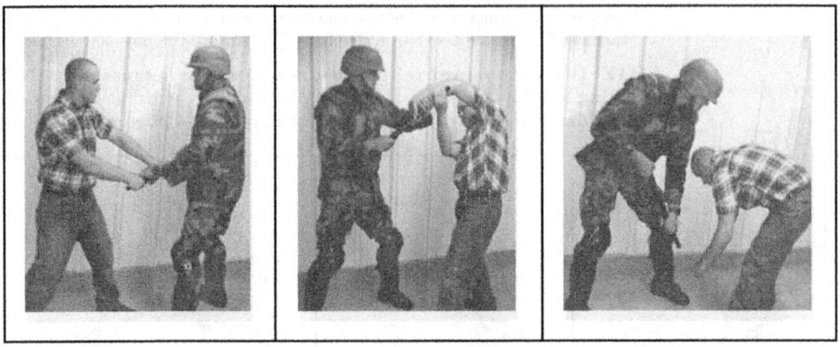

Figure 4-19. Riot Baton Retention

4-61. Once the soldier reaches the top of the C and the aggressor is tied up, it is impossible for the aggressor to maintain hold of the riot baton. Further actions such as a strike may or may not be necessary. Soldiers must be able to assess the situation and react accordingly.

WORKING AS A TEAM

4-62. Soldiers never have to rely solely on their own skills or abilities in riot control operations. Although individual skills were discussed in this chapter, it is important to remember that well-trained soldiers employing offensive, defensive, riot shield, and riot baton techniques will perform the collective task of crowd control well.

4-63. Riot control operations are dynamic and often chaotic. Soldiers armed with riot shields provide a first line of defense while those armed with riot batons and firearms provide overwatch as the formation moves toward engaging a hostile crowd. To work effectively and in concert, soldiers must be properly trained, drilled, and exercised. If an aggressor approaching a line of skirmishers violently grabs a riot shield, help is required to maintain control (see *Figure 4-20*, page 4-22). A soldier armed with a 36-inch wooden riot baton or an expandable riot baton (extended to 36 inches) can be of assistance. The soldier simply moves forward into a gap between the riot shields where he can quickly strike the top or side of the riot shield, slicing the long end across or straight down onto the hands and fingers of the aggressor (see *Figure 4-18*).

4-64. Riot batons are not the only way to effectively knock an aggressor's hands from a riot shield. Soldiers armed with 12-gauge shotguns or M16/M203s can just as effectively smash the hands or fingers of a persistent aggressor with the barrel of their weapons using the technique described above.

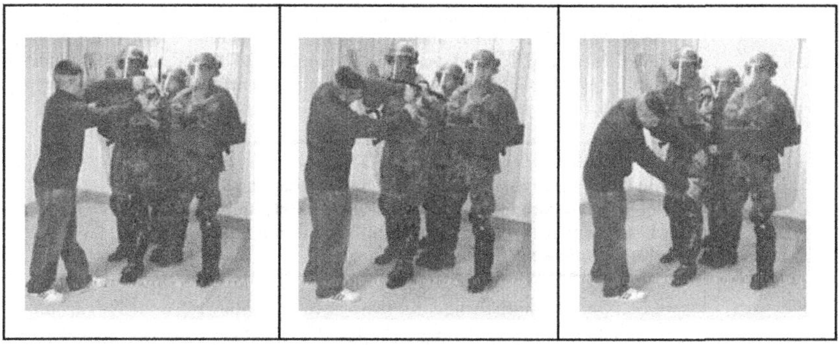

Figure 4-20. Clearing a Riot Shield

NONLETHAL MUNITIONS

4-65. With the inclusion of NL munitions, commanders now have additional tools that afford excellent standoff capabilities. This is critical because commanders now have intermediate options for dealing with a hostile crowd. No longer will a commander issue a proclamation, have it ignored by the crowd, and then be forced to use lethal munitions to backup the proclamation. Now he can use NL munitions to gain the same standoff distance but without the catastrophic results of lethal munitions.

4-66. NL munitions include (but are not limited to) a 12-gauge shotgun and 40-millimeter rounds. These blunt munitions have the capability of effectively engaging targets as close as 5 meters and as far as 50 meters. Range is dependent upon each round. These munitions are used from a distance to dissuade, discourage, and disperse the crowd. If the crowd disperses from 50 meters and soldiers are in close contact with the crowd, then the mission has been successful.

TAP-DOWN TECHNIQUE

4-67. Commanders must be familiar with the characteristics and capabilities of all rounds to use them effectively. When 12-gauge and M203 weapons are used within a line of skirmishers (usually located behind the riot shields), it is imperative that a method is developed, trained, and practiced to effectively operate as a team. Operators of these weapons may encounter problems in riot control situations when trying to identify and effectively engage legitimate targets. The tap-down technique is used to overcome target identification and engagement with NL munitions. If ordered to engage targets or the situation authorizes an engagement, soldiers armed with NL munitions may step forward, level their weapons, and fire upon legitimate targets (see *Figure 4-21*).

4-68. A soldier that is armed with a NLW and is using the tap-down technique must first tap on the shield holder's shoulder. Tapping on the shield holder's shoulder alerts him that the gunner is preparing to fire. In response to the tap, the shield holder drops to one knee while keeping the riot shield firmly affixed to their front for protection. The gunner takes a step forward and fills the gap. He then leans into the riot shield bearer's back with his knee and fires. This technique ensures that the long end of the weapon is extended beyond the riot shield.

WARNING

The long end of the weapon must be projected beyond the riot shield before firing. This is a critical safety step, which will prevent NL projectiles from hitting the riot shield and injuring soldiers.

4-69. Once soldiers firing munitions have fired, they raise their weapon back to a high port arms. The soldiers clear their weapons and tap the riot shield bearer a second time, signaling them to rise to a standing position.

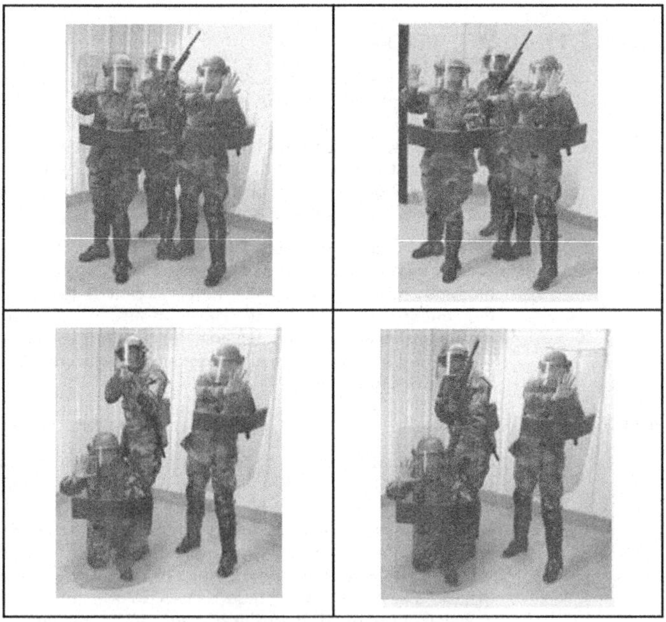

Figure 4-21. Tap-Down Technique

Chapter 5

Nonlethal Capabilities Set and Employment Considerations

During the past several decades, US forces have regularly performed peacekeeping and peace enforcement missions at a greater tempo than in the past. In the peacekeeping and peace enforcement environment, the use of conventional firearms or the threat of their use may not be the solution to a situation where US forces must separate two belligerent, hostile ethnic groups or prevent a similar group from entering an area that is off-limits to them. However, a show of force using NLW and NL munitions will cause crowds to disperse, separate, or leave the area with minimal casualities. The NLCS is a well-rounded, versatile package of both commercial off-the-shelf (COTS) and government off-the-shelf (GOTS) NL equipment and munitions. NLCS contents are divided into four distinct categories: personnel protectors, personnel effectors, mission enhancers, and training devices.

PERSONNEL PROTECTORS

5-1. Personnel protectors provide the individual soldier with added protection to the sensitive and vital areas of his body. They provide excellent protection for the individual soldier from trauma often inflicted by thrown objects (such as rocks, bricks, sticks, and bottles).

NONBALLISTIC RIOT FACE SHIELD

5-2. A nonballistic riot face shield (see *Figure 5-1,* page 5-2) has two adjustable positions and is constructed of hard durable plastic that securely attaches to the Kevlar® helmet. It may be worn with an M17 or M40 protective mask and is designed to protect the soldier's head, face, and neck from thrown objects. Nonballistic riot face shields will not protect from attacks with ballistic weapons, such as firearms. All soldiers operating in a civil disturbance environment should be outfitted with riot face shields.

NONBALLISTIC FULL-LENGTH SHIELD

5-3. A nonballistic full-length riot shield (see *Figure 5-2,* page 5-2) is a body shield constructed of hard durable plastic. Left- and right-handed personnel can easily use the nonballistic full-length riot shield. It is designed to protect the soldier's face, torso, and upper legs from thrown objects. Nonballistic full-length riot shields will not provide protection from ballistic weapons. Only the front line of skirmishers and those providing lateral protection to both flanks of the formation will be outfitted with nonballistic full-length riot shields.

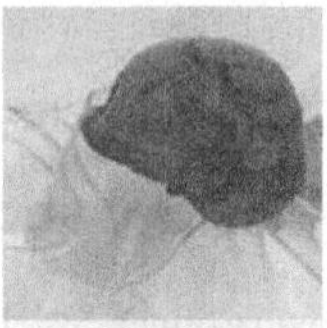

Figure 5-1. Nonballistic Riot Face Shield

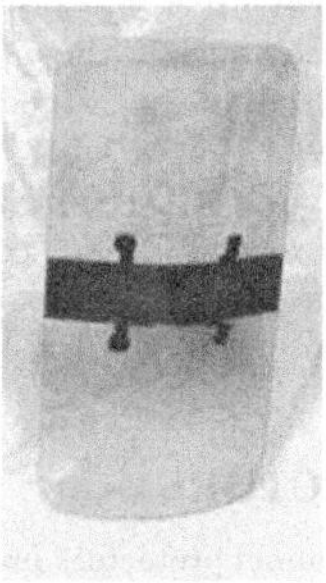

Figure 5-2. Nonballistic Full-Length Riot Shield

NONBALLISTIC SHIN GUARD

5-4. The nonballistic shin guards (see *Figure 5-3*) are constructed of hard durable plastic and are easily fastened to a soldier's legs. They are designed to protect the soldiers, feet, shins, and knees from thrown objects. They will not provide protection from ballistic weapons, such as firearms. All soldiers operating in a civil disturbance environment should be outfitted with shin guards.

BALLISTIC RIOT FACE SHIELD

5-5. The ballistic riot face shield (see *Figure 5-4*) has two adjustable positions constructed of acrylic and bullet-resistant materials that provide the wearer with Level IIIA protection. The ballistic riot face shield easily attaches to the Kevlar helmet and is capable of being worn with either an M17 or M40 protective mask. It will protect the soldier from hand-thrown objects. Because of its excessive weight (3.4 pounds), the ballistic riot face shield should not be worn for standard riot control operations. Ballistic riot face shields are best suited for use within specialized teams associated with riot control operations; for example, teams responsible for the search and apprehension of identified criminals or building- and alley-clearing operations.

Figure 5-3. Nonballistic Shin Guards

Figure 5-4. Ballistic Riot Face Shield

BALLISTIC FULL-LENGTH RIOT BODY SHIELD WITH LIGHT KIT

5-6. The ballistic full-length riot body shield (see *Figure 5-5,* page 5-4) is constructed from Spectra Shield® bullet-resistant material that provides a soldier with Level IIIA protection. It will protect the soldier from thrown objects. Because of its excessive size (18 pounds and 20 by 36 inches), the ballistic full-length riot body shield should not be used as part of a standard riot control operation. Ballistic full length riot body shields are best suited for use within specialized teams associated with riot control operations; for example, teams responsible for the search and apprehension of identified criminals or building- and alley-clearing operations.

Figure 5-5. Ballistic Full-Length Riot Body Shield

BALLISTIC SHIN GUARDS

5-7. Ballistic shin guards (see *Figure 5-6*) are constructed of Kevlar KM2 materials and provide a soldier with Level IIIA protection to the feet, shins, and upper legs. They protect the soldier from hand-thrown objects. There are three different sizes, and because of their excessive weight (7 to 10 pounds), ballistic shin guards should not be deployed as part of a standard riot control operation. Ballistic shin guards are best suited for use within specialized teams associated with riot control operations; for example, teams responsible for the search and apprehension of identified criminals or building- and alley-clearing operations.

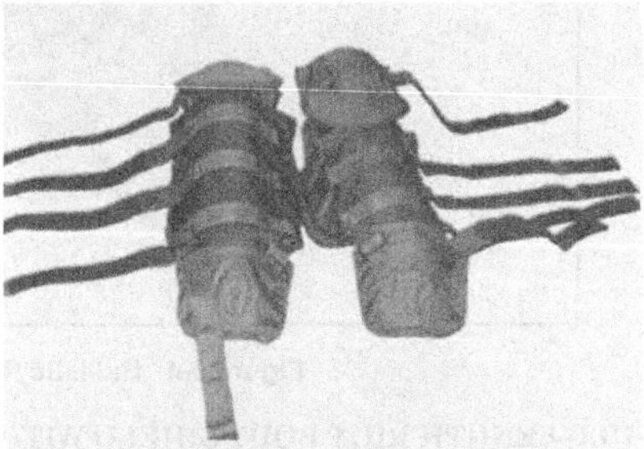

Figure 5-6. Ballistic Shin Guards

PERSONNEL EFFECTORS

5-8. Personnel effectors are those items that provide soldiers with an NL response without the necessity of always coming into direct physical contact with aggressors. Personnel effectors are 12-gauge and 40-millimeter ammunition that inflict blunt trauma. These munitions are compatible with the shotgun and the M203 grenade launcher already in the Army inventory. There are also diversionary and rubber ball grenades, which distract and inflict blunt trauma.

WOOD RIOT BATON WITH BELT RING (36 INCHES)

5-9. The 36-inch wood riot baton (see *Figure 5-7*) is made from solid oak. When held properly, the riot baton thong provides greater retention capabilities. Intended primarily for self-defense, the 36-inch wood riot baton can be used effectively as an offensive tool if it becomes necessary to keep rioters out of arm's reach of soldiers conducting crowd control operations.

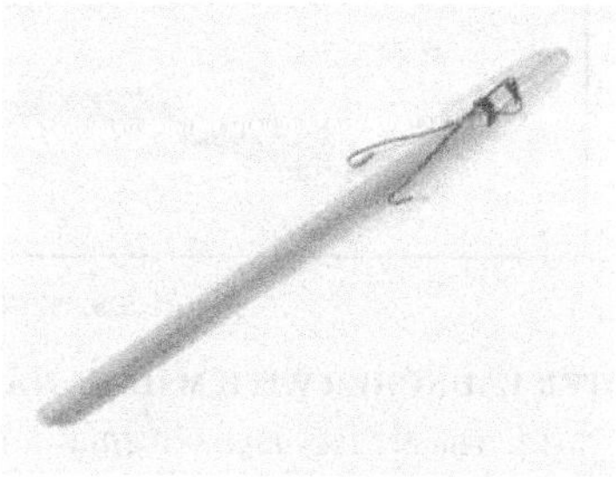

Figure 5-7. Wood Riot Baton

EXPANDABLE RIOT BATON WITH CARRIER (24 TO 36 INCHES)

5-10. The expandable riot baton (see *Figure 5-8*) provides the soldier with an NL means of crowd control and self-defense. The riot baton comes with a mounting device that attaches the riot baton to a belt. The riot baton is 24 to 36 inches in length and is primarily intended for self-defense. The expandable riot baton can be used effectively as an offensive tool if it becomes necessary to keep rioters out of arm's reach of soldiers conducting crowd control operations.

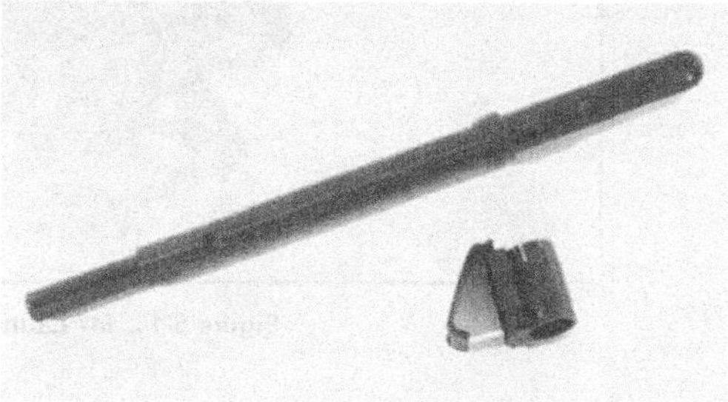

Figure 5-8. Expandable Riot Baton and Carrier

500 (3 INCH) 12-GAUGE SHOTGUN

5-11. The 12-gauge shotgun (see *Figure 5-9*) is a pump action shotgun currently in NLCS inventory. The pump action shotgun is chambered to take up to 3-inch shells. The 3-inch chamber allows for the use of M1012 and M1013 NL munitions. This shotgun also provides a visually distinct alternative to standard military weapons that may be desired based on mission considerations (force signature and acceptability).

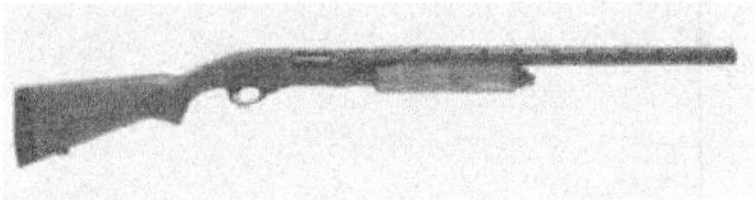

Figure 5-9. 12-Gauge Shotgun

M7 66-MILLIMETER LAUNCHER WITH M315 INSTALLATION KIT

5-12. The M7 (see *Figure 5-10*) is a 66-millimeter vehicle-mounted, NL, grenade-launching device that is mounted on a HMMWV. It is an indirect fire support system that can deliver the M98 distraction grenade that creates a flash-bang effect, L96A1 antiriot grenade, or M99 blunt trauma grenade that creates a sting-ball effect. The M315 installation kit is used to install an M7 discharger on the turret ring of appropriate HMMWV variants. An adjustable bracket allows the launch angle to be depressed for engaging targets at ranges of 50, 75, and 100 meters. The system enforces standoff distances and deters potential threats.

Figure 5-10. M7 Launcher

MODIFIED CROWD CONTROL MUNITION-GROUND EMPLACEMENT

5-13. Similar in nature to the claymore mine, the modified crowd control munition-ground emplacement (MCCM-GE) (see *Figure 5-11*) is an NL munition that can deliver six hundred .32-caliber rubber pellets and a flash-bang effect. The MCCM-GE will stop, confuse, disorient, and/or temporarily incapacitate area targets and/or personnel at ranges out to 30 meters.

WARNING

Avoid employing the MCCM-GE within 10 meters of personnel as it may cause serious bodily harm or death.

Figure 5-11. MCCM-GE

12-GAUGE, FIN-STABILIZED POINT, NONLETHAL CARTRIDGE (M1012)

5-14. The M1012 (see *Figure 5-12*) is a single projectile round made of hard rubber that is shaped like a bomblet and designed to be fired a at a single target. With a muzzle velocity of 500 feet per second, the M1012 has the effective range of no closer than 5 meters and no further than 30 meters. Engagement inside 5 meters could cause serious injury or death. Beyond 30 meters the kinetic energy dissipates to the point where the round becomes ineffective.

Figure 5-12. M1012

12-GAUGE, CROWD DISPERSAL, NONLETHAL CARTRIDGE (M1013)

5-15. The M1013 (see *Figure 5-13*) is a multiple projectile round with .23 caliber hard rubber pellets that is designed to be fired at and employed with the purpose of affecting multiple targets. With a muzzle velocity of 900 feet per second, the M1013 has an effective range of no closer than 5 meters and no further than 30 meters. Engagement inside 5 meters could cause serious injury or death. Beyond 30 meters, the kinetic energy dissipates to the point where the rubber pellets become ineffective.

Figure 5-13. M1013

40-MILLIMETER SPONGE GRENADE (M1006)

5-16. The M1006 (see *Figure 5-14*) is a single projectile round made of pliable foam rubber with a hard plastic back. The M1006 is designed to be aimed and fired at a single target. With a muzzle velocity of 265 feet per second, the M1006 has an effective range of no closer than 10 meters and no further than 50 meters. Engagements inside 10 meters could cause serious injury or death. Beyond 50 meters, the kinetic energy dissipates to the point where the round becomes ineffective.

Figure 5-14. M1006

40-MILLIMETER CROWD DISPERSAL CARTRIDGE

5-17. The crowd dispersal cartridge (CDC) (see *Figure 5-15*) is a multiple projectile round with .48 caliber hard rubber balls that is designed to be fired and employed with the purpose of affecting multiple targets. With a muzzle velocity of 450 feet per second, the CDC has an effective range of no closer than 10 meters and no further than 30 meters. Engagement inside 10 meters could cause serious injury or death and beyond 30 meters the kinetic energy dissipates until the rubber balls become ineffective.

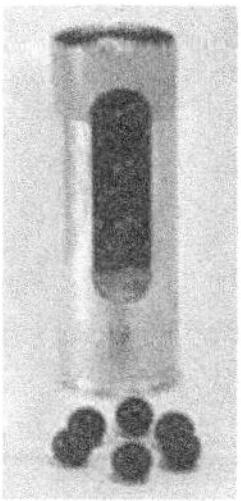

Figure 5-15. CDC

RUBBER BALL GRENADE (GG04)

5-18. The GG04 (see *Figure 5-16,* page 5-10) is a multiple projectile, flash-bang grenade with 100 .25-caliber hard rubber pellets. Each grenade has a fuse delay of 2.8 to 3 seconds with a flash measuring approximately 1 million CP and 180 decibels at 3.5 feet. At detonation, rubber pellets are dispatched at 360° with an effective range of 2 to 3 meters and a maximum engagement range of 15 to 20 meters. The rubber ball grenade is designed to be hand-thrown or muzzle-launched from a 12-gauge shotgun.

RUBBER BALL GRENADE LAUNCHING CUP AND AA30 12-GAUGE LAUNCH ROUND

5-19. The launch cup (see *Figure 5-17,* page 5-10) firmly attaches to the 12-gauge shotgun without any additional tools. It works in concert with the AA30 12-gauge launch round to propel the rubber ball grenade approximately 100 meters. The shooter simply inserts the full body and the safety lever of the grenade into the launching cup, pulls the pin, loads a 12-gauge round, and fires (approximately) at a 30° angle. The desired effect is to have the grenade explode approximately 7 to 10 feet above the target.

RELOADABLE FLASH-BANG GRENADE (M84)

5-20. The M84 (see *Figure 5-18,* page 5-10) is a reloadable flash-bang thrown grenade. Each grenade has a 1- to 2-second fuse delay, with a flash capability of 1.5 to 2.5 million CP and a bang of 168 to 175 decibels. Although it is part of the NLCS, use of the M84 is not recommended for use in crowd control situations. Specialty teams, such as security response teams, and search teams may use this device effectively.

Figure 5-16. GG04

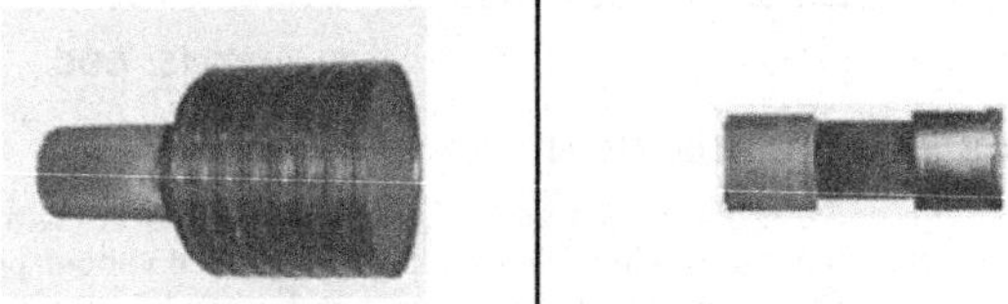

Figure 5-17. Launching Cup and AA30

Figure 5-18. M-84

FLEX-CUFS

5-21. Flex-Cufs consist of a tough pliable plastic band with a self-locking mechanism in the center for each end. When the Flex-Cuf (see *Figure 5-19*) is threaded the restraint bands circle around the wrists or ankles, impeding movement and securing the individual. They are lightweight and disposable and require a cutting tool or other instrument to remove.

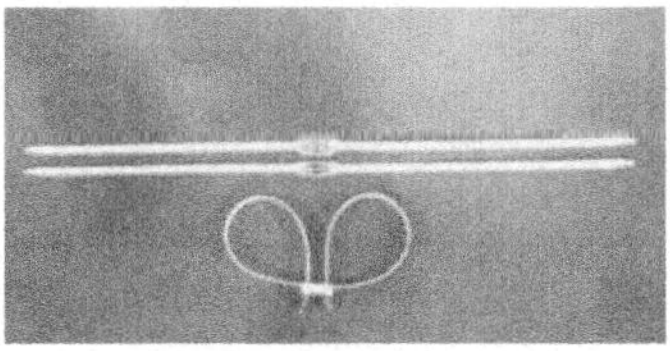

Figure 5-19. Flex-Cufs

INDIVIDUAL RIOT CONTROL AGENT DISPERSER (M36)

5-22. The M36 (see *Figure 5-20*) contains a Dibenz (F, -1 4-oxazepine [CR] solution. It can deliver 25 one-second bursts out to 12 feet. Individual disposable RCA dispersers are intended primarily for self-defense or to keep rioters out of arm's reach of soldiers conducting crowd control formations or soldiers engaged in missions where a noncombatant exists.

Figure 5-20. M36

MIDSIZE RIOT CONTROL AGENT DISPERSER (M37)

5-23. The M37 (see *Figure 5-21,* page 5-12) is the size of a standard fire extinguisher that uses compressed air to force the RCA out to a range of 30 feet. It has the capacity to employ 18 bursts for 3 seconds per charge. It is excellent for providing a wide coverage of RCAs onto a hostile crowd while maintaining excellent standoff capability. The M37 can be refilled and is rechargeable. It can be filled with CR solution (liquid agent) or CS1 (dry agent). For training purposes, CR can be substituted with water and CS1 can be substituted with talcum powder.

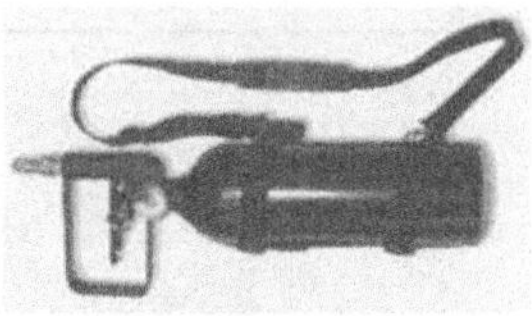

Figure 5-21. M37

SQUAD RIOT CONTROL AGENT DISPERSER (M33A1)

5-24. The M33A1 (see *Figure 5-22*) is designed to provide crowd control and protection at the squad level. It is capable of projecting a ballistic stream of RCAs beyond 25 feet in up to 25 half-second bursts. It consists of a frame and harness assembly, compressed-gas cylinder (agent container assembly), air pressure assembly, gun and hose assembly, multijet spray unit, and check valve assembly. The M33A1 can be refilled and is rechargeable. For training purposes, CR can be substituted with water and CS1 can be substituted with talcum powder.

Figure 5-22. M33A1

MISSION ENHANCERS

5-25. Mission enhancers aid the commander in various tactical situations. Mission enhancers include various types of devices used for illumination, vehicle denial, traffic control missions, and force protection enhancement.

PORTABLE BULLHORN

5-26. The portable bullhorn (see *Figure 5-23*) provides squad leaders with the capability to project their voices above the noise and commotion created by crowds and mobs. The megaphone has a built-in siren for alarm purposes that allows the operator to use the device as an ordinary megaphone and siren alarm during emergencies. The bullhorn is portable and easy to carry. The acoustic range of the megaphone for a clear voice output is ¾ to 1 mile. The power output of the megaphone is rated at 15 watts with a maximum rating of 20 watts. The dimensions of the megaphone are 14.5 inches with a 9-inch diameter bell. It weighs 3.5 pounds without batteries. Battery requirements for the megaphone are eight dry-cell size C batteries.

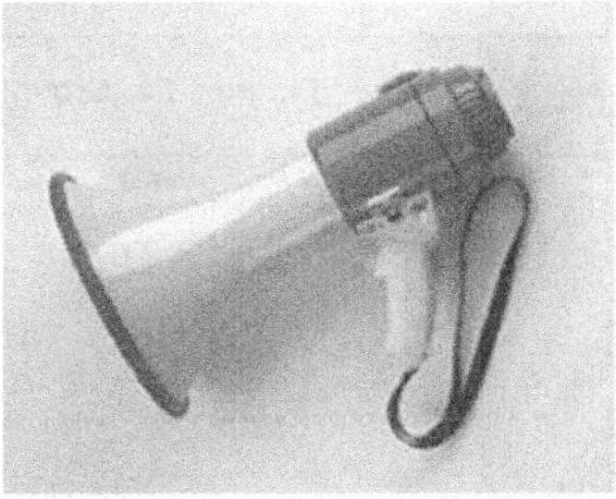

Figure 5-23. Portable Bullhorn

GROUND-MOUNTED BULLHORNS

5-27. Ground-mounted bullhorns (see *Figure 5-24,* page 5-14) are a critical communication enhancement device for conducting crowd control tactics. The bullhorn can facilitate communication with the crowd in conjunction with PSYOP support and assistance with the communication of commands to troops engaged in the crowd control process. The power output of the ground-mounted bullhorn is rated at 22 watts. The dimensions of the bullhorn are 13.75 by 9 by 14.75 inches, and it weighs 15 pounds.

INDIVIDUAL VOICE AMPLIFICATION SYSTEM (M7)

5-28. The M7 (see *Figure 5-25,* page 5-14) is a critical communication enhancement device for conducting crowd control tactics using RCAs while wearing the M40 protective mask. The M7 is fitted to the M40 protective mask to facilitate oral communication and increase the user's ability to communicate using radios and other devices.

Figure 5-24. Ground-Mounted Bullhorns

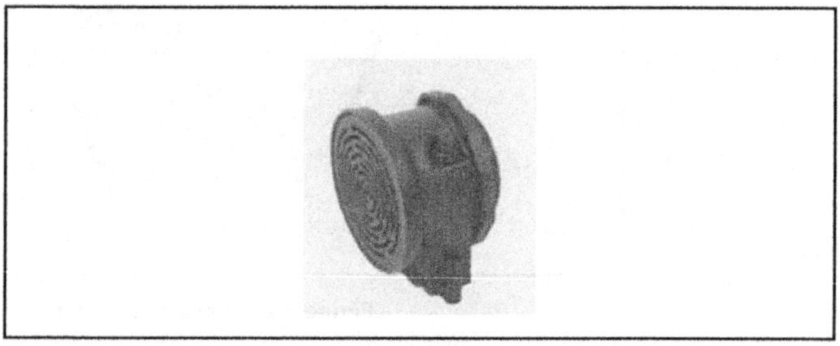

Figure 5-25. Individual Voice Amplification System

INDIVIDUAL HIGH-INTENSITY SEARCHLIGHT WITH HOLSTER

5-29. The individual high-intensity searchlight (see *Figure 5-26*) is used primarily for illumination in crowd control operations at night. An individual can use it for general illumination of the operational area to pinpoint agitators and threats, reduce the ability of the rioters to see troop formations and actions, enhance tactical deception and actions, and enhance tactical deception techniques for units conducting crowd control operations.

HIGH-INTENSITY LIGHT

5-30. A high-intensity light (see *Figure 5-27*) is intended for use in low light or night conditions. A high-intensity light can project a beam that will enable the user to identify an individual person up to 1,900 yards away. It has an adjustable beam spread of 1° to 15° and will run continuously at maximum power for up to 45 minutes. The internal power supply requires 4 hours for a full charge.

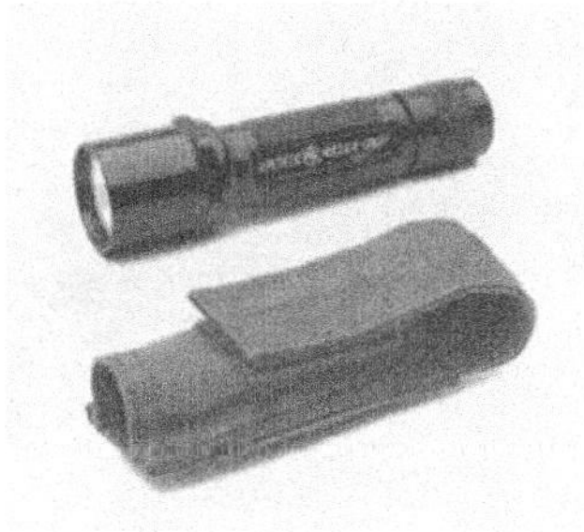

Figure 5-26. Individual High-Intensity Searchlight with Holster

Figure 5-27. High-Intensity Light

HIGH-INTENSITY LIGHT SYSTEM WITH 12-GAUGE SHOTGUN ATTACHMENT KIT

5-31. This 12-guage high-intensity light system (see *Figure 5-28,* page 5-16) is used to illuminate possible targets in crowd control situations where there is little or no light. It serves to blur or blind the aggressor's ability to see. It mounts easily to the end of the 12-gauge shotgun barrel.

40-MILLIMETER CARRYING POUCH

5-32. The 40-millimeter carrying pouch is made from durable canvas material, (see *Figure 5-29,* page 5-16) and can be easily affixed to load-bearing equipment (LBE) and/or a load-bearing vest (LBV) using Velcro straps. It is capable of carrying up to six 40-millimeter rounds.

Figure 5-28. 12-Gauge High-Intensity Light System With 12-Gauge Shotgun Attachment Kit

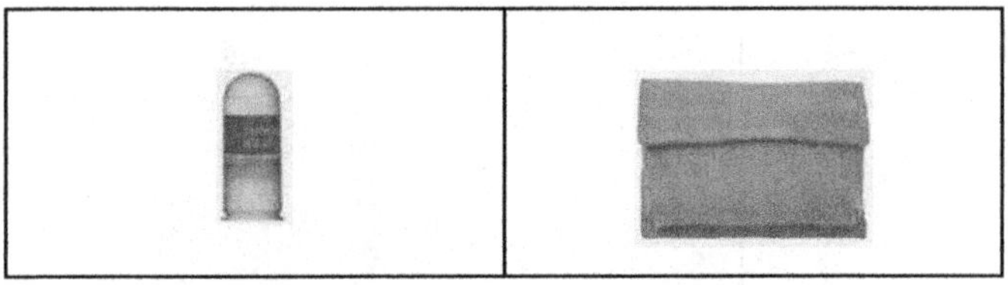

Figure 5-29. 40-Millimeter Carrying Pouch

12-GAUGE UTILITY POUCH

5-33. The 12-gauge utility pouch is made from durable canvas material (see *Figure 5-30*). It can be easily affixed to an LBE and/or LBV using metal clasps. The 12-gauge utility pouch is capable of carrying up to twenty-five 12-gauge shotgun rounds.

Figure 5-30. 12-Gauge Utility Pouch

12-GAUGE BUTTSTOCK CUFF

5-34. The 12-gauge buttstock cuff (see *Figure 5-31)* firmly attaches to the shotgun by sliding over the stock. Elastic straps aid in preventing the buttstock cuff from sliding or rolling. It holds up to seven 12-gauge cartridges, which give the soldier quick access to additional rounds.

Figure 5-31. 12-Gauge Buttstock Cuff

DIVERSIONARY AND/OR RUBBER BALL GRENADE POUCHES

5-35. The diversionary and/or rubber ball grenade pouch (see *Figure 5-32*) is an ammunition-carrying pouch made of durable canvas material. It is partitioned in a manner to provide the soldier with a means of holding up to six rubber ball grenades or six flash-bang grenades in separate compartments.

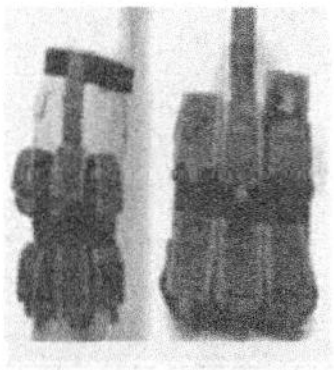

Figure 5-32. Diversionary and/or Rubber Ball Grenade Pouch

CALTROPS

5-36. The caltrop (see *Figure 5-33,* page 5-18) is a field-expedient area denial system. Securing approximately 50 caltrops to 550-type military cord allows a soldier to quickly toss and recover the system. Caltrops are used to deny vehicle and pedestrian movement in designated areas. The four prongs are approximately 2 inches tall and will puncture tires or boots. A soldier can easily employ the caltrop by scattering it in a designated area hidden by camouflage.

Figure 5-33. Caltrop

PORTABLE VEHICLE ARREST BARRIER

5-37. A portable vehicle arrest barrier (PVAB) (see *Figure 5-34*) is a lightweight, portable, manually-emplaced and recoverable-barrier entrapment device that is used (on command) to create a barrier that will stop vehicles being driven by known or suspected hostile forces with minimal damage to the vehicle and its occupants. Two individuals can easily emplace it in approximately 4 to 8 hours. It has the capability of stopping a light truck (up to 7,500 pounds) at speeds of up to 45 miles per hour with a stopping distance of less than 200 feet. It is not to be used as a stand-alone system. The PVAB may be used at dismount points, checkpoints, and roadblocks to prevent unauthorized access by wheeled vehicles into or out of areas.

Figure 5-34. PVAB

TRAINING DEVICES

5-38. Training devices that commanders can use to effectively prepare their soldiers are included in the NLCS. Training devices allow soldiers to learn and build confidence in the development of personal defensive skills. These are critical pieces of equipment because they allow soldiers to train at almost full speed, which provides reinforcement of realistic motor skills. Training in this manner has proven to be more effective than just striking out at the air.

IMPACT TRAINING SUIT

5-39. Impact training suits (see *Figure 5-35*) are made of closed-cell shock-absorbent foam that allows soldiers to hone their riot baton skills through realistic hand-to-hand training. Impact training suits absorb strikes inflicted by the current expandable and wooden riot batons. They provide protection for the head, face, hands, and legs without significant degradation of the wearer's mobility.

Figure 5-35. Impact Training Suits

TRAINING STRIKE BAGS

5-40. Training strike bags (see *Figure 5-36*) give the soldier a durable, yet absorbing surface to strike or kick for training. It also helps develops proficiency in open-hand control and riot baton techniques.

Figure 5-36. Training Strike Bag

PRACTICE GRENADE BODY (GG05)

5-41. The GG05 is blue in color and is easily identified as a practice grenade (see *Figure 5-37*). It is designed to replicate the live GG04 rubber ball grenade. It is made of durable rubber and can be hand-thrown or muzzle-launched from the 12-gauge shotgun. It can be reloaded.

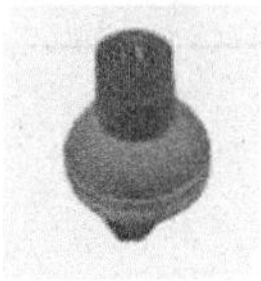

Figure 5-37. GG05

PRACTICE GRENADE FUSE

5-42. The practice grenade fuse (see *Figure 5-38*) is identified by its blue spoon and brown stripe. It easily screws into the body of the GG05. Once used, unscrew the practice grenade fuse from the body and reload it with a new fuse.

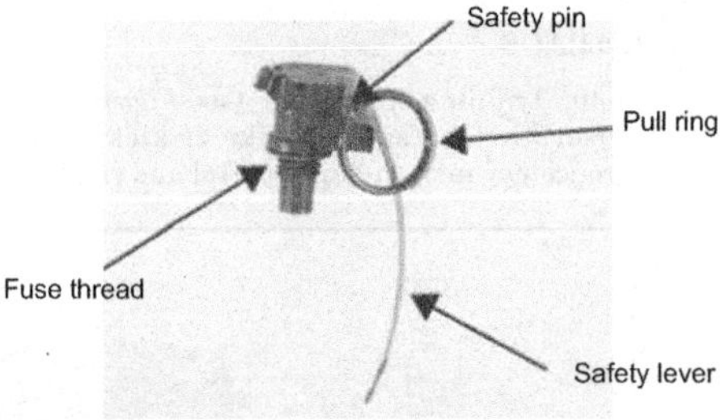

Figure 5-38. Practice Grenade Fuse

CORE CAPABILITIES

5-43. Core capabilities are those fundamental competencies that enable the commander to achieve the desired operational outcome. In the case of NLW, this includes providing a flexible means of response in order to protect friendly forces and/or influence the actions of potential adversaries and noncombatants. With the availability of the NLCS (see *Figure 5-39*) these goals can be achieved without resorting to lethal force and in a manner that will minimize collateral damage. The core capabilities associated with NL effects fall into the following two major categories:

NOTE: There are nine platoon equivalents within a light infantry battalion.

PLT Organization	Personnel and Weapons	Nonlethal Weapons & Equipment
PLT	7 personnel (5 M16/M4 and 2 M60/M240)	3 riot face shields 3 sets of shin guards 3 sets of full-body restraints 3 individual voice amplification systems 3 individual RCA dispensers with carriers 1 portable bullhorn 1 ground-mounted bullhorn 1 high-intensity searchlight 1 set of caltrops 3 ballistic riot face shields 3 sets of ballistic shin guards 3 ballistic riot body shields 1 riot baton training suit 1 riot baton strike pad
SQD SQD SQD	3 personnel (3 M16/M4)	3 riot face shields 3 sets of shin guards 3 riot body shields 3 expandable riot batons with carriers 9 sets of cinch-strap restraints 3 individual voice amplification systems 3 individual RCA dispensers with carriers 3 individual high-intensity search lights with carriers (see *Figure 5-26*)
TM TM TM TM TM TM	24 personnel (12 M16/M4 and 6 M16/M4 with M203 6 SAW)	24 riot face shields 24 sets of shin guards 100 sets of disposable flex-cuffs 24 individual voice amplification systems 24 individual RCA dispensers with carriers 6 expandable riot batons with carriers 11 grenade pouches 12 forty-millimeter grenade pouches 6 twelve-gauge ammunition pouches 6 twelve-gauge buttstock ammunition carriers

Figure 5-39. NLCS Distribution

- **Counterpersonnel.** NL counterpersonnel capabilities enable the application of military force with a reduced risk of fatalities or serious casualties among noncombatants or, in some instances, among enemy forces. There are several specific NL counterpersonnel capabilities to be explored. These include the means to influence the behavior and activities of a potentially hostile crowd, and the capability to bring a mob engaged in a riot under control. While there are many similarities in these two scenarios, each involves unique challenges, which may require radically different solutions. For more information, see *FM 3-22.40.*
- **Countermaterial.** NL countermateriel capabilities enhance operations by rendering equipment and facilities unusable without complete destruction. An NL countermateriel capabilities enable the

application of military force to defuse potentially volatile situations under circumstances in which more destructive conventional military means might prove counterproductive. For example, pre-emptive strikes against troublesome and aggressive nations may be politically unacceptable when only conventional weapons with the high risk of casualties are involved. With NL countermateriel capabilities, the enemy's ability to threaten its neighbors could be curtailed with less political risk by attacking only their weapons of war and their supporting infrastructure. For more information, see *FM 3-22.40.*

ADDITIONAL CAPABILITIES

5-44. NLW are only NL when fired within their parameters.

DANGER

If NLW are fired outside their parameters, they may cause serious injury or death.

5-45. If a situation turns from NL to lethal, an NL gunner only needs to change his point of aim and distance to inflict serious bodily harm and/or death to the threat. The closer the soldier is to the target, the greater the kinetic energy and the more severe the effects are to the threat.

EMPLOYMENT CONSIDERATIONS

5-46. NL munitions and equipment are not completely NL; they are NL by intent. The term NL does not guarantee zero mortality or nonpermanent damage. NLW can add flexibility to combat operations and enhance force protection by providing an environment in which friendly troops can engage threatening targets with a reduced risk of noncombatant casualties and collateral damage.

5-47. If the tactical situation dictates a NL response, a soldier who levels a weapon loaded with NL munitions must be trained in distance to the target and appropriate points of aim. Based on the soldier's understanding and training on the characteristics and capabilities of the round, they can easily discern and engage targets to the desired effect. If the tactical situation dictates a lethal response, soldiers can easily adjust the point of aim to a more vulnerable target location on the body and move in closer to the target. NL munitions can be used as a lethal response if the situation dictates and the intent of the soldier is to inflict death.

NONLETHAL WEAPONS CAPABILITIES IN FORMATIONS

5-48. The mix of NLW within the formation is based on METT-TC. For example, the first rank of the formation will have riot shields, riot batons, and their personal weapon (slung across their back [left to right] with the butt up and muzzle down or holstered). The second row will have a mix of shotguns and M203s. Shotguns and M203s may be used as area denial or point target weapons depending on the specific munitions used. The shotgun provides cover for the slower loading M203. The commander may move these weapons within the formation to meet the needs of the mission and to create distance between the formation and the crowd (see *Appendix D*).

Chapter 6

Civil Disturbance Formations

Control force formations, when properly employed and executed against a crowd, are some of the most practical methods of crowd control. The riot baton has been the weapon of choice for the control force. With the development of NLW, the commander now has more options at his disposal. Before NLW, the only other choices were chemical irritants, MWDs, and lethal force. Now commanders can create a standoff distance with NLW that reduces the risk of close fighting with the mob. During any crowd control operation, the commander must protect the control force with lethal overwatch. This overwatch may be in the form of marksmen and observer teams (sniper with observer) and a response force with lethal capabilities.

OVERVIEW

6-1. With advances in NLW, it is possible to create an effective standoff distance of 15 to 100 meters between the crowd and the control force formation. It is now possible to advance and disperse the crowd without coming into direct contact with it. Some NLW may also be used to assist in the identification and apprehension of instigators.

6-2. Control force formations are used to disperse, contain, and/or block a crowd. These formations are more effective in urban areas than in open areas, such as parks or fields. When control force operations are employed in urban areas, it is easier to split a crowd into smaller segments, isolate instigators, or funnel the crowd into the desired location by using buildings and other man-made structures. Commanders must be mindful of the fact that well-organized crowds may attempt to move to the flanks of a formation and/or get behind the formation to gain a tactical advantage.

6-3. Commanders must be aware of the limitations of control force formations. They are not the answer to all civil disturbance situations. Do not expose the formation to sniper fire and/or unnecessary violence simply for a show of force. Control force formations should be used when it is decided that a crowd poses a threat and must be dispersed or moved to a specified area. Before dividing a large crowd, the commander must consider that this might not solve the problem. It may worsen the problem by creating smaller elements that may engage the control force in small-mob tactics, such as sniping, looting, burning, and attempting to envelop the control force. Commanders must anticipate that breakaway crowds may engage in small-mob tactics. These tactics can be defeated by area control measures, such as building searches, saturation patrolling, and other tactics. After dispersing the crowd, the control force must not assume that there will be an immediate return to peaceful activities. The use of formations is only part of the total dispersal effort.

6-4. Commanders must assess and secure the areas of operations (AOs). This may be accomplished with the use of helicopter overflights and by visually securing rooftops, high elevations, and advantage points. Designated marksmen teams may be used to secure the rooftops and provide lethal overwatch for the control formation. The control force commander must be aware that designated marksmen teams are positioned on rooftops.

6-5. Crowd control formations and their support teams have more capabilities than just crowd dispersal. They also have the capability to apprehend and detain certain members of the crowd that the commander feels may instigate the crowd to further violence. Commanders must analyze the threat, determine the mission of the control force, and decide which control force formations will accomplish the mission according to METT-TC.

6-6. During CONUS operations, there are usually no enemy forces. However, commanders must analyze and prepare for any potential opposition. The formations discussed in this chapter are guidelines and may be adapted to fit any mission or situation. Whatever the modification, soldiers must stay in position and on line. Only through training and rehearsals will soldiers become proficient in basic formation movements and be able to adjust to changing situations. Soldiers must also be prepared to employ NLW. *Figure 6-1* is used as a key to identify soldiers within crowd control formations.

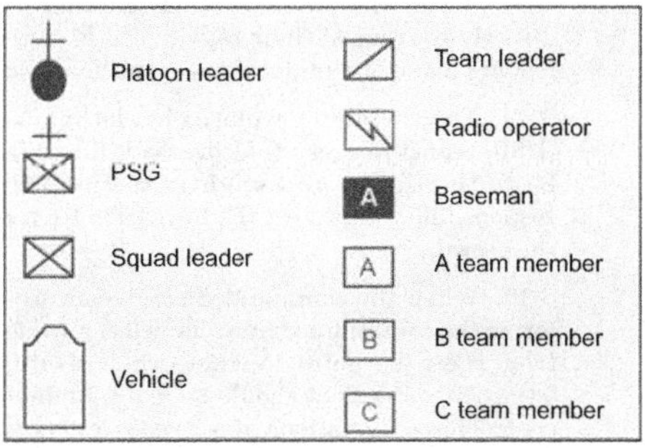

Figure 6-1. Symbols for a Formation Element

ACTIONS BEFORE MOVEMENT

6-7. Before moving the control force to the civil disturbance area and into a potentially hostile urban environment, the commander and his staff must review the preincident plans and expand them (as needed) based on current information. Because of the employment of forces against ambiguous threats,

IPB becomes critical. Information is the key to developing an appropriate response to civil disturbances.

6-8. Civil disturbance training must be part of the annual unit training plan. If a soldier is deployed to an area where civil disturbances are a recurring event, sustainment training must occur more frequently. Before moving the control force to the rally point, an intense rehearsal must be conducted by the leadership and the control force. A top-down review of the rules of engagement (ROE/rules for the use of force (RUF) must be accomplished. The ROE/RUF must be part of the unit training plan and each soldier must be trained to standard (see *Chapter 8*). The use of NLW (type specific) must be granted to the on-site commander from the approval authority.

6-9. Information is the key to developing plans for appropriate responses to a civil disturbance, and IPB is crucial to the success of the mission. During the planning phase (see *Chapter 2*), the information provided by the IPB will aid the staff in selecting various routes and rally points. The selection of ingress and egress routes must include a variety of ways in and out of the area. Direct and indirect routes and those with cover and concealment are necessary. The rally point must be cleared and secured by an advanced party capable of controlling the area.

RALLY POINT ACTIONS

6-10. Once at the rally point, the commander makes contact with the local police, civil authorities, or military officer. He then assesses the situation, decides the next COA, and issues orders, as appropriate. The platoon sergeant (PSG) forms the platoon into a column formation and places team leaders directly in front of their teams (see *Figure 6-2,* page 6-4). Team and squad leaders make last minute checks of their soldiers and await orders.

6-11. The commander places his lethal overwatch (designated marksmen [DM] teams) in areas that overlook the control force and the mob. They may be to the flanks and slightly to the front of the formation. It is their responsibility to protect the formation from lethal fire by constantly scanning the crowd.

6-12. When the commander receives word that DM teams are in place, he forms the control force into a crowd control formation and moves them quickly from the rally point to where the crowd is assembled. Depending on the situation, movement should be in a column formation or one of the three basic control force formations. Reserve forces are left at the rally point until ordered forward as additional lethal overwatch or to reinforce the formation.

CONTROL FORCE FORMATIONS

6-13. Control force formations, when properly employed and executed against a crowd of limited size, are one of the most practical methods of crowd control. Experience has indicated that the most frequently used control force

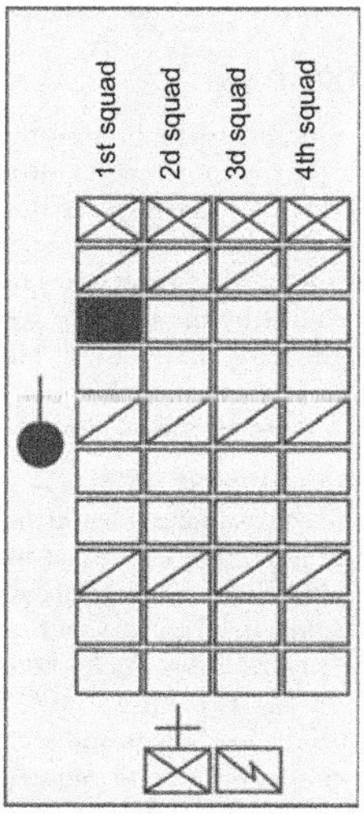

Figure 6-2. Column Formation

formations are the line, echelon, and wedge. The two lesser-used formations are the diamond and circle. Descriptions of all the formations are as follows:

- **Line formation.** The line is the basic formation and it is used more often because of its offensive and defensive applications. As an offensive formation, the line is used to push or drive crowds straight back, across an open area, or up a city street. As a defensive formation, the line is used to hold the crowd or to deny access to areas.
- **Echelon formation.** The echelon is an offensive formation used to turn or divert groups in open or built-up areas and to move crowds away from buildings, fences, and walls.
- **Wedge formation.** The wedge is an offensive formation that is used to penetrate and split crowds into smaller groups.
- **Diamond formation.** The diamond, when used as an offensive formation, is used to enter a crowd and is the formation of choice for extraction teams. As a defensive formation, the diamond is used when all-around security is required in open areas. The decision to use this formation is based on the conformation of the crowd.
- **Circular formation.** The circular formation is used for the same purpose as the diamond formation. The decision to use this formation is based on the conformation of the crowd.

6-14. There are many suitable variations of control force formations that may be employed, but appropriate commands and signals must be created to execute the formations. Because of their somewhat complicated nature and the coordination required for these formations, new variations must be trained, practiced, and rehearsed before they are used in a civil disturbance.

VEHICLES AND FORMATIONS

6-15. Vehicles may be employed with troops in control force formations, especially when a large riotous crowd is on hand. Although vehicles add strength to formations, certain precautions must be taken. Covering the windshield with sturdy, close mesh fencing and the standard safety glass will offer some level of protection to the occupants. Shields and/or mobile barriers may be built by mounting a wooden or metal frame strung with barbed wire across the front of a vehicle. Members of the formation should walk as near to the front corners of each vehicle as possible to keep rioters from attacking the sides and rear of the vehicles.

6-16. Up-armored HMMWVs, ASVs, or armored vehicles may be employed with the control force formation when determined by METT-TC. They present a strong psychological effect and offer protection for the occupants. Wheeled armored vehicles are best because they do not deface the pavement. Whenever vehicles are used in a crowd control formation, the commander must ensure that he is still able to see and control the formation. This may require him to occupy a position in a similar vehicle behind the formation or command subordinate leaders to move away from the front line and into line-of-sight (LOS) control. This will require additional hand-and-arm signals to allow the commander and subordinate leaders to communicate. Commanders can effectively use vehicles as primary communication tools by using horns and lights as signals to organize movement or actions. However vehicles are used, it takes a good measure of training and practice to be effective.

ELEMENTS OF A CONTROL FORCE FORMATION

6-17. Four elements make up the basic crowd control formation. They are as follows:

- **Base element.** The base element is made up of two ranks. The first rank is shield holders, while the second rank contains the NLW. This is the front line of the formation.
- **Support element.** The support element forms up in a column formation behind the base element and has a variety of uses. It may be used to replace base element members as needed, provide lateral or direct support, or perform extraction team operations. The support element helps the base element by performing the following three essential tasks:
  - **General support.** The general-support element is formed from an uncommitted squad in the platoon (generally the 2d squad). When a company is tasked as the control force element, one of the platoons becomes the general-support element. The element is in a column formation centered on and behind the main formation. From this formation, the general-support element can move to lateral or direct support, as needed.
  - **Lateral support.** The lateral-support element is used to protect the flanks of the formation. This is done by moving a set number of teams forward from the general-support element or by using the

end teams in the formation; however, this will make your formation smaller. Once in position, these teams become part of the formation, with their riot shields facing the flanks of the formation.

  - **Direct support.** When direct support is ordered, the general-support element moves forward as they form their own formation. While moving forward, the squad and team leaders from the control force formation step back and allow the riot shield holders from the support element to step behind riot shield holders of the original formation. This allows the support riot shield holders to form up between and behind the riot shield holders of the original formation. Now there are two lines of riot shield holders between the crowd and NLW firers. This formation is the strongest of the formations and requires more planning and practice to master.

- **Command element.** The command element contains several different members. A general configuration for the command element is the platoon leader, the PSG, a radio operator, a video recorder operator (if required), and an interpreter (if required). This element does not have a fixed location within the formation and moves about as needed.
- **Lethal overwatch element.** The lethal overwatch element is a team formed from reserve security forces.

6-18. The reserve support element is not part of the control force formation until it is brought forward from the rally point to join the formation. It remains until released by the formation commander. Lethal overwatch teams are formed from the reserve force. Once lethal overwatch teams are deployed and in position, they are under the control of the formation commander. Communication between the lethal overwatch team and the formation commander is a priority because he approves target selection and engagement.

WEAPON POSITIONS

6-19. At the commander's discretion, members of the control force formation may carry their weapons in any of the positions prescribed in *FM 3-21.5*. However, the primary method of carrying weapons in the control force formation is at the safe-port position (see *Figure 6-3*). It allows the soldier to control both ends of the weapon while moving in and out of the formation and advancing on the crowd.

6-20. High-port position (see *Figure 6-3*) is a position of complete readiness. It is used whenever troops are in contact with a crowd that is showing resistance or not withdrawing. This is a tiring position and is hard to maintain for extended periods. Commanders must rest their troops at every opportunity by using less tiring positions.

Safe-port arms | High-port arms

Figure 6-3. Weapons Positions

COMMANDS

6-21. Commands are given to the control force formation orally or with hand-and-arm signals. Oral commands for the on-guard position are given in one count. All other commands are given in two counts, a preparatory command followed by a command of execution. However, oral commands cannot be relied on completely, so commanders must plan to use hand-and-arm signals too. When vehicles are part of the formation, radios become an alternate method for relaying commands. To improve communication, commanders may use hand-and-arm signals (see *Figure 6-4,* page 6-8) with oral commands. The basic commands used in control force formations are outlined in *Table 6-1,* page 6-9. These commands can be added to or combined to form more complex commands, therefore forming more complex formations.

MOVEMENTS

6-22. Control force formations move in the same manner as regular formations. This allows the commander to more precisely maneuver the formation by commanding the number of steps it should move and in what direction, such as "Five Steps, Forward March." The normal rate of march for entering and leaving a control force formation is double-timing. The half-step march is used when the formation is in direct contact with the crowd. This slows the formation down and allows for better command and control. The squad and team leaders echo preparatory commands and provide cadence while in control force formations.

INTERVAL AND DISTANCE

6-23. Interval is the lateral space between elements. Distance is the space between elements in a column. The usual interval and distance between soldiers in control force formations is 30 inches. The interval and distance can be adjusted based on METT-TC. If the crowd has to be physically pushed back, a close interval is preferred. This allows the riot shields to be overlapped, creating a stronger wall. During peaceful demonstrations, the interval could increase to double arm width, allowing the formation to cover more area and still control the crowd. The interval and distance could change several times during the course of operations before the crowd is dispersed.

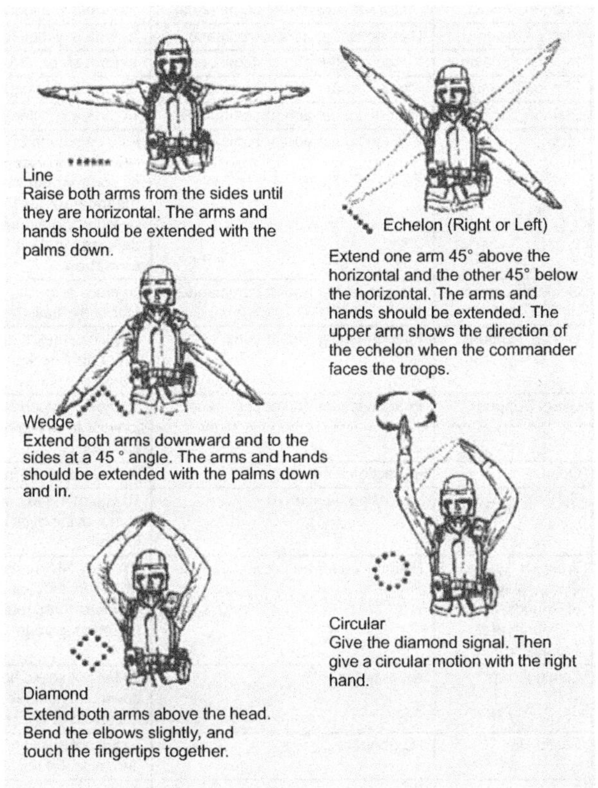

Figure 6-4. Hand-and-Arm Signals

Table 6-1. Basic Commands

Commands for	Command	Given by	Purpose
Formations	Platoon On Line	Platoon leader and/or commander	To form a line formation.
	Platoon Wedge	Platoon leader and/or commander	To form a wedge formation.
	Platoon Echelon	Platoon leader and/or commander	To form an echelon formation.
	Platoon Diamond	Platoon leader and/or commander	To form a diamond formation.
	Platoon Circle	Platoon leader and/or commander	To form a circular formation.
	Move	Platoon leader and/or commander	As a command of execution. **NOTE: The platoon leader and/ or commander also identifies the location for the formation by pointing his arm to the desired location for the formation.**
Support	General Support	Platoon leader and/or commander	To place a specified unit in the rear of the base element.
	Lateral Support	Platoon leader and/or commander	To place support elements on the left or right flank of the base element.
	Direct Support	Platoon leader and/or commander	To move support elements forward to strengthen the base element.
	Open	Extraction team leader	To open a space in the formation.
	Up	Extraction team leader	To inform the extraction team to start moving back to the main formation.
Fire	Weapon System Gunners, Number of Rounds, Type of Round, and Prepare to Fire	Platoon leader and/or commander	To ready NL gunners to fire weapons; for example, M203 gunners, three rounds, area target, prepare to fire.
	Shield Down	NL gunners	To tell riot shield holders to get down on one knee and lower the riot shield to the ground.
	Shield Up	NL gunners	To tell riot shield holders to stand and raise the riot shield.
Weapons	Port Arms	Platoon leader and/or commander	To use when not in contact with the crowd.
	High Port	Platoon leader and/or commander	To tell NL gunners what position to hold their weapons in while the formation is in physical contact with the crowd. This allows for rapid targeting and firing of NL munitions.

EXTRACTION TEAM OPERATIONS

6-24. The extraction team is a squad (see *Figure 6-5*) that is generally formed from the support element, but could also be formed from reserve security forces. Once the team is formed, it is OPCON to the formation commander until the team's mission is complete and it returns to the rear of the formation. This team provides the riot control formation with the means to employ NL and lethal cover forward of the formation. It also provides the formation with the ability to remove and detain rioters downed by munitions.

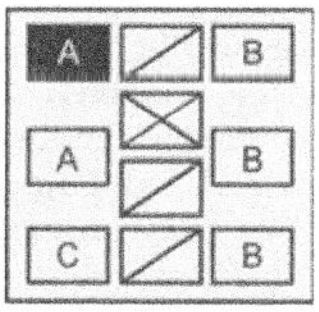

Figure 6-5. Extraction Team Formation

The extraction team—

- Extracts vehicles or personnel that are in immediate danger from the crowd.
- Detains and escorts downed rioters to the rear of the formation.
- Restrains and searches subjects.
- Covers confined or congested areas where a full riot control formation cannot be inserted.

6-25. The extraction team is usually a squad. While the recommended distance for deploying the extraction team from the control force formation is no farther than 10 meters, this may change based on METT-TC. The control formation protects the extraction team and provides immediate support, if necessary.

6-26. The extraction team may deploy from anywhere within the formation. When given orders to deploy, the squad leader of the extraction team gives the formation commands, identifies at least two personnel in the squad to conduct the search and apprehension of the subject, and leads the squad to the base line of the formation. He then extends his arm between the two riot shield holders and commands, "Open," while tapping the riot shield holders on the side of the shoulder. The two riot shield holders take one step backward and one step to the right or left. This clears a path for the extraction team. The extraction team leader then states the number of personnel leaving the formation, such as "Ten security personnel leaving." As the last extraction team member exits the formation he states, "Last Man." The two riot shield holders then return to their original position in the formation. The adjacent base element squad and/or team leaders pass the number of personnel in the team through the formation.

6-27. As the extraction team approaches an instigator or incapacitated demonstrator, the riot shield holders envelop the target and face in the direction of the nearest threat. NL gunners on the left and right cover their respective areas. The search and apprehension team immediately controls and restrains the target. To avoid confusion, one team member concentrates on controlling the subject while another team member applies a restraining system. When the search and apprehension team has the subject under control, the extraction squad leader sounds off with, "Up." Upon hearing the "Up" command, it is repeated by all extraction team members. Team members then grab a shoulder of the team member in front of them to backtrack into the formation. As the extraction team approaches the formation, the nearest base element leader will extend his arm and command, "Open," while tapping the riot shield holders on the side of the shoulder. The riot shield holders move one step backward and to the right or left to open the formation. As he enters the formation, the extraction team leader sounds off with the number of personnel entering the formation, such as "Ten security personnel and one demonstrator." The adjacent base element leader counts the number of personnel as they enter the formation. When the last man enters the formation, he states, "Last Man." The riot shield holders then return to their original places in the formation. The adjacent squad and team leaders in the base element pass information through formation that the extraction team has returned.

LETHAL OVERWATCH TEAMS

6-28. During a NL engagement, the use of a DM team provides confidence and safety to those facing the mob. If a lethal threat is presented, the DM team in an overwatch position (armed with a standard infantry rifle that is mounted with a high-powered scope) can scan the crowd, identify agitators and riot leaders for apprehension, or fire lethal fire if so ordered or warranted. It is also ideally suited for flank and countersniper operations.

6-29. The DM team sets the security overwatch and provides real-time intelligence and reporting that is vital to mission success. It provides coverage during the entire approach to the crowd, and its ability to select positions and provide cover fire (if and when warranted) is critical to the safety of the control force. Successful execution requires training, practice, and rehearsal with the control force formation.

6-30. The DM team is organized from a standard three- or four-man team. Each individual on the team is equipped with a standard-issue weapon. One individual has an infantry rifle with a scope, another has a set of binoculars, and two other individuals provide security for the team. Each team is equipped with a radio for communication with the HQ element of the control force formation.

6-31. The teams must have a visual advantage over the crowd to provide lethal protection to the formation. This is best accomplished by placing them on nearby rooftops, in the upper floors of buildings, and on hilltops. They must have the optical equipment to identify mob leaders, instigators, and/or individuals with weapons. Team integrity must be maintained. In organizations with three-man teams, one individual becomes the marksman, another the observer, and another the security. When teams have a 4th member, that individual becomes the video camera operator. This will help identify individuals for later apprehension and provides a record of events.

6-32. Two teams should be deployed to cover the flanks of the formation. Each team should be deployed to a position that is in front of the formation and in an overwatch position. As the formation moves forward, the teams leapfrog forward from the formation.

6-33. The reserve forces, located at the rally point, can provide another lethal force of some size that can be brought forward, as needed. Each soldier in the control force formation should have a sidearm (9-millimeter pistol) or an M16 or M4 rifle. The rifle is slung diagonally across his back with the rifle butt over his left shoulder and the muzzle below his right hip. Lethal munitions are in a separate ammunition pouch that is isolated from NL munitions. This is to prevent the mixing of NL and lethal ammunition.

6-34. Once the lethal overwatch teams are deployed, they are under the OPCON of the formation commander. The lethal overwatch element must have direct and secure communications with the formation commander. The number of teams deployed is based on METT-TC. Teams will leapfrog to keep ahead of the advancing control force formation with one team in the overwatch position at all times.

RESERVE FORCES

6-35. A large reserve of soldiers should be maintained during civil disturbance operations. Knowing that a large reserve force is available provides confidence and safety within the control force and helps prevent them from overreacting to provocative acts with disorderly and criminal elements in the crowd. When determining the number of reserve forces required, it should be resolved in favor of a large number. Lethal overwatch teams, forces to augment the control force, and apprehension teams come from the reserve forces.

6-36. Apprehension teams are a must when it is likely that a large number of people will be apprehended. These teams provide an organized response to what could be a chaotic situation. The teams can be organized at the team or squad level, depending on the number of apprehensions expected. These teams usually operate behind the base element. During peaceful demonstrations, these teams can be sent forward to remove demonstrators who refuse to disperse.

6-37. Each team consists of an apprehension, movement, and security element (added if the control force cannot provide security); a recorder; and a video or still photographer. The apprehension element is an officer or NCO and a civilian police officer (CONUS or OCONUS) or US marshal. It decides who will be apprehended, informs the subject of the reasons, supervises the handling of the subject, and ensures that the apprehension is properly conducted and documented. The apprehension team is responsible for the overall apprehension process, limiting the number of individuals appearing in court to document the apprehension. The movement element helps the apprehension element handle the subject. They move, restrain, and search the individuals being apprehended under the supervision of the apprehension element. The security element prevents the crowd from interfering with the mission. The recorder documents the apprehension on *DD Form 2708*. When possible, the video and/or still photographer recorder photographs each subject with the apprehension element. This provides the apprehension element with the necessary documentation for use in court. One individual may need to conduct many of the duties of an apprehension team based on METT-TC.

6-38. The video and/or still photographer record the situation before, during, and after the apprehensions. These images may also be used as pictorial documentation for use in court. It is possible that the presence of video and still cameras actively recording the situation can cause crowd members to be less prone to unruly or criminal acts.

6-39. The behavior of the selected individual to be apprehended is observed to decide if the subject is violent, cooperative, or passive. Unless there is good reason to believe otherwise, the subject must be considered dangerous. The mood of the crowd is observed to estimate what it might do. Is it only curious onlookers or does it support the individual targeted for apprehension? Will it interfere?

6-40. When apprehending an individual, the actions must be kept as simple as the situation allows. The element of surprise can give the apprehension team an advantage over the apprehended individual. It is best not to use surprise if the individual might panic and injure someone. Select an area for the apprehension that is to the advantage of the team and not the individual. Avoid apprehensions in crowded areas as these areas provide many avenues of escape and contain supporters of the apprehended individual.

6-41. All verbal commands given to the apprehended individual must convincing authority with short, clear, and distinct commands. Voice, attitude, and deportment must convey an expectation that the apprehended individual will comply.

6-42. The individual should be searched immediately; however, circumstances (violence or sympathizers for the individual being apprehended) may dictate temporarily postponing the search until the individual can be moved to a safer area. Flex-Cufs or hand irons may be applied.

SQUAD FORMATIONS

6-43. Civil disturbance training is a collective task. Using team and squad concepts, team leaders instruct their teams and squad leaders instruct their squads. Safety is key when using force on force training for the control force formation. Soldiers must understand that serious injuries may occur if they do not follow guidelines set by the commander.

6-44. The smallest formation is a squad formation, which is used in back of the main formation to cover side streets. Squad members must know their positions regardless of which control force formation they are in, squad to company.

6-45. Training must enforce soldiers' understanding of the need to stay informed for individual protection. Individual soldiers are much more vulnerable to attack when they break ranks and chase after crowd members. When an individual soldier breaks ranks he not only puts himself in danger, but the entire formation.

6-46. When moving a squad into a crowd control formation from a column formation, the squad leader takes one or more steps to the right of the squad and faces it. He gives the command and hand-and-arm signal for the formation he wants. Then he indicates where he wants the formation located by pointing in that general direction. The A team leader commands, "Follow Me." He places the baseman at the position indicated by the squad leader. The squad forms on the baseman position.

LINE FORMATION

6-47. The command for forming a squad in a line formation is "Squad Line Formation, Move." On the command of execution, the A team leader leads the baseman to the place indicated by the squad leader. The squad members align themselves in sequence with the baseman at normal intervals (see *Figure 6-6*). Based on the situation, the squad leader may designate a specific interval. He does this when he gives the preparatory command. If no interval is specified, the squad automatically forms using the normal interval.

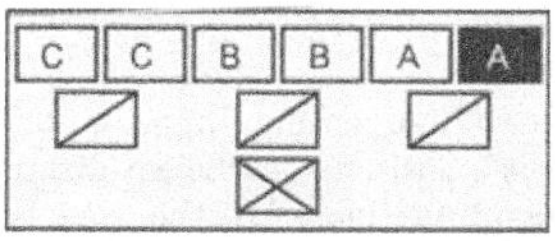

Figure 6-6. Squad Line Formation

ECHELON FORMATION

6-48. The command for having a squad form in an echelon formation is "Squad Echelon Right (or Left), Move" (see *Figure 6-7*). On the command of execution, the A team leader places the baseman at the location indicated by the squad leader. The squad members align themselves in sequence with the baseman, one pace to the side and one pace to the rear.

WEDGE FORMATION

6-49. The command for moving a squad formation into a wedge formation is "Squad Wedge, Move." On the command of execution, the A team leader moves the baseman to the place indicated by the squad leader. The A team lines up to the left of the baseman, one pace to the left and one pace to the rear of each preceding man. B team members align themselves with the baseman, one pace to the right and one pace to the rear of each preceding man. C team members align themselves in the same way to the left of the baseman (see *Figure 6-8*).

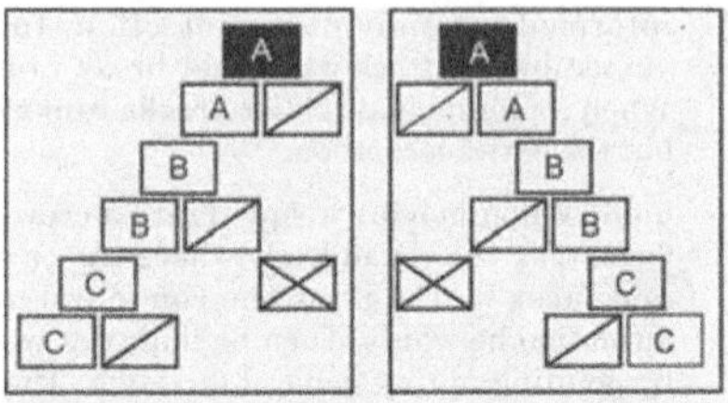

Figure 6-7. Squad Echelons Left and Right Formations

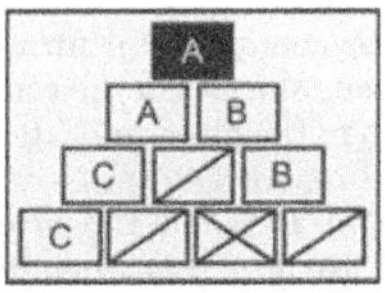

Figure 6-8. Squad Wedge Formation

ASSEMBLY

6-50. When assembling a squad from the line formation, the squad leader takes a position a sufficient distance to the rear of the squad and commands, "Squad Assemble." At the same time, he raises his right hand in the air and makes a circular motion. Squad members, who hear the command, automatically come to the port arms position. The A team leader and the baseman do an about-face movement.

6-51. The other squad members face toward the baseman. On the command of execution "Move," the squad leader points to the place where he wants the squad to assemble. The A team leader double-times to the designated spot, and the other members of the squad follow him. Team leaders fall into the file in their designated positions as they move to the AA. To assemble from a squad echelon formation, the steps are the same as a line formation.

6-52. To assemble from a squad wedge formation, the squad leader takes the same steps as for the line and echelon formations. B team members do a half left face, and the A team and C team members do a half right face. The C team pauses at the baseman position and allows the B team to clear the formation before moving to the AA.

DIAMOND AND CIRCLE FORMATIONS

6-53. The diamond and circle formations are used during extraction team operations. They are small formations used to penetrate the crowd or cover small areas. The decision on which formation to use is based on METT-TC and the ability of the squad to perform.

6-54. Commands for forming a squad in a diamond or circle formation is "Squad Diamond (or Circle), Move." On the command of execution, the A team leader moves the baseman to the place indicated by the squad leader. The A team lines up to the left of the baseman, covering the 8 to 12 o'clock positions. B team members align themselves to the right of the baseman, covering the 12 to 4 o'clock positions. C team members align themselves covering the 4 to 8 o'clock positions (see *Figure 6-9*).

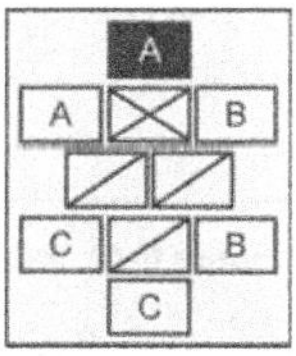

Figure 6-9. Squad Diamond or Circle Formation

6-55. To assemble from a squad diamond or circle formation, the squad leader takes the same steps as he would for a line formation. The A team leads the squad to the AA followed by the B team and then the C team. Teams fall into a column formation as they file back to the AA.

THREE-SQUAD PLATOON FORMATIONS

6-56. The platoon HQ consists of the following individuals: the platoon leader, the PSG, and a messenger or radio operator. When directed by the platoon leader, other individuals such as the selected marksman, interpreter, RCA disperser operators, and firefighters may augment the platoon HQ. In forming all control force formations from the column, the platoon leader moves to the right of the platoon and faces them when giving his commands. He gives the hand-and-arm signal and the preparatory command. On the command of execution "Move," he points to the approximate location where he wants the platoon to form. The platoon leader pauses between the preparatory command and the command of execution to allow each squad leader to issue a preparatory command to his squad.

Platoon Line Formation

6-57. The command for forming a platoon line is "Platoon on Line, Move." Immediately following the platoon leader's preparatory command, the squad leaders of the 1st and 3d squads give the command, "Follow Me." The leader of the 2d squad gives the command, "Stand Fast." On the platoon leader's command of execution, the 1st and 3d squads move forward to the point designated by the platoon leader. The 1st team member of the 1st squad is the baseman for the platoon formation. The 1st squad forms a squad line to the left of the baseman while the 3d squad forms a squad line to the right of the baseman and dresses with the 1st squad. After the 1st and 3d squads have cleared the platoon, the 2d squad leader gives the command, "Follow Me." The 2d squad splits and the A and B teams form a line to the right of the 3d squad.

The C team (with the squad leader) forms a line to the left of the 1st squad (see *Figure 6-10*).

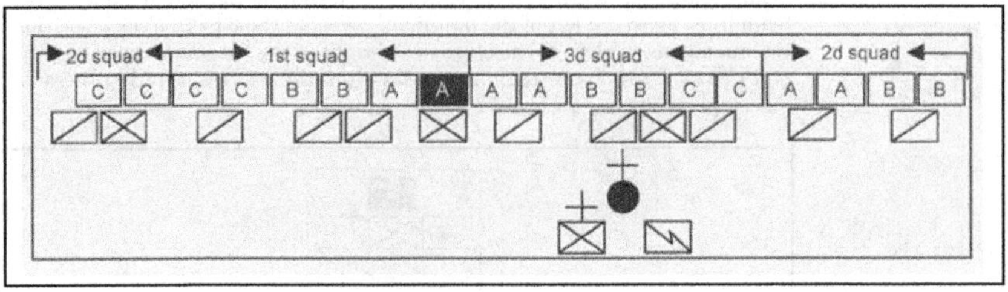

Figure 6-10. Platoon Line Formation

Assembling From a Platoon Line Formation

6-58. When assembling the platoon from the platoon line formation, the platoon leader and his HQ personnel take a position to the rear of the platoon. He then gives the command, "Platoon Assemble," while raising his right arm and making a circular motion above his head. Platoon members automatically assume the port arms position. The 1st squad leader and the baseman do an about face. The 3d squad leader moves to the head of his squad. All other members do an about-face movement toward the baseman. On the platoon leader's command of execution, the platoon leader points to where he wants the platoon to assemble. The 1st squad leader gives the command, "Follow Me," and moves his squad toward the AA. The squad moves to make a single file line to the baseman position followed by the C team of the 2d squad. As the 1st squad clears the baseman position, the 2d squad leader comes to a halt at the baseman position. The 3d squad leader gives the command, "Follow Me," and leads his squad to the AA followed by the A and B teams of the 2d squad. The A team leader of the 2d squad comes to a halt at the baseman position. Once the 3d squad has cleared the baseman position, the 2d squad leader gives the command, "Follow Me," and leads his squad to the AA. The 2d squad reassembles into their original column formation as they move to the AA. As the 1st squad comes to a halt (facing the crowd) in the AA, the 3d squad dresses with the 1st squad and leaves space for the 2d squad to return to its place in the formation. Once the platoon is formed into a column formation, HQ personnel take their position in the formation. The PSG then gives the command, "Countercolumn March," and moves the platoon to the rally point.

Platoon Line Formation With General Support

6-59. Whenever the command for establishing a crowd control formation contains the phrase "In Support" (without modification), it means that the support element is to remain in general support.

6-60. The command for forming a platoon line with the 2d squad in general support is "Platoon Line, 2d Squads in Support, Move." The 1st and 3d squads of the platoon line with general support execute the line, while the 2d squad remains in the column (see *Figure 6-11*). To assemble a platoon with a squad in general support, the procedure is the same as the platoon line, except that the 1st and 3d squads dress with the 2d squad.

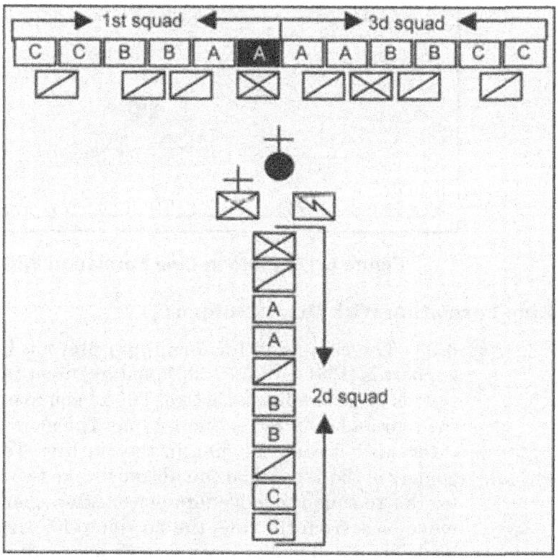

Figure 6-11. Platoon Line Formation With General Support

Platoon Line Formation With Lateral Support

6-61. The command for forming a platoon line with one squad in lateral support is "Platoon Line, 2d Squad in Lateral Support, Move." The 1st and 3d squads execute the line as before, and the 2d squad stands fast. After the line has been formed, the 2d squad leader gives the command, "Follow Me." The A and B teams move to the right flank while the squad leader and the C team move to the left flank of the formation. The 2d squad forms columns behind the last men on the line formation. If no direction is given, the 2d squad splits to cover both flanks. A squad may be used to cover one flank by giving the command, "2d Squad Right (or Left) Lateral Support" (see *Figure 6-12*).

6-62. The 2d squad may be moved from general to lateral support at any time by the platoon leader. The platoon leader commands, "2d Squad, Lateral Support, Move." To have the 2d squad join the line from either support position, the platoon leader commands, "2d Squad, Extend the Line, Move." The 2d squad leader commands, "Follow Me." On the command of execution, the squads set up individual lines, closing and dressing on the existing line. To assemble the platoon from a platoon line with lateral support, follow the same steps as for the platoon line.

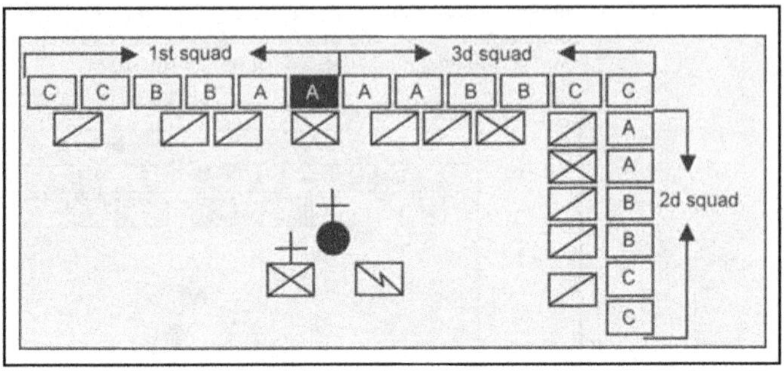

Figure 6-12. Platoon Line Formation With Lateral Support (Right)

Platoon Line Formation With Direct Support

6-63. The command for forming a platoon line with one squad in direct support is "Platoon Line, 2d Squad in Direct Support, Move." The 1st and 3d squads execute a line as before. The 2d squad executes a squad line directly to the rear and centered on the 1st line. The men in the supporting line cover the intervals between the men in the 1st line. To do this, the squad and team leaders of the 1st and 3d squads must take two steps to the rear to allow space for the 2d squad to move into place. After seeing the squad and team leaders move back from the line, the 2d squad leader commands, "Follow Me," and leads his squad into position (see *Figure 6-13*).

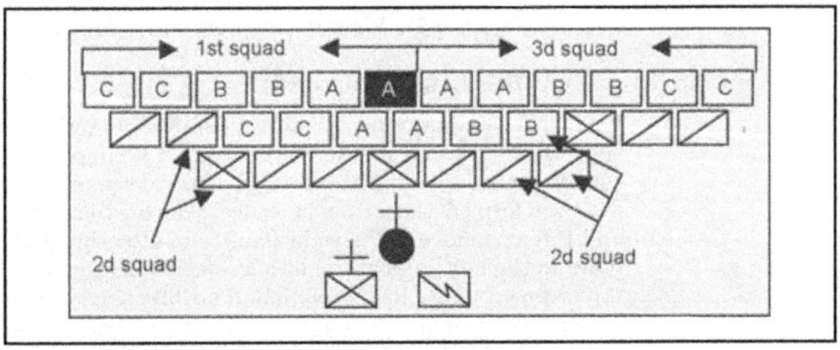

Figure 6-13. Platoon Line Formation With Direct Support

6-64. To assemble a platoon with direct support, the procedure is the same as the platoon line, except that the 2d squad moves to the AA first. To assemble the support squads from any position to general support, the platoon leader commands, "2d Squad in Support, Move." The 2d squad then returns to a column behind the line formed by the other two squads.

Platoon Echelon Right and Left Formations

6-65. The command for forming a platoon echelon right is "Platoon Echelon Right, Move." After the platoon leader's preparatory command, the squad leader of the 1st squad commands, "Follow Me." The squad leaders of the 2d and 3d squads command, "Stand Fast." On the command of execution, the 1st squad moves out and executes an echelon right at the location designated by the platoon leader. As the 1st squad clears the column, the 3d and 2d squads extend the echelon. The 2d squad performs support to this formation in the same ways it supports a line formation.

6-66. The command for forming a platoon echelon left is "Platoon Echelon Left, Move." The platoon echelon right is formed in the same manner as the echelon left (see *Figure 6-14*). *Figures 6-15; 6-16,* page 6-22; and *6-17,* page 6-23 show support formations in a left echelon.

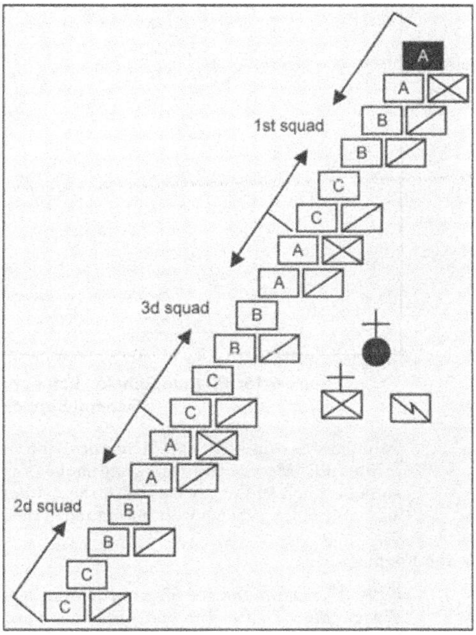

Figure 6-14. Platoon Echelon Left Formation

6-67. To assemble the platoon, the commander takes his position to the rear of the formation with the platoon HQ. On the preparatory command, "Platoon Assemble," the 1st squad leader does an about-face movement while all other

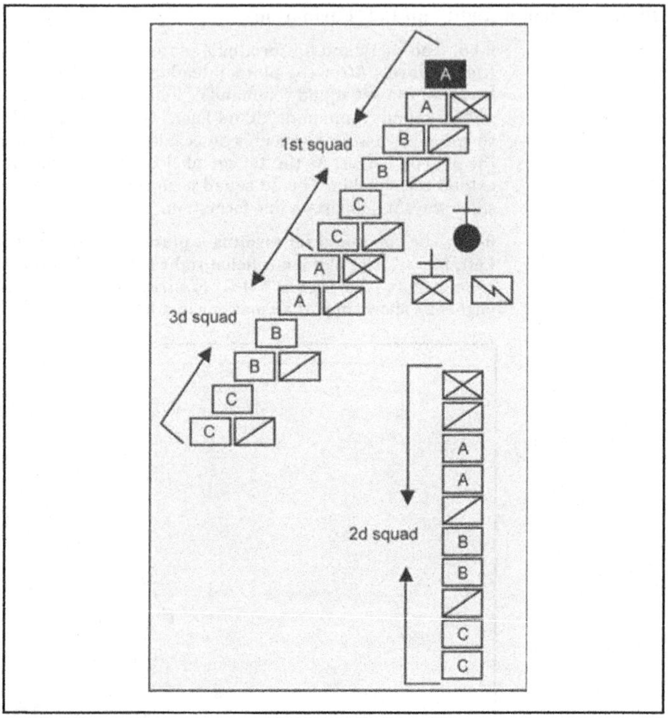

Figure 6-15. Platoon Echelon Left Formation With 2d Squad in General Support

members assume the port arms position and face the baseman. On the command, "Move," the 1st squad moves first, followed by the 3d and 2d squads. Each squad dresses with the 1st squad as it reaches the AA. The assembly of the echelon right formation is the same as the echelon left.

Platoon Wedge Formation

6-68. The command for forming the platoon wedge formation is "Platoon Wedge, Move." After the platoon leader's preparatory command, the squad leaders of the 1st and 3d squads command, "Follow Me." At the same time, the 2d squad leader commands, "Stand Fast." On the platoon leader's command of execution, the 1st and 4th squads move directly to the front. The 1st squad executes an echelon left on the baseman while the 3d squad executes an echelon right. When the last man of the 1st and 4th squads clear the 2d squad, the 2d squad leader commands, "Follow Me," and moves out to the left and right. The 2d squad splits and the A and B teams go to the right and the C

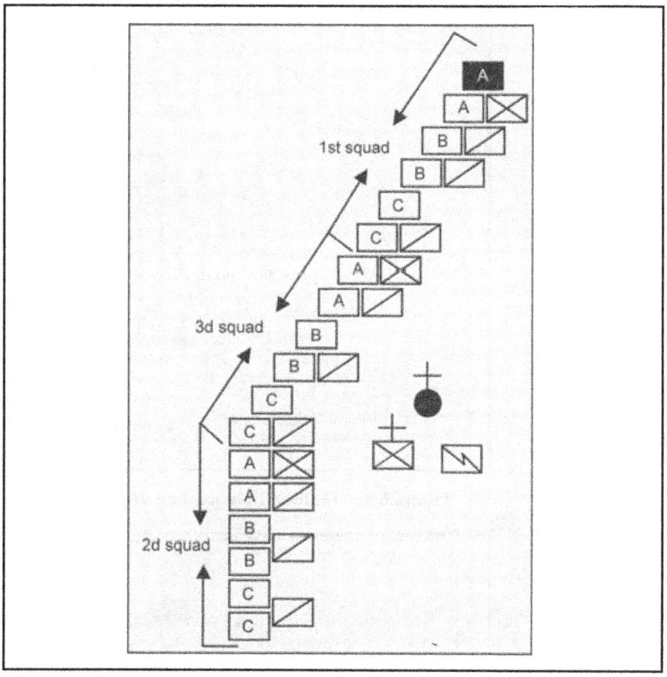

Figure 6-16. Platoon Echelon Left Formation With 2d Squad in Lateral Support

team goes to the left to extend the formation (see *Figure 6-18*). Assembly of the wedge formation is the same as the platoon line.

Platoon Wedge Formation With General Support

6-69. The command for forming a platoon wedge formation with support is "Platoon Wedge, 2d Squad in Support, Move." The 1st and 3d squads execute the wedge while the 2d squad remains in a column formation (see *Figure 6-19,* page 6-24).

Platoon Wedge Formation With Lateral Support

6-70. The command for forming a platoon wedge formation with lateral support is "Platoon Wedge, 2d Squad in Lateral Support, Move." The 1st and 3d squads execute the wedge while the 2d squad stands fast. After the 1st and 3d squads have cleared the column, the 2d squad leader commands, "Follow Me." If no direction is given, the 2d squad splits to cover both flanks. A squad

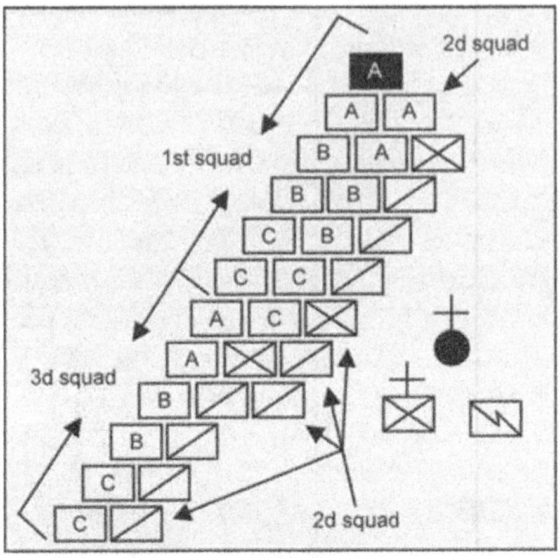

Figure 6-17. Platoon Echelon Left With 2d Squad in Direct Support

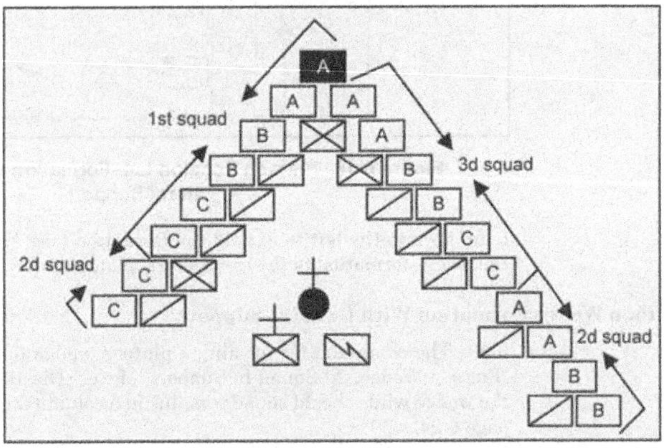

Figure 6-18. Platoon Wedge Formation

may be used to cover one flank by commanding, "2d Squad Right (or Left) Lateral Support" (see *Figures 6-20* and *6-21,* page 6-25).

6-71. To move the 2d squad from general support to lateral support, the platoon leader commands, "2d Squad, Lateral Support, Move." To have the 2d squad join the wedge from general or lateral support, the platoon leader

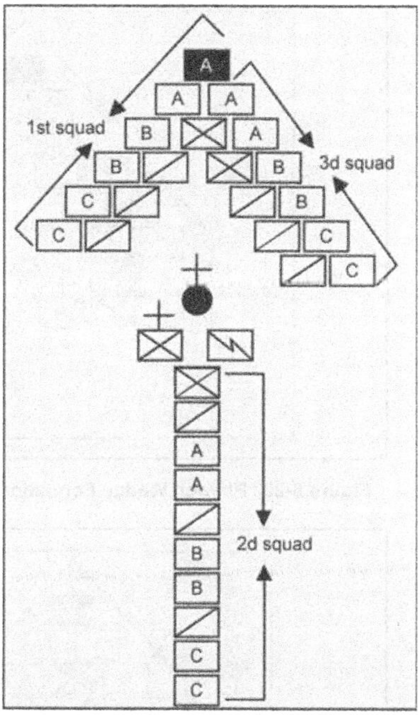

Figure 6-19. Platoon Wedge Formation With 2d Squad in General Support

commands, "2d Squad, Extend the Wedge, Move." To extend the wedge, the 2d squad splits and the teams move to the left and right, respectively.

Platoon Wedge Formation With Direct Support

6-72. The command for forming a platoon wedge formation with the 2d squad in direct support is "Platoon Wedge, 2d Squad in Direct Support, Move." The 1st and 3d squads execute a wedge. The 2d squad waits for the squad and/or team leaders to move to the rear, as in the line formation, before executing a wedge directly behind and centered on the main formation. The individuals in the supporting wedge cover the intervals between individuals in the leading wedge (see *Figure 6-22,* page 6-26).

Platoon Diamond and Circle Formations

6-73. These formations are used in open areas or where the crowd may be able to envelop the formation. The diamond and circle formations are primarily defensive in nature and difficult to maneuver. The decision about which formation to use is based on the ability of the platoon to perform them.

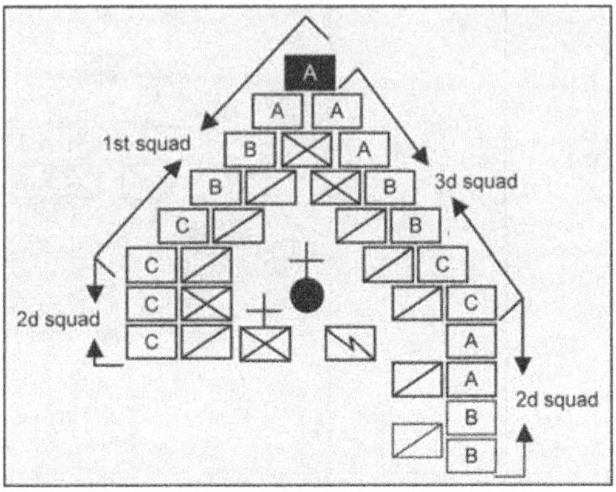

Figure 6-20. Platoon Wedge Formation With 2d Squad in Lateral Support

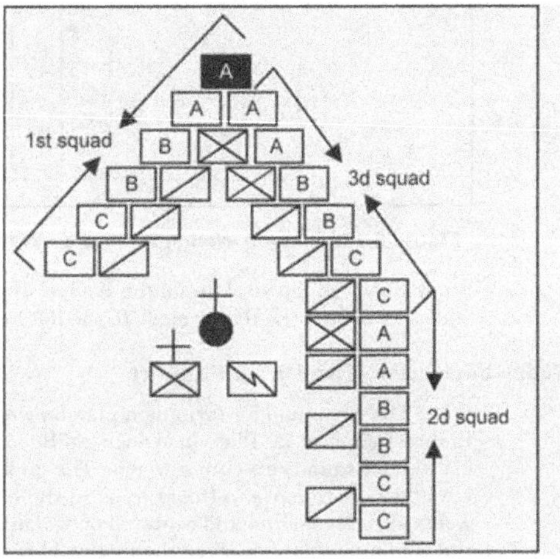

Figure 6-21. Platoon Wedge Formation With 2d Squad in Right Lateral Support

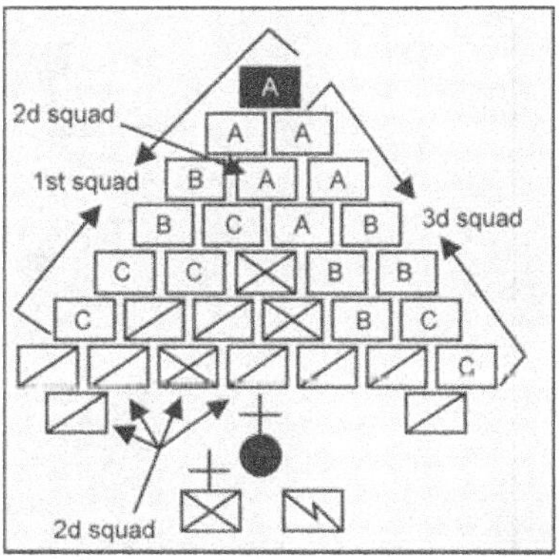

Figure 6-22. Platoon Wedge Formation With 2d Squad in Direct Support

The command for having a platoon form in a diamond or circle formation is "Platoon Diamond (or Circle), Move." On the command of execution, the 1st squad leader moves the baseman to the place indicated by the squad leader. The baseman becomes the 12 o'clock position while the 1st squad lines up to the left of the baseman covering the 8 to 12 o'clock positions. The 3d squad aligns itself to the right of the baseman, covering the 12 to 4 o'clock positions and the 2d squad aligns itself, covering the 4 to 8 o'clock positions (see *Figures 6-23* and *6-24*).

6-74. To assemble from a platoon diamond or circle formation, the HQ element takes the same steps as that of a line formation. The 1st squad leads the platoon back to the AA followed by the 2d and 3d squads. Squads fall into a column formation as they file back to the AA.

RELIEF ELEMENTS IN PLACE

6-75. During crowd control operations, the platoon leader may rotate his squads to give them a rest. He is not limited to using only those squads specified in the preceding examples as the lead elements of the formations. He can replace the base elements by commanding direct support to the 1st squad and then directing the 1st squad into general support. The support element (the 2d squad) then passes through the 1st squad. Once the 2d squad is formed directly behind the 1st squad, the 1st squad leader commands, "Right Face." The members of the 1st squad do a right-face movement and the 2d squad steps quickly though the 1st squad and forms the base formation. The 1st squad leader then commands, "Follow Me," and leads his squad to the general support position. This procedure is often necessary when protective masks are used and chemical irritants employed. The 1st squad then replaces the 3d squad in the same manner. Platoons participating in company operations are relieved in the same manner.

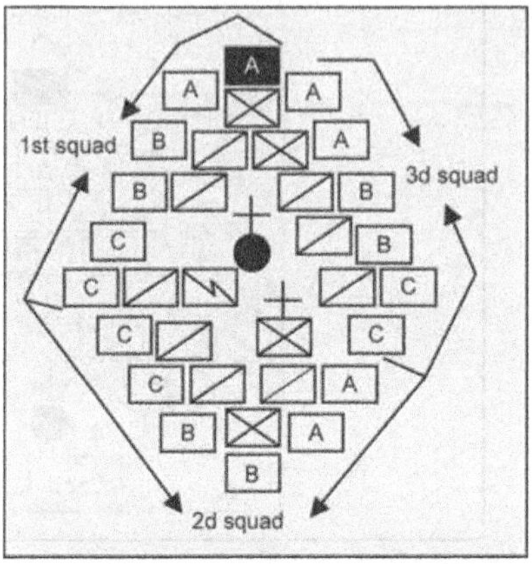

Figure 6-23. Platoon Diamond Formation

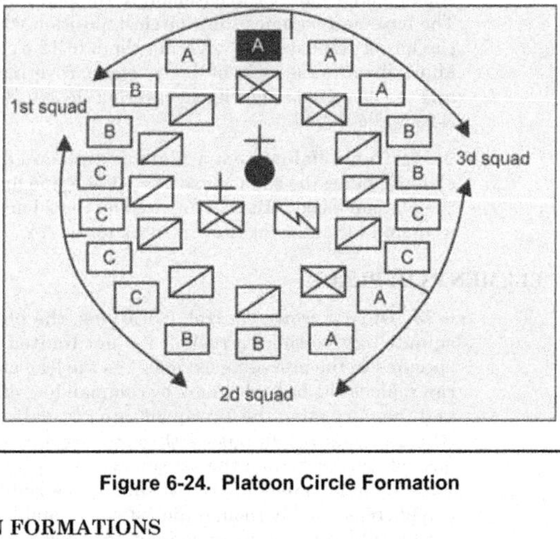

Figure 6-24. Platoon Circle Formation

FOUR-SQUAD PLATOON FORMATIONS

6-76. When forming control force formations from the column, the platoon leader moves to the right of the platoon and faces it when giving his commands. He gives the hand-and-arm signal and the preparatory command. On the command of execution, "Move," he points to the approximate location where he wants the platoon to form. The platoon leader pauses between the preparatory command and the command of execution to allow each squad leader to issue a preparatory command to his squad.

Platoon Line Formation

6-77. The command for forming a platoon line formation is "Platoon Line, Move." Immediately following the platoon leader's preparatory command, the squad leaders of the 1st and 4th squads command, "Follow Me." The squad leaders of the 2d and 3d squads command, "Stand Fast." On the platoon leader's command of execution, the 1st and 4th squads move forward to the point designated by the platoon leader. The 1st A team member of the 1st squad is the baseman for the platoon formation. The 1st squad forms a squad line to the left of the baseman and the 4th squad forms a squad line to the right of the baseman with all lines dressed to the right. After the 1st and 4th squads have cleared the platoon, the 2d and 3d squad leaders command, "Follow Me." The 2d squad forms a squad line to the left of the 1st squad. The 3d squad forms a squad line to the right of the 4th squad. All squads dress with the 4th squad (see *Figure 6-25*).

Platoon Line Formation Assembly

6-78. When assembling from the platoon line formation, the platoon leader and his HQ personnel take a position to the rear of the platoon. He then gives the command, "Platoon Assemble," while raising his right arm and making a circular motion above his head. Platoon members automatically assume the port arms position. The 1st squad leader and the baseman do an about-face movement. All other platoon members do an about-face movement toward the baseman. On the platoon leader's command of execution, he points to where he wants the platoon to assemble. He commands, "Follow Me," and moves his squad toward the AA. The 1st squad files to the baseman position, followed by the 2d squad. As the 2d squad clears the baseman position, the 4th squad leader then commands, "Follow Me," and leads his squad to the AA, followed by the 3d squad. As the 1st squad comes to a halt facing the crowd in the AA, the 2d squad assumes its position to the right of the 1st squad. As the 4th squad assembles, it dresses on the 1st squad, leaving space for the 3d squad to return to its place in the formation. Once the platoon is formed into a column formation, HQ personnel take their position in the formation. The PSG then commands, "Countercolumn, March," and moves the platoon to the rally point.

Platoon Line Formation With General Support

6-79. Whenever the command for establishing a crowd control formation contains the phrase "In Support Without Modification," it means that the support element is to remain in general support. The command for forming a platoon line with two squads in general support is "Platoon Line, 2d and 3d Squads in Support, Move." The 1st and 4th squads execute the line, while the 2d and 3d squads remain in the column (see *Figure 6-26,* page 6-30). To assemble a platoon with two squads in general support, the procedure is the same as that of the platoon line, except that the 1st and 4th squads dress on with the 2d and 3d squads.

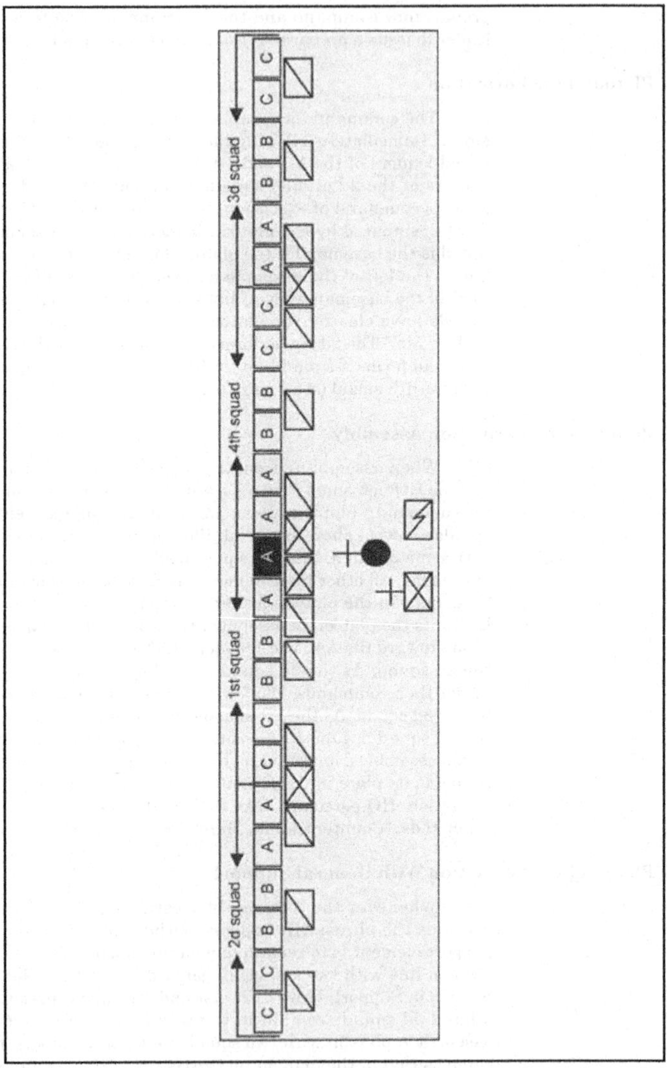

Figure 6-25. Platoon Line Formation With Four Squads

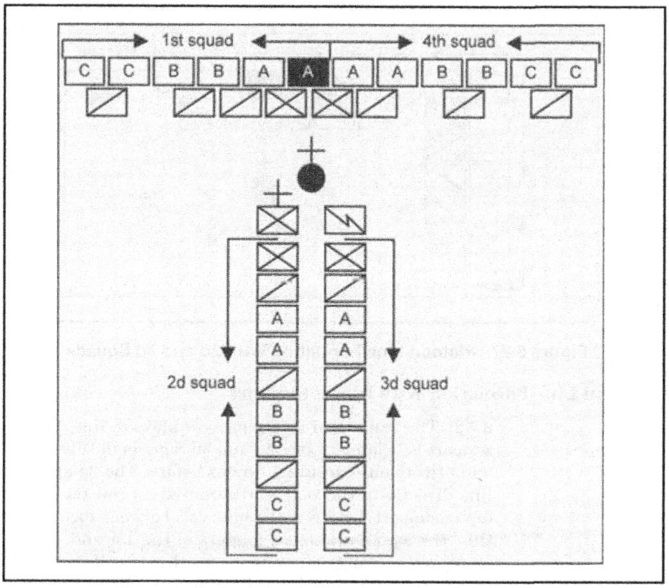

Figure 6-26. Platoon Line Formation With d and 3d Squads in General Support

Platoon Line Formation With Lateral Support

6-80. The command for forming a platoon line formation with two squads in lateral support is "Platoon Line, 2d and 3d Squads in Lateral Support, Move." The 4th and 1st squads execute the line as before, and the 2d and 3d squads stand fast. After the line has been formed, the squad leaders of the 2d and 3d squads command, "Follow Me." The 2d squad forms a column behind the last man of the 1st squad. The 3d squad forms a column behind the last man of the 4th squad (see *Figure 6-27*).

6-81. Squads may be moved from general support to lateral support at any time by the platoon leader. He commands, "2d and 3d Squads, Lateral Support, Move." To have the 2d and 3d squads join the line from either support position, the platoon leader commands, "2d and 3d Squads, Extend the Line, Move." The 2d and 3d squad leaders command, "Follow Me." On the command of execution, the squads set up individual lines, closing and dressing with the existing line. To assemble the platoon from a platoon line with lateral support, the procedure is the same as that of the platoon line, except lateral support squads face forward until they move to the AA.

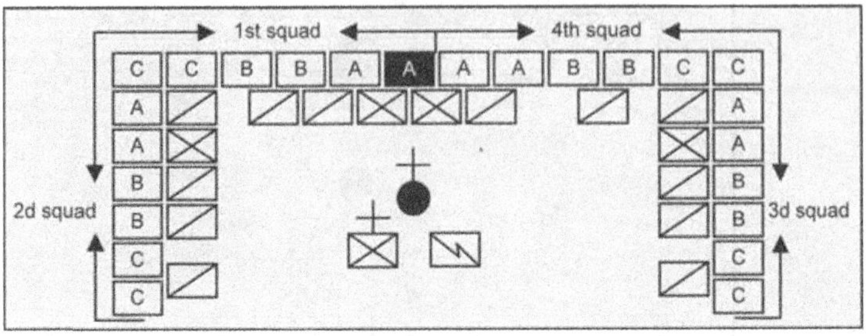

Figure 6-27. Platoon Line Formation With 2d and 3d Squads in Lateral Support

Platoon Line Formation With Direct Support

6-82. The command for forming a platoon line with two squads in direct support is "Platoon Line, 2d and 3d Squads in Direct Support, Move." The 1st and 4th squads execute a line as before. The 2d and 3d squads execute a 2d line directly to the rear and centered on the 1st line. The members in the direct support line cover the intervals between members in the baseline. To do this, the squad and team leaders of the 1st and 4th squads must take two steps to the rear to allow space for the 2d and 3d squads to move into place. After seeing the squad and team leaders move back from the line, 2d and 3d squad leaders command, "Follow Me," and lead their squads into position (see *Figure 6-28*).

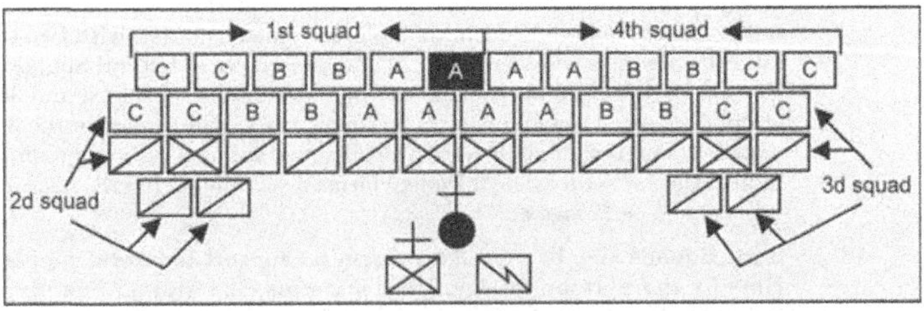

Figure 6-28. Platoon Line Formation With 2d and 3d Squads in Direct Support

6-83. To assemble the support squads from any position to general support, the platoon leader commands, "2d and 3d Squads in Support, Move." The 2d and 3d squads then return to a column behind the line formed by the 1st and 4th squads. To assemble the entire platoon, the procedure is the same as that of a platoon line, except that the 2d squad moves to the AA first, followed by the 3d, 1st, and 4th squads.

Platoon Line Formation With One Support Squad

6-84. The command for forming a platoon line formation with one squad in support is "Platoon Line, 3d Squad in Support, Move." The 1st squad moves out and executes a squad line. The 2d and 4th squads form a line on the side of the left and rights sides of the 1st squad. The 3d squad remains in column formation. The support squad may be used in lateral support on one or both sides of the formation or in direct support of any segment of the formation. To assemble the platoon, the procedure is the same as that of a platoon line, except that all squads dress with the 3d squad.

Platoon Echelon Right and Left Formations

6-85. The command for forming a platoon echelon right formation is "Platoon Echelon Right, Move." After the platoon leader's preparatory command, the squad leader of the 1st squad commands, "Follow Me." The squad leaders of the 2d, 3d, and 4th squads command, "Stand Fast." On the command of execution, the 1st squad moves out and executes an echelon right at the location designated by the platoon leader. As each squad clears the column, the next successive squad moves out and extends the echelon that has been formed. To assemble the platoon, the procedure is the same as that of a platoon line, except that all squads file back in order.

6-86. The command for forming a platoon echelon left formation is "Platoon Echelon Left, Move." The platoon echelon left is formed in the same manner as the echelon right, but in reverse order (see *Figure 6-29*). To assemble the platoon, the procedure is the same as that of an echelon right. See *Figures 6-30,* page 6-34*;* *6-31,* page 6-35*;* and *6-32,* page 6-36, for examples of echelon formations with support.

Platoon Wedge Formation

6-87. The command for forming the platoon wedge formation is "Platoon Wedge, Move." After the platoon leader's preparatory command, the 1st and 4th squad leaders command, "Follow Me." At the same time, the 2d and 3d squad leaders command, "Stand Fast." On the platoon leader's command of execution, the 1st and 4th squads move directly to the front. When the last members of the 1st and 4th squads clear the 2d and 3d squads, the 2d and 3d squad leaders command, "Follow Me," and move out to the left and right. The 1st squad executes an echelon left on the baseman and the 4th squad executes an echelon right. The 2d squad forms an echelon left on the last man of the 1st squad. The 3d squad forms an echelon right on the last man of the 4th squad. This completes the wedge formation (see *Figure 6-33,* page 6-36). To assemble the platoon, the procedure is the same as that of a platoon line.

Platoon Wedge Formation With General Support

6-88. The command for forming a platoon wedge formation with two squads in support is "Platoon Wedge, 2d and 3d Squads in Support, Move." The 1st and 4th squads execute the wedge while the 2d and 3d squads remain in column formation (see *Figure 6-34,* page 6-37).

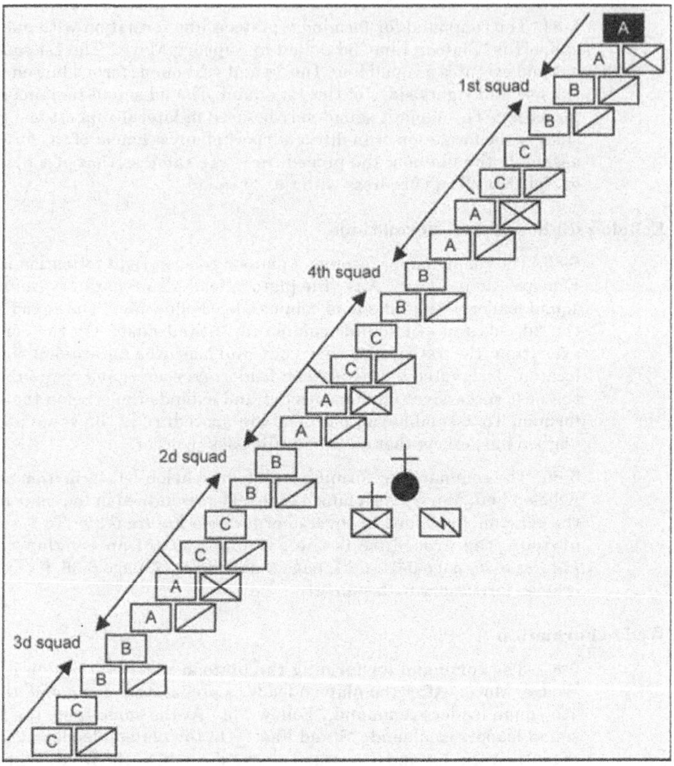

Figure 6-29. Platoon Echelon Left Formation With Four Squads

Platoon Wedge Formation With Lateral Support

6-89. The command for forming a platoon wedge formation with lateral support is "Platoon Wedge, 2d and 3d Squads in Lateral Support, Move." The 1st and 4th squads execute the wedge while the 2d and 3d squads stand fast. After the 1st and 4th squads have cleared the column, the 2d and 3d squad leaders command, "Follow Me," and move out to the left and right. The 2d squad forms a column formation behind the last man of the 1st squad. The 3d squad forms a column formation behind the last man of the 4th squad (see *Figure 6-35,* page 6-38).

6-90. To move the 2d and 3d squads from general to lateral support, the platoon leader commands, "2d and 3d Squads, Lateral Support, Move." To have the 2d and 3d squads join the wedge from general or lateral support, the

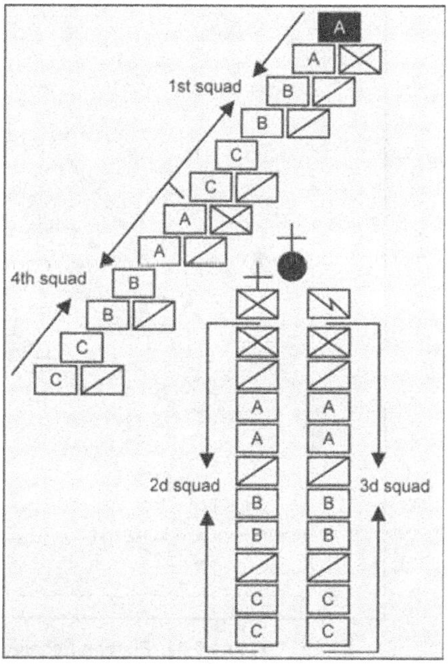

Figure 6-30. Platoon Echelon Left Formation With 2d and 3d Squads in General Support

platoon leader commands, "2d and 3d Squads, Extend the Wedge, Move." The 2d and 3d squad leaders command, "Follow Me," and move out to the left and right to extend the wedge on the 1st and 4th squads.

Platoon Wedge Formation With Direct Support

6-91. The command for forming a platoon wedge with two squads in direct support is "Platoon Wedge, 2d and 3d Squads in Direct Support, Move." The 1st and 4th squads execute a wedge. The 2d and 3d squads execute a wedge and close in on the leading wedge. The members in the supporting wedge cover the intervals between members in the leading wedge. To do this, the squad and team leaders of the 1st and 4th squads must take two steps to the rear to allow space for the 2d and 3d squads to move into place. After seeing the squad and team leaders move back from the line, the 2d and 3d squad leaders command, "Follow Me," and lead their squads into position (See *Figure 6-36,* page 6-38). To move the 2d and 3d squads back to general support, the platoon leader commands, "2d and 3d Squads Support, Move." Then the 2d and 3d squads move to their general support positions. To

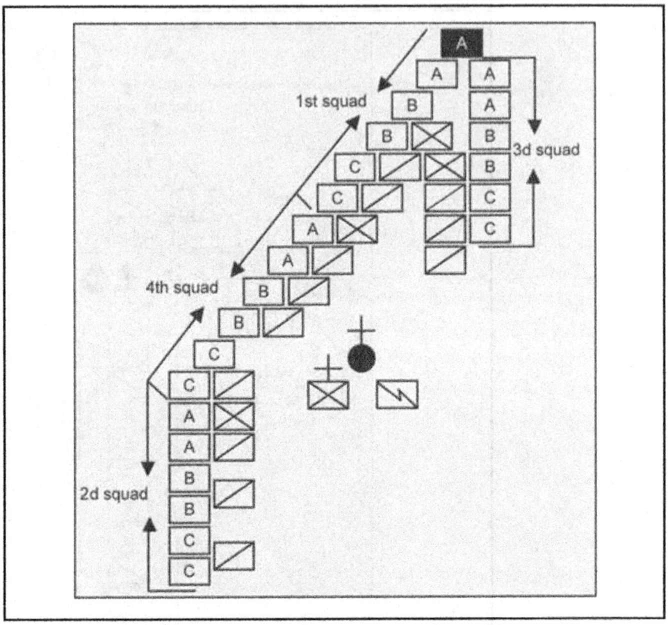

Figure 6-31. Platoon Echelon Left Formation With 2d Squad in Lateral Support

assemble a platoon from a platoon wedge with direct support, the procedure is the same as that of a platoon wedge, except that the 2d squad assembles first, followed by the 3d, 1st, and 4th squads.

Platoon Wedge Formation With One Support Squad

6-92. The command for forming a platoon wedge formation with one squad in support is "Platoon Wedge, 3d Squad in Support, Move." The 1st squad moves out and executes a squad wedge. The 2d and 4th squads form echelons left and right on the 1st squad. The 3d squad remains in the column formation.

Platoon Diamond and Circle Formations

6-93. The platoon diamond and circle formations are used in open areas or where the crowd may be able to envelop the formation. They are primarily defensive in nature and hard to maneuver. The decision concerning which formation to use is based on the ability of the platoon to perform them. The command for having a platoon form in a diamond or circle formation is "Platoon Diamond (or Circle), Move." On the command of execution, the 1st squad leader moves the baseman to the location indicated by the squad leader. The baseman becomes the 12 o'clock position. The 1st squad lines up to the

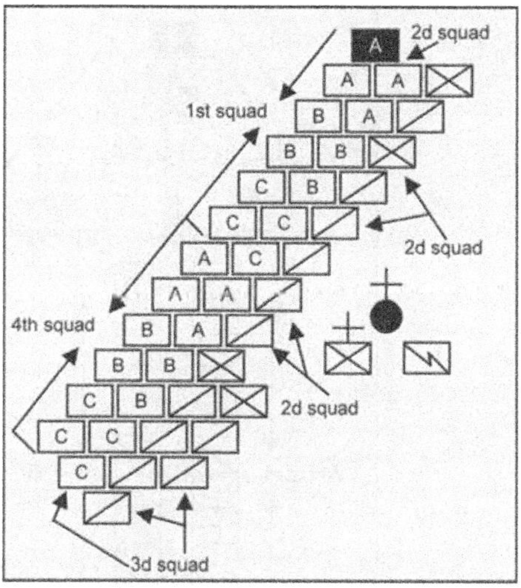

Figure 6-32. Platoon Echelon Left Formation With 2d and 3d Squads in Direct Support

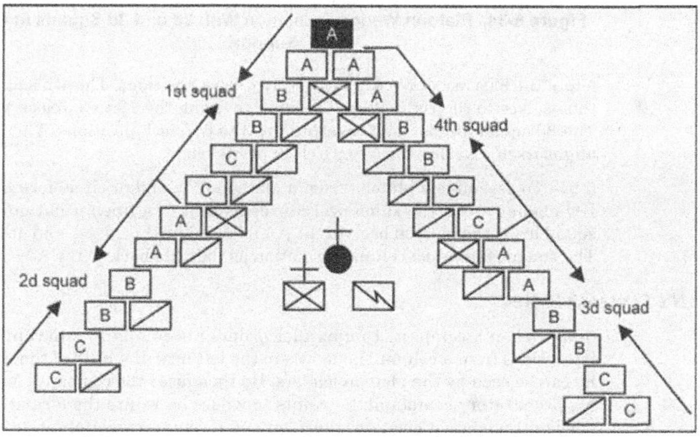

Figure 6-33. Platoon Wedge Formation With Four Squads

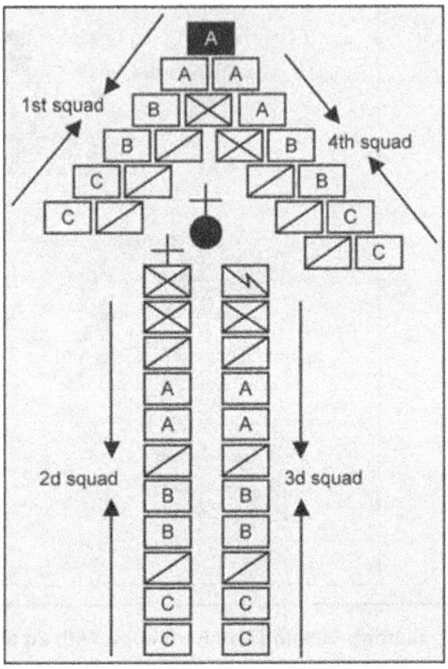

Figure 6-34. Platoon Wedge Formation With 2d and 3d Squads in General Support

left of the baseman covering the 9 to 12 o'clock positions. The 4th squad aligns themselves to the right of the baseman covering the 12 to 3 o'clock positions. The 3d squad aligns itself covering the 3 to 6 o'clock positions. The 2d squad aligns itself covering the 6 to 9 o'clock positions.

6-94. To assemble a platoon from a platoon diamond or circle formation, the HQ element uses the same procedures as that of a line formation. The 1st squad leads the platoon back to the AA followed by the 2d, 3d, and 4th squads. The squads fall into a column formation as they file back to the AA.

COMPANY FORMATIONS

6-95. When a company commander orders his company into control force formations from a column, he moves to the left near the head of the column so he can be seen by the platoon leaders. He then faces the company. As he gives his preparatory command, he points to where he wants the formation. If he does not indicate a location, the company forms immediately in front of the lead platoon.

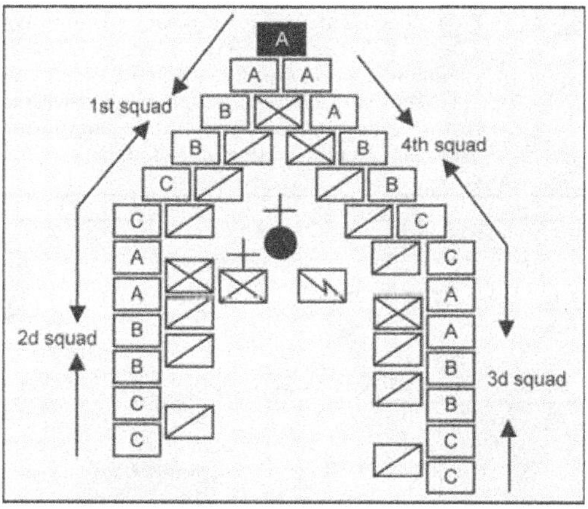

Figure 6-35. Platoon Wedge Formation With 2d and 3d Squads in Lateral Support

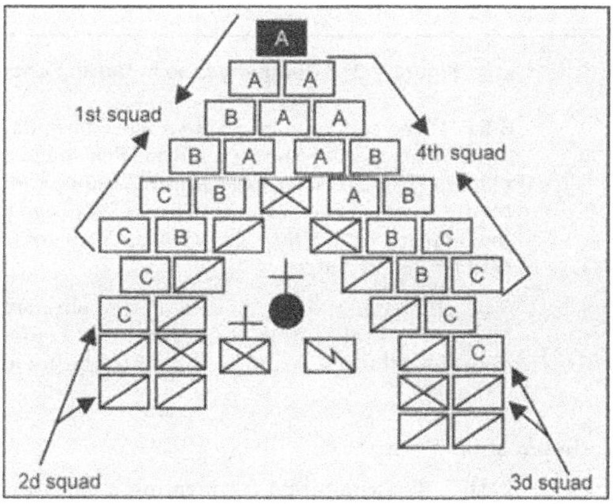

Figure 6-36. Platoon Wedge Formation With 2d and 3d Squads in Direct Support

Relieving Elements in Place

6-96. During crowd control operations, the platoon leader may rotate his squads to give them a rest. See *paragraph 6-76,* page 6-27.

Company Line in Depth Formation

6-97. The command for forming a company line in depth formation (see *Figure 6-37*) is "Company Line in Depth, Move." After the company commander's preparatory command, each platoon leader gives the command, "Platoon Line." The squad leaders follow with their commands to their squads to form the platoon line. On the command of execution, each platoon establishes a platoon line immediately to its front.

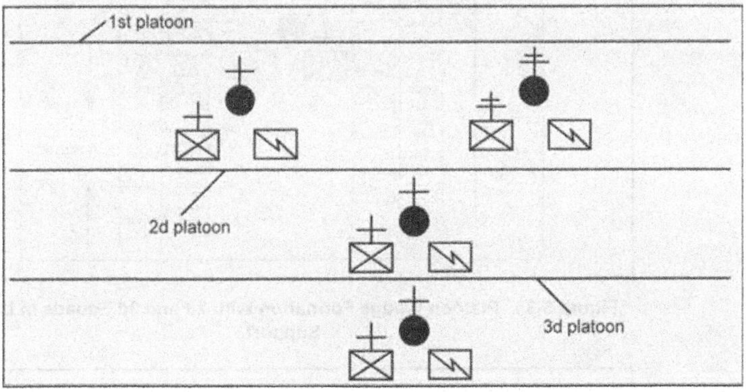

Figure 6-37. Company Line in Depth Formation

6-98. If the commander wants a more formidable formation, he commands, "2d Platoon, Direct Support, Move." The squad and team leaders from the 1st platoon take two steps backward. The 2d platoon moves forward and covers the intervals between the members of the lead platoon. The platoon leader of the 3d platoon then moves his platoon forward to occupy the position formerly held by the 2d platoon.

6-99. If the 3d platoon is called on to support the company line, it moves forward and takes up a position directly behind the 2d platoon in the same manner. Platoon leaders and PSGs help each other in controlling the company.

Company Line in Mass Formation

6-100. The command for forming a company line in mass formation is "Company Line in Mass, Move" (see *Figure 6-38*, page 6-40). The company line in mass is the same as a company line with both platoons in direct support. Instead of forming it in stages, the company commander indicates his desire for this formation in his initial command. The platoons each form a line, and the 2d and 3d platoons close on the 1st platoon without further command.

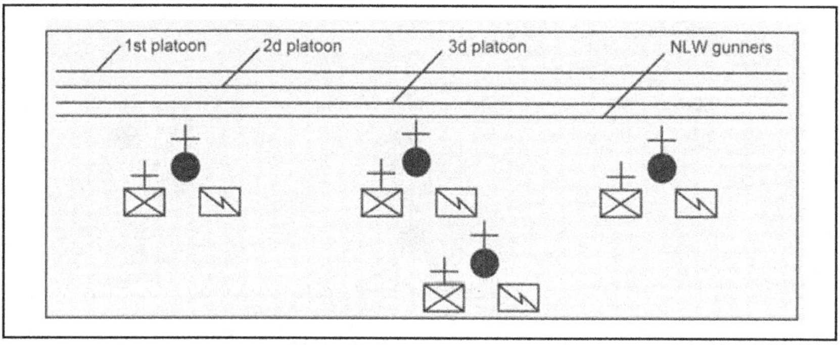

Figure 6-38. Company Line in Mass Formation

Company Line Formation With Support

6-101. In company formations, the 1st platoon usually forms the lead element and the 2d and 3d platoons provide support (see *Figures 6-39; 6-40;* and *6-41*). The support platoons can be employed in the same manner as the support squads in platoon formations. One platoon can be relieved from the lead element by another platoon while in formation by having a support platoon pass through the lead platoon.

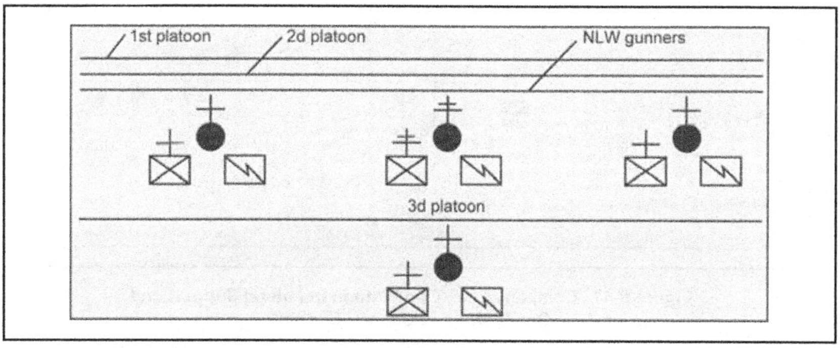

Figure 6-39. Company Line in Mass Formation With One Platoon in Depth

Company Echelons and Wedge Formations

6-102. Company echelons (see *Figures 6-42,* page 6-42*; 6-43,* page 6-43*; 6-44,* page 6-43*;* and *6-45,* page 6-44) and company wedge formations (see *Figures 6-46,* page 6-45; *6-47,* page 6-46; *6-48,* page 6-46; and *6-49,* page 6-47) are formed in the same manner and with the same variations as the company line formation. A company assembles from a crowd control formation in the same manner as a platoon or a squad does. However, due to the size of company

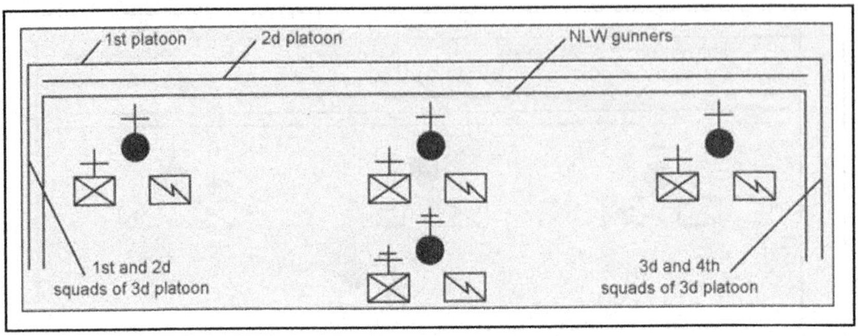

Figure 6-40. Company Line in Mass Formation With One Platoon in Lateral Support

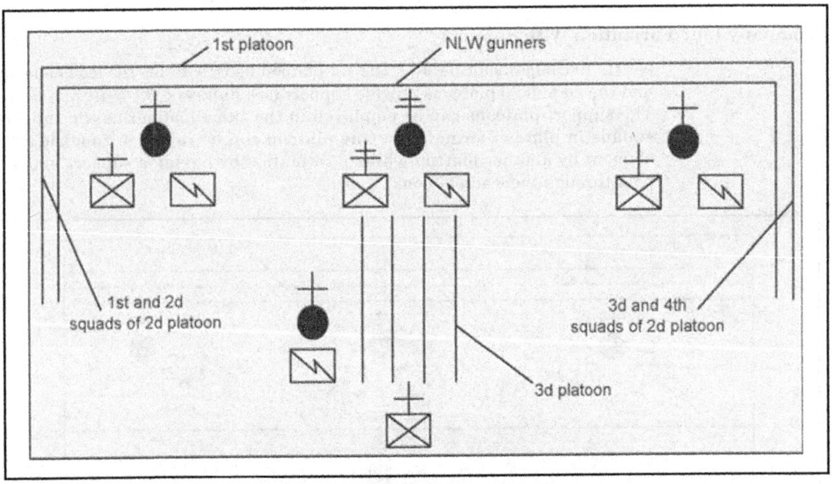

Figure 6-41. Company Line, One Platoon in Lateral Support and One Platoon in General Support

formations, the commander must consider the area that the company is operating in when he assembles the company. If he follows the guidelines of platoon assembly, he will have no difficulty. When assembling the company, the commander takes a position to the rear of the formation and commands, "Company Assemble." The platoon leaders take charge of their respective platoons. Usually, the 3d platoon assembles first, just behind the formation. After the 3d platoon is assembled, the 2d platoon assembles in front of the 3d platoon and the 1st platoon assembles in front of the 2d. All platoons assemble facing the commander. The 2d and 3d platoons then dress with the 1st platoon. As each platoon is assembled, they are ordered to port arms by the platoon leader to await further orders from the commander. If four platoons are used, the 4th platoon assembles first.

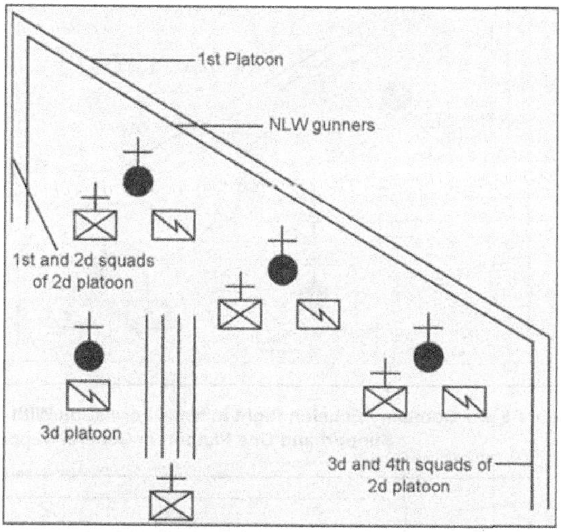

Figure 6-42. Company Echelon Right Formation With One Platoon in Lateral Support and One Platoon in General Support

Company and Platoon Formations With Vehicles

6-103. The commands for vehicles and foot troops are the same as those for foot troops alone. Because of the additional noise and distances involved, the company commander gives the hand-and-arm signal for the line. The command and the signal apply to the lead platoon only. The other platoons remain in general support until further directed (see *Figures 6-50,* page 6-48; *6-51,* page 6-49; *6-52,* page 50; *6-53,* page 6-50; *6-54,* page 6-51; *6-55,* page 6-52; and *6-56,* page 6-53).

6-104. The motor section moves out with vehicle number two moving to the right. Vehicle number two passes the lead vehicle and establishes the position for the center of the line. At the same time, the 3d and 4th vehicles swing out to the left and right. They form to the left and right and slightly to the rear of vehicle number two. The 1st and 4th squads of the lead platoon then move forward and form lines to the left and right, on the front end of vehicle number two. As soon as the troops are in position, the 3d and 4th vehicles close in on the squads and tighten the formation. The 2d and 3d squads of the lead platoon then form a line to the left and right of the 3d and 4th vehicles to complete the company line. The command vehicle (vehicle number one) takes

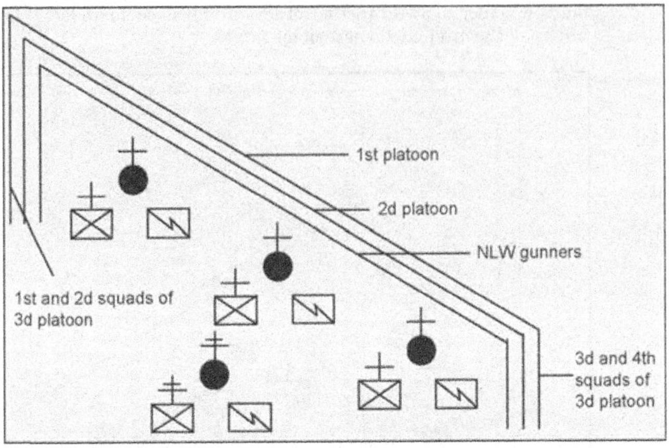

Figure 6-43. Company Echelon Right in Mass Formation With One Platoon in Lateral Support and One Platoon in General Support

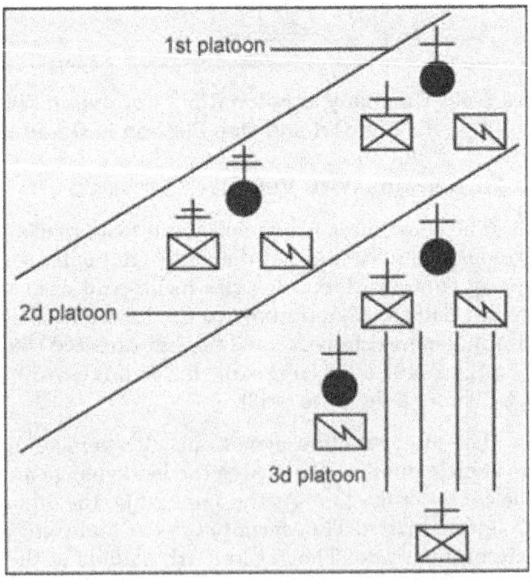

Figure 6-44. Company Echelon Right in Mass Formation With One Platoon in Lateral Support

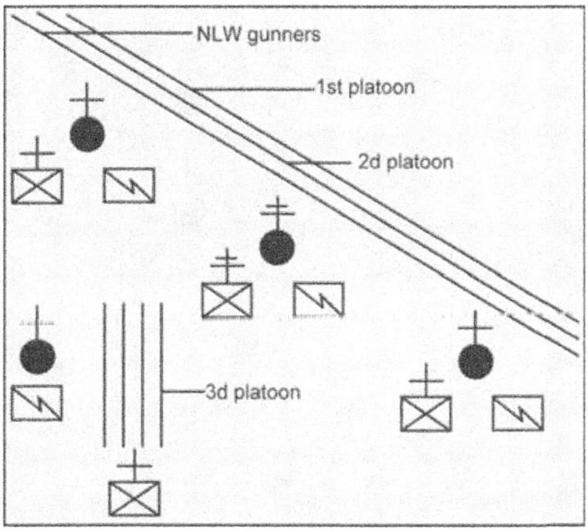

Figure 6-45. Company Echelon Right in Mass Formation With One Platoon in General Support

up a position behind the line where the commander can direct and control the unit. The 2d and 3d platoons move forward and are used in general, lateral, or direct support. In some instances, foot troops move into position first and vehicles join the formation, as needed.

6-105. To form a company echelon or a company wedge with vehicles, the company follows the same procedures used for company formations without vehicles. Vehicles are moved into the appropriate positions using procedures similar to those used for a company line.

6-106. To assemble troops and vehicles, the command vehicle moves to a position behind the formation. The commander faces the formation and gives the vehicle section the signal for assembly. Immediately, the vehicles return to their proper positions in column formation behind the command vehicle while the troops stand fast. The second in command then assembles the troops in the usual manner.

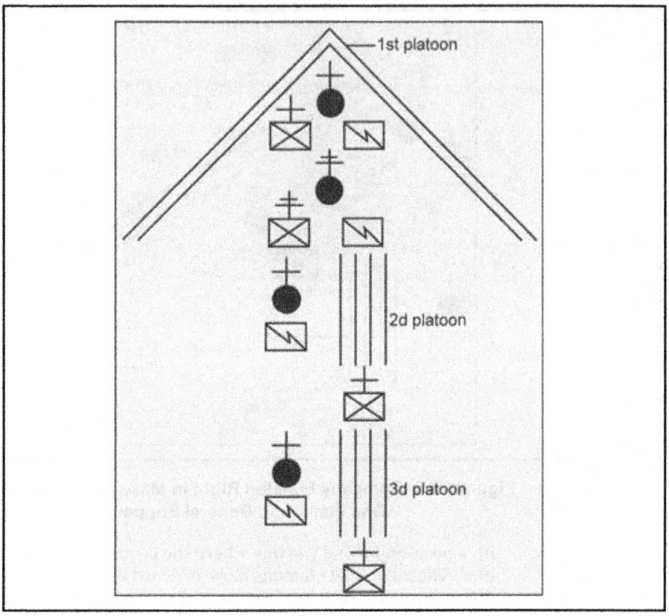

Figure 6-46. Company Wedge Formation With Two Platoons in General Support

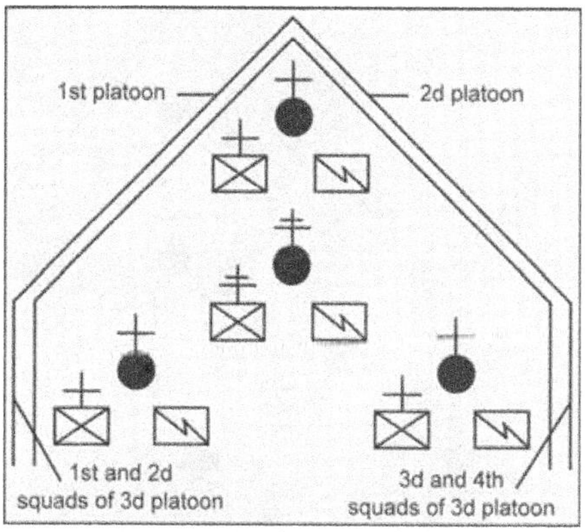

Figure 6-47. Company Wedge Formation With One Platoon in Lateral Support

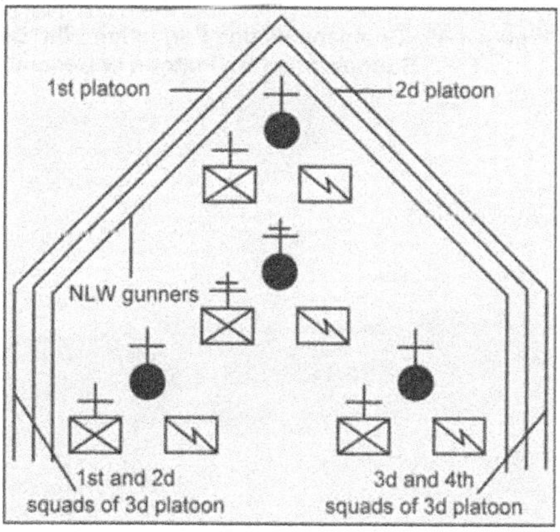

Figure 6-48. Company Wedge Formation With One Platoon in Lateral Support and One Platoon in General Support

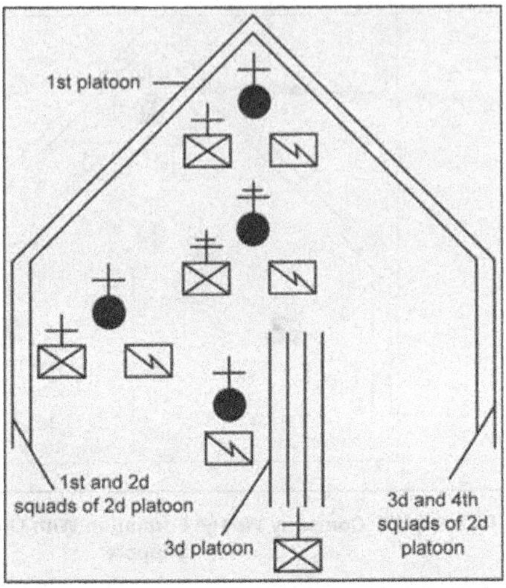

Figure 6-49. Company Wedge Formation With One Platoon in Lateral Support and One Platoon in General Support

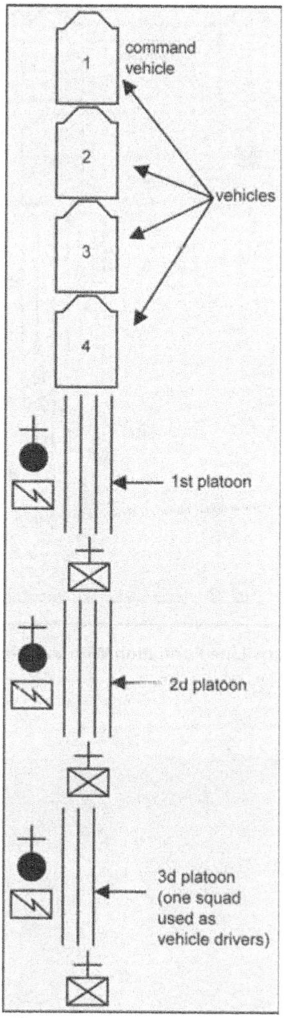

Figure 6-50. Company Column Formation With Vehicles

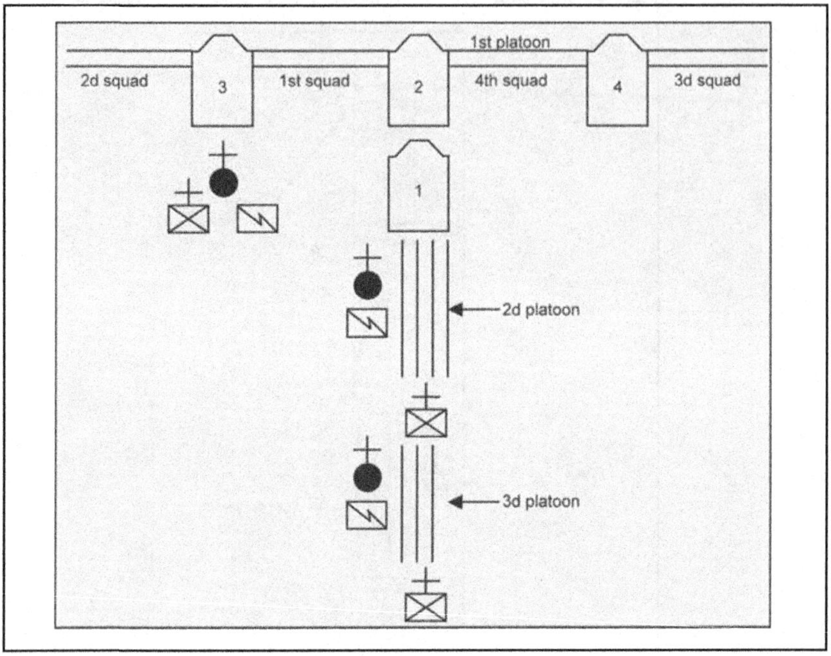

Figure 6-51. Company Line Formation With Vehicles and Two Platoons in General Support

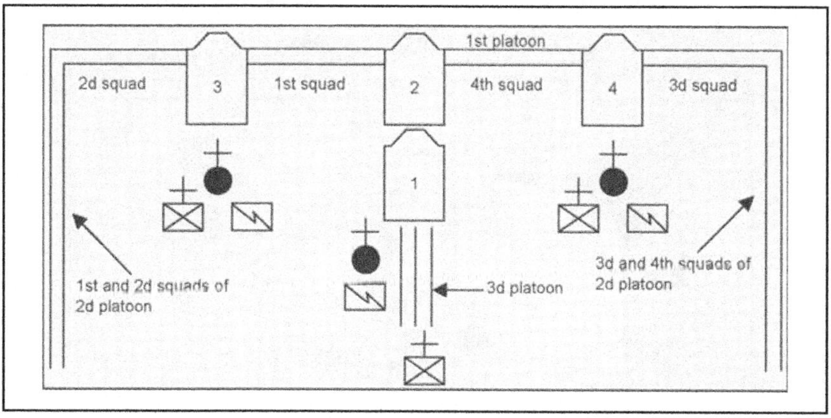

Figure 6-52. Company Line Formation With Vehicles, One Platoon in Lateral Support, and One Platoon in General Support

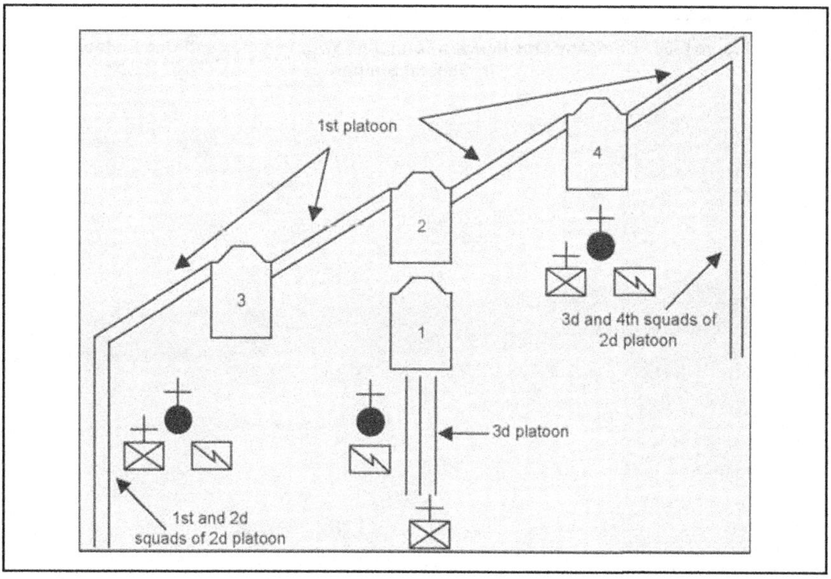

Figure 6-53. Company Echelon Right Formation With Vehicles, One Platoon in Lateral Support, and One Platoon in General Support

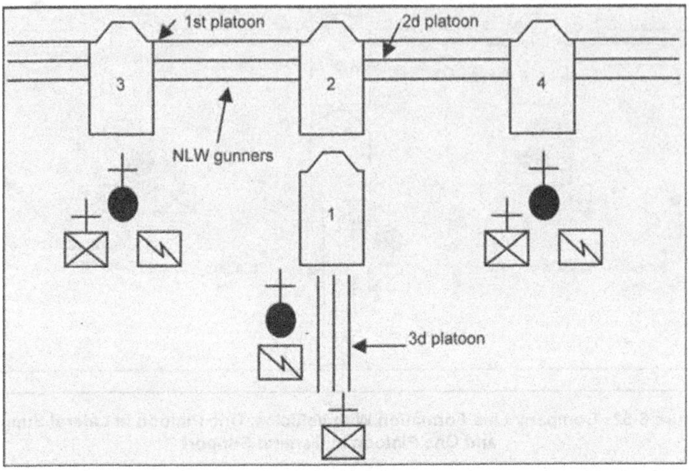

Figure 6-54. Company Line in Mass Formation With Vehicles and One Platoon in General Support

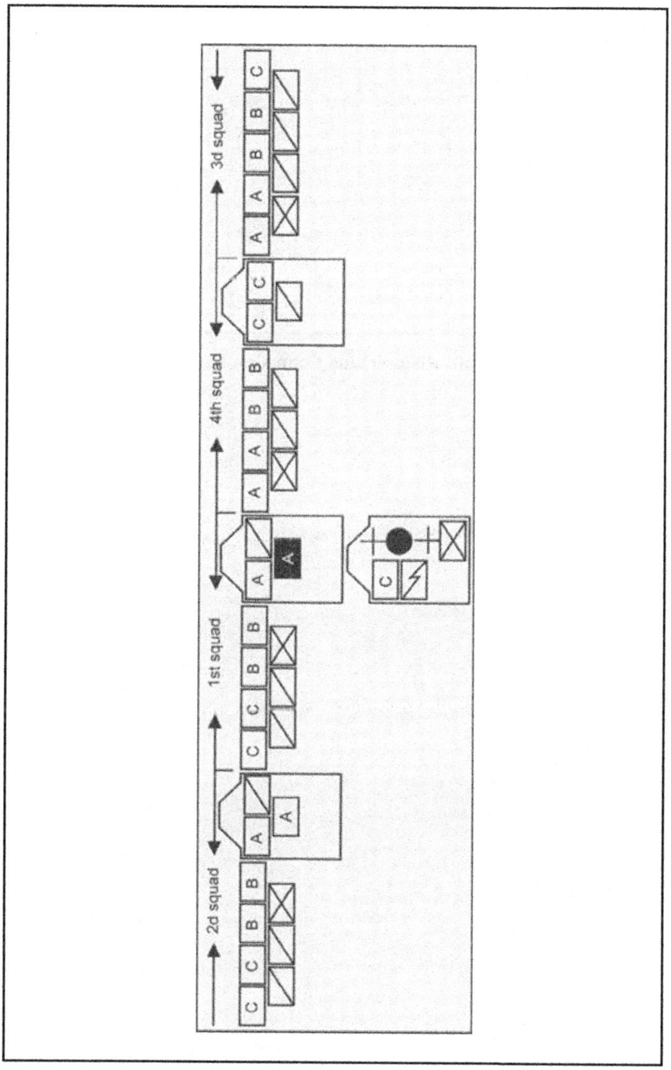

Figure 6-55. Platoon Line Formation With Vehicles

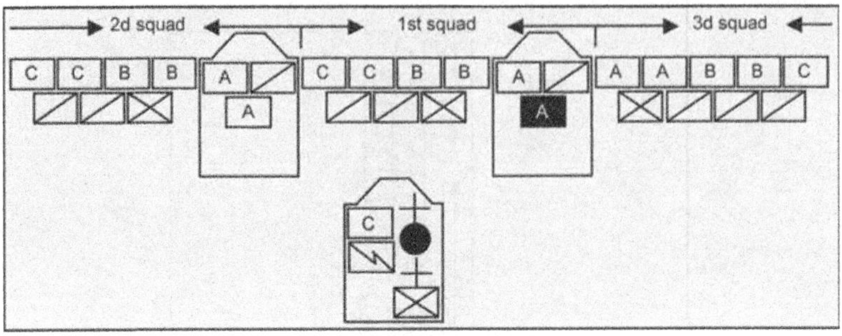

Figure 6-56. Platoon Line Formation With Direct Support and Vehicles

Chapter 7

Civil Disturbance Operations in Confinement Facilities

Civil disturbance doctrine is usually written for quelling disturbances that occur in large open areas where the size of the responding force or the type of formation employed is not hampered by a lack of space. However, in the military confinement environment, inmate disturbances often occur in small, closed areas, such as an inmate's housing unit, a gymnasium, a dining facility, or a chapel where maneuver space and equipment limit the size of the element that may be employed. The layout of each facility is different and presents its own unique problems when employing forces. This chapter covers the use of small military police teams, forced cell move teams (FCMTs), and the formations used to move an unruly inmate from one cell to another. Small generic riot control formations are discussed to provide the commander with a basic employment option that he can tailor to his facility to enable him to regain control of the affected area. The use of NL munitions and RCAs are incorporated into this discussion.

CROWD DYNAMICS WITHIN THE CONFINEMENT FACILITY

7-1. In confinement facility disturbances, inmates employ a number of tactics to resist control or to achieve their goals. Tactics may be unplanned or planned and nonviolent or violent. When a disturbance carries many purposes, it is likely that their tactics are well planned.

NONVIOLENT TACTICS

7-2. Nonviolent tactics range from name-calling to building barricades. Inmates may attempt to distract control force members by shouting at or ridiculing them or by using abusive language, obscene remarks, taunts, and jeers. The inmates' goals are to anger and demoralize the control force. They also want authorities to take actions that may later be exploited as acts of brutality.

VIOLENT TACTICS

7-3. Violent crowd tactics used by inmates are often extremely destructive and can include physical attacks on cadre, other inmates, and property. Use of violent tactics is limited only by the attitudes and ingenuity of the inmates, the training of their leaders, and the materials available to them. Inmates often commit violence with crude homemade weapons. If unplanned violence occurs, inmates will use mops, brooms, chairs, beds, or whatever else is on hand as weapons of violence. During planned violent disturbances, inmates can easily conceal makeshift weapons or tools to use against the control force. Inmates often erect barricades to impede troop movement and/or to prevent the control force from entering certain areas or buildings. Inmates can be expected to vent their emotions on individuals (cadre or inmates), the control force, their equipment, and/or the institution's property. Inmates direct dangerous objects like carts, barrels, liquids, and burning shoe polish at the control force.

GRIEVANCE PROTEST

7-4. A grievance protest may be organized as a riot. However, under normal circumstances, a riot of this type is not of an extremely violent nature. It may turn violent when leaders attempt to exploit any successes or weaknesses found in the control force.

UNORGANIZED RIOTS

7-5. Unorganized riots are characterized as being spontaneous in nature. However, they could be exploited and diverted by leaders at any subsequent stage of the riot, turning it into a different type. They may begin as an isolated assault against authority figures, or acts of civil disobedience. Under determined leadership, an unorganized riot could change to an organized riot.

MULTIPLE RIOTS

7-6. Multiple riots may happen when a disturbance occurs in one area of the confinement facility, causing others to riot in their areas. To control this situation with only one riot control force available, the force commander should first subdue the most violent riot. At the same time, he should employ some of his control forces to contain the other riots until the main control force is able to move to those locations to subdue them.

CADRE BEHAVIOR

7-7. The cadre is susceptible to crowd behavior. They become emotionally stimulated during tense confrontation with unruly and violent inmates. Commanders must be aware of the mood and attitude of the crowd and its effect on the control force cadre. To counteract the effects of crowd behavior on the cadre, the commanders must institute rigorous training and firm and effective leadership. This training must include a complete awareness and understanding of the use of force and the commander's intent. These are all necessary to offset the effect of crowd contagion upon the control force. Even with the best training and leadership, control force members must exercise individual and collective restraint.

PLANNING

7-8. Preplanning is the preparation and training conducted before a crisis occurs. The purpose of preplanning is to have plans and standing operating procedures (SOPs) in place so the cadre may react to an emergency and successfully contain and neutralize the situation. The preplanning process includes plan development, intelligence gathering, and training.

NOTE: Intelligence gathering is essential for the successful containment, neutralization, and/or prevention of a riot.

7-9. Once the control force team has been alerted to a disturbance within the confinement facility, leaders further develop preplans to fit the situation much the same way that DOD develops war plans.

7-10. During the planning process, leaders must be aware of the potential risks involved in quelling disturbances within a confinement facility. Careful planning minimizes collateral damage and risks to control force members and hostages.

7-11. The use of NLW and NL munitions must also be considered when developing plans. There must be strict accountability and control of RCAs, their employment, and other appropriate NL means (such as high-pressure water hoses). When using RCAs, plans must be flexible enough to accommodate changes in the situation and weather because it impacts the effectiveness of some agents. Planners must also consider the—

- Location of the disturbance.
- Estimated number of rioters.
- Accessibility of weapons, tools, or cleaning supplies.
- Inmate's military training.
- Hostage situation.

7-12. Based on an analysis of these factors, the commander makes an estimate of the situation. The estimate must be as thorough as time permits. Using the estimate, the commander considers COAs, selects RCAs, and determines the need for engineer support. In choosing a COA, consider the—

- Number of hostages.
- Number of inmates holding the hostages.
- Leader.
- Attitude and demeanor of the rioters.
- Location of the riot.
  - Inside areas of the confinement facility, such as the dining facility, work site, or housing wing.
  - Outside areas of the confinement facility, such as in exercise or work areas.
- Accessibility of weapons or material to make weapons.
- Accessibility of chemicals, such as cleaning supplies, cooking oils, and shoe polish.

USE OF CHEMICAL IRRITANTS

7-13. The use of chemical irritants can be a valuable NL tool for control force leadership to consider during the planning phase. These chemical irritants can drive a threat from an established, enclosed position or deny the rioters access to a certain area without long-lasting effects to those involved. The proper use of chemical irritants may prevent the control force from having to enter a dangerous area; however, improper use can cause injury, death, or property damage.

USE-OF-FORCE GUIDELINES

7-14. The installation commander authorizes the use of chemical irritants after the control force leader determines the type and dosage. The SOP outlines the procedures for securing authorization and provides guidance to help the control force leader make his decisions. The confinement facility commandant should inform the installation commander of the situation, since the installation commander is ultimately responsible. Team members must be aware of how chemical irritants affect personnel and plan accordingly. Inmates not involved must be removed from the area and the chemical cloud's path. The fire department and emergency medical services (EMS) should be on standby during operations involving chemicals.

PEPPER IRRITANT

7-15. The pepper irritant receives its name from varieties of the capsicum plant that is it made from. The active ingredient, Oleoresin Capsicum (OC), is extracted from the plant and micropulverized to make the irritant. It is neither a chemical nor a gas, but an all-natural organic substance. It is effective against individuals who are emotionally disturbed or under the influence of drugs or alcohol. It is effective against domestic and wild animals. OC is often used to divert the threat from making or continuing an assault. OC does not cause permanent damage. Its effects last 30 to 40 minutes after individuals are placed in fresh air.

7-16. OC can have various effects on people. Some effects are—

- Swelling of mucous membranes and the upper respiratory system; however, it does not shut down the system.
- Burning (intense) and discoloration (bright red) of exposed skin.
- Dilating the capillaries.
- Swelling of the eyelids.
- Burning and involuntary closing of the eyes.
- Coughing uncontrollably.
- Gagging.
- Gasping for air.
- Losing strength and coordination (temporarily).

OC has some disadvantages. They are that—

- The canister depressurizes over time if not checked regularly.
- Inmates may gain access to OC and use it against SRT members.
- The canister must be shaken on a regular basis.

DISSEMINATION OF CHEMICAL IRRITANTS

7-17. There are six methods for disseminating chemical irritants.

Pyrotechnic Method

7-18. In the pyrotechnic method of dissemination, the chemical irritant is placed in a canister with an inert material that is ignited when the device is activated. The agent is then carried into the air on the smoke particles of the inert material.

7-19. There are some disadvantages when using a pyrotechnic device indoors. These disadvantages are that the device—

- May cause a fire.
- Has a very slow saturation time (40 to 45 seconds to burn completely).
- Can be thrown back by the threat, even while burning.

Bursting Method

7-20. In the bursting method of dissemination, the irritant and an inert powder are placed in a saw-toothed canister. After a small detonation, the canister splits and expels the irritant in a cloud.

7-21. The bursting method has its advantages and disadvantages. Some advantages are that—

- There is no risk of fire.
- There is no risk of the canister being thrown back.
- The contents disseminate quickly.
- The canister is effective, easy to carry, and easy to control.

Some disadvantages are that—

- There is a slight possibility of fragmentation when detonated.
- The fuse head may separate from the canister.
- Only 50 to 90 percent of the irritant may be disseminated.
- The direction of the cloud formation depends on where the device is detonated.

Aerosol Method

7-22. In the aerosol method of dissemination, the irritant is suspended in an inert liquid that is located in the rear portion of the device. When the projectile penetrates the structure, the rear portion splits open and the irritant is disseminated in a mist. The aerosol method is most appropriate for tactical operations when chemical irritants are used before entry.

7-23. The aerosol method has its advantages and disadvantages. Some advantages are that it—

- Is easily carried.
- Is an excellent deployment system (uses 40-millimeter, 37-millimeter, or 12-gauge ammunition).
- Disperses the contents quickly.
- Is easily controlled and accurate (fin-stabilized).

NOTE: Practice firing is always necessary.

Some disadvantages are the possibility of—

- Injury to individuals if it is fired from close range.
- Damage to the structure if it strikes a weak portion.

Projectiles Method

7-24. Chemical-containing projectiles can be very effective when used correctly. A soldier must know which projectile to use in each unique situation.

7-25. **Thirty-Seven-Millimeter Projectile.** A 37-millimeter projectile can incapacitate anyone in an enclosed area of 4,500 cubic feet. It can penetrate the following:

- A sheet of ¾-inch-thick plywood at 25 yards.
- An automobile windshield at 33 yards.
- A hollow-core door at 58 yards.

7-26. **Forty-Millimeter Projectile.** A 40-millimeter projectile is more effective at greater ranges than a 37-millimeter projectile because of its rifled bore. It can incapacitate anyone in an enclosed area of 4,500 cubic feet. It can penetrate the following:

- A 1-inch-thick plywood at 10 yards.
- An automobile windshield at 55 yards.
- A hollow-core door at 109 yards.

7-27. **Twelve-Gauge Projectile.** A 12-gauge projectile can be loaded into a shotgun with no modifications to the weapon. It can be fired as a single round, or it can be magazine-fed. The 12-gauge projectile can incapacitate anyone in an enclosed area of 1,000 cubic feet. The angle at which the projectile strikes the barrier can affect its penetration capability. It can penetrate the following:

- A sheet of ¾-inch plywood at 33 yards.
- An automobile windshield at 33 yards.
- A hollow-core door at 100 yards.

Fogging Method

7-28. The irritant is suspended in a fogging solution and dispersed with a fogging machine. It is not recommended for tactical operations.

7-29. The fogging method has its advantages and disadvantages. Some advantages are that it—

- Produces enough chemical irritant to cover an enclosed area of 100,000 cubic feet in 26 seconds.
- Disseminates quickly.
- Provides high-volume capabilities.
- Is effective for riot control or other crowd dispersal situations.

Some disadvantages are that it is—

- Noisy.
- Cumbersome.

Multipurpose Grenade Method

7-30. A multipurpose grenade can be hand-thrown or launched from a shotgun with an adapter. It has an extended shelf life of 6 years and an adjustable fuse delay of 2 to 5 seconds. The irritant is located in the cylinder portion of the device and is forced out the bottom of the grenade upon detonation.

WARNING
A multipurpose grenade deployed incorrectly could result in injury. Safety rules should be observed at all times.

RECORD OF EVENTS

7-31. Reporting procedures for serious incidents should be included in the plans. A record of events must be initiated to provide a basis for the preparation and submission of a formal report to higher HQ. As a minimum, include the—

- Time the incident was reported and by whom.
- Time the incident was reported to the facility commander.
- Time the control force was assembled.
- Time the control force entered the facility.
- Weather conditions as they relate to the use of RCAs.
- Number of cadre and inmates injured or killed, including how they were injured or killed and the medical attention given to them.
- Time the operation was completed.
- Time the riot control force restored order.

TRAINING

7-32. Guard units and associated teams must train on a regular basis in the five basic riot control formations. There must be a continuous training program established, to include—

- Interpersonal communication.
- Self-defense.
- Use of force.
- Use of a riot baton.
- Fatal areas of the body to avoid when using the riot baton.
- First aid.
- Emergency plans.
- Intelligence and counterintelligence techniques.
- The use of RCAs and the various methods of dispersing them.
- The use of NLW.
- The use and effects of high-pressure water.
- Riot control formations.

EQUIPMENT

7-33. The recommended basic riot gear used by the confinement riot control force is as follows:

- Riot baton.
- Riot shield.
- Helmet with face shield.
- Groin protector.
- Flak vest.
- Leather gloves.
- Shin protection.

RIOT BATON

7-34. The riot baton is an invaluable weapon in crowd control situations. A riot baton in the hands of a well-trained control force member is the most appropriate weapon, except with extremely violent crowds. The riot baton is not meant to replace NL munitions, but is instead used in situations in which NL munitions are not needed. Losing a riot baton to the crowd does not present a serious threat.

7-35. Use of the riot baton is based on the commander's appraisal of the situation and his choice of force option. Control forces committed with the riot baton must have teams equipped with RCAs, NLW, and lethal coverage. If the confrontation increases in intensity, the commander has these force options available and may rapidly engage the control force.

7-36. A riot baton is a versatile offensive weapon that varies in length from 26 to 42 inches with the optimum length being 36 inches. It is made of hickory and does not shatter or break easily. Both ends of the riot baton are rounded to prevent unnecessary injury. It is approximately 1¼ inches in diameter. Each riot baton has a hole drilled at the grip end with a leather thong threaded through the hole. The thong helps secure the riot baton to the user's hand.

RIOT SHIELD

7-37. The riot shield (see *Figure 7-1*) is an excellent piece of defensive equipment. It is used by control force personnel in the front and on the flanks of the formation.

7-38. The riot shield has its advantages and disadvantages. Some advantages are that—

- OC pepper spray or high-pressure water can be used to back the control force.
- It provides protection from thrown objects.
- It absorbs some of the impact of rioters hurling themselves at the control force.

The disadvantages of the riot shield are that it—

- Severely limits the use of the riot baton.
- May crack if hit with a heavy object.

Figure 7-1. Riot Shields

RIOT CONTROL FORMATIONS

7-39. The five basic formations for civil disturbance operations are the line, wedge, echelon, diamond, and circle. In the correctional environment the basic formations for disturbances are the line, wedge, or echelon with apprehension teams in support. Generally, the diamond and circle formations are not used in the correctional environment. The minimum size force that the commander should consider is an augmented platoon, with a two-to-one advantage over the rioters.

7-40. Correctional facility cadre must be proficient in all riot control formations. These are usually trained at the squad level and above, but performed by a platoon or company-sized unit. Squad line, echelon left, echelon right, and wedge formations form the basis for platoon and company formations. Each squad must be adept in the basic formations before practicing in platoon-sized or larger formations. For more detailed information on formations, see *STPs 19-95B1-SM* and *19-95B24-SM-TG*. In the correctional environment, the support platoon plays a key role by supporting the control formation with apprehension and equipment teams. See *Chapter 6* for an in-depth discussion on civil disturbance formations.

APPREHENSION TEAMS

7-41. Apprehension teams give the control force the ability to extract inmates. Inmates identified as leaders or agitators and inmates that are injured or no longer want to resist can be extracted from the crowd. Apprehension teams can also be used to enter confined areas where the formation cannot operate.

7-42. The apprehension team will consist of eight members and one leader. Each team will have four shield holders, two nonlethal gunners and two apprehension/search members. The squad leader controls the movement of the team and assists the team in exiting the control formation. He maintains communications with the formation and coordinates the team's operations with the formation. The squad leader also ensures that the team uses the minimum amount of force necessary to extract inmates.

7-43. The apprehension team may deploy through the middle of the formation or from around either of the flanks.

7-44. The apprehension team should be deployed within a distance of approximately 10 meters so that the formation can still provide immediate support. If the apprehension team is deployed farther than 10 meters, the formation must advance to close the gap.

7-45. The team will exit the formation in two columns; shield holders in the first and second ranks, nonlethal gunners in the third rank, apprehension/search members in the fourth rank and the leader last. Team members grip the shoulder or equipment of the member in front of them. The shield holders will envelop the inmate that is to be extracted and form a protective barrier. Once the shield holders have enveloped the inmate, the nonlethal gunners will cover their sides with their weapons while the apprehension/search members control and restrain the inmate. When the team has the inmate under control, the squad leader will give the order for the team to withdraw. The team withdraws in reverse order (apprehension/search members with the inmate, followed by nonlethal gunners and shield holders). Team members then move backwards, face the crowd and grip the shoulder or equipment of the member in front of them.

TEAM EQUIPMENT

7-46. The apprehension team uses the following equipment:

- Flex-Cufs.
- Restraint cutters.
- Pepper spray.
- Helmets with face shields.
- Groin protectors.
- Flak vests.
- Elbow pads.
- Leather gloves.
- Shin protectors.
- Knee pads.
- A video camera.

TEAM COMPOSITION

7-47. The apprehension team is composed of five team members. This team consists of four members and a squad leader. Team member one is the team leader.

Squad Leader

7-48. The squad leader assists the team in exiting the control formation and ensures that the team uses the minimum force necessary to subdue inmates. He is also responsible for signaling the PSG if more help is needed.

Team Member One (Team Leader)

7-49. Team member one's responsibility is to make initial contact with the inmate and use pepper spray and apply Flex-Cufs, as needed. He also carries a set of restraint cutters. Once the apprehension team has returned to the relative safety of the control formation, team member one searches the inmate.

7-50. Team member one receives the command to move forward and secure the inmate from the squad leader. This command is usually given by a signal and verbal command. Team member one then notifies the main force commander where his team exits the control force. Once the team is in position to exit the main formation, team member one signals the member of the formation in front of him. This is usually a prearranged signal, such as a tap on the right shoulder. The control force opens and the team exits the formation and makes contact with the assigned inmate. Team members then subdue and control the inmate. If pepper spray is needed, team member five shouts, "Pepper." This gives the other team members time to react to the command and spray the inmate. Team member one then applies Flex-Cufs to the inmate's wrists and ankles. Depending on the mission, the inmate may be left for follow-on security forces or carried behind the control force to be secured by security forces.

7-51. Should the mission require that the inmate be moved, team member one shouts, "Prepare to Move," this allows each member to grab an arm and position the inmate toward the control force formation. When the team is ready to move, team member two shouts, "Ready," alerting team member one that the team is ready for the last command. Team member one then commands, "Move." The team enters the control formation with the inmate. The inmate is searched by team member one and turned over to the security team.

Team Member Two

7-52. Team member two stands directly behind team member five with his hands on team member five's waist. He protects the left side of team member five by moving to team member five's left and helps to secure the inmate after team member one makes initial contact. He repeats team member five's commands and controls the inmate's head after the inmate is on the ground.

Team Member Three

7-53. Team member three stands behind team member two with his hands on team member two's waist. He protects the right side of team member five by moving to team member five's right and helps to secure the inmate after initial contact.

Team Member Four

7-54. Team member four stands behind team member three with his hands on team member three's waist. His task is to help subdue the inmate in any way possible. He usually concentrates on the inmate's body below the knees. He acts as the eyes and ears for the team during control of the inmate.

Team Member Five

7-55. Team member five gives commands to control the inmate.

FORCED CELL MOVE TEAMS

7-56. The intent of a forced cell move is to move an unruly and/or uncooperative inmate from one cell to another. This is a difficult task and must not be taken lightly. A forced cell move must be thoroughly planned, rehearsed, and resourced to ensure the safety of the team members and the inmate.

7-57. A number of reasons may compel the commander to consider a forced cell move. The inmate may refuse to eat, take medication, accept medical attention, get a haircut, keep himself clean, work, or move or rotate cells when ordered to. These are some of the reasons that the commander may order a forced cell move.

TEAM COMPOSITION AND EQUIPMENT

7-58. The FCMT is composed of six corrections cadre and two support cadre from military police and medical sections specially trained in forced cell movement. Primarily, the team consists of the following personnel:

- One officer in charge (OIC) (E7 or above).
- Five corrections cadre.
- One military police investigator.
- One medic.

7-59. The FCMT uses the following standard equipment:

- Helmets with face shields.
- Groin protectors.
- Flak vests.
- Elbow pads.
- Leather gloves.
- Shin protectors.
- Knee pads.
- A pinning shield.
- Restraints (hand or leg irons or Flex-Cufs).
- A video camera.

ACTIONS PRIOR TO TEAM EMPLOYMENT

7-60. The OIC gives an operations order (OPORD) to team member five (the team leader) of the FCMT. As a minimum, the OPORD should answer the following questions about the incident leading up to the move:

- Did the inmate fail to comply with orders?
- Did the inmate assault cadre or other inmates?
- Is the objective of the move to force a shower or haircut?
- Is OC pepper spray authorized for use?
- Has the inmate been moved before?

- What is the condition of the area where the inmate is currently located?
- Has the inmate tested positive for human immunodeficiency virus (HIV)?
- Does the inmate have a weapon or the materials needed to make one?
- Is the area barricaded?
- Has he smeared himself with body fluids or other slippery substances?
- Can you approach the inmate without risk to him and/or others?

7-61. The team is assembled for inspection by the OIC, and the team members are required to brief the OIC on their individual duties and responsibilities. A line formation is the customary formation for this inspection. It should take place out of site of the inmates.

7-62. Each soldier takes one step forward and briefs the OIC as shown in the following example: *"I am (rank and name), team member one. My mission is to pin the inmate using the minimum amount of force necessary. I carry the pinning shield."* He then faces so the camera can record the number printed on the back of his flak jacket and holds up the pinning shield [see *Figure 7-2*]).

Figure 7-2. Recording Team Members' Duty Position

INDIVIDUAL TEAM MEMBER RESPONSIBILITIES

7-63. Each individual team member has his own responsibilities during a forced cell move. Each team member must do his designated job while working as a team.

Team Member One

7-64. Team member one is the eyes and ears of the team. His responsibility is to carry the pinning shield up to the cell door and describe the scene with as much detail as possible (especially including anything that may cause harm to the team). The pinning shield is used to protect the team from body fluids thrown at the them by the inmate.

7-65. Team member one describes the scene in the cell to his teammates by turning his head to the right and speaking in a loud voice so that all team members can hear him. A slap on the right shoulder by team member two alerts him that all team members have heard him. If he receives a slap on the left shoulder from team member two, he repeats the situation report (SITREP) in a loud, clear voice. The decision to enter the inmate's area with or without the pinning shield is team member one's decision, unless the inmate is known to have a weapon. Should he decide to enter without the pinning shield, team member one then enters with both hands held approximately chest high. When team member one is unable to determine if the inmate has a weapon, the pinning shield is used. Charging through the open cell door, team member one drives the inmate to the bunk or ground with the shield. Once pinned, team member one secures the inmate's head and ensures that it is not twisted or turned in relation to its natural form. He may have to use the various pressure points on the inmate's head to help subdue him.

Team Member Two

7-66. Team member two steps forward and briefs the OIC as follows: *"I am (rank and name), team member two. My mission is to secure the inmate's right arm using the minimum amount of force necessary. I carry one set of hand irons."* He then holds the hand irons up for video camera taping (see *Figure 7-3*) and says, *"I also carry one canister of OC pepper spray to be used at the direction of the team OIC."* He then holds up the canister for video camera taping and faces so the camera can record the number on the back of his flak jacket (see *Figure 7-2,* page 7-13.)

Figure 7-3. Recording Team Members' Equipment

7-67. Team member two lines up directly behind team member one. He keeps contact with team member one until he can secure the inmate's strong arm. Team member two taps team member one on his right or left shoulder after being tapped by team member three. Once inside the cell, team member two secures the prisoners strong hand with wrist restraints and announces, "*Hands secure*," to team member five.

Team Member Three

7-68. Team member three steps forward and briefs the OIC as follows: *"I am (rank and name), team member three. My mission is to secure the inmate's left arm using the minimum amount of force necessary."* He then faces so the camera can record the number on the back of his flak jacket (see *Figure 7-4*).

Team member three lines up directly behind team member two. Team member three keeps contact with team member two until he can grasp a portion of the inmate's body. He then concentrates on securing the inmate's weak hand and assisting team member two in applying hand restraints. Team member three taps team member two on his right or left shoulder after being tapped by team member four.

Figure 7-4. Recording Number on Flak Jacket

Team Member Four

7-69. Team member four steps forward and briefs the OIC as follows: *"I am (rank and name), team member four. My mission is to secure the inmate's right leg using the minimum amount of force necessary. I carry one set of leg irons."* He holds the leg irons up for the camera to record (see *Figure 7-2,* page 7-13). He then faces so the camera can record the number on the back of his flak jacket (see *Figure 7-2*). Team member four lines up directly behind team member three. He taps team member three on his right or left shoulder after being tapped by team member five. Team member four keeps contact with team member three until he can secure a portion of the inmate's body. Team member four concentrates on securing the right leg of the inmate. It is his responsibility to apply the leg irons. Once the legs are secure, he announces, "*Legs secure*," to team member five.

Team Member Five (Team Leader)

7-70. Team member five steps forward and briefs the OIC as follows: *"I am (rank and name), team member five. My mission is to secure the inmate's left leg using the minimum amount of force necessary. I carry a restraint key (hold up for the camera to view). I am also the team leader for this mission and it is my responsibility to ensure that the team uses the minimum amount of force necessary to subdue the inmate."* He then faces so the camera can record the number on the back of his flak jacket (see *Figure 7-4*).

7-71. Team member five taps team member four on the right shoulder when he has heard the SITREP from team member one. If he does not hear the report or needs to hear it again, he taps team member four on the left shoulder. Team member five keeps in contact with team member four until he can secure a portion of the inmate's body. He secures the left leg and assists team member four in applying the leg irons. Once he has heard that the hands and legs are secure, team member five signals for the team to conduct an equipment check. The team members check the equipment in place.

NOTE: At no time is the inmate to be released by anyone.

Medic

7-72. The medic briefs the OIC in the following manner: *"I am (rank and name), the medic. My mission is to treat all injuries that may occur during the move.*" He follows the instructions given to him by the OIC/NCOIC. His mission is to observe the inmate for signs of injury before the forced cell move and assess the inmate's condition during and after the move.

Military Police Investigator

7-73. The military police investigator (MPI) is responsible for videotaping the entire forced cell move. This includes the OIC/NCOIC briefing the OPORD, each team member briefing his responsibilities to the OIC, and any medical treatment rendered to the inmate.

ACTIONS DURING THE FORCED CELL MOVE

7-74. There are many actions performed during a forced cell move. Each member of the FCMT must complete his task successfully to accomplish the forced cell move.

Officer in Charge or Noncommissioned Officer in Charge

7-75. The OIC, an E7 or above, is responsible for ensuring that the team uses the minimum amount of force necessary to subdue the inmate. He is also responsible for ensuring that all necessary precautions have been taken to ensure the safety of the team and the inmate. The OIC gives the inmate the initial orders to comply and informs him of the consequences of noncompliance. He does this by placing himself at the front of the inmate's cell (see *Figure 7-5*). The orders are not given until it can be enforced by the FCMT. The following prepared statement is read to the inmate by the OIC: "*I am (rank and name), OIC of this FCMT. At this time I am ordering you to lie face down on your bunk (or other designated location) with your head to the rear of the cell. Cross your hands behind your back and cross your feet as the team enters the cell and uses the minimum amount of force necessary to place you in restraints and move you to (give location). Failure on your part to comply with these instructions or to resist the FCMT will result in the use of force (OC pepper spray if applicable). Do you understand these instructions? If so, lie down on the bunk.*"

7-76. After reading the instructions to the inmate, the OIC faces the camera and states whether the inmate complied. He then waits 60 seconds and motions for the team to position itself in front of the inmate's cell (see *Figure 7-6*). The OIC directs the opening of the cell when the team is positioned at the entrance to the cell and prepared to enter. He positions himself so he can observe team entry to and departure from the cell. If OC spray is authorized, the OIC tells the team when to spray the inmate with OC spray. This task is done in silence. Communication between team members should be done through hand-and-arm signals, unless noted otherwise.

Figure 7-5. OIC Reading Compliance Statement to Inmate

Figure 7-6. OIC and FCMT

Team Member One

7-77. Once the team is positioned in front of the inmate's cell, team member one starts his report. At a minimum, he must answer the following questions and report them to the team:

- Does the inmate have a weapon, if so, what kind?
- Does the inmate appear dry?
- Is the inmate clothed or wearing extra clothing?
- Has the inmate smeared anything on his body?
- Can you see the inmate's hands?
- Is the cell barricaded or visibly booby-trapped?

- Is the cell dry?
- Are cell furnishings intact?

7-78. Based on the answers to the above questions, team member one must decide whether to enter the cell with or without the shield (see *Figure 7-7*). Team member one's ultimate goal is to pin the inmate on the bunk or floor.

Figure 7-7. FCMT Preparing to Enter a Cell

Team Member Two

7-79. Team member two keeps contact with team member one until he can secure the inmate's strong arm. If he cannot immediately secure the inmate's arm, he secures a portion of the inmate's body and progresses to the inmate's strong arm. Once the inmate's hands are secure, he announces, "*Hands secure*," to team member five (see *Figure 7-8*).

Team Member Three

7-80. Team member three keeps contact with team member two until he can grasp a portion of the inmate's body. Team member three concentrates on securing the inmate's weak hand and assists team member two in applying Flex-Cufs. Team member three maintains control of the inmate's hands and looks at team member five for further instructions (see *Figure 7-8*).

Team Member Four

7-81. Team member four keeps contact with team member three until he can secure a portion of the inmate's body. Team member four concentrates on securing the inmate's right leg. He applies leg irons and when the inmate's legs are secure, he announces, "*Legs secure*," (see *Figure 7-8*).

Team Member Five (Team Leader)

7-82. Team member five keeps contact with team member four until he secures a portion of the inmate's body. Team member five concentrates on securing the inmate's left leg and assists with the application of leg irons. When he is informed that the hands and legs are secure, the team searches the inmate for weapons and contraband. Once the search is complete, team member five commands, "Prepare to Lift."

Figure 7-8. FCMT Securing an Inmate

Movement of the Inmate

7-83. Team member one positions his hands to control the inmate's head during the move, turning the inmate's head toward the wall. Team members two and three position themselves on the inmate's right and left sides. They grasp the wrist or forearm of the team member opposite them to form a platform for the inmate to lie on. Team member four positions himself on the inmate's right side even with the inmate's knees. This places him in position to control the inmate's legs. Team member five's position is on the inmate's left side and enables him to control the inmate's lower legs and feet. When in position, team member five ensures that all team members are in place before commanding, "Ready, Lift."

7-84. On the command "Lift," all team members simultaneously stand with the inmate. They should sandwich the inmate between them to facilitate control. Team member five then commands, "Prepare to Turn, Ready—Turn." The inmate is always turned toward the bunk. If a bunk is not available, team member five should specify which direction to turn the inmate. The inmate is then removed from the cell (see *Figure 7-9,* page 7-20).

7-85. Once the inmate is on the tier, team member five commands, "Prepare to Lower, Team—Lower." The team simultaneously lowers the inmate to the floor with the team keeping positive control of the inmate. The medic asks the inmate, *"Do you have any injuries at this time?"* (see *Figure 7-10,* page 7-20). If the inmate has no injuries, the medic looks at the camera and states, *"Inmate (inmate name) appears to have sustained no injuries."* If the inmate has sustained injuries, the medic treats those injuries and completes a sworn statement. If OC was employed, the medic decontaminates the inmate using water and paper towels. The medic instructs the inmate to tilt his head to the side and flush his eyes with water for approximately 5 minutes. The medic then dries the inmate's face.

Figure 7-9. Moving the Inmate

Figure 7-10. Treating Injuries

Placement of the Inmate in a Cell

7-86. Team member five orders an equipment check before returning the inmate to a cell. He then commands, "Prepare to Lift, Team—Lift." Team members position themselves exactly where they were when they removed the inmate from the cell. Next, team members simultaneously lift the inmate and return him to the predetermined cell. They move the inmate into the cell head first with his face toward the bunk. Once inside the cell, team member five commands the team to lower the inmate onto the bunk (see *Figure 7-11,* page 7-20). Team members then remove the inmate's restraints, keeping positive control of the inmate at all times. Team members use a nonverbal signal to notify team member five when the restraints are removed. He signals another equipment check. Team members then account for all equipment in place and prepare to exit the cell. The team exits the cell in the same manner as they entered, but in reverse order.

Figure 7-11. Placing Inmate in Cell

7-87. Team member four moves into position where he can control both of the inmate's legs while team member five moves toward the cell door. When team member four is ready to exit the cell, team member three moves to control the inmate's legs, team member two moves to where he can control both of the inmates arms, and team member one maintains control of the inmate's head. Team member four maintains contact with team member three at all times. When team member three is ready to exit the cell, team member two moves into a position where he can control the inmate's lower body and team member one moves into position to control the inmate's upper body. Using a prearranged, nonverbal signal, the team pulls team members one and two from the cell. The FCMT is prepared to reenter the cell if the inmate attempts to attack team members one and two as they exit the cell. The FCMT remains ready to reenter until the cell door is secured.

Chapter 8

Civil Disturbance Training

Training requires planning and preparation. Unit commanders structure their training program according to the mission essential task list (METL) (see *FM 7-1)* and expected contingencies. These training plans include prior training and the experience and proficiency of their unit. Leaders agree that the operational tempo and expected contingencies for the next 10 to 15 years will likely involve low-intensity military operations other than war (MOOTW). Therefore, civil disturbance training must be included in the commander's annual training program and must cover all aspects of that type of operation. To effectively function as a member of a control force, individual soldiers must be trained in civil disturbance operations. Training must be intensive, realistic, and sustained. Any lack of training (from basic soldier skills to leadership skills) can have an adverse effect to the point of endangering lives, especially the lives of soldiers. Leader involvement in the training process is essential for mission success in crowd control operations.

UNIT TRAINING STRATEGY DEVELOPMENT

8-1. Developing a unit training strategy begins with the mission analysis portion of the MDMP and identifying the GRM. Missions that require soldiers and units to enforce treaties or accords, protect the lives of civilians in uncertain or hostile environments (such as NEOs), or provide large-scale humanitarian assistance demand some sort of graduated-response criteria to maintain order and prevent uncertain environments from becoming hostile.

8-2. When developing training strategies, planners should focus training and evaluation on the echelon level above the target element (for example TF) or the unit identified as needing training.

8-3. Planners must agree on the intent of the GRM. The GRM is for most a training and rehearsal tool. It provides leaders with likely vignettes that can be incorporated into COA analysis, predeployment training, and rehearsals. The GRM becomes a handy reference during situations that require graduated responses.

TRAINING PLAN DEVELOPMENT

8-4. In developing a training plan for civil disturbances, many considerations should be addressed in terms of available resources. Considerations include the number of personnel to train, the equipment needed and its availability, the time available to train to standard, the location of where training will be executed, and the incorporation of SOPs (based on the theater of operations

[CONUS or OCONUS]). A personnel assessment is necessary to determine the training and experience level of soldiers. It is critical to identify and use soldiers with previous real-world experience and leaders that have graduated from the Interservice Nonlethal Individual Weapons Instructor Course (INIWIC) as primary instructors.

8-5. Commanders should structure the unit training program based on the METL, expected contingencies, prior training, and the proficiency of the unit. In developing a training program, use the seven-step approach. For in-depth instructions on training development, see *FMs 7-0* and *7-1*.

8-6. The objective is to train and evaluate units in the performance of civil disturbance procedures. Proficiency in performing civil disturbance procedures is reflected by a unit's ability to quickly and deliberately execute the disturbance control procedures required to prevent loss of life and property or to regain control of a civil disturbance in its assigned area of responsibility.

8-7. Training should focus on, but is not limited to—

- Instructors.
- Individual and collective drills.
- Leaders.
- Equipment.
- Weapons.
- Munitions.
- Live-fire range exercises.
- Qualification and familiarization.
- Soldier training exercises (STXs).
- Culminating exercises.

8-8. There are several successful methods to evaluate trained leader skills and trained soldier skills using scenarios, vignettes, lanes, or STXs. Flexibility is an important key in training plan development due to time restraints.

RANGE SETUP

8-9. The key to setting up a successful range for NL munitions is to ensure that they are in compliance with installation safety regulations. Kinetic energy is the force that pushes the projectiles down range. In most cases, kinetic energy will only carry these NL munitions projectiles 75 meters or less. This gives trainers flexibility as to the locations where a live-fire exercise can be conducted, especially in a field-expedient situation. *Figure 8-1* is an example for conducting a NL live fire range setup.

8-10. As illustrated in *Figure 8-1*, mark off 5-, 10-, 20-, 30-, 40-, and 50-meter distances from the targets with markers. Each of the NL munitions has minimum and maximum effective ranges within the distances of 5 to 50 meters. Impact power and accuracy are greater at close distances and depleted at farther distances. Targets can be anything from the standard E-type to more realistic three-dimensional manikins. Evenly divide your shooters into three groups as shown in *Figure 8-1*. Group one will initially be shield holders, group two will be shooters, and group three will be grenadiers.

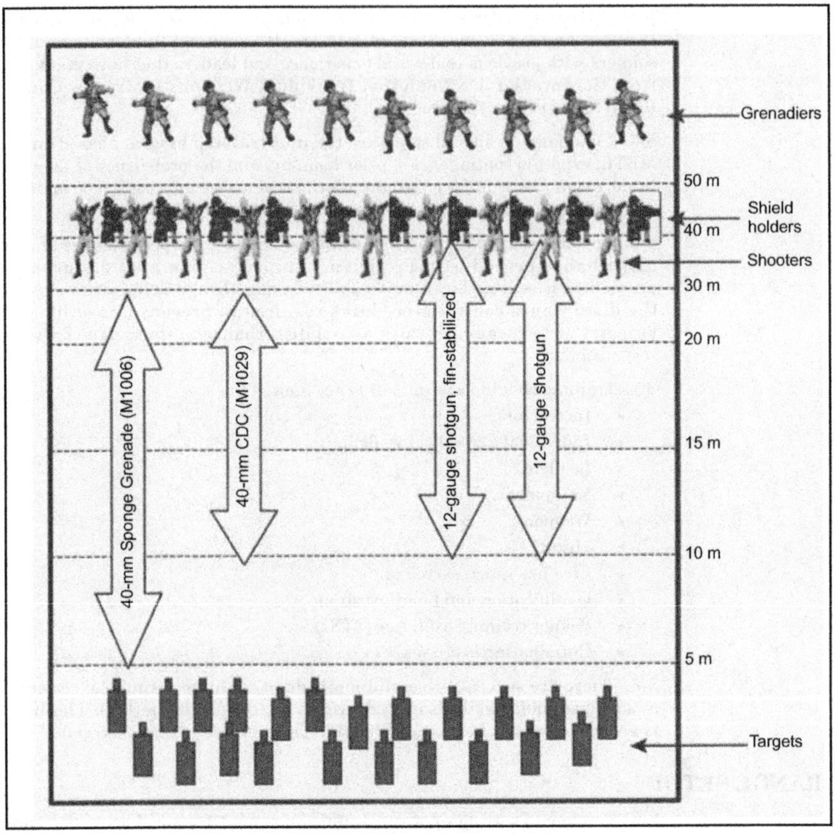

Figure 8-1. NL Munitions Range Setup

8-11. The formation should begin at the farthest effective range firing line (50 meters). After each volley of fire, the formation moves closer to the targets, but stays outside the closest effective range firing line (5 meters). For safety reasons, ensure that the formation stays online throughout each iteration.

8-12. When moving forward to each firing point, it is recommended that the formation always come to a halt before the fire command is given. It is also important to use the tap down drill method (see *Figure 8-1*) rather than having the shooters move forward and fire between the shield holders.

8-13. Range instructors can use verbal commands, such as, "Ready," or "Fire," for each volley. Shooters can be briefed to move forward, tap down the shield holder, and engage targets at the first explosion of distraction grenades thrown by grenadiers. In that instance, the command could simply be "Grenadiers Prepare to Throw, Throw."

8-14. The success of any civil disturbance operation is through effective communication, especially when incorporating the use of NL munitions. For effective training, use verbal commands in conjunction with effective, predetermined hand-and-arm signals during live-fire exercises.

8-15. Each iteration requires two phases of fire. During the first phase, half of the shooters will be firing NL, 40-millimeter rounds using an M203. The others will be firing NL, 12-gauge rounds using a 12-gauge shotgun. In the second phase, shooters switch weapons (safely). It is critical that line safeties are alert during this transition between phases.

8-16. Upon completion of each iteration, clear the weapons and the range. Establish a safe method and return all groups back to the farthest firing line. Once that is accomplished, group one rotates to become group three, group three rotates to become group two, and group two rotates to become group one for the next iteration. Using this method affords a safe and efficient way to get all shooters through the proficiency course. *Figure 8-2* is an example of how to fire each volley based on its characteristics, effectiveness, and capabilities.

Distance	Command	Subject	Actions
10 Meters	12-Gauge Shotgun, Two Point	Shotgun shooters	Load two M1013 rounds.
	M203, Two Point	M203 shooters	Load one M1006 round and have one readily available.
	Ready, Fire	Shotgun shooters	Tap down shields and engage designated targets.
		M203 shooters	Tap down shields, engage designated targets, reload, and reengage.
	Clear Range	Range safeties	Ensure that weapons are clear and safe.
	Safeties, Clear All Weapons	All shooters	Face to right, keep weapons pointed up and downrange.
	Shooters, Right Face, Move	All personnel	Step off shield holder, then shooters, and grenadiers on the command, "Move." Grenadiers reform at 50 meters.

Figure 8-2. Sample Live-Fire Exercise

Distance	Command	Subject	Actions
15 Meters	Shotgun, Two Point	Shotgun shooters	Load two M1013 rounds.
	M203, Two Point	M203 shooters	Load one M1006 round and have another available.
	Grenadiers, Prepare to Throw	Grenadiers	Make distraction grenade ready.
	Throw	Grenadiers	Throw grenade.
		Grenadiers	Throw distraction grenade.
	Ready, Fire	Shot gun shooters	Tap down shields and engage designated targets.
		M203 shooters	Tap down shields, fire at designated targets, reload with M1006 rounds, and engage designated target.
	Clear Range	Range safeties	Ensure that weapons are clear and safe.
	Three Steps Forward	All shooters and grenadiers	Echo preparatory command.
	Move	All shooters and grenadiers	Stay on line and step off three steps, counting each step aloud.
	Clear Range	Range safeties	Ensure that weapons are clear and safe.
	Seven Steps Forward	All shooters and grenadiers	Echo preparatory command.
	Move	All shooters and grenadiers	Stay on line and step off seven steps, counting each step aloud.

Figure 8-2. Sample Live-Fire Exercise (Continued)

Distance	Command	Subject	Actions
20 Meters	Shotgun, Three Point	Shotgun shooters	Load three M1013 rounds.
	M203, One Area, Two Point	M203 shooters	Load one M1026 round and have two M1006 rounds available.
	Grenadiers, Prepare to Throw	Grenadiers	Make distraction grenade ready.
	Throw	Grenadiers	Throw grenade.
		Grenadiers	Throw distraction grenade.
	Ready, Fire	Shotgun shooters	Tap down shields and engage designated targets.
		M203 shooters	Tap down shields, fire area at targets, reload with M1006 rounds, and engage designated target.
	Clear Range	Range safeties	Ensure that weapons are clear and safe.
	Shotgun, Two Point	Shotgun shooters	Load two M1013 rounds.
	M203, Two Point	M203 shooters	Load one M1006 round and have another available.
	Grenadiers, Prepare to Throw	Grenadiers	Make distraction grenade ready.
	Clear Range	Range safeties	Ensure that weapons are clear and safe.
	Three Steps Forward	All shooters and grenadiers	Echo preparatory command.
	Move	All shooters and grenadiers	Stay on line and step off three steps, counting each step aloud.

Figure 8-2. Sample Live-Fire Exercise (Continued)

Distance	Command	Subject	Actions
30 Meters	Shotgun, Two Area, Two Point	Shotgun shooters	Load two M1012 rounds and two M1013 rounds.
	M203, Two Area, Two Point	M203 shooters	Load one round M1026 and have one M1026 and two rounds M1006 available.
	Grenadiers, Prepare to Throw	Grenadiers	Make distraction grenade ready.
	Throw	Grenadiers	Throw grenade.
		Grenadiers	Throw distraction grenade.
	Ready, Fire	Shotgun shooters	Tap down shields, engage targets, and engage designated target with M1013 rounds.
		M203 shooters	Tap down shields, fire area at target, reload with M1006 rounds, and engage designated target.
	Clear Range	Range safeties	Ensure that weapons are clear and safe.
	Seven Steps Forward	All shooters and grenadiers	Echo preparatory command.
	Move	All shooters and grenadiers	Stay on line and step off seven steps, counting each step aloud.

Figure 8-2. Sample Live-Fire Exercise (Continued)

Distance	Command	Subject	Actions
40 Meters	Shotgun, Two Area, One Point	Shotgun shooters	Load two M1012 rounds and one round M1013.
	M203, One Area, One Point	M203 shooters	Load one M1026 round and have one M1006 round available.
	Grenadiers, Prepare to Throw	Grenadiers	Make distraction grenade ready.
	Throw	Grenadiers	Throw grenade.
		Grenadiers	Throw distraction grenade.
	Ready, Fire	Shotgun shooters	Tap down shields and engage targets with M1013 rounds.
		M203 shooters	Tap down shields, fire area at target, reload with M1006 rounds, and engage designated target.
	Clear Range	Range safeties	Ensure that weapons are clear and safe.
	Seven Steps Forward	All shooters and grenadiers	Echo preparatory command.
	Move	All shooters and grenadiers	Stay on line and step off seven steps, counting each step aloud.
50 Meters	Shotgun, One Area	Shotgun shooters	Load one M1012 round.
	M203, One Area, One Point	M203 shooters	Load one M1026 round and have one M1006 round available.
	Grenadiers, Prepare to Throw	Grenadiers	Make distraction grenade ready.
	Throw	Grenadiers	Throw grenade
		Grenadiers	Throw distraction grenade.
	Ready, Fire	Shotgun shooters	Tap down shields, engage targets, and engage designated target with M1013 rounds.
		M203 shooters	Tap down shields, fire area at the target, reload with M1006 rounds, and engage the designated target.
	Clear Range	Range safeties	Ensure that weapons are clear and safe.
	Seven Steps Forward	All shooters and grenadiers	Echo preparatory command.
	Move	All shooters and grenadiers	Stay on line and step off seven steps, counting each step aloud.

Figure 8-2. Sample Live-Fire Exercise (Continued)

NOTE: Shield holders are referred to as group one, shooters are referred to as group two, grenadiers are referred to as group three.

8-17. Currently, there are no qualification standards established in the Army for NL munitions. Area rounds, which contain multiple projectiles designed to affect multiple targets when engaged, only require approximately five rounds. This affords shooters an opportunity to realize the characteristics and capabilities of these rounds.

8-18. Single projectile rounds like the M1013 and the M1006 require a greater degree of proficiency. If situations call for the engagement of targets with these rounds, it is essential that trained shooters be able to hit what they are aiming at. For example, firing 10 rounds from various ranges can provide soldiers with the confidence to be effective shooters. A standard can be established that requires shooters to hit targets with at least 6 out of the 10 rounds used in the example to be considered proficient.

8-19. *Figures 8-3; 8-4,* page 8-11; *8-5,* page 8-21; and *8-6,* page 8-22, give examples of plans to follow if in a situation similar to the following scenario (see *Army Training and Evaluation Program (ARTEP) 19-313-10-MTP*):

A company commander is providing security patrols in support of a joint task force (JTF). Besides providing security patrols for a sector of responsibility, one platoon must always be designated as a quick reaction force (QRF) for civil disturbance operations. For the last 2 weeks the sector has been quiet and the commander has ordered all platoons to conduct civil disturbance training on a rotational basis.

A new order has been sent down to provide a squad-sized element to provide security for the future polling site for an upcoming election. This polling site has yet to open, but intelligence indicates potential problems as the election approaches.

On the day prior to the scheduled election, a crowd of approximately 25 to 30 demonstrators aggressively move toward the polling site, chanting slogans and yelling toward the security forces posted there. While some demonstrators begin throwing whatever they can pick up, the majority are simply yelling, locking arms, and blocking roadways. Hundreds of additional people, not part of the crowd, begin to gather to watch the activities as they unfold.

In an assessment of the demonstrators, it is determined that they intend to stay through the night into the morning of the election with the intention to block access to the polls. It is essential that the crowd is dispersed and that access to the polls is unrestricted in time for the election to begin the next morning.

<table>
<tr><th colspan="5">Evaluation Scheme</th></tr>
<tr><td colspan="5">The evaluation will consist of five lanes conducted over a 3-day period. Platoons will execute 24-hour operations during the evaluation. Any retraining will be done immediately following the execution of the lane if time permits.
The following tasks will be evaluated:</td></tr>
<tr><td>Lane 1</td><td colspan="4">Cordon and search operations</td></tr>
<tr><td>Lane 2</td><td colspan="4">Civil disturbance QRF</td></tr>
<tr><td>Lane 3</td><td colspan="4">Security and defense of a critical site (polling site)</td></tr>
<tr><td>Lane 4</td><td colspan="4">Night civil disturbance QRF</td></tr>
<tr><td>Lane 5</td><td colspan="4">Night LP/OP (PIO)</td></tr>
<tr><th colspan="5">Evaluation Schedule</th></tr>
<tr><th>Date and Time</th><th>1st Platoon</th><th>2d Platoon</th><th>3d Platoon</th><th>4th Platoon</th></tr>
<tr><td></td><td>Lane 1</td><td>Lane 2</td><td>Lane 3</td><td>Lane 4</td></tr>
<tr><td></td><td>Lane 2</td><td>Lane 3</td><td>Lane 4</td><td>Lane 5</td></tr>
<tr><td></td><td>Lane 3</td><td>Lane 4</td><td>Lane 5</td><td>Lane 1</td></tr>
<tr><td></td><td>Lane 4</td><td>Lane 5</td><td>Lane 1</td><td>Lane 2</td></tr>
<tr><td></td><td>Lane 5</td><td>Lane 1</td><td>Lane 2</td><td>Lane 3</td></tr>
</table>

Figure 8-3. Evaluation Scheme

Unit Training Schedule and Firing Tables

Unit Training Schedule 1

Training Day 1

Time	Subject	Location	Uniform	Instructor	Notes
0800	Administrative time	Classroom	U1		
0830	GRM	Classroom	U1		
1000	ROE and use of force	Classroom	U1		
1200	Chow				
1300	MOOTW	Classroom	U1		
1500	Effective communication	Classroom	U1		

Administration:

Uniforms:

U1: BDU

U2: BDU and LBE with harness and canteens

U3: BDU and LBE with harness, canteens, flak jacket, and Kevlar helmet

U4: BDU and LBE with harness, canteens, flak jacket, Kevlar helmet, and assigned weapon

Transportation:

Weapons:

Ammunition:

Required Equipment for Training:

1. Projection screen
2. Video projector with CPU
3. Television
4. VCR

Miscellaneous:

1. All personnel are required to have a notebook and a pencil or pen.
2. Handouts must be prepared and passed out.

Figure 8-4. Sample Two-Week Plan

Training Day 2

Time	Subject	Location	Uniform	Instructor	Notes
0800	NLCS	Classroom	U1		
0900	PPCT and empty hand	Classroom	U1		
1000	PPCT and empty hand (Practical application)	PT field	U2		
1200	Chow				
1300	PPCT and empty hand (Practical application)	PT field	U2		
1700	COB				

Administration:

Uniforms:

U1: BDU

U2: BDU and LBE with harness and canteens

U3: BDU and LBE with harness, canteens, flak jacket, and Kevlar helmet

U4: BDU and LBE with harness, canteens, flak jacket, Kevlar helmet, and assigned weapon

Transportation:

Weapons:

Ammunition:

Required Equipment for Training:

1. Projection screen
2. Video projector with CPU
3. Training strike bags
4. Flex-Cufs

Miscellaneous:

1. All personnel are required to have a notebook and a pencil or pen.
2. Handouts must be prepared and passed out.
3. Medical personnel must be on location.

Figure 8-4. Sample Two-Week Plan (Continued)

Training Day 3					
Time	**Subject**	**Location**	**Uniform**	**Instructor**	**Notes**
0800	PPCT and empty hand (Practical application)	PT field	U2		
1200	Chow				
1300	PPCT and empty hand (Practical application)	PT field	U2		
1700	COB				

Administration:

Uniforms:

U1: BDU

U2: BDU and LBE with harness and canteens

U3: BDU and LBE with harness, canteens, flak jacket, and Kevlar helmet

U4: BDU and LBE with harness, canteens, flak jacket, Kevlar helmet, and assigned weapon

Transportation:

Weapons:

Ammunition:

Required Equipment for Training:

1. Projection screen
2. Video projector with CPU
3. Training strike bags
4. Flex-Cufs

Miscellaneous:

1. All personnel are required to have a notebook and a pencil or pen.
2. Handouts must be prepared and passed out.
3. Medical personnel must be on location.

Training Day 4					
Time	**Subject**	**Location**	**Uniform**	**Instructor**	**Notes**
0800	Riot baton training	Classroom	U1		
0900	Riot baton training (Practical application)	PT field	U2		
1200	Chow				

Figure 8-4. Sample Two-Week Plan (Continued)

1300	Riot baton training (Practical application)	PT field	U2		
1700	COB				

Administration:

Uniforms:

U1: BDU

U2: BDU and LBE with harness and canteens

U3: BDU and LBE with harness, canteens, flak jacket, and Kevlar helmet

U4: BDU and LBE with harness, canteens, flak jacket, Kevlar helmet, and assigned weapon

Transportation:

Weapons:

1. Expandable riot baton with carrier
2. Training riot batons

Ammunition:

Required Equipment for Training:

1. Projection screen
2. Video projector with CPU
3. Training strike bags
4. Flex-Cufs

Miscellaneous:

1. All personnel are required to have a notebook and a pencil or pen.
2. Handouts must be prepared and passed out.
3. Medical personnel must be on location.

Training Day 5

Time	Subject	Location	Uniform	Instructor	Notes
0800	Riot baton training (Practical application)	PT field	U2		
1000	Crowd dynamics	Classroom	U1		
1130	Chow				
1230	Crowd dynamics	Classroom	U1		
1330	CD formations	Classroom	U1		

Figure 8-4. Sample Two-Week Plan (Continued)

1430	RCAs and OC	Classroom	U1		
1700	COB				

Administration:

Uniforms:

U1: BDU

U2: BDU and LBE with harness and canteens

U3: BDU and LBE with harness, canteens, flak jacket, and Kevlar helmet

U4: BDU and LBE with harness, canteens, flak jacket, Kevlar helmet, and assigned weapon

Transportation:

Weapons:

1. Expandable riot baton with carrier
2. Training riot baton

Ammunition:

Required Equipment for Training:

1. Projection screen
2. Video projector with CPU
3. Training strike bags
4. Flex-Cufs

Miscellaneous:

1. All personnel are required to have a notebook and a pencil or pen.
2. Handouts must be prepared and passed out.
3. Medical personnel must be on location.

Training Day 6

Time	Subject	Location	Uniform	Instructor	Notes
0800	RCAs and OC (Practical application)	PT field	U2		
1200	Chow				
1300	RCAs and OC (Practical application)	PT field	U2		
1700	COB				

Figure 8-4. Sample Two-Week Plan (Continued)

Administration:

Uniforms:

U1: BDU

U2: BDU and LBE with harness and canteens

U3: BDU and LBE with harness, canteens, flak jacket, and Kevlar helmet

U4: BDU and LBE with harness, canteens, flak jacket, Kevlar helmet, and assigned weapon

Transportation:

Weapons:

Ammunition:

1. MK inert OC
2. MK 4 live OC disperser
3. M36 live CR disperser

Required Equipment for Training:

1. Five large clean trash cans with liners
2. Two water hoses with running water

Miscellaneous:

Medical personnel must be on location.

Training Day 7

Time	Subject	Location	Uniform	Instructor	Notes
0800	NL munitions employment	Classroom	U1		
1030	Chow				
1115	Safety brief	Range	U4		
1130	NL munitions (live fire)	Range	U4		
TBD	Evening chow				
TBD	Safety brief	Range	U4		
TBD	NL munitions night fire exercise	Range	U4		
TBD	Clean up and COB				

Figure 8-4. Sample Two-Week Plan (Continued)

Administration:

Uniforms:

U1: BDU

U2: BDU and LBE with harness and canteens

U3: BDU and LBE with harness, canteens, flak jacket, and Kevlar helmet

U4: BDU and LBE with harness, canteens, flak jacket, Kevlar helmet, and assigned weapon

Transportation:

Yes

Weapons:

1. M16A2 with M203
2. 12-gauge shotgun

Ammunition:

1. Fin-stabilized M1012
2. M1013 multipellet (rubber)
3. M1006 40-millimeter sponge grenade
4. M1026 40-millimeter multiball (rubber)
5. M84 diversionary grenade

Required Equipment for Training:

1. Fifteen to twenty human form targets
2. Eye and hearing protection
3. Fifteen to twenty riot shields
4. Two MREs per student
5. Video recorder
6. Three to five high-intensity lights
7. All students should be issued face shields, shin guards, and ammunition pouches

Miscellaneous:

1. Medical personnel must be on location.
2. Range safety officers must be available.
3. Primary and secondary means of communication with range control must be available.
4. Operational risk assessments must be conducted.

Figure 8-4. Sample Two-Week Plan (Continued)

Training Day 8					
Time	**Subject**	**Location**	**Uniform**	**Instructor**	**Notes**
0800	CD formations (Practical application)	TA 1	U4		
1200	Chow				
1300	CD formations (Practical application)	TA 1	U4		
1700	COB				

Administration:

Uniforms:

U1: BDU

U2: BDU and LBE with harness and canteens

U3: BDU and LBE with harness, canteens, flak jacket, and Kevlar helmet

U4: BDU and LBE with harness, canteens, flak jacket, Kevlar helmet, and assigned weapon

Transportation:

Yes

Weapons:

1. Training weapons
2. Training riot batons

Ammunition:

Required Equipment for Training:

1. Eye protection
2. Fifteen to twenty riot shields
3. Video recorder
4. Three to five high-intensity lights
5. Face shields, shin guards, and ammunition pouches

Miscellaneous:

1. Medical personnel must be on location.
2. Primary and secondary means of communication must be available.
3. Operational risk assessment must be conducted.
4. Twenty to thirty OPFOR personnel must be available.

Figure 8-4. Sample Two-Week Plan (Continued)

Training Day 9

Time	Subject	Location	Uniform	Instructor	Notes
0800	Introduction to M5, PVAB, and M315	Classroom 1	U1		
1200	Chow				
1300	M5, PVAB, and M315 (Practical application/LFE)	Range	U4		
1700	COB				

Administration:

Uniforms:

U1: BDU

U2: BDU and LBE with harness and canteens

U3: BDU and LBE with harness, canteens, flak jacket, and Kevlar helmet

U4: BDU and LBE with harness, canteens, flak jacket, Kevlar helmet, and assigned weapon

Transportation:

Yes

Weapons:

Two complete M315s and installation kits

Ammunition:

1. Ten live MCCMs
2. Ten live M315 rounds of each

Required Equipment for Training:

1. Eye protection
2. Ten inert MCCMs
3. Two complete PVAB systems

Miscellaneous:

1. Medical personnel must be on location.
2. Primary and secondary means of communication must be available.
3. Operational risk assessment must be conducted.

Figure 8-4. Sample Two-Week Plan (Continued)

Training Day 10

Time	Subject	Location	Uniform	Instructor	Notes
0800	Tactical consideration	MOUT classroom	U1		
0930	Safety brief	MOUT classroom	U4		
1000	Tactical exercise	MOUT Facility	U4		

Administration:

Uniforms:

U1: BDU

U2: BDU and LBE with harness and canteens

U3: BDU and LBE with harness, canteens, flak jacket, and Kevlar helmet

U4: BDU and LBE with harness, canteens, flak jacket, Kevlar helmet, and assigned weapon

Transportation:

Yes

Weapons:

1. Assigned weapons
2. Inert RCA/OC individual dispersers
3. Training riot batons
4. Fifty practice grenades

Ammunition:

Required Equipment for Training:

1. Eye protection
2. Ten inert MCCMs
3. Face shields, shin guards, and ammunition pouches
4. Fifteen to twenty full-length riot shields

Miscellaneous:

1. Medical personnel must be on location.
2. Primary and secondary means of communication must be available.
3. Operational risk assessment must be conducted.
4. A video recorder must be available.

Figure 8-4. Sample Two-Week Plan (Continued)

Sample Unit Training Schedule				
Training Day 1				
Time	**Event**	**Location**	**Instructor**	**Notes**
0730-0830	Force continuum	Classroom		L
0830-0930	Management of aggressive behavior	Classroom		L
0930-1030	ROE	Classroom		L
1030-1130	Crowd control	Classroom		L
1130-1230	Chow			
1230-1700	PPCT and empty hand	PT field		L, PA
Training Day 2				
Time	**Event**	**Location**	**Instructor**	**Notes**
0730-1130	Riot baton training	PT field		L, PA
1130-1230	Chow			
1230-1530	Riot baton training	PT field		PA
1530-1700	CD formations	Classroom		L
Training Day 3				
Time	**Event**	**Location**	**Instructor**	**Notes**
0730-1130	CD formations	PT field		PA
1130-1230	Chow			
1230-1700	NL munitions, NLW	Range		L, PA
Training Day 4				
Time	**Event**	**Location**	**Instructor**	**Notes**
0730-1130	RCAs and OC	Classroom and PT field		L, PA
1130-1230	Chow			
1230-1700	RCAs and OC	PT field		PA

Figure 8-5. Sample One-Week Plan

Training Day 5				
Time	**Event**	**Location**	**Instructor**	**Notes**
0730-1700	Tactical exercise	MOUT facility		OPS
NOTE: Appointed student leaders ensure that troops are fed and rotated for rest.				

Figure 8-5. Sample One-Week Plan (Continued)

Sample Unit Training Schedule				
Training Day 1				
Time	**Event**	**Location**	**Instructor**	**Notes**
	Shield and riot baton techniques	MOUT site		Individual and collective tasks
	NLCS and munitions	Range Classroom		Leader training
	Formation training	MOUT site		Individual and collective tasks
	Chow			
	Formation training	MOUT site		Individual and collective tasks
Training Day 2				
Time	**Event**	**Location**	**Instructor**	**Notes**
0730-1930, 12-hour operations	Live-fire exercise	Range		M203/12-gauge
	QRF exercise	MOUT site		Individual and collective tasks
NOTE: Once weapons and munitions familiarization is complete, shooters join the QRF exercise to be used. For live-fire range setup, see *Figure 8-1*.				

Figure 8-6. Sample Two-Day Plan

Appendix A

Metric Conversion Chart

This appendix complies with current Army directives which state that the metric system will be incorporated into all new publications. *Table A-1* is a metric conversion chart.

Table A-1. Metric Conversion Chart

US Units	Multiplied By	Metric Units
Cubic feet	0.0283	Cubic meters
Feet	0.3048	Meters
Feet per second	0.3048	Meters per second
Inches	0.0254	Meters
Inches	25.4001	Millimeters
Miles per hour	1.6093	Kilometers per hour
Pounds	0.4536	Kilograms
Yards	0.9144	Meters
Metric Units	**Multiplied By**	**US Units**
Cubic meters	35.3146	Cubic feet
Kilograms	2.2046	Pounds
Kilometers per hour	0.6214	Miles per hour
Meters	3.2808	Feet
Meters	39.3701	Inches
Meters	1.0936	Yards
Meters per second	3.2808	Feet per second
Millimeters	0.0394	Inches

Appendix B

Operations and Legal Considerations in the Continental United States

Providing military assistance to state and local governments to assist them in quelling a civil disturbance or riot requires close coordination through a host of state and federal agencies. It requires a thorough briefing of soldiers at all levels on what they can and cannot do with respect to law enforcement. Civil authorities must be briefed on the restrictions placed on federal forces by the *Constitution of the United States* and federal statutes and laws. The national guard (NG), when not in a federal status, operates under the control of the state governor and the adjutant general (AG). The NG has historically been the first military responder during emergencies.

FEDERAL INTERVENTION AND AID

B-1. Under the *Constitution of the United States* and *United States Code*, the President is empowered to direct federal intervention in civil disturbances to—

- Respond to state requests for aid in restoring order.
- Enforce the laws of the United States.
- Protect the civil rights of citizens.
- Protect federal property and functions.

B-2. Under the *Constitution of the United States*, each state is responsible for protecting life and property within its boundaries. State and local governments use their LEA forces to maintain law and order and to quell civil disturbances. However, if civil disturbance support requirements exceed the resources of a state, federal troops may be requested to help restore and maintain law and order.

B-3. The *Constitution of the United States* and federal statutes authorize the President to direct the use of federal armed troops within the 50 states and territories and their political subdivisions. The President is also empowered to federalize the NG of any state to suppress rebellion and enforce federal laws.

B-4. Federal assistance is provided to a state when the state has used all of its resources, including its NG, to quell a disorder and finds the resources insufficient. Usually, active duty federal forces are used to augment the NG of the requesting state. However, the President may choose to federalize the NG of another state and use them alone or with other forces to restore order.

B-5. The President may employ federal troops to ensure the execution of US law when a state opposes or obstructs US law or impedes the course of justice under those laws. The President may employ armed federal troops to suppress insurrection, domestic violence, unlawful assemblies, and conspiracy if such acts deprive the people of their constitutional rights and privileges and the civil authorities of a state cannot or will not provide adequate protection.

B-6. The president is also authorized to use armed federal troops to protect federal property and functions when the need for protection exists and local civil authorities cannot or will not give adequate protection. The right of the United States to protect all federal property and functions regardless of their locations is an accepted principle of our government.

B-7. As a temporary measure, federal military equipment and facilities may be loaned to law enforcement agencies of state and local governmental bodies for use during civil disturbances. These resources may also be loaned to state NG and non-DOD federal agencies. The requesting agencies are expected to provide enough resources of their own to minimize the need for US military resources. In addition, the loan of the resources must not conflict with US military needs.

ROLES AND RESPONSIBILITIES OF VARIOUS AGENCIES

B-8. Several agencies within the federal and state governments play a key role in supporting the request of local authorities for support during a civil disturbance.

SECRETARY OF DEFENSE

B-9. The Secretary of Defense retains approval for federal support to civil authorities involving the use of DOD forces, personnel, and equipment. Approval may also involve DOD support that will result in a planned event with the potential for confrontation with specifically identified individuals and groups or will result in the use of lethal force.

SECRETARY OF THE ARMY

B-10. The Secretary of the Army, as an executive agent for the Secretary of Defense, is the approval authority for federal emergency support in response to natural and man-made disasters (except weapons of mass destruction [WMD]). Military forces respond to directions and guidance from the director of military support (DOMS) (the action agent for the Secretary of the Army).

DIRECTOR OF MILITARY SUPPORT

B-11. The DOMS plans and executes the DOD domestic support mission to civil authorities. He is a general officer appointed by the Secretary of the Army. The DOMS is the DOD primary contact for all federal departments and agencies during DOD involvement in most domestic support operations.

STATE COORDINATING OFFICER

B-12. The governor appoints the state coordinating officer (SCO) to coordinate disaster operations for the state. The SCO is the primary point of contact for the federal coordinating officer (FCO) in facilitating disaster assistance.

DIRECTOR OF THE FEDERAL EMERGENCY MANAGEMENT AGENCY

B-13. The director of the Federal Emergency Management Agency (FEMA), on behalf of the President, appoints the FCO. The FCO coordinates the timely delivery of federal disaster assistance to the affected state and local governments and disaster victims. He works closely with the SCO to determine state requirements and to coordinate national level issues with the catastrophic disaster response group. There must be close and continuous coordination with FEMA to ensure the proper use of military assets in support of civil authorities.

DEFENSE COORDINATING OFFICER

B-14. The defense coordinating officer (DCO) operates under the authority of the President and is the DOD representative designated to coordinate on-the-scene activities with the FCO. FEMA and other federal agencies that request support from DOD go through the DCO for validation and resourcing.

NATIONAL GUARD

B-15. The NG, in state status (see *Title 32, United States Code {USC}),* is the governor's primary military response organization for emergencies and disasters. The NG responds and operates under state laws unless they are federalized, when they operate under the same constraints and laws as federal units.

UNITED STATES ARMY RESERVE

B-16. The US Army Reserve is capable of providing military support. This assistance and support may include the use of equipment and other resources, including units and individuals.

STATE AND LOCAL GOVERNMENT RESPONSIBILITIES

B-17. State and local government officials, operating under the authority granted by state constitutions and the *Constitution of the United States,* are responsible for daily safety and security issues that impact the quality of life of their citizens. State and local officials have primary responsibility for emergency preparedness.

STATE RESPONSIBILITIES

B-18. Many state government agencies are responsible for coming to the aid of state citizens.

Governor

B-19. A state governor is empowered by each state constitution and the *Constitution of the United States* to execute the laws of the state and to command the state NG while it is serving in state status. Governors are also responsible for issuing executive orders or emergency proclamations, such as declaring states of emergency and ensuring that state agencies plan for potential disasters and civil disorder.

B-20. Once a man-made or natural disaster occurs, the governor assesses its extent and determines if local government requests for assistance should be honored. If appropriate, the governor declares a state of emergency, activates the state emergency response plan, and may order the NG into state active duty. The governor assigns the NG its mission through the office of emergency service (OES)/emergency medical dispatching (EMD).

Plans, Operations, and Military Support Officer

B-21. Each state NG has a plans, operations, and military support officer (POMSO) who prepares and coordinates contingency plans for military support to civil authorities (MSCA) during disaster response and recovery operations. The POMSO coordinates training plans and exercises between the state NG and federal, state, and local emergency management agencies. The POMSO serves as the NG point of contact with DOD officials during a declared federal emergency or disaster.

Office of Emergency Services

B-22. Each state has a specific agency and/or office that coordinates emergency preparedness and serves as the governor's coordinating agency during an emergency. The titles of these offices vary from state to state. Some examples are the Division of Emergency Government, Emergency Management Agency, Department of Public Safety, or Office of Emergency Preparedness. This manual refers to this office using the generic term OES/EMD.

B-23. Generally, the OES/EMD is organized as a stand-alone office under the governor or aligned under the AG or the state police, and in some instances, the AG is the OES/EMD director. It operates the state emergency operations center and coordinates with state agencies and federal officials for assistance.

Adjutant General

B-24. The state NG is the governor's primary military response force in an emergency. The AG, through the state area command (STARC) (specifically the POMSO), develops and coordinates emergency response plans for disasters and emergencies. The AG is in command of state NG forces ordered into state active duty.

State Government Agencies

B-25. State government departments and agencies prepare emergency preparedness plans for their areas of specialization in support of the governor's plan. They also participate in emergency preparedness exercises.

LOCAL RESPONSIBILITIES

B-26. Mayors, city managers, local police, fire protection officials, county executives, sheriffs, prosecuting attorneys, and public health officials are some of the people responsible for law enforcement, safety, health, and fire protection on a daily basis. They are responsible for developing appropriate emergency response plans and responding to emergencies within their jurisdictions. Most local jurisdictions have an OES/EMD to plan and coordinate actions in an emergency. In many cases, local jurisdictions have mutual aid agreements with other jurisdictions that allow for firefighting, medical, and police assistance across jurisdictional boundaries. Once local officials determine that an emergency is beyond the scope of their resources or ability to respond, the senior local official is responsible for requesting additional assistance from the next agency in the assistance request process.

LEGAL CONSIDERATIONS AND CONSTRAINTS

B-27. The *Constitution of the United States*, laws, regulations, policies, and other legal issues limit the use of federal military personnel in domestic support operations. Any Army involvement in civil disturbance operations involves many legal issues requiring comprehensive legal reviews.

B-28. Under the *Constitution of the United States*, Congress has the authority to raise and support an army, provide and maintain a navy, and make rules for governing and regulating the land and naval forces. The *Constitution of the United States* places the military under civilian control and designates the President as commander-in-chief. Statutes provide for civilian leadership in the office of the secretary of defense, service secretaries, and various other civil authorities. Under the *Constitution of the United States*, the civilian government must preserve public order and carry out governmental operations within its territorial limits. The *Constitution of the United States* allows the use of the federal military to execute or enforce the law when necessary to protect federal or civilian property and functions. However, significant restrictions exist on employing federal military forces within the United States.

B-29. The unique capabilities of the military enable it to support federal, state, or local civilian agencies. In most circumstances, the DOD is one of many federal agencies reacting to a domestic emergency or crisis, playing a subordinate, supporting role to a lead civilian agency.

LAW

B-30. In the United States, civilian agencies provide for the needs of citizens. Civilian, federal, state, and local government and law enforcement agencies execute US laws. Laws governing the use of the military in domestic operations are complex, subtle, and constantly changing. Commanders involved in civil disturbance operations should staff plans, policies, programs, exercises, funding, operations, constraints, and limitations with their local SJA to ensure conformity with legal requirements. They should scrutinize each request for aid (whether it is for equipment, individuals, or training) to ensure that it conforms to statutory requirements. The following references are a snapshot of the law on military support to civilian authorities:

- *Section 331, Chapter 15, Part 1, Subtitle A, Title 10 USC.*
- *Section 332, Chapter 15, Part 1, Subtitle A, Title 10 USC.*
- *Section 333, Chapter 15, Part 1, Subtitle A, Title 10 USC.*
- *Section 334, Chapter 15, Part 1, Subtitle A, Title 10 USC.*
- *Section 1385, Chapter 67, Part 1, Title 18 USC.*
- *DOD Directive 3025.12.*

- *DOD Directive 3025.15*
- *DOD Directive 5525.5*
- *DOD Civil Disturbance Plan 55-2*
- *NGR 500-1/ANGI 10-8101*
- *FM 3-19.15*

INFORMATION OPERATIONS

B-31. Information superiority helps forces anticipate problems and requirements. It allows commanders to control situations earlier and with less force, creating the conditions necessary to achieve the end state. Public affairs, PSYOP, and civil military operations are activities that will allow the commander to control situations earlier and with less force.

B-32. Information is available from a multitude of sources, the primary being open sources, law enforcement, and the military. A diversity of sources is the best approach because it prevents biased behavior.

Open sources include—

- Libraries.
- Newspapers and news periodicals.
- Radio and television.

Law enforcement sources include—

- Local law enforcement agencies.
- National law enforcement agencies.

Military sources include—

- DOD intelligence community (most restrictive source).
- Local MI field offices.

B-33. The DOD intelligence community operates under limitations imposed by regulations and executive orders. Attempts to skirt these restrictions may violate regulations or federal statutes.

RESTRICTIONS

B-34. Strict limits are placed on collecting information related to a civil disturbance in order to protect the civil rights of people and organizations not affiliated with the DOD. Civil disturbance plans and materials must not include lists of groups or people not affiliated with the DOD. Information on civilians and civilian organizations can be collected only with specific authorization from the Secretary or the Under Secretary of the Army. Conditions for collecting information include the existence of threats against Army personnel, functions, or property (see *ARs 380-10* and *380-13).* Civil disturbance information (available in public documents) or open-source information may be collected. However, specific rules regarding its storage must be followed. Commanders must coordinate with SJA, MI, and USACIDC before collecting any such information.

B-35. Civilian law enforcement agencies may limit access to law enforcement intelligence. In such circumstances, the military may request that agencies release the unclassified version.

B-36. If the Department of Justice (DOJ) determines that federal intervention into a civil disturbance is likely, then information relating to the disturbance is provided to the Army Assistant Chief of Staff for Intelligence. The information is analyzed and then passed through channels to the TF commander for planning purposes.

B-37. Military intelligence collection efforts, except liaison, may begin only when DA orders. The orders will come through the CSA's personal liaison officer and the TF commander. Covert operations to gather information on non-DOD individuals and groups must be approved on an operation-by-operation basis by the Undersecretary of the Army.

B-38. Once DA approves the collection effort, MI elements establish and maintain liaison with the appropriate local, state, and federal authorities. Then they collect information pertaining to the incident and general situation. Their estimate of the situation is passed to the TF commander and DA.

POSSE COMITATUS

B-39. Generally, federal military forces may not give law enforcement assistance to civil authorities without conflicting with the *Posse Comitatus Act*. However, constitutional and statutory exceptions to this prohibition do exist. The recent emphasis on drug interdiction has led to an increase in those exceptions.

B-40. The *Judiciary Act of 1789* allowed US marshals to call upon the military as a posse comitatus. This continued until after the Civil War, when the federal government used the Army to execute reconstruction era policies. The southern states regarded the use of the military for this purpose as abusive and repressive, and in 1878, President Rutherford B. Hayes signed the original bill ending the practice.

B-41. The *Posse Comitatus Act* prescribes criminal penalties for the use of the US Army or Air Force to execute the laws of or to perform civilian law enforcement functions within the United States. DOD policy extends this prohibition to the US Navy and Marine Corps. Prohibiting the military from executing the laws means that military personnel may not participate directly in—

- An arrest; a search and seizure; a stop and frisk; or an interdiction of vessels, aircraft, or vehicles.
- A surveillance or pursuit.
- A civilian legal case or any other civilian law enforcement activity as informants, undercover agents, or investigators.

B-42. The *Posse Comitatus Act* does not apply to the following:

- Members of the NG when in state service.
- Members of a Reserve Component when not on active duty or active duty for training.
- DOD civilians, unless under the direct command and control of an active duty officer.

- A soldier when off duty and acting only in a private capacity.
- Soldiers taking action for the primary purpose of furthering a military or foreign affairs function of the United States; for example, enforcing military justice, maintaining law and order on military installations, and protecting classified materials.

CONSTITUTIONAL EXCEPTIONS TO THE POSSE COMITATUS ACT

B-43. Under its inherent authority, the US government is responsible for preserving public order and carrying out governmental operations within its territorial limits, by force if necessary. Under the *Constitution of the United States*, exceptions allow the use of military force to execute or enforce the law. Some of the exceptions are to—

- Protect civilian property and functions.
- Ensure preservation of public order and carry out government operations, by force if necessary.
- Protect civil rights where local authorities do not or cannot protect them.
- Protect federal property and functions.
- Quell civil disturbances and labor strife that rises to the level of civil disorder.

B-44. The President may order the armed forces to support state civil authorities suffering from an insurrection or civil disturbance. He must act personally by first issuing a proclamation calling on insurgents to disperse and retire peacefully within a limited time. This occurred when federal forces were called in to counter the Los Angeles riots in 1992.

Protect Civilian Property and Functions

B-45. A sudden and unexpected civil disturbance, disaster, or calamity may seriously endanger life and property and disrupt normal governmental functions to such an extent that local authorities cannot control the situation. At such times, the federal government may use military force to prevent loss of life or wanton destruction of property and to restore government functions and public order. This exception has rarely been used.

Protect Federal Property and Functions

B-46. The federal government may use military force to protect federal property and federal government functions when local authorities cannot or decline to provide adequate protection.

ARMED FORCES

B-47. Basic guidance for the interaction of the military, the Reserve Component, and civilian LEAs is given in *10 USC Chapter 18*. Guidelines on reimbursement and restrictions, direct participation in law enforcement activities, the use of information collected during military operations, and the use of military equipment and facilities are some of the topics covered. Title 10 prohibits the military from directly participating in arrests, searches, seizures, or other similar activity unless authorized by law (such as arrests on military property). The *Fiscal Year 1989 Act* and subsequent *National Defense Authorization Acts* have authorized the DOD to provide more support to LEAs in the counterdrug effort.

NATIONAL GUARD

B-48. For a description of how the secretary of defense may provide funds to state governments (including the District of Columbia, Commonwealth of Puerto Rico, and US territories) for civil disturbance operations performed by the National Guard when not in federal service, see 32 *USC*. The *Posse Comitatus Act* does not apply to the NG when not in federal service. Each state specifies how NG may be employed.

SUPPORT TO CIVIL LAW ENFORCEMENT

B-49. It is DOD policy to cooperate with civilian law enforcement officials to the greatest practical extent. However, cooperation must consistently meet the needs of national security and military preparedness, the historic tradition of limiting direct military involvement in civilian LEAs, and the requirements of applicable law. US military forces are never placed under the command of civilian law enforcement officers or nonfederalized NG. DOD policy concerning the provision of military support to LEAs, including personnel and equipment, are contained in *DOD Directive 5525.5.*

B-50. The Army assists civilian law enforcement by providing personnel, equipment, training, and expert advice within the limits of applicable law. Army NG units in state status (see 32 *USC*) provide the primary source of military assistance. They may be able to provide assistance to civil authorities in instances when federal units are precluded due to the restrictions of the *Posse Comitatus Act.*

ROLES OF THE NATIONAL GUARD AND FEDERAL FORCES

B-51. The preservation of law and order in the civilian community is the responsibility of state and local governments and law enforcement authorities. When a civil disturbance in a civilian community turns to widespread rioting that includes arson, looting, and acts of violence, the civil authorities may decide that they do not have the resources to quell the riot. They may then turn to the NG and federal forces to support the civil authorities in restoring law and order. Civil disturbances in any form are prejudicial to public law and order.

NATIONAL GUARD FORCES

B-52. The NG (as a state organization) responds to the governor according to state law for civil disturbance operations. NG regulations direct planning and training for the civil disturbance mission. During most civil disturbance situations, the NG will be the first military responder and will usually remain in state active duty status (according to 32 *USC*) throughout the operation. In extreme circumstances, the NG can be brought on federal service for civil disturbance operations when ordered to under the appropriate federal statute by the President. During the 1992 Los Angeles civil disturbance, the President federalized the California NG. Once the NG has been deployed to assist the civil authorities with a civil disturbance, it is too far into the deployment cycle to start determining what tasks the NG can do as a state or as a federally controlled force. It is the commander's responsibility not to accept missions that the unit is not adequately trained to do. The following are examples of appropriate missions for the NG:

- Manning traffic control points.
- Providing building security.
- Providing area security and patrols.
- Providing security at custody facilities.
- Providing security and escort for emergency personnel and equipment.
- Protecting sensitive sites.
- Transporting law enforcement personnel.
- Providing show of force.
- Dispersing crowds.
- Employing riot control munitions.
- Providing very important person (VIP) protection and escort.
- Providing quick reaction and reserve force.

B-53. Examples of inappropriate taskings:

- Negotiating hostage situations.
- Barricading suspects.
- Participating in criminal investigations.
- Acting as federal Army forces.

B-54. The commitment of federal troops to deal with domestic civil disturbances must be viewed as a drastic measure of last resort. Their role should never be greater than what is absolutely necessary under the circumstances. Commanders should take every measure to avoid the perception of an invading force. A JTF designated to respond to a civil disturbance should project the image of a restrained and well-disciplined force whose sole purpose is to help restore law and order with minimal harm to the people and property and with due respect for all law-abiding citizens.

B-55. The role of federal Army forces is to assist civil authorities in restoring law and order when the magnitude of the disturbance exceeds the capabilities of local and state LEA, including the NG. Under the provisions of the *Constitution of the United States* and selected federal statutes, the President may order the employment of the federal armed forces to aid local and state civil authorities to protect the Constitutional rights of citizens. Federal military forces may also protect federal facilities and installations in any state, territory, or possession.

B-56. The *DA Civil Disturbance Plan 55-2* gives guidance for Army forces directed to quell civil disturbances. In addition, *AR 500-50* outlines responsibilities, prescribes policies, and provides guidance for the use of Army resources in response to civil disturbances. It outlines statutory restrictions and designates staff and command responsibilities for planning and executing civil disturbance operations.

B-57. The JTF commander exercises control of all federal military forces (including NG in federal status) committed to assist civil authorities. Federal military forces remain under the military chain of command during civil disturbance operations. Federal forces will not be placed under the command of either civil officials or NG commanders in a nonfederal status. Civil authorities retain control of their state and local law enforcement agencies. The JTF commander establishes liaison with the senior representative of the attorney general (SCRAG) and other appropriate federal, state, and local civil authorities.

B-58. When a JTF is being assembled it must deploy with a provost marshal (PM) who will be the POC for situation assessment, operational information, and security matters. He also serves as a liaison with civilian and law enforcement agencies.

B-59. Federal military forces must be tailored to the specific civil disturbance situation. Sufficient combat support and combat service support units will be required to sustain the force throughout the deployment. Coordination with civil officials may allow the force to draw on resources available from state and local agencies. Close and continuous coordination between federal military forces and LEAs will provide the commander with the detailed and timely information required to employ and protect the force effectively.

B-60. In supporting *DA Civil Disturbance Plan 55-2,* intelligence personnel may conduct close and continuous liaison with LEAs and military police to ensure that their units receive the information needed to allow the commander to adequately protect the force. The JTF commander should staff intelligence support missions with his senior intelligence officer and legal counsel before approving the mission.

B-61. Federal military forces must be employed in tasks or missions appropriate to their organization and training; they must not be employed in ways that violate the legal restrictions in effect. Military forces may be used to disperse unlawful assemblies and to patrol disturbed areas to prevent unlawful acts. They may be used to assist in the distribution of essential goods and the maintenance of essential services. Forces may also establish traffic control points, cordon off areas, release smoke and obscurants, and serve as security or quick reaction forces. Certain types of missions (such as gathering intelligence on civilians) are always inappropriate for military forces during civil disturbance operations.

B-62. Requests for specific military missions are typically passed through a single state or federal law enforcement coordinating officer, as approved by the SCRAG. Validated requests are transmitted to the JTF commander and his HQ for staffing and coordination. Approved missions are assigned through the military chain of command to the appropriate element or unit for execution. Units and soldiers will not accept taskings or missions directly from law enforcement or civilian officials, except in a direct support relationship as approved and ordered through the military chain of command.

B-63. It is important that a military liaison be provided to each LEA HQ generating requests for support. This liaison can assist LEA officials in determining the types and quantities of military support to be requested. JTF HQ can facilitate this mission assignment process by providing LEAs with a detailed listing of the types of missions military forces may conduct.

B-64. The AO for a deployed unit should coincide with the jurisdiction or subdivision boundaries of the LEA it supports. This arrangement facilitates liaison and coordination between law enforcement and military chains of command.

LEAD AGENCY CONCEPT AND ROLE OF MILITARY

B-65. The DOJ is the lead federal agency for civil disturbance operations. The Attorney General's on-scene representative is the SCRAG. The DOD has designated DA as their executive agent for military assistance for civil disturbance (MACDIS) (see *FM 27-100.*)

JOINT TASK FORCE

B-66. Military support to civil authorities in disasters and domestic emergencies is a DOD (not a service component) responsibility. The supported commander may decide that the severity and scope of a disaster require a joint response. In civil assistance, as in operational level warfighting, the commander-in-chief (CINC) uses the different and complementary capabilities of each service to accomplish the mission. Such use requires knowledge of both the capability and the availability of all service component assets, to include their agencies and installation.

B-67. The CINC may establish a JTF to provide comprehensive military support. Establishing a JTF may provide the best DOD response to civil emergencies. A JTF is established to execute a specific mission limited in scope and duration. The JTF objective in civil emergencies is to deploy forces to the area rapidly to assist immediately in saving lives and safeguarding property and to continue providing assistance required by the federal response plan (FRP) and the particular situation.

B-68. The JTF is configured for each specific mission. In some civil emergencies, the JTF may require a greater proportion of combat service support type units and capabilities than in typical warfighting deployments. The JTF must be able to provide emergency assistance across all lines of support.

JOINT TASK FORCE CONTROL ELEMENT

B-69. Responsibility for controlling the civil disturbance resides with the civil authorities. The JTF is organized to support and lend assistance to them in their effort to return civil order to the community.

B-70. The control element consists of the command group and the crisis management team (CMT). The command group is made up of city, county, state, and military command personnel. They set policy and issue directives, coordinate the activities of civil and military support agencies, supervise the CMT, and coordinate with outside agencies. The CMT is made up of representatives of civilian and military staff sections. The CMT advises the command group and coordinates operations and support for the action element of the TF. The control elements are located with the emergency operations center (EOC) to facilitate information processing, resource management, and operational control. The JTF command element should colocate with the civilian command element at the EOC. This will expedite the exchange of information. If an EOC has not been set up, the commander may establish one while making provisions for including civil authorities to ensure a unity of effort.

B-71. Not all CMT members are located at the EOC. Some key people may use liaisons to represent them at the EOC. Key personnel can then research and discuss ideas freely, away from the confusion associated with crisis management. The liaison can transmit guidance and answers to the EOC. Some agencies may not be needed in the EOC at first, but they must be prepared for inclusion. The decision to include an agency in the EOC is based on the likelihood that they may have to take an action or a support role on the importance of the agency to the mission.

Appendix C

Urban Terrain Analysis

Of all the environments in which the Army may need to conduct operations, the urban environment confronts commanders with a combination of difficulties rarely found in other environments. The distinct characteristics of the environment result from the combination of man-made features and supporting infrastructure superimposed on the existing natural terrain and the density of civilians in close proximity to combat forces. The human dimension it represents is potentially the most important and perplexing for commanders and their staffs to understand and evaluate. Commanders at all levels must make extraordinary efforts to assess and understand their particular urban environment to plan, prepare for, and execute effective civil disturbance operations.

SIGNIFICANT FACTORS

C-1. Although urban areas throughout the world possess similar characteristics, no two are alike. The sprawl of Los Angeles, for example, bears little physical resemblance to New Delhi. The most significant factors affecting the uniqueness and complexity of each area are its societal characteristics. While the terrain can be extremely complex, information about this aspect, its potential effects on operations, and how it changes over time may be determined with some degree of certainty. However, the human dimension is more difficult to understand and assess, particularly its effect on military operations. The urban environment is not neutral—it helps or hinders. The commander who can best understand and exploit the effects of the urban environment has the best chance of success.

C-2. Whether a metropolitan area or a village, each urban environment has an identifiable system of urban components that constantly change and interact. This system of components consists of the terrain, the society, and the infrastructure that provides the critical link between the two (see *Figure C-1,* page C-2).

C-3. These systems are not separate and distinct categories but instead are overlapping and interdependent. A thorough analysis of these systems contributes to the commander's situational understanding and allows him to develop COAs that apply appropriate resources to control a crowd and prevent a major civil disturbance. This understanding allows commanders to engage decisive points critical to maintaining peace or restoring normalcy to the urban environment.

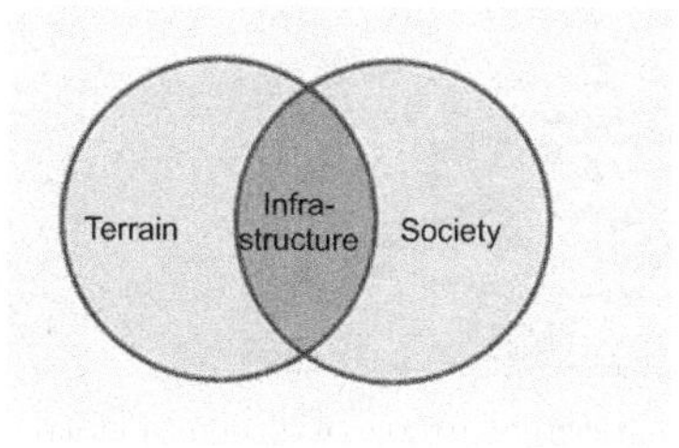

Figure C-1. Keys to Understanding the Urban Environment

C-4. The social aspects and the integrating infrastructure are the most difficult for commanders to understand and assess. These issues require greater dependence on LEAs and local governmental agencies for their information, knowledge, and expertise.

URBAN TERRAIN TYPES

C-5. The physical terrain is the most basic and easily recognizable aspect of the urban area; however, truly understanding it requires comprehending its multidimensional nature, urban and street patterns, urban form and function, and size. The infinite ways in which these factors (see *paragraph C-15)* can intertwine make it impossible to describe a typical urban area. However, these factors provide a framework for understanding the complex terrain within an urban environment. Although the terrain consists of both natural and man-made features, the latter dominates the physical characteristics of the urban environment and has significant effects on military systems and soldiers, tactics, and operations.

URBAN PATTERNS

C-6. The combinations of urban areas, lines of communications (LOCs), and natural terrain result in the identification of four urban patterns (see *Figure C-2*) that will influence civil disturbance operations. An integral part of each urban pattern is the hub or the central urban area around which outlying urban areas radiate. In SOSO, the hub is important, as it is often the economic, political, or cultural center of the surrounding area.

Satellite Pattern

C-7. This pattern consists of a central hub surrounded by smaller, dependent urban areas. LOCs tend to converge on the hub. The natural terrain throughout this pattern is relatively homogenous. Outlying or satellite urban areas often support the principal urban area at the hub.

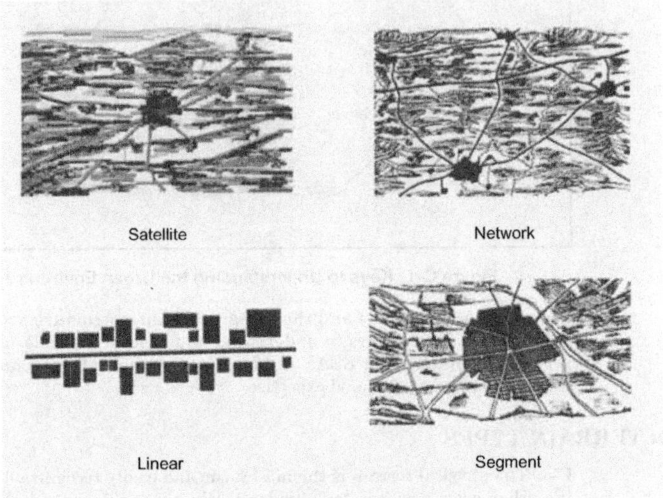

Figure C-2. Urban Patterns

Network Pattern

C-8. This pattern represents the interlocking of the primary hubs of subordinate satellite patterns. Its elements are more self-sufficient and less supportive of each other, although a dominant hub may exist. Major LOCs within a network are more extensive than in a satellite pattern and take a rectangular rather than a convergent form. The natural terrain within a network may vary more than in a single satellite array.

Linear Pattern

C-9. This pattern is a potential subelement of the previous two patterns. It may form one ray of the satellite pattern or be found along connecting links between the hubs of a network. Most frequently, however, this pattern results from the string of minor urban areas aligning a confined natural terrain corridor such as an elongated valley, a body of water, or a man-made communications route.

Segment Pattern

C-10. When dominant natural terrain (such as a river or man-made features [canals, major highways, or railways] divides an urban area, it creates a segmented pattern). This pattern makes it easier to assign areas of responsibility to units. However, if operations are in support of the city LEA, then units should be assigned to a police precinct to avoid crossing several precinct lines, as would happen when assigning units by dominant terrain features.

STREET PATTERNS

C-11. Patterns within the urban area result from the arrangement of the streets, roads, highways, and other thoroughfares. They are a result of the influences of natural terrain, the original designer's personal prejudices, and the changing needs of the inhabitants. The commander must understand the street patterns in the area in which his control force will operate. This will offer force protection, preventing the control formation from being outmaneuvered, flanked, or surrounded by a hostile crowd. Urban areas can display any of the three basic patterns or any combination of the three (see *Figure C-3).*

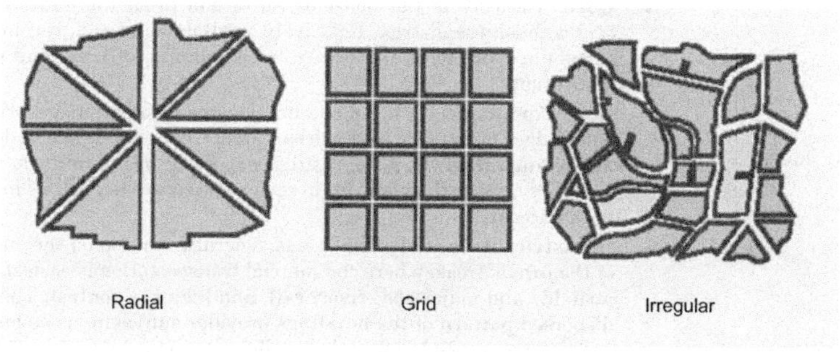

Figure C-3. Basic Internal Street Patterns

Radial

C-12. Societies of highly concentrated religious or secular power often construct urban areas with a radial design with all primary thoroughfares radiating out from the center of power. Terrain permitting, these streets may extend outward in a complete circle or may form a semicircle or arc when a focal point abuts a natural barrier such as a coastline or mountain. To increase mobility and traffic flow, societies often add concentric loops or rings to larger radial patterns.

Grid

C-13. The most adaptable and universal form for urban areas is the grid pattern (lines of streets at right angles to one another forming blocks similar to the pattern of a chessboard). A grid pattern can fill in and eventually take over an original radial pattern. Most urban areas, regardless of the original intent, plan, or vision, emerge from successive plans overlaid on one another.

Areas are well planned to fit with previous plans and others a haphazard response to explosive urban growth. The results may mix different patterns, blend patterns in symmetric combinations, or have no discernable geometric pattern (the irregular pattern).

Irregular

C-14. Irregular patterns can be specifically designed for aesthetic reasons (as in many suburban housing developments). They are sometimes used to conform to marked terrain relief (see *FMs 3-06* and *3-06.11*).

URBAN FORM AND FUNCTION

C-15. Throughout the world, urban areas have similar form and function. In form, urban areas contain like characteristics, readily divisible into distinct sections or areas. Functionally, they tend to be the center of population, finance, politics, transportation, industry, and culture.

- **Core.** The core is the heart of the urban area, the downtown, or central business district. Relatively small and compact, it contains a large percentage of the urban area's shops, offices, and public institutions.
- **Core Periphery.** As the name implies, the core periphery forms the outer edges of the core of the urban area. Generally, it has undergone less change than the core, resulting in buildings of uniform height (generally two to three stories in towns and five to ten stories in larger urban areas).
- **Industrial Area.** Industrial areas generally develop on the outskirts of the urban areas, where commercial transportation is easiest (along airfields and major sea, river, rail, and highway routes). The open/ dispersed pattern of the buildings provides sufficient space for large cargoes, trucks, and materials-handling equipment.
- **High-Rise Area**. High-rise areas consist of multistoried apartments, commercial offices, and businesses separated by large open areas such as parking lots, parks, and individual one-story buildings. The automobile, mass transit systems, and improved road networks encourage these areas to grow and function further from the urban core.
- **Residential Area.** Residential areas can be found dispersed throughout the urban area; however, large suburban areas (or sprawl) normally form on the outskirts. The combined population of surrounding suburban areas often outnumbers that of the urban area. Proper and specific suburbs tend toward homogeneity based on ethnicity, religion, economics, or some other social aspect.
- **Commercial Ribbon Area**. Commercial ribbon areas are rows of stores, shops, and restaurants built along both sides of major streets that run through urban areas. These same types of areas often develop along the roads that connect one urban area to another (strip areas).

URBAN SOCIETY

C-16. Although complex, understanding the urban terrain is relatively straightforward in comparison to comprehending the multifaceted nature of urban society. As urban areas increase in size, they generally lose homogeneity; therefore, commanders must understand and account for the characteristics of a diverse population whose beliefs may vary. The behavior of civilian populations within an urban area is dynamic and poses a special challenge to commanders.

C-17. The center of gravity during an urban operation may be the civilian inhabitants themselves. The side that enjoys the support of the population has many advantages. To gain and retain this support, commanders must first understand the complex nature and character of the urban society. Second, they must understand and accept that every military action (or inaction) may influence the relationship between the urban population and Army forces, and by extension, mission success.

C-18. The identification of groups helps commanders focus on specific segments of the urban society to determine their beliefs and needs and how those beliefs and needs motivate them to future action (or inaction). Groups may be categorized based on race, religion, national origin, tribe, clan, class, party affiliation, education level, union memberships, occupation, age, or any other significant social grouping.

C-19. Commanders should consider political, economic, and historical factors in this analysis. These factors usually affect all groups to some extent and often provide the basis for many of their beliefs, needs (actual or perceived), and subsequent behavior. In most cases, training and discipline, grounded in cultural understanding and sensitivity, will help mitigate many of the potential adverse effects resulting from military and civilian interaction.

URBAN INFRASTRUCTURE

C-20. Urban infrastructures are those systems that support the inhabitants, their economy, and their government. They provide the link between the physical terrain and the urban society. Hundreds of systems may exist. Each system has a critical role in the smooth functioning of the urban area. All systems fit into the following five broad categories: communications and information, transportation and distribution, energy, commerce, and human services.

Appendix D

Practical Application

It is always best to derive lessons learned from practical operations, in the field. These are operations where a measure of success was acquired, along with important lessons learned. The following reflects an actual TF application of the doctrine highlighted in the contents of this manual.

OVERVIEW

D-1. The purpose of this manual is to provide TF units and elements with the latest guidance on providing crowd control and preventing civil disturbances.

D-2. Local civilians often gather in towns or along strategic locations throughout an AO to demonstrate grievances or other causes (for example, protesting the apprehension of a fellow citizen, policies of US forces, and policies of government services). Although civilians have the right to assemble, crowd gatherings are predominantly ethnically based, motivated, and influenced. Local civilians can apply for a permit to legally conduct a rally 72 hours in advance as long as their assembly is peaceful and unobtrusive to government services, freedom of movement, or the rights of others. On occasion, intelligence collection can indicate a possibility for violence. In these instances or during spontaneous crowd formations, the TF must provide forces in support of a safe and secure environment or to assist civilian police in preventing unlawful acts. If unlawful acts are committed, detain violators for adjudication under the legal system.

PROCEDURES

D-3. Take precautions to avoid becoming a catalyst to the civil disturbance. During lawful assemblies, the TF must make every effort to treat HN individuals with respect. This means that the TF must engage in peaceful gatherings using interpersonal communication skills. This ensures that demonstrators understand that TF personnel respect their right to lawfully assemble during peaceful demonstrations. When possible, TF engagement of a peaceful assembly should be unobtrusive.

D-4. Do not display civil disturbance equipment or weapons, overtly prepare defenses against a crowd, or aggressively violate crowd space, unless presented with information, intelligence, or negative indicators of pending violence.

NEGATIVE INDICATORS

D-5. Any response to negative indicators should be sequential, measured, and focused on deescalating rather than escalating the crowd.

NEGATIVE INDICATORS OF A POTENTIAL CIVIL DISTURBANCE

D-6. Predictive intelligence is the best indicator of a potential civil disturbance and may provide insight on internal planning, external influences, environmental conditions, or catalyst events that should influence TF planning and response.

D-7. Other indicators of a potential civil disturbance are people beginning to assemble during a scheduled or spontaneous rally. When people begin to assemble, leaders should pay particular attention to crowd dynamics. As the crowd demeanor or dynamics change from peaceful assembly to unruly behavior or unlawful acts, negative indicators should provide an advance warning and prompt a measured response. Negative indicators of a potential change in crowd dynamics and unlawful activities should prompt immediate countermeasures (see *Figure D-1*).

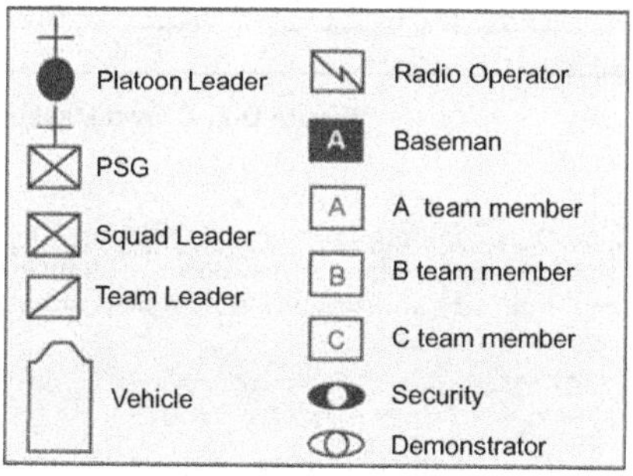

Figure D-1. Symbols for a Formation Element

Orientation

D-8. Crowd dynamics change from people milling around and talking among themselves in isolated groups to a more collective focus concentrated on a single objective (for example, TF personnel, specific agitators or magnets, a government figure, or other activity).

Massing

D-9. The crowd begins to mass or tighten into a large contiguous body from a loose formation to a static congregation located in a more concentrated area. Unchecked, massing can provide anonymity and a collective feeling of invulnerability. This change in group dynamics could provide ideal conditions for a civil disturbance (see *Figure D-2*).

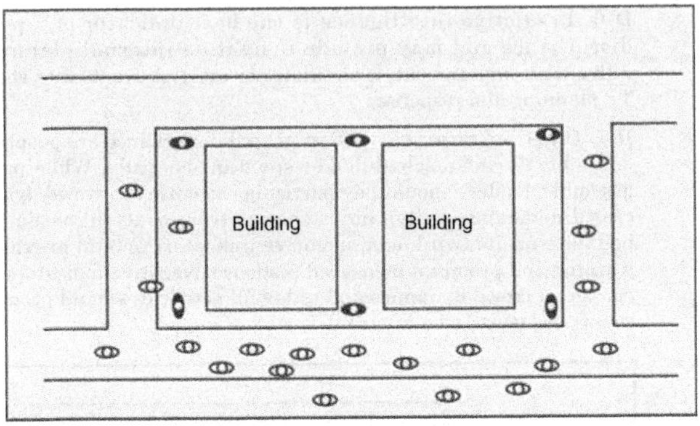

Figure D-2. Crowd Massing

Chanting

D-10. People may begin chanting slogans or yelling ethnically charged rhetoric or obscenities. Because chanting can indicate an increase in crowd intensity, interpreters should be positioned in strategic locations to interpret crowd rhetoric.

Signaling

D-11. Signaling occurs when members of the crowd position themselves to communicate and synchronize crowd actions tied to a plan with specific objectives. Signaling may indicate the intent to escalate the message through the use of violent or unlawful acts.

Blocking

D-12. The crowd may attempt to block public thoroughfares (for example, entrances and exits to public buildings, MSRs, community streets, or other traffic areas that may degrade freedom of movement or similar actions). This escalates the level of attention given to the crowd's message or prevents freedom of movement to opposing persons or groups.

Arming

D-13. Any signs that the crowd is arming itself with objects (for example, bottles, rocks, and sticks) is a clear indication of impending danger to the TF and other civilians. In fact, during a recent civil disturbance, some members of the crowd were seen handing out bottles to others in the crowd. Arming demonstrates intent toward violence is unlawful.

Outside Influence

D-14. The presence of people from outside the community may indicate more sophisticated planning and resources than would otherwise be expected from local community members. Outside people may feel insulated from responsibility for local actions and, therefore, provide a dangerous catalyst for a violent civil disturbance.

Agitators and Instigators

D-15. Agitators and instigators are common elements associated with civil disturbances. They often provide a catalyst to incite the crowd toward violence. Whether spewing rhetoric or committing unlawful acts, their continued presence will fuel potential strife.

Absence of Children

D-16. Recent civil disturbances have shown that even when children were initially present at a rally, they were absent before the civil disturbance erupted with violence. While the presence of children is not a guarantee of a peaceful assembly, the removal of children from a rally may provide early warning for a pending escalation of hostilities or a full-blown riot.

PREVENTION OF CROWD ESCALATION

D-17. If TF personnel believe, based on intelligence or negative indicators, that a crowd is planning to, or may spontaneously, riot they should immediately engage crowd participants using the preventive TTP measures.

D-18. The force executing these preventive TTP can include a number of civil disturbance teams and elements capable of executing the tasks shown in *Figures D-2,* page D-3; *D-3; D-4,* page D-6; and *D-5,* page D-8. These figures demonstrate a full range of crowd control teams and elements, to include camera teams with interpreters, checkpoints, QRF response, riot control formation, NLW teams, snatch and detain teams, blocking teams, and detainee evacuation, as required. *Figures D-2, D-3, D-4,* and *D-5,* illustrate the evolution of a crowd formation and the recommended positioning of teams and elements to engage people as they congregate. Included with each security team is a designated camera or video recorder.

D-19. Depending on the situation and crowd dynamics, it may not be necessary to establish all crowd control teams and elements, but they should be included in planning. Regardless of task organization, teams and elements must be prepared to execute the crowd control and civil disturbance TTP. The order to execute the following TTP should be tailored based on local conditions.

Early Contact

D-20. Engaging people as a crowd forms is essential to prevention. Whether based on intelligence or negative indicators, people should be engaged early, before the crowd masses as shown in *Figure D-2,* rather than waiting until after the crowd masses as shown in *Figure D-3.* Early contact will assist TFs in personalizing their interaction and ensuring that people understand their presence is known, documented, and can be linked to their actions.

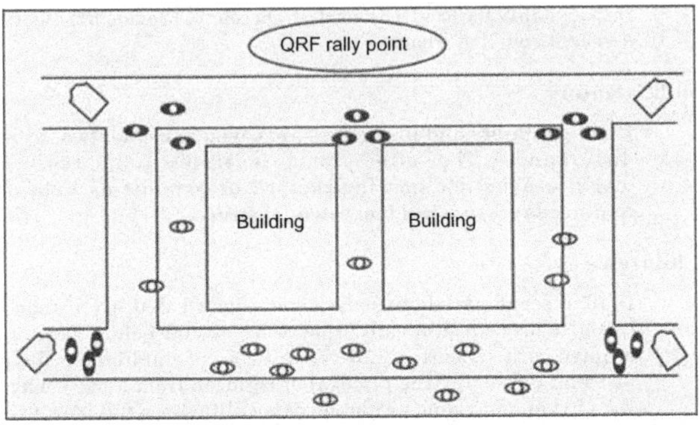

Figure D-3. Early Contact and Unobtrusive Crowd Control Measures

D-21. Cameras are one of the best tools for preventing civil disturbances. The use of cameras to document crowd formations, particularly lens-size groups (small enough to fit in a single photo), will personalize contact and eliminate feelings of anonymity among the crowd. People whose identity has been documented will think twice before committing unlawful acts. Even without documenting names, photo analysis can identify people associated with the civil disturbance, especially those who unlawfully participated as crowd instigators, agitators, or magnets or may commit unlawful acts.

D-22. As people begin to gather in and along streets, camera teams should be deployed as unobtrusively as possible to engage people before they congregate in mass. As depicted in *Figure D-3*, camera teams should consist of two soldiers with an interpreter and be employed in a widely dispersed pattern. Such small, dispersed camera teams are less likely to intrude on lawful rallies or create invasive friction that might become a catalyst to a civil disturbance.

D-23. As the crowd begins to demonstrate negative indicators for a possible civil disturbance (*Figure D-4,* page D-6), camera teams should become more invasive by actively engaging groups to take their pictures. Camera teams should focus cameras on instigators, agitators, or magnets (personnel that garner respect or compliance from the crowd). If a person is purposely avoiding the camera, even if his activities are lawful, the team should take special measures to obtain his picture. Each time a camera team takes a picture of a person or group of people, they must immediately engage them using the warning outlined in *paragraph D-26.*

D-24. Once the crowd has massed and displays negative indicators or begins to riot, camera teams should immediately move behind the riot formation line (see *Figure D-4*). It is essential, however, that camera teams continue to document the event, focusing on people who are committing illegal acts such as arming themselves, throwing objects, and assaulting TF personnel.

D-25. Immediately following a crowd control or civil disturbance operation, photos must be consolidated, processed, and submitted to the TF analysis and control element (ACE) for analysis. A near real-time analysis will provide immediate intelligence for ongoing operations to prevent subsequent civil disturbances or to detain violators from the previous operation.

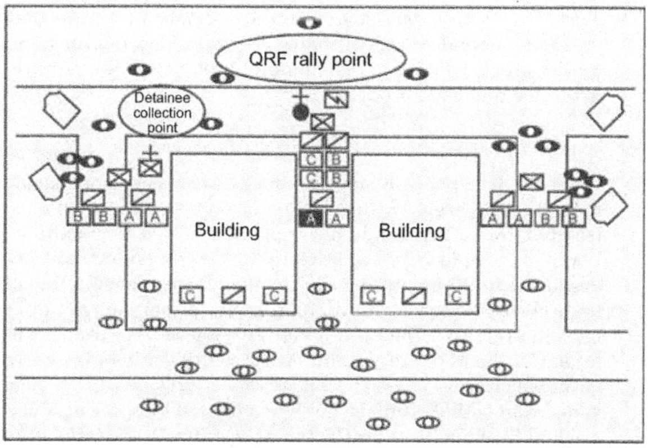

Figure D-4. More Obtrusive Engagement as the Crowd Demonstrates Negative Indicators

Crowd Warning

D-26. Once the team has taken a picture of a person or group of people, they should warn the crowd that as individuals they would be held accountable for their actions. Using an interpreter, the camera team leader will make solid eye contact with the person or group and state the following warning: *"We have your picture. You will be held accountable for your actions. If you commit an unlawful act, you will be detained, if not today, tomorrow. Blocking roads or access to public buildings is unlawful."*

Limit Crowd Size

D-27. Persons seeking to conduct a rally must apply for a permit through civilian authorities at least 72 hours prior to the meeting. If a rally has not been lawfully scheduled through civilian authorities, or if it has, but is demonstrating negative indicators and is turning into a civil disturbance, TF units and elements should attempt to limit the crowd or separate gatherings under 10 people (see *Figure D-3,* page D-5). Walking patrols can assist camera teams in breaking up large groups. Walking patrols should operate in conjunction with camera teams. This will ensure that camera teams can accurately document and provide the aforementioned warning to prevent further escalation of crowd size or behavior.

Clear Routes

D-28. Routes should be considered key terrain and must be controlled by the TF to retain the initiative. Note how a QRF divides in order to clear side streets between protected buildings in *Figure D-4.* Blocking routes or public thoroughfares is not only illegal, but severely restrictive to civil disturbance operations. Routes must remain open to allow movement of TFs and elements, whether it is for camera teams during early contact or for the QRF to respond during an actual civil disturbance. Because blocking routes is illegal, camera teams must document responsible persons for immediate or subsequent detention and legal processing.

Isolate the Area

D-29. As depicted in *Figure D-3,* page D-5, checkpoints should be established to isolate the area. While checkpoints can prevent people from entering the isolated area, they should never prevent people from exiting the area. When trapped, people may escalate the level of violence out of fear, anger, or desperation. When placing checkpoints, leaders should find a location that is both close enough to limit the number of persons being isolated, but far enough away that they will not draw a crowd from people within the isolated area. Checkpoints are essential to preventing others from joining the already gathered crowds. These checkpoints must provide storage room for processing vehicles in both directions; but when limited by space or other considerations, it should simply block traffic trying to enter the isolated area. Improve force protection measures in and around checkpoints as necessary. Strategically placed checkpoints can provide numerous countermeasures to prevent and control a civil disturbance or to identify and detain persons who may have committed unlawful acts during the riot. Use checkpoints to prevent potential participants from joining the crowd or civil disturbance. Checkpoints should also prevent any activity that may commingle ethnic groups. The checkpoint must process all people departing the area by documenting personal and vehicle information with the pictures of all occupants in front of or adjacent to the vehicle.

Establish Lethal Overwatch

D-30. The mentality of "taking the high ground" is just as important in civil disturbance operations as in any other type of operation. Although QRF assets on the ground have both NL and lethal capabilities, it is critical to ensure their safety by deploying marksman/observer teams as shown with the C-team designates on top of the buildings in *Figures D-4* and *D-5,* page D-8. Detain persons committing unlawful acts. The detention and legal processing of persons who commit unlawful acts, against TFs or local laws sends a clear message that all people will be held accountable for their actions. Immediately detain persons who are committing unlawful acts and are isolated or easily accessible, unless the detain operation would create a catalyst event or endanger the apprehension team. If it is not feasible to immediately detain them, a picture with relevant information should be forwarded for analysis and inclusion on the detain list for their detention at a later date and time. Remember, as stated in the warning, they will be detained, if not today, tomorrow. Detainees should be quickly relocated to the forward processing site for detainee evacuation (see *Figure D-5*). Detainee evacuation sites should be behind the control formations to prevent any interference from the local population. Sites should be positioned at two separate locations, such as at the end of a town or located along an MSR. All people who are detained based on

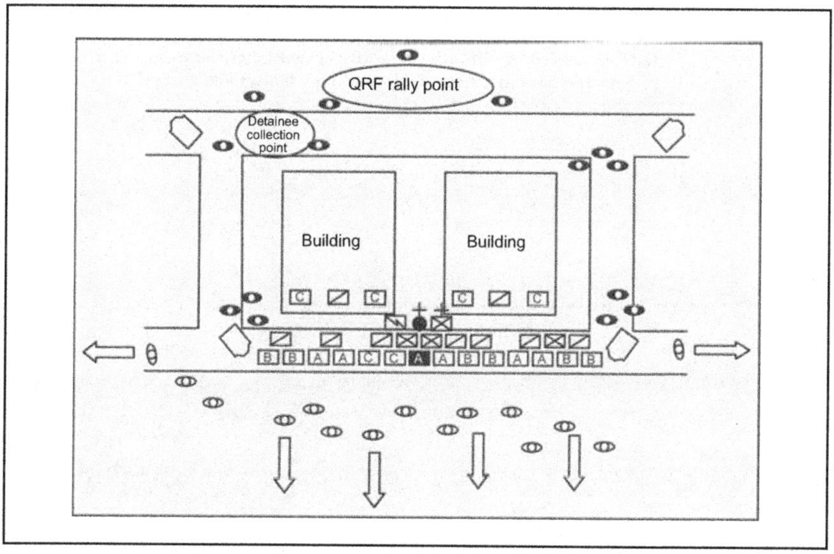

Figure D-5. Full Engagement as Crowd Unlawfully Demonstrates

solid evidence of a crime (for example, pictures of them armed with dangerous objects, blocking a road, assaulting TF members) will be turned over to local authorities for legal processing.

Detain List

D-31. When executing operations based on the detain list, do not show people the list or separate pictures from the list.

Quick Reaction Force Position

D-32. The QRF should be positioned with easy access to decisive points to interdict formations or gathering of crowds, support checkpoints, or provide sustained operations in support of early contact teams. Notice in *Figure D-1*, page D-2, how the QRF is positioned out of direct sight of the crowd until employment becomes necessary. QRF personnel are readily organized, equipped, and easily moved into position to move between buildings and through obstacles (see *Figure D-4,* page D-6).

D-33. At this point, if the crowd ignores verbal orders, NL munitions from appropriate standoff distances could be used to influence and motivate the crowd to comply. Use of NL munitions could prevent the necessity for QRF personnel to come into direct physical contact with the crowd. Early positioning of the QRF could become a catalyst event that may instigate a civil disturbance.

D-34. Assembling in mass (see *Figure D-5*), directly in front of the protected buildings, illustrates to the crowd that there is strength in numbers and the QRF is determined to disperse the crowd from the area. As a reminder, remove the flanking checkpoints that may block the dispersal of the crowd.

Glossary

A2C2	Army airspace command and control
AA	assembly area
AFL-CIO	American Federation of Labor-Congress of Industrial Organizations
AG	Adjutant General
AO	area of operations
ARTEP	Army training and evaluation program
AS	area security
ASV	armored security vehicle
ATTN	attention
BDU	battle dress uniform
CA	civil affairs
CD	combat development
CDC	crowd dispersal cartridge
CFR	Code of Federal Regulations
CINC	commander-in-chief
CMT	crisis management team
COA	course of action
COB	close of business
CONUS	continental United States
COTS	commercial off-the-shelf
CP	candle power
CPU	central processing unit
DA	Department of the Army
DCO	defense coordinating officer
DD	Department of Defense
DM	designated marksmen
DOD	Department of Defense
DOJ	Department of Justice
DOMS	director of military support
EMD	emergency medical dispatching
EMS	emergency medical services
EOC	emergency operations center
FCMT	forced cell move team
FCO	federal coordinating officer
FEMA	Federal Emergency Management Agency
FM	field manual

FRP	federal response plan
FSE	fire support element
FSO	fire support officer
G2	Assistant Chief of Staff, G3 (Intelligence)
G3	Assistant Chief of Staff, G3 (Operations and Plans)
G5	Assistant Chief of Staff, G5 (Civil Affairs)
G6	Assistant Chief of Staff, G6 (Information Management)
GOTS	government off-the-shelf
GRM	graduated response matrix
HIV	human immunodeficiency virus
HMMWV	high-mobility multipurpose wheeled vehicle
HN	host nation
HQ	headquarters
I/R	internment and resettlement
INIWIC	Interservice Nonlethal Individual Weapons Instructor Course
IO	information operations
IPB	intelligence preparation of the battlefield
JMC	joint military commission
JTF	joint task force
L&O	law and order
LBE	load-bearing equipment
LBV	load-bearing vest
LEA	law enforcement agency
LOC	lines of communication
LOS	line of sight
LP	listening post
MACDIS	military assistance for civil disturbance
MANSCEN	Maneuver Support Center
MCCM-GE	modified crowd control munition-ground emplacement
MCM	Manual for Courts-Martial
MDMP	military decision-making process
METL	mission essential task list
METT-TC	mission, enemy, terrain, troops, time available, and civilian considerations
MI	military intelligence
mm	millimeter
MO	Missouri
MOOTW	military operations other than war
MOUT	military operations on urbanized terrain
MPI	military police investigator
MRE	meal, ready to eat

MS	mobility support
MSCA	military support to civil authorities
MTP	mission training plan
MWD	military working dog
NCO	noncommissioned officer
NEO	noncombatant evacuation operations
NG	National Guard
NL	nonlethal
NLCS	nonlethal capabilities set
NLW	nonlethal weapons
OC	Oleoresin Capsicum
OCONUS	outside continental United States
OES	office of emergency services
OIC	officer in charge
OP	observation post
OPFOR	opposing forces
OPORD	operations order
PA	public address
PAO	Public Affairs Office
PIO	police intelligence operations
PM	provost marshal
PLT	platoon
PMO	Provost Marshal Office
POMSO	plans, operations, and military support officer
PPCT	pressure point control tactic
PPE	personal protective equipment
PSG	platoon sergeant
PSYOP	psychological operations
PT	physical training
PVAB	portable vehicle arrest barrier
QRF	quick reaction force
RCA	riot control agent
ROE	rules of engagement
RUF	rule for the use of force
S2	Intelligence Officer (US Army)
SCO	state coordinating officer
SCRAG	senior representative of the attorney general
SITREP	situation report
SJA	Staff Judge Advocate
SOFA	status of forces agreement

SOMA	status of mission agreement
SOP	standing operating procedure
SOSO	stability operations and support operations
SQD	squad
STARC	state area command
STP	soldier training publication
STX	soldier training exercise
TA	Training Area
TF	task force
TLP	troop leading procedure
TM	team
TPT	tactical psychological operations team
TRADOC	United States Army Training and Doctrine Command
TSOP	tactical standing operating procedure
TTP	tactics, techniques, and procedures
UCMJ	Uniform Code of Military Justice
UN	United Nations
UNGA	United Nations General Assembly
UNHCR	United Nations High Commissioner for Refugees
UNSC	United Nations Security Council
US	United States
USAMPS	United States Army Military Police School
VCR	video cassette recorder
VIP	very important person
WMD	weapons of mass destruction
WTO	World Trade Organization

Bibliography

AR 380-10. *Foreign Disclosure and Contacts With Foreign Representatives.* 6 June 2003.

AR 380-13. *Acquisition and Storage of Information Concerning Nonaffiliated Persons and Organizations.* 30 September 1974.

Code of Federal Regulations.

Constitution of the United States.

DA Form 2028. *Recommended Changes to Publications and Blank Forms.*

DA Form 4137. *Evidence/Property Document.*

DD Form 2708. *Receipt For Inmate or Detained Person.*

DOD Civil Disturbance Plan 55-2. *Operation Garden Plot.* 15 February 1991.

DOD Directive 3025.12. *Military Assistance for Civil Disturbances.* 4 February 1994.

DOD Directive 3025.15. *Military Assistance to Civil Authorities.* 18 February 1997.

DOD Directive 5525.5. *DOD Cooperation With Civilian Law Enforcement Officials.* 15 January 1986.

Executive Order 12333. *United States Intelligence Activities.* 4 December 1981.

Fiscal Year 1989 Act.

FM 3-06. *Urban Operations.* 1 June 2003.

FM 3-06.11. *Combined Arms Operations in Urban Terrain.* 28 February 2002.

FM 19.10. *Military Police Law and Order Operations.* 30 September 1987.

FM 3-19.4. *Military Police Leaders' Handbook.* 4 March 2002.

FM 3-19.40. *Military Policy Internment/Resettlement Operations.* 1 August 2001.

FM 3-21.5. *Drills and Ceremonies.* 7 July 2003.

FM 3-22.40. *Multiservice Tactics, Techniques, and Procedures (MTTP) For Tactical Employment of Nonlethal Weapons (NLW).* 15 January 2003.

FM 7-0. *Training the Force.* 22 October 2002.

FM 7-1. *Battle Focused Training.* 15 September 2003.

FM 27-100. *Legal Support to Operations.* 1 March 2000.

Judiciary Act of 1789.

MCM.

National Defense Authorization Acts.

NGR 500-1/ANGI 10-8101. *Military Support to Civil Authorities.* 1 February 1996.

Posse Comitatus Act. 1878.

STP 19-95B1-SM. *Soldier's Manual for MOS 95B Military Police Skill Level 1.* 14 January 2003.

STP 19-95B24-SM-TG. *Soldier's Manual and Trainer's Guide for MOS 95B, Military Police, Skill Levels 2/3/4.* 2 December 2002.

STP 19-95C1-SM. *MOS 95C, Corrections Specialist, Skill Level 1 Soldier's Manual.* 30 October 2003.

STP 19-95C24-SM-TG. *MOS 95C, Corrections Specialist, Skill Levels 2/3/4, Soldier's Manual and Trainer's Guide.* 30 October 2003.

Uniform Code of Military Justice.

USC, Title 32. *National Guard.*

USC, Title 10, Chapter 18. *Military Support for Civilian Law Enforcement Agencies.*

USC, Title 10, Subtitle A, Part I, Chapter 15, Section 331. *Federal Aid for State Governments.*

USC, Title 10, Subtitle A, Part I, Chapter 15, Section 332. *Use of Militia and Armed Forces to Enforce Federal Authority.*

USC, Title 10, Subtitle A, Part I, Chapter 15, Section 333. *Interference With State and Federal Law.*

USC, Title 10, Subtitle A, Part I, Chapter 15, Section 334. *Proclamation to Disperse.*

USC, Title 18, Part I, Chapter 67, Section 1385. *Use of Army and Air Force as a Posse Comitatus.*

Index

Made in the USA
Monee, IL
05 November 2026

67820229R00334